# NanoScience and Technology

Springer
*Berlin*
*Heidelberg*
*New York*
*Barcelona*
*Hong Kong*
*London*
*Milan*
*Paris*
*Tokyo*

**Physics and Astronomy**

ONLINE LIBRARY

http://www.springer.de/phys/

# NANOSCIENCE AND TECHNOLOGY

*Series Editors:* P. Avouris   K. von Klitzing   H. Sakaki   R. Wiesendanger

The series NanoScience and Technology is focused on the fascinating nano-world, meso-scopic physics, analysis with atomic resolution, nano and quantum-effect devices, nano-mechanics and atomic-scale processes. All the basic aspects and technology-oriented developments in this emerging discipline are covered by comprehensive and timely books. The series constitutes a survey of the relevant special topics, which are presented by leading experts in the field. These books will appeal to researchers, engineers, and advanced students.

**Sliding Friction**
Physical Principles and Applications
By B. N. J. Persson

**Scanning Probe Microscopy**
Analytical Methods
Editor: R. Wiesendanger

**Mesoscopic Physics and Electronics**
Editors: T. Ando, Y. Arakawa, K. Furuya, S. Komiyama,
and H. Nakashima

**Biological Micro- and Nanotribology**
Nature's Solutions
By M. Scherge and S. N. Gorb

**Semiconductor Spintronics and Quantum Computation**
Editors: D. D. Awschalom, N. Samarth, and D. Loss

**Semiconductor Quantum Dots**
Physics, Spectroscopy and Applications
Editors: Y. Masumoto and T. Takagahara

**Nano-Optoelectronics**
Concepts, Physics and Devices
Editor: M. Grundmann

Series homepage – http://www.springer.de/phys/books/nst/

Y. Masumoto   T. Takagahara  (Eds.)

# Semiconductor Quantum Dots

## Physics, Spectroscopy and Applications

With 278 Figures

Springer

Professor Yasuaki Masumoto
Institute of Physics
University of Tsukuba
Tennoudai 1-1-1
Tsukuba, Ibaraki 305-8571, Japan

Professor Toshihide Takagahara
Department of Electonics
and Information Science
Kyoto Institute of Technology
Hashigami-cho, Matsugasaki
Sakyo-ku, Kyoto 606-8585, Japan

*Series Editors:*
Professor Dr. Phaedon Avouris
IBM Research Division, Nanometer Scale Science & Technology
Thomas J. Watson Research Center, P.O. Box 218
Yorktown Heights, NY 10598, USA

Professor Dr., Dres. h. c. Klaus von Klitzing
Max-Planck-Institut für Festkörperforschung, Heisenbergstrasse 1
70569 Stuttgart, Germany

Professor Hiroyuki Sakaki
University of Tokyo, Institute of Industrial Science, 4-6-1 Komaba, Meguro-ku
Tokyo 153-8505, Japan

Professor Dr. Roland Wiesendanger
Institut für Angewandte Physik, Universität Hamburg, Jungiusstrasse 11
20355 Hamburg, Germany

ISSN 1434-4904
ISBN 3-540-42805-4 Springer-Verlag Berlin Heidelberg New York

Library of Congress Cataloging-in-Publication Data: Semiconductor quantum dots : physics, spectroscopy
and applications / Y. Masumoto, T. Takagahara (eds.) p. cm. – (Nanoscience and technology) Includes
bibliographical references and index. ISBN 3540428054 (alk. paper) 1. Quantum dots. I. Masumoto, Y.
(Yasuaki), 1948– II. Takagahara, T. (Toshihide), 1950– III. Series. TK7874.88.S457 2002 621.3815'2–dc21
2002021118

Springer-Verlag Berlin Heidelberg New York
a member of BertelsmannSpringer Science+Business Media GmbH

http://www.springer.de

© Springer-Verlag Berlin Heidelberg 2002
Printed in Germany

Typesetting: Verlagsservice Ascheron, Mannheim
Cover concept: eStudio Calamar, Barcelona
Cover design: *design& production*, Heidelberg

Printed on acid-free paper       SPIN: 10785741       57/3141/ba - 5 4 3 2 1 0

# Preface

Semiconductor quantum dots are very small three-dimensional systems whose dimension ranges from nanometers to tens of nanometers. Their size is smaller than the de Broglie wavelength of slow electrons, therefore quantum effects are manifest in the dots. As a result of quantum confinement, the energy states for electrons, holes, and excitons (a pair composite of an electron and a hole interacting mutually via Coulomb force) are composed of discrete series like those in atoms. The energy levels depend on the size; the size effect is called the quantum size effect. A typical quantum dot involves 1,000 to 100,000 atoms. This number of atoms is much smaller than that constituting a bulk crystal, a quantum well, or a quantum wire, but is larger than atoms, molecules, and clusters. Because of the small size of a quantum dot, the wavefunction extends coherently throughout the quantum dot and the squared transition dipole moment of excitons increases with the volume of the quantum dot. This is an advantageous feature for optical device applications. Furthermore, the surface-to-volume ratio of the number of atoms in a very small dot is as large as 20%. An important consequence of the large surface-to-volume ratio is the manifestation of surface-related phenomena in quantum dots.

Characteristics of optical properties of quantum dots are reduced to the quantum confinement effect and the surface effect. However, they are often masked by the inhomogeneous broadening of the optical spectra. To see the unique properties of quantum dots, site-selective laser spectroscopy and single-dot spectroscopy are successfully applied to the inhomogeneously broadened quantum dot system. The main scope of this book is to review the recent developments in these areas and the associated advancements in the theoretical understanding of the underlying physics of carrier relaxation, excitonic optical nonlinearity, and multiexciton states. This book covers the physics and spectroscopy of semiconductor quantum dots. Also described are recent developments in the formation of self-assembled quantum dots and their device applications.

Chapter 1 reviews the growth and real-time monitoring of quantum dots, including in situ ellipsometry. Since the angle of polarization is determined by the refractive index of the surface, changes in the monolayer are observed in ellipsometry by selecting a wavelength sensitive to the material covering the surface. An inflection point indicates the formation of quantum dots. In situ ellipsometry is effective for the precise real-time monitoring of growth of an atomic layer in metal organic vapor-phase epitaxy (MOVPE). We also discuss one-dimensional $In_{0.45}Ga_{0.55}As$ quantum dot arrays that were successfully fabricated on a GaAs (311)B substrate by MOVPE.

Chapter 2 describes characteristic electronic structures and optical properties of quantum dots. We deal with the excitonic energy levels and their oscillator strengths in spherical dots embedded in dielectric materials and clarify the dielectric confinement effect, which appears most pronounced in zero-dimensional systems. In the weak-confinement regime, the exciton oscillator strength is proportional to the volume of a quantum dot. This is due to the formation of a large transition dipole moment as a result of the coherent superposition of the atomic dipole moments within a quantum dot. But this proportionality does not persist over quantum dot sizes that exceed the wavelength of a photon corresponding to the exciton transition energy. This intriguing property of the size- and temperature-dependent radiative decay rate of excitons in quantum dots is clarified on the basis of the nonlocal response theory. Exciton fine structures arising from the electron–hole exchange interaction are discussed, taking into account the degenerate valence band structures. Quantitative comparison with recent results of single-dot spectroscopy is made. The prerequisite to enhance the excitonic optical nonlinearity is a large transition dipole moment. This can be achieved by formation of a regular array of quantum dots. Each quantum dot corresponds to an atom with a large oscillator strength. Through dipole–dipole coupling among quantum dots, the coherent superposition of the dipole moments of quantum dots can be accomplished in analogy to the Frenkel-type exciton. These features are investigated in detail.

Quantum dots are usually embedded in a medium with different elastic constants and different optical constants. Thus the acoustic and the polar optical lattice vibrations of quantum dots are highly sensitive to the boundary conditions at the interface. Chapter 3 presents the theoretical formulation of the acoustic and the polar optical phonon modes in quantum dots and their interaction with electrons. The interaction of acoustic phonon modes with electrons occurs through deformation potential coupling and piezoelectric coupling. The polar optical phonons primarily interact with electrons through Fröhlich coupling. The characteristic size dependence of the electron–phonon coupling strength is discussed.

Spectroscopy of a single quantum dot is a recent development. An ensemble of quantum dots shows inhomogeneous broadening due to the distribution of sizes, shapes, and the surrounding environment. Single quantum dot spectroscopy can reveal the optical spectrum of an individual dot and is free from these distributions. Chapter 4 reviews methods of single-dot spectroscopy and their application to a single InP quantum dot formed in a $Ga_{0.5}In_{0.5}P$ matrix. Results exhibit sharp emission lines from excitons and biexcitons in a quantum dot. These lines are changed to broad emission bands due to multi-exciton states as the excitation intensity is raised. Interface or surface effects of a quantum dot are observed in the optical spectra of a quantum dot: it was found that the luminescence spectra of an InP quantum dot embedded in a $Ga_{0.5}In_{0.5}P$ layer varied with the observation direction due to interface effects. Occasionally, one of several hundred dots in this experiment blinks at an interval of milliseconds to seconds due to defects near the quantum dot. The blinking mechanism is attributed to a fluctuating local electric field induced by carrier trapping in the host near a quantum dot.

Chapter 5 reviews the persistent hole-burning phenomena of quantum dots. The presence of persistent hole-burning phenomena in quantum dots shows that the inhomogeneous broadening of quantum dots does not come from the distribution of size and shape alone, but comes from various ground state configurations of the dot–matrix system. It means that the quantum dots are highly sensitive to the surrounding host. Persistent hole burning in quantum dots takes place as a result of photoionization. A variety of spatial arrangements induce inhomogeneous broadening in addition to the inhomogeneous broadening produced by the size and shape distribution. The persistent hole burning of quantum dots is applied to site-selective spectroscopy. It can reveal the electronic or excitonic excited quantum states and phonons confined in quantum dots. Application of persistent hole burning to optical multiple storage is demonstrated. Photoionization of quantum dots leads to photostimulated luminescence. In fact, the photostimulated luminescence of quantum dots is observed.

Understanding of carrier relaxation by phonons in quantum dots is very important, because slow relaxation leads to low luminescence yields as a result of competition between relaxation to luminescing states and nonradiative decay to non-luminescing states. The luminescence yield is a very important parameter for the application of quantum dots to optical devices. So far, slow carrier relaxation is predicted for phonons in the discrete energy spectra of quantum dots and was referred to as the phonon bottleneck effect. Chapter 6 describes electron relaxation in quantum dots by longitudinal optical phonons and acoustic phonons. Phonon sidebands have been observed as clear structures in the luminescence spectra of site-selectively excited InP as well as InGaAs quantum dots under an electric field. Carrier relaxation is also studied by time-resolved luminescence spectroscopy. Acoustic phonon-mediated relaxation was faster than expected from the phonon bottleneck effect, which indicates the irrelevance of the phonon bottleneck effect in quantum dots.

Chapter 7 describes resonant two-photon spectroscopy of quantum dots. It includes application of two-photon excited resonant luminescence, resonant second-harmonic scattering, and resonant hyper-Raman scattering to quantum dots. Because selection rules for two-photon electronic and phonon-assisted electronic transitions differ from those for one-photon transitions, two-photon spectroscopy provides information on electronic and phonon states that is unobservable in one-photon spectroscopy. Unique features of two-photon spectroscopy are discussed, including application to determination of orientation of quantum cubes and single quantum dot spectroscopy.

Exciton dephasing and dynamics in quantum dots are discussed in Chaps. 8 and 9. Chapter 8 reviews the experimental aspects of exciton dephasing in quantum dots. The dephasing time is inversely proportional to the homogeneous width and is proportional to the optical nonlinearity. The theoretical aspects are reviewed in Chap. 9. The population decay or longitudinal relaxation of excitons arises from the transition to other states. On the other hand, the exciton dephasing represents the decay of the excitonic transition dipole moment to which the population decay as well as the phase fluctuation of the exciton wavefunction contribute. We present a general formulation of the exciton dephasing due to the exciton–phonon interactions that are discussed in Chap. 3. The magnitude and the temperature dependence of the exciton dephasing rate are investigated and compared quantitatively with relevant experimental results.

The size dependence of the excitonic optical nonlinearity of quantum dots is a very intriguing subject. As mentioned in Chap. 2, the excitonic oscillator strength increases with the volume of a quantum dot, and we expect a corresponding increase in the optical nonlinearity. But experimentally the optical nonlinearity saturates rapidly with increasing dot size, even before the nonlocality effect discussed in Chap. 2 sets in. This puzzling feature was resolved by the discovery of the antibonding biexciton states, namely weakly correlated exciton pair states. Basic physics shows that the contribution to the optical nonlinearity from the excitonic states is partially compensated by that from the biexciton states. This compensation is dependent on the quantum dot size and the material parameters. Details are described in Chap. 10. The experimental observation of the antibonding biexciton states and the triexciton states in CuCl quantum dots by a time-resolved size-selective pump-and-probe technique are also presented.

Chapter 11 discusses multiexciton states in quantum dots. A characteristic feature of semiconductor quantum dots, in striking contrast to atoms and molecules, is the possibility of creating multiple electron–hole pair excitations in a quantum dot. Such multiparticle complexes exhibit various intricacies of the interparticle correlation induced by the Coulomb interaction. Recent advances in fabrication and spectroscopy of semiconductor quantum dots have made it possible to investigate such a quantum-confined electron–hole plasma in a well-controlled way. In Chap. 11, energy levels and optical spectra in highly excited quantum dots are calculated at varying levels of sophistication, based on the local density approximation and the configuration interaction calculation. As a consequence, the spin structure of triexciton states is clarified and configuration-crossing transitions from the biexciton and triexciton states are predicted.

Device applications of quantum dots are discussed in Chap. 12. Applications of quantum dots to low-power-consumption laser diodes, temperature-insensitive laser diodes, and low-noise and long-wavelength vertical-cavity surface-emitting lasers are discussed with a special emphasis on the advantages of self-assembled quantum dots. This chapter reviews the present status of the application of quantum dots to lasers. We discuss the improved temperature characteristics, wavelength control performance, reduction of threshold current density for lasing, and demonstration of vertical-cavity surface-emitting lasers. Further future prospects for quantum dot devices are also discussed.

Many chapters of this book are based on the research results of the Masumoto Single Quantum Dot Project in the ERATO (Exploratory Research for Advanced Technology) program of the Japan Science and Technology Corporation (JST). The editors would like to express their sincere gratitude to JST. The Masumoto Quantum Dot Project has greatly extended our understanding of the basic properties of quantum dots, while opening up a variety of new experimental and theoretical investigations in this field. Quantum dots have thus been established as fundamental and important objects that will certainly find useful applications in the future, while also extending our knowledge of basic physics and materials science.

Tsukuba and Kyoto,                                    *Yasuaki Masumoto*
January 2002                                          *Toshihide Takagahara*

# Contents

## 6   Dynamics of Carrier Relaxation in Self-Assembled Quantum Dots

## 7   Resonant Two-Photon Spectroscopy of Quantum Dots

## 8   Homogeneous Width of Confined Excitons in Quantum Dots – Experimental

# List of Contributors

**Alexander Baranov**
S.I. Vavilov State Optical Institute
St. Petersburg 199034
Russia
*and*
Single Quantum Dot Project,
ERATO, JST
Japan
baranov1@online.ru


**Ivan V. Ignatiev**
Institute of Physics
St.-Petersburg State University
St. Petersburg 198504
Russia
*and*
Single Quantum Dot Project,
ERATO, JST
Japan
ivan@paloma.spbu.ru


**Igor E. Kozin**
Institute of Physics
St.-Petersburg State University
St. Petersburg 198504
Russia
*and*
Single Quantum Dot Project,
ERATO, JST
Japan

**J.-S.Lee**
Nippon EMC
Higashimurayama
Tokyo 189-0011
Japan
*and*
Single Quantum Dot Project,
ERATO, JST
Japan
lee@n-emc.co.jp


**Yasuaki Masumoto**
Institute of Physics
University of Tsukuba
Tsukuba 305-8571
Japan
*and*
Single Quantum Dot Project,
ERATO, JST
Japan
shoichi@sakura.cc.tsukuba.ac.jp


**Selvakumar V. Nair**
Energenius Centre
for Advanced Nanotechnology
University of Toronto
170 College Street
Toronto M5S 3E3
Canada
*and*
Single Quantum Dot Project,
ERATO, JST
Japan
selva.nair@utoronto.ca

**Kenichi Nishi**
Photonic and Wireless Devices
Research Laboratories
NEC Corporation
Tsukuba 305-8501
Japan
*and*
Single Quantum Dot Project,
ERATO, JST
Japan
k-nishi@cj.jp.nec.com

**Mitsuru Sugisaki**
Energenius Centre
for Advanced Nanotechnology
University of Toronto
170 College Street
Toronto M5S 3E3
Canada
*and*
Single Quantum Dot Project,
ERATO, JST
Japan
mitsuru.sugisaki@utoronto.ca

**Toshihide Takagahara**
Department of Electronics and
Information Science
Kyoto Institute of Technology
Matsugasaki
Sakyo-ku
Kyoto 606-8585
Japan
takaghra@hiei.kit.ac.jp

# 1  Growth of Self-Organized Quantum Dots

J.-S. Lee

## 1.1  Introduction

In recent decades, considerable efforts have been devoted to the realization of semiconductor hetero-structures that provide carrier confinement in all three directions and behave as electronic quantum dots (QDs). Initially, mainly techniques like lithographic patterning and etching of quantum well (QW) structures have been employed. However, the feature sizes of the structures made using electron beam lithography and focused ion beam lithography, the most developed approaches, were still larger than the desirable level. More recently, substrate encoded epitaxy, which allows the fabrication of small nanostructures from a much larger template, has been conducted. QDs in tetrahedral-shaped pyramids were fabricated.

In another approach, cleaved edge overgrowth has been used to fabricate electronic T-shaped quantum wires (QWRs) and electronic QDs at the juncture of three orthogonal QWs. In these cases, the distance between QDs is usually large, and the structures contain additional emission from sidewalls and QW or QWR structure.

Self-organized QD formation without any artificial patterning is the main focus of this chapter. It realizes the highly dense arrays of QDs and shows high optical performance. Ordering of QD shape, size, and position for the ultimate realization of QD advantages is also summarized.

## 1.2  Fabrication Techniques of Quantum Dots

### 1.2.1  Quantum Dot Fabrication by Lithographic Techniques

By the end of the 1980s the fabrication of QDs by patterning of QWs was considered the most straightforward way of QD fabrication. Patterning has several advantages and still attracts much attention because QD size and spatial arrangement can be realized.

Lithographic techniques comprise many kinds of methods. Optical lithography, based on excimer lasers (resolution below 0.2 µm) and on ultraviolet optics and resists with steep photosensitivity curves (resolution below 0.1 µm) is the most conventional and popular method. X-ray lithography has the advantage of much shorter wavelengths and can be used for nanostructure

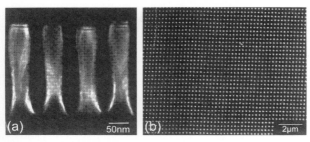

**Fig. 1.1.** (a) Cross-sectional and (b) plan-view SEM images of dry etched Al-GaAs/GaAs columns (after Scherer et al. [1])

fabrication. However, there are some problems to be solved. For example, high-resolution X-ray lenses are presently not available and an additional process is required to fabricate the mask.

For direct lateral patterning, the most developed approaches are electron beam (EB) lithography and focused ion beam lithography (resolution below 50 nm). Scherer et al. reported freestanding QD fabrication using EB lithography and BCl$_3$/Ar reactive ion etching [1]. Figure 1.1 shows a regular array of columns that has been etched from an AlGaAs/GaAs superlattice. Lateral column dimensions ranging from 30 to 40 nm were achieved.

Although electron beam lithography is widely used, the feature sizes of these structures and their size fluctuations tend to be larger than the desirable level. In addition, lithography and subsequent processes often lead to contamination and defect formation. New attractive methods forming high-quality 10-nm-scale QDs without direct patterning by lithography have been studied.

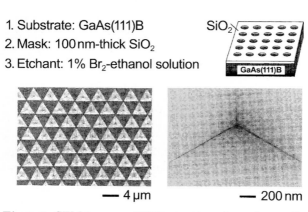

**Fig. 1.2.** SEM images of TSR structure formed on a GaAs (111)B substrate after removing SiO$_2$ mask (after Sugiyama et al. [2])

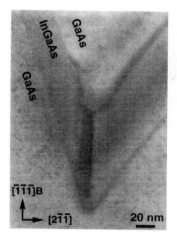

**Fig. 1.3.** Cross-sectional TEM image of the bottom region of a TSR (after Sugiyama et al. [2])

A useful approach for preparing semiconductor quantum nanostructures with well-defined positions, potential wells, and surrounding heterostructure consists in growth on patterned substrates. This approach has been employed successfully to produce high quality QWRs in heterostructures grown in V-grooves. More recently, a similar method has been developed for growing GaAs/AlGaAs and InGaAs/GaAs QDs in tetrahedral-shaped recesses (TSRs) [2]. The TSR consists of threefold (111)A formed by selective chemical etching of a (111)B-oriented GaAs substrate with a $SiO_2$ mask (Fig. 1.2). The tetrahedral GaAs surrounded by (111)A, (111)B, and high indices planes was observed to form a QD structure. The cross-sectional transmission electron microscopy (TEM) image at the bottom region of a typical TSR along the $(01\bar{1})$ face is shown in Fig. 1.3. The interfaces correspond to the darker lines. At the bottom of the TSR, a darker region exists in the InGaAs layer that is elongated in the vertical direction indicating the In-rich region. The In-rich region can be distinguished as small enough to act as a QD. In that way, it is possible to obtain a well-defined self-organized QD at the bottom of

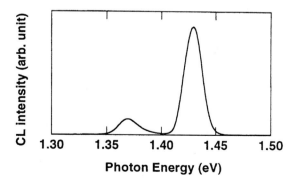

**Fig. 1.4.** CL spectrum of TSRs at 20 K (after Sugiyama et al. [2])

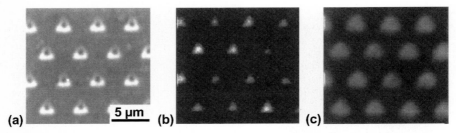

**Fig. 1.5.** (a) Secondary electron image and monochromatic CL images of TSRs taken at (b) 1.37 eV and (c) 1.43 eV at 20 K (after Sekiguchi et al. [3])

each TSR. Figure 1.4 shows a cathode luminescence (CL) spectrum of TSRs. The main peaks are at 1.37 and 1.43 eV representing QD and QW luminescence, as shown by the spatially resolved monochromatic CL images in Fig. 1.5. The 1.37 eV luminescence clearly originated in the quantum structure at the bottom of the TSRs. The 1.43 eV luminescence forms a triangular pattern, as can be expected for the QW formation on the {111}A sidewalls [3].

Some groups reported the fabrication of an array of mesas on an As-terminated GaAs (111)B substrate. An array of resist patterns aligned in the [110] direction was defined by lithography, followed by wet chemical etching. Growth was stopped when each mesa consisted of a truncated triangular pyramid Truncated triangular pyramid. The mesa top was a (111)B face and the three side facets were of the {110} type. The size of a mesa top can be decreased from the initial pattern-defined size, eventually down to practically zero depending on the duration of etching. Even if the size of the mesa top is not sufficient to provide efficient lateral confinement, one can reduce it further by overgrowth, due to lateral shrinking of the mesa top. Figure 1.6 illustrates this process [4]. A precondition of the shrinkage process is a higher growth rate on the mesa top than on the sidewalls. Using this or similar approaches QD structures with lateral dimensions as small as 50 nm can be created. Figure 1.7a,b shows scanning electron microscopy (SEM) images of the GaAs tetrahedral structures demonstrated by Fukui et al. [4]. Uniform

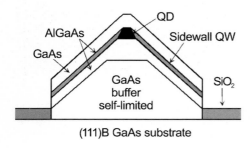

**Fig. 1.6.** Schematic view of cross-section of GaAs tetrahedral QD structure (after Fukui et al. [4])

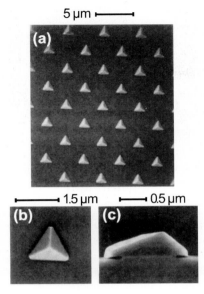

5 µm ├────┤

(a)

├──────┤ 1.5 µm     ├──────┤ 0.5 µm

(b)          (c)

**Fig. 1.7.** (a) and (b) SEM images of GaAs tetrahedral structure. (c) SEM image of cross-section of GaAs tetrahedral structures buried in AlGaAs (after Fukui et al. [4])

size of tetrahedral QDs with 1.5-µm sides are obtained, and no deposition occurred on the SiO$_2$-masked area. A cross-sectional SEM image of tetrahedral GaAs QD structure buried in AlGaAs (Fig. 1.7c) shows a GaAs tetrahedron and smooth AlGaAs overlayers. Although room temperature CL shows clear images of an array of QDs, the wavelength of 882 nm corresponding to the GaAs band edge indicates no carrier confinement.

Similar GaAs pyramids were fabricated on (100) substrates with square openings. SiN$_x$ mask patterns aligned in the $\langle 100 \rangle$ direction were defined by photolithography and wet chemical etching (Fig. 1.8) [5]. The pyramidal

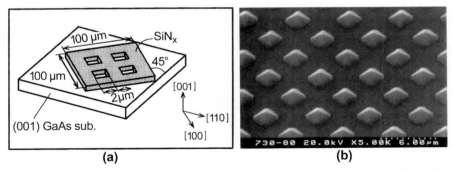

(a)                              (b)

**Fig. 1.8.** (a) Schematic view of SiN$_x$ mask pattern for selective area MOVPE growth; and (b) SEM image of pyramidal structures grown on (a) (after Kumakura et al. [5])

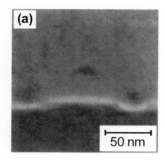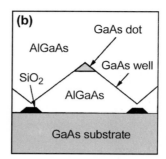

**Fig. 1.9.** (a) A cross-section at the center of the GaAs dot structure surrounded by AlGaAs and (b) its illustration (after Nagamune et al. [6])

structures having four {110} facets were formed by metal organic vapor-phase epitaxy (MOVPE). After the AlGaAs pyramids were completely formed, no growth occurred on the tops or sidewalls of the pyramids. The maximum size of all pyramids became almost the same and saturated at about 50 nm because the shape and maximum size of the pyramids are determined by thermal equilibrium. Namely, the width of the tops of the pyramids becomes very uniform due to the "self-limited growth" mechanism. On the pyramids, GaAs QDs were formed under different growth conditions. Sharp photoluminescence (PL) emission (full-width at half-maximum (FWHM): 22 meV at 18 K) from the QD array was observed.

Nagamune et al. also used MOVPE to fabricate pyramid-shaped GaAs QDs on GaAs (100) substrates of dimension $25 \times 25 \times 15$ nm, surrounded by (100) and {110} facets (Fig. 1.9) [6]. Photoluminescence spectra (8.7 K) of the sample shown in Fig. 1.10 have several peaks from sidewall QW, GaAs

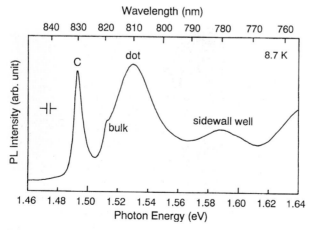

**Fig. 1.10.** PL spectra of the sample shown in Fig. 1.9 (after Nagamune et al. [6])

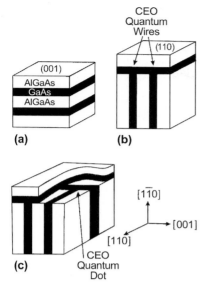

**Fig. 1.11.** (a) Schematic representation of twofold cleaved edge overgrowth; (b) standard process used for fabrication of QWRs; (c) the second cleavage and succeeding growth on the top of the (1$\bar{1}$0) plane allows the fabrication of QDs (after Grundmann et al. [7])

bulk, and GaAs dot. The blueshift value of the PL peak position of the GaAs dots from that of the GaAs bulk is about 19 meV.

Cleaved edge overgrowth has been used to fabricate electronic T-shaped QWRs, which develop at the juncture of two QW planes (Fig. 1.11a,b). Grundmann et al. predicted theoretically that electronic QDs form at the juncture of three orthogonal QWs, which could be fabricated with twofold cleaved edge overgrowth (Fig. 1.11c) [7]. In Fig. 1.12, the electron excitonic wave function is shown in different views. The extension of the wave func-

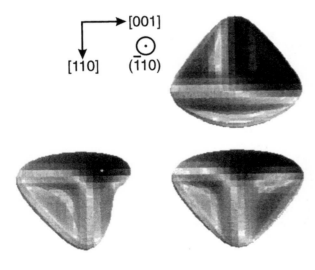

**Fig. 1.12.** The electron excitonic wave function for a 2CEO GaAs/AlGaAs QD at the junction of three QWs. The same orbital is shown in different angles (after Grundmann et al. [7])

tion into the three QW layers gives the orbital its characteristic shape. The Coulomb interaction leads to a general shrinking of the dot wave function, which also becomes more symmetric. They found a further enhancement of exciton binding energy for the QD. The localization energies of QDs have been predicted to be typically 10 meV for the AlGaAs/GaAs system.

### 1.2.2  Self-Organized Quantum Dot Fabrication

When conventional lithography techniques are used, the lateral density of QDs is defined by the pattern. The QD-to-QD distance is usually large, preventing the realization of dense arrays of QDs. Moreover, TSR and pyramids contain additional emission from sidewalls and QWR structures. The evolution of an initially two-dimensional (2D) growth into a three-dimensional (3D) corrugated growth front is a well-known phenomenon and has been frequently

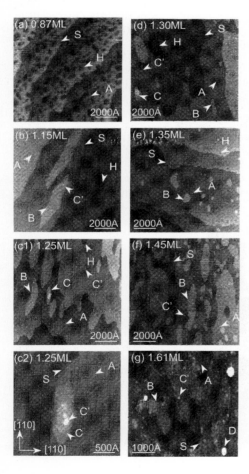

**Fig. 1.13.** STM images showing the evolution of InAs morphology on GaAs (100) for different thicknesses: (**a**) 0.87 ML; (**b**) 1.15 ML; (**c1**) and (**c2**) 1.25 ML in different positions; (**d**) 1.3 ML; (**e**) 1.35 ML; (**f**) 1.45 ML; and (**g**) 1.61 ML. The labels in the figure denote small 2D clusters (A), large 2D clusters (B), small 3D clusters (C′), large 3D clusters (C), 3D islands (D), 1 ML high steps (S), and 1 ML deep holes (H) (after Heitz et al. [8])

observed in the presence of strain. In a paper by Stranski and Krastanow, the possibility of island formation on an initially flat heteroepitaxial surface was proposed. Self-assembled islands formed in the Stranski–Krastanow (S–K) growth mode have attracted particular attention as QD structures. In the S–K mode, island structures are self-formed on a 2D wetting layer as a result of the transition of the growth mode, namely, from 2D to 3D at a certain layer thickness. The formation of coherent, i.e., defect free, islands as a result of S–K growth of strained heterostructures is systematically exploited for the fabrication of QDs.

Representative examples are In(Ga)As on GaAs, Ge on Si, and InP on InGaP. In the case of InAs on GaAs, the first monolayer of InAs grows in the form of a fully strained 2D layer, but 3D QDs are formed when the thickness of the InAs layer exceeds 1.5 monolayers (ML). Figure 1.13 shows typical STM images for InAs supply from 0.87 to 1.61 ML during molecular beam epitaxy (MBE) at a temperature of 500°C [8]. The image reveals 1-

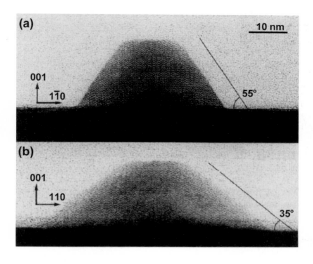

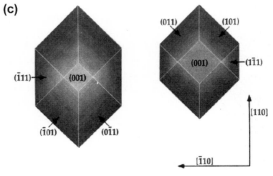

**Fig. 1.14.** (a) Cross-sectional HRTEM of uncapped InP/InGaP SK islands on GaAs (100) along the [011] and (b) [01$\bar{1}$] directions, together with (c) a schematic model of the island shape (after Georgsson et al. [9])

ML-high steps (labeled S) corresponding to 200–400-nm-wide terraces, and a high density of clusters a few nm wide and 1 ML high, referred to as small 2D clusters (labeled A). The first InAs layer is incomplete up to 1.35 ML InAs supply, showing holes 1 ML deep (labeled H). Further InAs deposition (1.15 ML) leads to the formation of clusters 1 ML high with lateral sizes up to hundreds of nanometers, referred to as large 2D clusters (labeled B) on top of the still incomplete InAs layer. At this deposition of 1.15 ML, the InAs surface starts to show features of small quasi-3D clusters 2–4 ML high and up to 20 nm wide (labeled C′). At 1.25 and 1.30 ML InAs supply, in addition to the small quasi-3D clusters, clusters of similar height but much larger lateral extension (> 50 nm large quasi-3D clusters, labeled C) are present. Remarkably, the quasi-3D clusters disappear as the InAs supply is increased to 1.35 ML, resulting again in a 2D-like surface. As the InAs deposition is further increased to 1.45 ML, only small quasi-3D clusters reappear (3D islands labeled D). When InAs deposition continues, QD density increased up to $10^{11}$ cm$^{-2}$. Study of these self-assembled QDs of InAs on GaAs has shown that the size fluctuation of dots is relatively small (< 10%) and that small dots are dislocation free and strained coherently with GaAs with sufficient material quality for use in lasers.

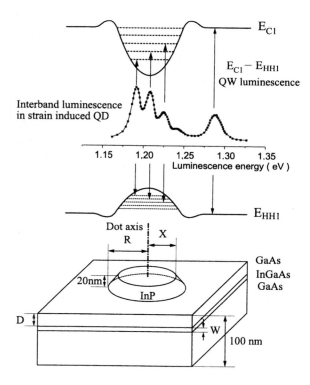

**Fig. 1.15.** The geometry of the strain-induced QD and a schematic presentation of the luminescence spectrum including the E1–HH1 transitions in the QW and QD. A schematic experimental photoluminescence spectrum is also shown (*center*) (after Tulkki et al. [10])

Georgsson et al. showed InP islands grown on InGaP/GaAs (100) sub-strates using MOVPE [9]. Figure 1.14 shows high-resolution TEM (HRTEM) images of the QD structure with 4 ML InP deposition at 580°C. A GaP cap 4 ML thick was grown on the top of the InGaP layer before InP growth was started in order to increase the density of fully developed islands and decrease the density of tiny InP islands. The size of InP QD is rather larger than In(Ga)As ones. The average width is found to be 40–50 nm, the av-erage length is 55–65 nm. Pyramidal islands develop whose side facets are low-index planes including {100}, {110}, and {111}.

Another method to achieve 3D confinement is lateral modulation of a QW band gap with local strain. The strain field has been introduced by form-ing stressors on the surface of the sample by lithographic techniques and/or nanoscale islands. The latter approach is favorable because it can make the stressors in situ without additional processes. Tulkki et al. have calculated the electronic structure of QDs induced by InP self-organized QDs acting as stressors for an InGaAs/GaAs QW [10]. They reported computational ana-lysis of the confinement effect in the quantum structures shown schematically

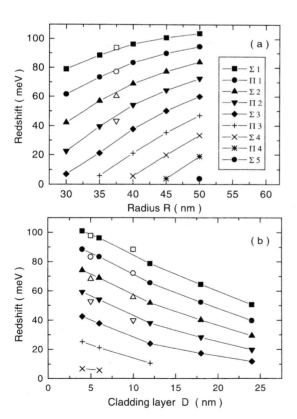

**Fig. 1.16.** (a) The redshifts of main allowed transitions in a strain-induced QD, mea-sured from the ground state transition in the unpatterned InGaAs/GaAs QW as a func-tion of the stressor dot base radius R. The InP stressor dot is assumed to be a flat cone with a top radius of X=R–10 nm. The QW thick-ness is 8 nm and the cap layer thickness is 6 nm. (b) Redshift as a function of cap layer thickness for fixed QW thickness of 8 nm and radius of 40 nm (after Tulkki et al. [10])

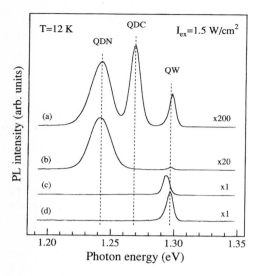

**Fig. 1.17.** PL spectra (**a**) of the as-grown stressor-induced QD sample; (**b**) sample etched for 20 s in diluted HCl; (**c**) sample etched for 3 min in fuming HCl; and (**d**) the reference QW sample. QDC and QDN denote coherent QD and incoherent QD, respectively (after Sopanen et al. [11])

in Fig. 1.15. The shifts of several transitions in the QDs from the QW ground state transition as a function of distance between QW and the surface and stressor radius was plotted in Fig. 1.16. The levels are essentially equidistant since the strain-induced potential has a prevalently parabolic shape.

Sopanen et al. reported 3D confinement of carriers to InGaAs/GaAs QW dots modulated by InP QD stressors [11]. The PL spectrum from the stress-

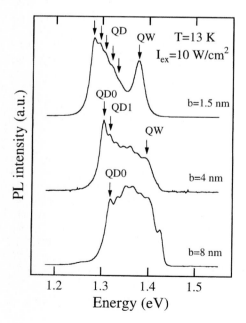

**Fig. 1.18.** PL spectra from samples with two 2-nm-thick $In_{0.25}Ga_{0.75}As$ QWs separated by GaAs barrier layers of different thicknesses b (after Sopanen et al. [12])

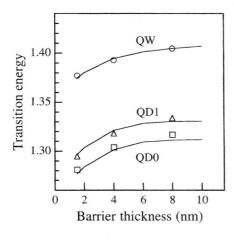

**Fig. 1.19.** Calculated transition energies (*lines*) for double $In_{0.25}Ga_{0.75}As$ QD samples having different barrier thicknesses and the measured peak positions for QW (*circles*), QD ground-state (*squares*), and the first excited QD state (*triangles*) peaks (after Sopanen et al. [12])

induced QD sample in Fig. 1.17a shows three distinct peaks, whereas the spectrum of the reference QW sample in Fig. 1.17d shows only one peak. The peaks at the lower energy side disappear when the top InP QD stressors were etched away. As a result, the QDC and QDN peaks are related to QDs produced by a strong field of the coherent islands and the larger, partially relaxed islands, respectively. They also fabricated the multi-QDs by using the strain of InP islands on top of a near-surface multi-QW structure and investigated the coupling of the electronic states [12]. Figure 1.18 shows the PL spectra from the sample with two 2-nm-thick InGaAs QWs separated by GaAs barriers of different thicknesses. As the barrier thickness decreases, the coupling is enhanced, which is observed as the redshift of the peaks and also as the narrowing of the peaks (from 15 meV to 13 meV). Comparison between the experimental and calculated transition energies of the QW, QD0, and the first excited state peak, QD1, as a function of the barrier thickness shows good agreement (Fig. 1.19).

## 1.3  Ordering of Three-Dimensional Islands

### 1.3.1  Structural Characterization of Quantum Dots

Quantum dots grown epitaxially introduce unavoidable fluctuations in size and shape. For example, just after the critical layer thickness is exceeded, the formed QDs are small, mostly do not show well-resolved crystalline shape, and exhibit large size dispersion.

The ultimate realization of the advantages of QDs requires a better understanding of QD electronic and optical properties. Moreover, it is important to learn more about how these properties may be engineered than what is presently known about QWs, because island structures depend on growth conditions and structural parameters. The lack of precise information on QD shape has been a major obstacle in furthering optoelectronics based on these structures. This knowledge is necessary to construct QD formation process models, which remain poorly understood. We must quantitatively model the electron structure, optical spectra, and relaxation dynamics – all key parameters in designing QD-based optoelectronic devices. Theoretical studies based on assumed shapes have described experimental PL spectra or explained observation of in-plane PL polarization anisotropy in QDs. However, these studies do not explain these phenomena quantitatively, leaving the precise shape of self-organized QDs unknown – knowledge critical to understanding the electronic structure of these technologically interesting nanostructures.

Since the shape of the QDs is quite different for different growth conditions such as growth temperature ($T_g$) and InAs deposition, one might question the existence of an equilibrium shape of the QDs. Many different shapes have been reported for InAs QDs grown on GaAs (100) substrates. Leonard et al. reported lens-shaped QDs without particular facets found in cross-section TEM analysis for InAs deposition at 530°C [13]. Plan-view TEM images of 2 ML and 4 ML InAs QDs grown at 480°C show square-shaped objects with case sides along the $\langle 100 \rangle$ directions.

On the other hand, Moison et al. found facets ranging from {140} to {110} for QDs deposited at 500°C in atomic force microscopy (AFM) investigations at room temperature [14]. Nabetani et al. concluded that large QDs formed on {311} facets based on reflection high-energy electron diffraction (RHEED) analysis for 13.3 nm upon 2 ML In deposition at 480°C [15]. From a detailed RHEED analysis, Lee et al. concluded that QDs formed on {136} facets for 1.68 ML of InAs deposited at 500°C [16].

Saito et al. investigated MBE-grown QD shape dependence on $T_g$ [17]. Figure 1.20 shows a series of AFM images of various growth temperatures ranging from 500 to 550°C. Small, dense QDs were formed at 500°C. Over 510°C, large QDs more than 9 nm high appear, and only large QDs with uniform size exist at 550°C. A clear difference was observed between small and large QDs. Figure 1.21 plots the QD height as a function of diameter. The different aspect ratios (QD height to diameter) in Fig. 1.21 and the RHEED patterns in Fig. 1.22 indicate the different QD shapes at high and low temperatures. The QDs grown at low temperatures (440–500°C) have a pyramidal

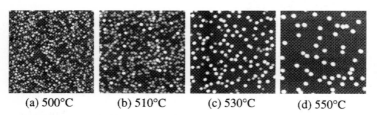

(a) 500°C      (b) 510°C      (c) 530°C      (d) 550°C

**Fig. 1.20.** (a–d) 1 μm×1 μm scan AFM images of InAs QDs grown at various temperatures (after Saito et al. [17])

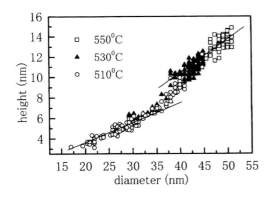

**Fig. 1.21.** Scatter plot of InAs QD height as a function of diameter. The QDs were grown at 510, 530, and 550°C (after Saito et al. [17])

shape with {136} facets deduced from RHEED patterns in Fig. 1.22 that show clear chevrons (also reported by Lee [16]). At high temperatures over 510°C, on the other hand, the RHEED pattern shows a new cross pattern along the [001] azimuth with a separation angle of 90°. It indicates the QD grown at high temperature has {110} facets. The emergence of a different facet orientation results in the different angle between the sidewall and the substrate surface (angle between (100) and (136): 27.8°, (100) and (110): 45°), leading to the different aspect ratios.

When QDs are grown on a high-Miller-index substrate, the islands are smaller and more uniform in size and show clear facet termination. The self-organized growth has led to the formation of quantum dashes (QDHs) as demonstrated by Guo et al., who observed the formation of InAs QDHs by growing InAs on GaAs (211)B substrates by MBE [18]. Figure 1.23a shows the 3D AFM image of QDHs for 12 ML ((211) face) of InAs grown at 500°C. Quantum dashes having an asymmetric hut-like geometry terminated by low-

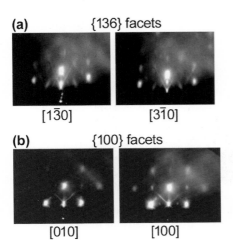

**(a)**   {136} facets

[1$\bar{3}$0]          [3$\bar{1}$0]

**(b)**   {100} facets

[010]          [100]

**Fig. 1.22.** RHEED patterns of InAs QDs grown at (**a**) 500°C and (**b**) 510°C (after Saito et al. [17])

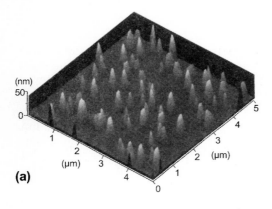

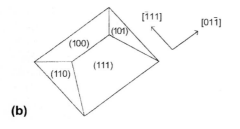

**Fig. 1.23.** (a) 5 μm×5 μm 3D AFM image for 12 ML of InAs deposited on GaAs (211)B substrate at 500°C; (b) schematic diagram of QDH structure (after Guo et al. [18])

index facets are observed (Fig. 1.23b). The clear facet termination was observed only when the $T_g$ is higher than 500°C. Islands grown at 450°C have vague shapes. Lacombe et al. reported formation of low-index facets in gas source MBE-grown InGaAs and InAs islands on InP (311)B substrates. In this case, relatively long growth interruption after InGaAs deposition results in smaller island size and clear facet termination [19]. As observed by HRTEM (Fig. 1.24), the islands are always bounded by a (111)B facet. On the other side, they are bounded by a (100) facet, or more frequently, by a steeper facet

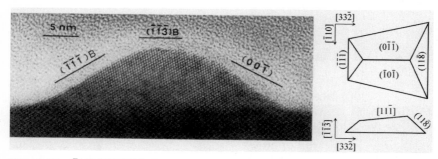

**Fig. 1.24.** [$\bar{1}$10] HRTEM observation of an InAs island and schematic view of InAs islands obtained with 30 s growth interruption (after Lacombe et al. [19])

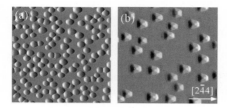

**Fig. 1.25.** $1\,\mu\text{m}\times1\,\mu\text{m}$ scan surface AFM images of (**a**) $In_{0.5}Ga_{0.5}As$ and (**b**) $In_{0.35}Ga_{0.65}As$ QD samples grown on (411)B substrates at 650°C (after Lee et al. [20])

forming a typical 35–38° angle with the substrate, corresponding on average to (h11)B facets where 6<h<8.

Similar facet-terminated QDs (FTQDs) were reported by Lee et al. [20]. Figure 1.25 shows AFM images of MOVPE-grown $In_{0.5}Ga_{0.5}As$ and $In_{0.25}Ga_{0.75}As$ QDs grown on (411)B substrates at 680°C with a nominal layer thickness of 20 Å. Lens-shaped QDs (LSQDs) 145 Å in height and 600 Å in diameter on the average are randomly arranged on the $In_{0.5}Ga_{0.5}As$ layer sample (Fig. 1.25a). Quantum dots 160 Å in height and 800 Å in base along [$2\bar{4}4$] on the average formed during the $In_{0.25}Ga_{0.75}As$ growth (Fig. 1.25b). An analysis of AFM profiles showed these islands to be bound by a large (100) facet and a small (111)B facet (along [$2\bar{4}4$] at tilt angles of about 19.5° and 35.3°) and by (110) and (101) facets (both at tilt angles of 33.5°).

Other substrate orientations also showed FTQDs, e.g., $In_{0.25}Ga_{0.75}As$ QDs grown on (711)B, (511)B, and (211)B substrates grown at 680°C (Fig. 1.26). Schematic images of QDs obtained from high-magnification AFM

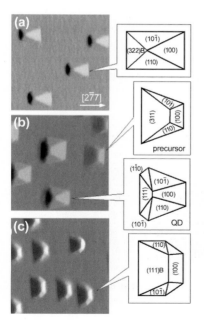

**Fig. 1.26.** 500 nm scan AFM images of QD structures grown on (**a**) (711)B; (**b**) (511)B; and (**c**) (211)B substrates. Schematic diagram of QD structures from AFM profiles are also shown (after Lee et al. [20])

profiles are also shown. All QDs consist of low-index facets. Islands consist of low-index facets that are energetically more favorable than high-index ones. Pehlke et al. reported that minimizing total energy derives the equilibrium shape of a strained coherent island due to competition between surface energy and elastic energy [21]. The variation with island size differs for these two contributions. While elastic energy scales linearly with volume $V$, total surface energy is proportional to the surface area, i.e., it scales with $V^{2/3}$. The strain effect thus becomes more important as volume increases, and QD shape depends mainly on the In composition (lattice mismatch). This agrees well with clear facet termination for the QD with lower In composition (Fig. 1.25).

An optimum island at equilibrium is comprised of {110}, {111}A, and {111}B facets and a {100} surface on top, on the (100) substrate. Formation of {110} and {111}B facets has also been reported in QDs grown by MBE, but high-index facets emerge in most actual QD growth because they are very steep (45° and 54°) against the (100) substrate [13, 14]. A steep angle of the QD sidewall against the substrate surface results in large QD height and large elastic energy. On the other hand, as was seen in Fig. 1.26, the emergence of low-index facets is enhanced on $(n11)$B ($n$=2–7) substrates because the $(n11)$ planes have small angles between low-index planes, therefore the complex surface structure of $(n11)$B results in high surface energy.

Quantum dot shape shows strong dependence on $T_g$. Figure 1.27 shows AFM images of $In_{0.25}Ga_{0.75}As$ QD structures grown on (711)B substrates at various values of $T_g$. At low $T_g$, 3D formation becomes vague and a mound-like structure 60 Å high and 600 Å wide elongated toward [011] without a clear facet formed. With increasing $T_g$, mound-like structures were cut into pieces along [011] because growth stops due to emergence of the {110} facet (Fig. 1.27b). This results in a shape transition from a mound to a clear facet-terminated island (Fig. 1.27c). Quantom dot structures 78 Å high with a 650 Å base along [$2\bar{7}7$] form on the surface for growth at 680°C. Mound-like structures (Fig. 1.27a, growth at 580°C) change to large QDs when the sample is annealed at 680°C for 5 min (Fig. 1.27d). Strong $T_g$ dependence of QD shape and facet formation at high $T_g$ suggest that the transition in surface reconstruction changes the surface structure. Transition

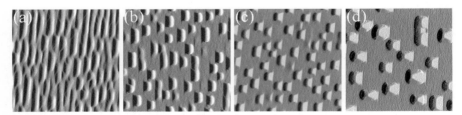

**Fig. 1.27.** Surface AFM images of $In_{0.25}Ga_{0.75}As$ QD structures grown on (711)B substrates at (**a**) 580°C; (**b**) 630°C; and (**c**) 680°C. (**d**) AFM image of an annealed QD sample ($T_g = 580$°C, annealed for 5 min at 680°C) is also shown (after Lee et al. [20])

from $(2 \times 2)$ to $(\sqrt{19} \times \sqrt{19})$ surface reconstruction on the GaAs (111)B surface, for example, occurs at a critical $T_g$ [22]. Since the surface energy depends on surface reconstruction, the facet arrangement minimizing total energy changes. Namely, the facet orientation preferentially emerging on the surface changes. For example, on the (111)B surface, the (110) facet prefers to emerge at high $T_g$. Therefore a simple $(1 \times 1)$ reconstruction of a (110) surface results in a lower surface energy than that of the complex surface structure of a $(\sqrt{19} \times \sqrt{19})$ reconstruction of a (111)B surface. Low growth rate of the (110) direction also enhances the emergence of the (110) facet. Transformation of the surface structure such as shape of step [23] and morphology [24] following changing surface reconstruction was observed. In the same manner, complex surface structure of the $(n11)$B surface having large surface energy and higher growth rate than that of low-index surfaces enhances the emergence of low-index facet termination of QDs.

In addition, strain energy relaxation contributes to the shape transition of 3D structures. Comparison of pyramids and prisms is presented in Fig. 1.28, where the volume elastic relaxation energy versus the tilt angle of facets is displayed for a square-based pyramid and for an infinitely elongated prism [25]. For the prism, the component of the strain field in the direction along the prism remains equal to lattice mismatch throughout the entire volume of the island, whereas for the pyramid, all components of the strain undergo relaxation and decrease with height. Thus, the figure demonstrates that the volume elastic relaxation is more efficient for pyramids (QDs) than for infinitely elongated prisms (QWRs), which explains the preferred pyramid-like shape of the islands. At low temperatures, wire-like structures are preserved because vague features of the structure are equivalent to the low facet angle (Fig. 1.27a). Clear facet formation of the structure at high $T_g$ results in the increase of facet angle and leads to enhancement of QD formation.

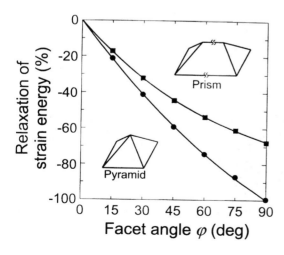

**Fig. 1.28.** Relaxation of elastic strain energy for 3D coherently strained islands versus the tilt angle of island facets for pyramids (*lower curve*) and for elongated prisms (*upper curve*). *Circles* and *squares* represent results of calculations by the finite element method, which are fitted by the *solid lines* (after Schukin et al. [25])

### 1.3.2    Ordering of Quantum Dot Position

**Artificial Alignment of Quantum Dots.** S–K growth mode apparently introduces unavoidable fluctuation in QD size and distribution. Considerable size fluctuation and broad emission may be mainly due to the nearly random initiation of islanding, making it difficult to reproducibly control QD size and density. Since the size and positional randomness of S–K islands limit device applications, it is important to control islanding initiation, including the onset of QD formation and distribution. Various attempts have been undertaken to improve site control for self-organized QDs, which is of importance for fabrication of devices incorporating QDs of homogeneous size. The basic concepts introduced are ordering of QDs on masked surfaces, along multiatomic steps, on corrugated surfaces, and so on. Further, given damage-free and contamination-free nanostructure fabrication, QD alignment by single step crystal growth is favorable and application of the substrate orientation is preferable.

In this section, various techniques for QD alignment are summarized. In the vicinity of a dislocation, the local strain is changed. Depending on the sign of change of strain, in- or out-diffusion of strained material is induced. Xie et al. introduced the dislocation network at the interface of the Si (100) substrate and a SiGe buffer layer (Fig. 1.29) [26]. The average dislocation distance is 100 nm, but dislocation spacing varies in a random fashion. Ge QDs were deposited on top of a thin Si cap. Ge dots were found to nucleate on the intersection of [110] and [1$\bar{1}$0] misfit dislocations. The use of relaxed templates eliminated random island nucleation. The typical Ge dot size is 200nm, but the dot size is far from uniform.

The lattice mismatched stress-induced 2D to 3D morphology transition is combined with interfacet adatom migration to selectively assembled parallel chains of InAs islands on the top of [1$\bar{1}$0]-oriented stripe mesas sub-100-nm

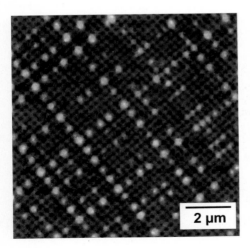

2 µm

**Fig. 1.29.** A 10 µm×10 µm AFM image of sample with 1.0 nm of Ge coverage grown at 750°C (after Xie et al. [26])

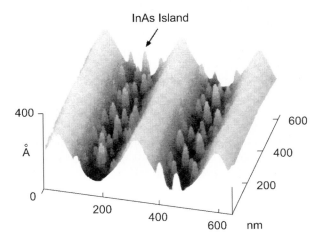

**InAs Island**

400

Å

0

200        400

600     nm

600

400

200

**Fig. 1.30.** AFM of a perspective view of the QD sample grown on V-groove aligned along [01$\bar{1}$] direction (after Mui et al. [27])

wide on GaAs (100) substrates. On such mesa stripes, prepared in situ via size-reducing epitaxy, deposition of InAs amounts subcritical for island formation on the planar GaAs (100) surface was performed. On the corrugated substrates, islands tend to nucleate at characteristic sites of the structure, such as edges, sidewalls, or trenches. Mui et al. reported the alignment of InAs islands grown on etched GaAs ridges along [110] and [1$\bar{1}$0] by MBE [27]. Rather large pitch islands are found to nucleate on the sidewalls for ridges along [110] and are found on the (100) top of the mesas or at the foot of the mesa for ridges along [1$\bar{1}$0]. In short pitch V-grooves, islands are only found at the bottom and sidewalls of grooves (Fig. 1.30). Alignment of InAs islands grown using chemical beam epitaxy in chains at the bottom of trenches was reported in Jeppesen et al. [28]. The islands densely fill the trenches in chains as long as the patterns, tens of microns long. They also reported preferred nucleation in cylindrical holes. There are one to four islands

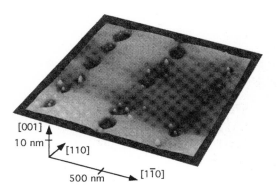

[001]

10 nm

[110]

500 nm        [1$\bar{1}$0]

**Fig. 1.31.** AFM tip height image of sample that shows islands in the holes (after Jeppesen et al. [28])

in each hole, apparently determined by the lithography, and the islands show very selective placement in the pockets (Fig. 1.31).

Jin et al. reported the ability to arrange Ge islands on patterned Si (100) substrates grown by gas source MBE [29]. The preferential location of Ge islands on top Si (100) ridges with {311} sidewalls. Regularly spaced one-dimensional (1D) arrays of the Ge islands are formed on the ridges of the Si stripe mesas (Fig. 1.32). All islands are dome shaped and have a common size of 70–90 nm. They also investigated the Ge islands formed on the square Si mesas. After the formation of Si square mesas in the exposed Si windows, four corners on the mesa are formed, which are the energetically preferred sites. Therefore, four Ge islands are formed on the square mesas with the base square oriented in the [110] directions, as seen in Fig. 1.33a. In contrast,

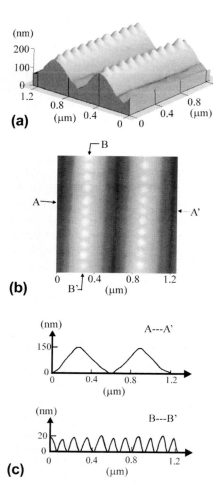

**Fig. 1.32.** (a) 3D AFM image of the self-organized Ge islands on the ⟨110⟩-oriented Si stripe mesas with a window width of 0.6 µm. Self-aligned and well-spaced 1D arrays of the Ge islands are formed on the ridges of the Si mesas after the deposition of 10 ML Ge. (b) 2D image of the island arrays in (a), along with the cross-sections of the mesas line (*line* AA′) and one array of the islands (*line* BB′), respectively. The sidewall facets are {311} facets (after Jin et al. [29])

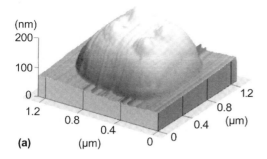

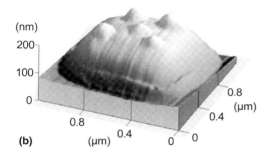

**Fig. 1.33.** (a) 3D AFM image with four Ge islands located at the corners on a square Si mesa with the base lines parallel to the ⟨110⟩ directions. The Ge thickness is 9 ML. (b) 3D AFM image with five Ge islands with 10 ML Ge. The fifth pyramidal island is formed in the central region. The average base size of the islands is about 140 nm (after Jin et al. [29])

the central region is free of Ge islands. This is because the sites in the central region are not preferred, and Ge adatoms have a sufficiently long diffusion length to migrate to the preferential corner sites. With increasing the Ge deposition thickness, the fifth island is formed in the central region of the square mesas (Fig. 1.33b). It is found that the central island is a pyramid with a square base, which is different from the other four dome-shaped islands at the corners. The central island is believed to be at the earlier stage of its evolution, i.e., it has not yet undergone a shape transformation of the four islands at the corners, changing the strain distribution, thus leading to the change of the energy distribution on the mesa. Compared with the corners, other sites, such as the center of the mesa, become the preferential sites. Then the excessive Ge forms the fifth island at the center.

**Spontaneous Lateral Alignment of Quantum Dots by Single Step Crystal Growth.** Quantum wires and boxes have been grown utilizing the periodic step structure on vicinal surfaces. Kitamura et al. demonstrated natural alignment of InGaAs QDs on the GaAs multiatomic step structure by MOVPE growth [30]. This growth technique results in spontaneously aligned InGaAs QDs without any preprocessing techniques prior to the growth. Figure 1.34 shows AFM images of the QDs grown on the (100) surface 2°-misoriented toward the [010] direction. The InGaAs QDs are aligned at the

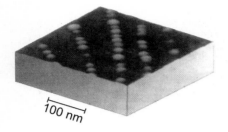

**Fig. 1.34.** 300 nm×300 nm size AFM image of InGaAs QDs aligned at multiatomic steps on GaAs (100) surface misoriented by 2° toward [010] (after Kitamura et al. [30])

multiatomic step edges. However, this technique cannot fulfill the condition of well-ordered QD arrays with high density. QD coalescence under high-QD density results in the formation of QWR-like structure. Moreover, it is difficult for the (100)-misoriented surface to fabricate uniform and straight multiatomic steps. On the other hand, the breakup of high-index surfaces into arrays of low-index facets to decrease the surface energy is also used for ordering of QDs. Formation of straight and uniform corrugations (or giant steps) on (211)B and (111) substrates has been reported by Lee et al. [23, 31].

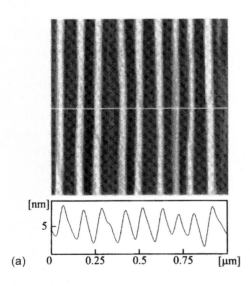

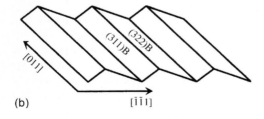

**Fig. 1.35.** AFM images and profiles of GaAs (211)B surfaces grown at (**a**) 650°C. The *line* across the AFM images shows the path taken for the profiles shown below. (**b**) Schematic diagram of stepped surface in Fig. 1.34 (after Lee et al. [31])

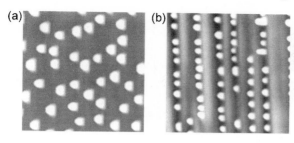

**Fig. 1.36.**      $1\,\mu m \times 1\,\mu m$ scan surface AFM images and oblique views of $In_{0.5}Ga_{0.5}As$ QD samples grown on (**a**) smooth surface and (**b**) saw-tooth corrugation (after Lee et al. [31])

Figure 1.35a shows an AFM image of multistep arrays on the (211)B substrate. Facets emerging on the surface correspond to the (311)B and (322)B surfaces from the angles of about $10°$ toward the [100] direction and $8°$ toward the [111]B direction, respectively, as was seen in the schematic diagram of a surface structure (Fig. 1.35b). Dramatic step feature improvement with increasing $T_g$ on the (211)B surface was related to facet formation accompanied by changes in surface reconstruction, similar to the shape transition from wire to QD structures. Spontaneous (111)B facet formation at step edges was observed when the surface reconstruction changes from $(2 \times 2)$ to $(\sqrt{19} \times \sqrt{19})$ on GaAs (111)A vicinal surfaces [23].

Quantum dot structures grown on the surface had a nominal $In_{0.5}Ga_{0.5}As$ layer thickness of 12 Å. Figure 1.36 shows the surface AFM images and the oblique views of $In_{0.5}Ga_{0.5}As$ QD structures grown on the planar and corrugated surfaces. The period of the corrugation was 1100 Å and its depth was 60 Å. On corrugated surface, QDs do not form on (322)B terraces, but do form on (311)B terraces. The low growth rate and large migration length of cation adatoms on the (322)B surface results in selective QD formation on the (311)B surface. For the sample seen in Fig. 1.36a, the width of (311)B terraces (450 Å) is comparable to or smaller than the base width of QDs. In this case, 1D QD alignment along the corrugation is well defined. Size uniformity is improved because the (311)B terrace width limits the lowest size of QDs.

R. Nötzel et al. reported the phenomena of self-organized formation of box-like microstructures on GaAs (311)B surfaces during epitaxial growth by MOVPE [32]. In the growth of strained InGaAs/AlGaAs heterostructures, they found that InGaAs films naturally arrange into homogeneous nanoscale disks directly covered with AlGaAs during growth interruption (Fig. 1.37). Unlike islanding on conventional (100) substrates, the system produces well-ordered, high-density arrays of AlGaAs microcrystals containing disk-shaped InGaAs box structures. Due to the absence of any artificial patterning procedure, disks exhibit very high uniformity and structural perfection that manifests itself in the well-resolved exciton resonances in photoluminescence excitation (PLE) spectra and in the smaller PL linewidth compared to that of conventional (100) QWs. Figure 1.38a,b shows the PL and PLE spectra at

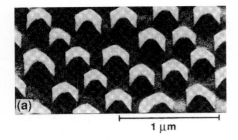

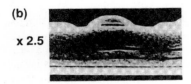

**Fig. 1.37.** (**a**) Scanning electron topographical image of $Al_{0.5}Ga_{0.5}As$ microcrystals containing 150-nm-diameter (311)B $In_{0.2}Ga_{0.8}As$ disks. (**b**) Cross-sectional image of the same structure after stain etching (after Nötzel et al. [32])

room temperature of the quantum disks and reference QW. The linewidth of the disks is 13 meV, compared to 27 meV for the reference QW. This narrow linewidth of the disks indicates smoothing and ordering of the interfaces during disk formation, which is most directly shown in the PL and PLE taken at

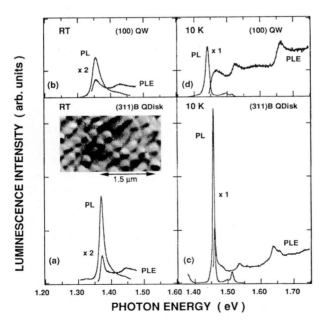

**Fig. 1.38.** (**a**) Room temperature PL and PLE spectra of coupled InGaAs quantum disks and (**b**) reference (100) QW. The *inset* shows topographical image of the modulated surface; (**c**) and (**d**) show PL and PLE taken at 10 K (after Nötzel et al. [32])

10 K from small Stokes shift of 3 meV and the 6 meV linewidth (Fig. 1.38c). The unusually small linewidth at room temperature, i.e., the reduced thermal broadening, can be attributed to efficient lateral localization of the excitons in the disks.

**Quantum Dot Self-Alignment Using Vertical Stacking.** Vertical stacking of layers containing QDs is important for most device applications in order to increase the filling factor of QDs in a given sample. Multistacking can also be used to control initiation of islanding. Upper islands tend to stack just on lower islands due to the strain field induced by the bottom islands. The stress on a crystal surface can provide a natural driving force for 3D island formation in lattice-mismatched growth. The first set of islands produces a tensile stress in the capping layer above them, whereas a region having little or no stress may exist on the wetting layer. Another effect of multistacking is enhanced islanding due to the integrated strain induced by repeated In-GaAs/GaAs growth. The strain induced by lower-layer islands or the In atom segregation from these islands affects subsequent islanding.

Figure 1.39 shows a representative [011] cross-sectional TEM image of a sample with five sets of InAs islands and GaAs spacers 36 ML thick (101.7 Å). Figure 1.40 summarizes the pairing probabilities as a function of the spacer thickness. Probability of an island pairing to the set just below is maintained at a level of 0.95 for all adjacent island sets. In pairing probability, there are three typical regimes depending on the spacer thickness: (1) a regime for small spacer thickness, probability is greater than 0.95, indicating a nearly completely correlated behavior; (2) a regime of gradual decrease in probability; and (3) a regime for large spacer thickness, probability saturating at a value corresponding to random overlapping of islands.

Based on the experimental data, Xie et al. have proposed the phenomenological model for vertical self-organization of coherent InAs islands separated by GaAs spacer layers along the growth direction [33]. The kinetic process giving the vertically self-organized growth behavior is depicted in Fig. 1.41. Islands in the first set produce a tensile stress in the GaAs above the islands (region I), whereas region II, having little or no stress, may exist depending on the average separation $l$ between the first set of islands and the GaAs spacer layer thickness-dependent range of the surface strain fields, $l_s$. The In atoms impinging in region I would then be driven by the strain field to

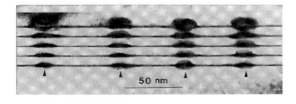

**Fig. 1.39.** Dark-field TEM image for sample having five sets of islands and 36 ML spacer layers (after Xie et al. [33])

50 nm

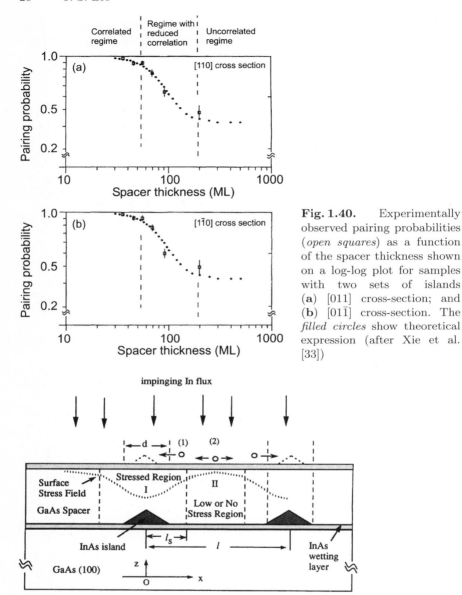

**Fig. 1.40.** Experimentally observed pairing probabilities (*open squares*) as a function of the spacer thickness shown on a log-log plot for samples with two sets of islands (**a**) [011] cross-section; and (**b**) [01$\bar{1}$] cross-section. The *filled circles* show theoretical expression (after Xie et al. [33])

**Fig. 1.41.** A schematic representation showing the two major processes for the In adatom migration on the stressed surface: (**1**) directional diffusion under mechanochemical potential gradient contributing to vertical self-organization and (**2**) largely symmetric thermal migration in regions from the islands contributing to initiation of new islands not vertically aligned with islands below (after Xie et al. [33])

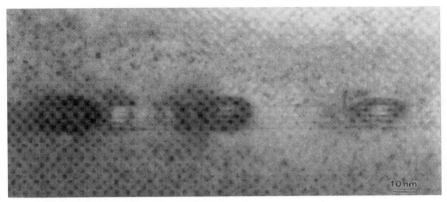

**Fig. 1.42.** A (011) cross-sectional TEM image of a 5-layer stacked InAs island grown at 2 nm intervals (after Nakata et al. [34])

accumulate on top of the lower islands, where the atoms can also achieve the lower energy thermodynamic state due to lower lattice mismatch of InAs with the GaAs in tension. On the other hand, the In atoms impinging in region II may initiate formation of islands in region II, an undesirable feature for achieving the most efficient vertical ordering. When the spacing layer is chosen appropriately, i.e., $2l_s > l$, island formation in region II is suppressed and vertical self-organization of islands occurs. A characteristic spacer thickness for vertically self-organized growth to occur inferred from the model was found to be consistent with the experimental data.

When QDs are closely stacked, the islands are electronically coupled and behave as a single QD. Nakata et al. stacked InAs QDs with GaAs intermediate layers of less than 3 nm thickness on (100) GaAs substrates by MBE [34]. Figure 1.42 shows a [011] cross-sectional TEM image of a 5-stacked QD structure grown at 2 nm intervals. The upper-layer QDs grew nearly on the lower-layer QDs, aligning vertically with the first layer QDs. These QDs were stacked almost in columns of about 22 nm in diameter and 13 nm in height, shown in the image as dark megaphone-like strained regions. QDs in an individual layer were seen to be spatially isolated in the vertical direction, but the distance between the bottom of the upper-layer QDs and the top of the lower-layer QDs was seen to be 3 to 4 ML or less, which is so thin that the wave functions can be distributed along the vertical column. Reflecting this, PL spectra show clear differences between the closely stacked and single QD layers. Figure 1.43 shows the PL spectra of single, 3-, and 5-layer QD structures grown at 3 nm intervals. The broad PL spectrum of a single QD layer transformed to sharper spectra as the number of stacked layers increased. These narrow linewidths a re attributed to the suppression of relative QD size fluctuation in the vertical direction. The peak energy shift to the lower

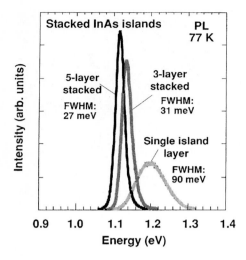

**Fig. 1.43.** PL spectra of single, 3-layer, and 5-layer stacked InAs island layer structures grown at 3 nm GaAs intervals as measured at 77 K (after Nakata et al. [34])

energy by about 90 meV implies the reduction of quantum size effects due to the increase in effective QD heights.

The alternating InAs/GaAs epitaxy based on the atomic layer epitaxy condition forms highly uniform microcrystals that are practically QDs [35]. Figure 1.44 shows plan-view and cross-sectional TEM images. A plan-view image shows uniform dot-like microstructures of about 20 nm in diameter. Cross-sectional dark field images indicated that the dots of spherical shape were formed within the short growth period layer. The dots were about 10 nm in height and surrounded laterally by a QW layer having the same thickness as the dots. Reflecting the uniform size of QDs, the narrow PL peak (FWHM~30 meV) was obtained around 1.35 μm at 300 K.

In addition to the vertical alignment of QDs by multistacking, the anisotropic crystallographic arrangement of high-Miller-index planes makes it possible to develop high-density and laterally self-aligned QDs. Lee et al. reported 1D self-alignment of multistacked QDs on $(n11)B$ ($n=2$–7) substrates [36, 37]. Figure 1.45 shows surface AFM images of QDs grown on (311)B sub-

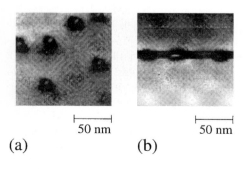

(a)                    (b)

**Fig. 1.44.** (a) Plan-view image and (b) cross-sectional dark-field image of QDs (after Mukai et al. [35])

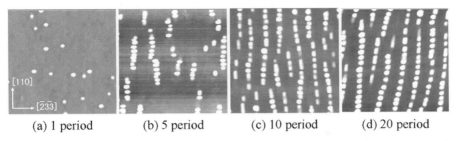

(a) 1 period        (b) 5 period        (c) 10 period        (d) 20 period

**Fig. 1.45.** 1 μm surface AFM images of QDs with (**a**) 1-, (**b**) 5-, (**c**) 10-, and (**d**) 20-periodic In$_{0.45}$Ga$_{0.55}$As/GaAs structures (after Lee et al. [36])

strates in structures including 1-, 5-, 10-, and 20-periodic In$_{0.45}$Ga$_{0.55}$As/GaAs multilayers grown at 530°C. Randomly distributed QDs then begin to align in the [011] direction as the number of stacking layers increases. With 5-periodic layer growth, the QDs form chains that lengthen with increasing number of stacking layers (Fig. 1.45b). The elongated QD chains then coalesce to form arrays (Fig. 1.45c,d). Individual QDs are clearly separated in the 20-periodic sample, even though the distance between them is very small (within 150 Å). The distance between the bottom of the upper-layer QDs and the top of the lower-layer QDs is so small (less than 30 Å) that the In atom segregation and strain effects are quite large. It is presumably this large strain that leads to the rapid increase in QD density (by one order of magnitude) in 1- to 20-periodic samples, although the same amount of In$_{0.45}$Ga$_{0.55}$As is supplied to each layer.

In ordinary multistacking QD growth, the island-induced evolving strain field drives island self-alignment in the vertical direction. Self-alignment of the QD array cannot be explained by the multistacking mechanism alone, however. The GaAs (311)B surface exhibits a smooth surface with monoatomic height steps [36]. The step density (a few steps per 1 μm$^2$) is much less than the QD array density (around 10 steps per 1 μm$^2$), indicating that the in-plane alignment was not caused by the surface morphology or steps. The direct origin for QD self-alignment is the anisotropy in the GaAs layer deformation, originating in the anisotropy in the crystallographic arrangement (Fig. 1.46a). The (311)B surface structure has hexagonally shaped unit cells. The bond density is the lowest in the diagonal lines of the unit cells, namely in the [130] direction, as indicated by the arrow in Fig. 1.46a. The largest deformation occurs along this direction when the epitaxial layer with a different lattice constant causes biaxial strain. The direction of alignment of the single InGaAs QD layer on the (311)B substrate matches this direction [38]. When local strain was caused by QD formation, adjacent atoms arranged along the [011] direction are deformed, as seen in Fig. 1.46a (only one In atom was described here for simplicity). The strain field around QDs is distributed elongated to the [011] azimuth because of the crystallographic

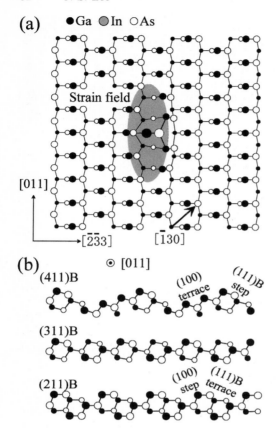

**(a)** ● Ga ◐ In ○ As

Strain field

[011]

[2̄3̄3]    [1̄30]

**(b)** ⊙ [011]

(411)B

(100) terrace    (111)B step

(311)B

(211)B

(100) step    (111)B terrace

**Fig. 1.46.** (a) Schematic top view of atomic geometry of GaAs of the (311)B surface. (b) Cross-sectional arrangement of Ga and As atoms of $(n11)$B surfaces viewed from the [011] direction (after Lee et al. [37])

anisotropy of the (311)B plane. In this case, a region with the highest probability of QD formation is also directly above the QDs in the lower layer. In a difference from the QD multi-stacking on a (100) substrate, however, the anisotropic strain field creates a region with the second highest probability of QD formation distributing to both sides of the QDs in the [011] direction. QD chains thus expand along the [011] azimuth as the number of stacking layers increases. This assumption is well supported by the high QD array density, when a GaAs spacing layer is thin enough to form a large strain field. On the other hand, low QD density featuring vague alignment is observed when the thick GaAs spacing layer is thick [36].

The multistacking QD structures grown on (411)B and (211)B substrates exhibit somewhat different characteristics, reflecting the presence of steps on the surface (Fig. 1.47). A (411)B ((211)B) surface consists of (111)B steps (terraces) and (100) terraces (steps), as seen in Fig. 1.46b. Straight and uniform multiatomic height steps are formed on the surface consisting of the (111)B terrace. The straight multistep arrays formed on the (211)B surface

GaAs          20-periodic QD

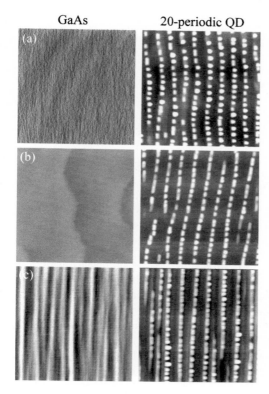

**Fig. 1.47.** 1 μm×1 μm AFM images of GaAs and QD samples with 20-periodic In$_{0.45}$Ga$_{0.55}$As/GaAs layers grown on (**a**) (411)B; (**b**) (311)B; and (**c**) (211)B substrates (after Lee et al. [37])

of the GaAs buffer layer (Fig. 1.47c) and the spontaneous QD formation on the step edges improve continuity and straightness in QD alignment. Superior straightness was seen in the QD arrays for the (211)B sample in comparison to samples in the other orientations. Meanwhile, the (411)B sample shows the best continuity in QD arrays. Existence of QD chains more than a few microns long was confirmed in a wider-area AFM scan. Although the image of the GaAs (411)B surface displays no structure due to poor lateral AFM resolution, there ought to be steps with high density in crystallography. It is considered that the growth proceeds in a step-flow mode owing to the densely distributed steps. In the step-flow growth mode, the incoming atoms impinge at the step edges, enhancing spontaneous QD formation in the step region. This may result in superior continuity in the QD arrays. Thus, introduction of steps on the surface using high Miller indices planes effects on improvement in the QD alignment.

The effect of the 1D QD arrangement is clearly shown in polarization-resolved photoluminescence (PRPL) spectra. Figure 1.48 shows the PRPL spectra for the QD sample with the surface morphology shown in Fig. 1.45d. PRPL measurement was performed with the electric field vectors parallel and

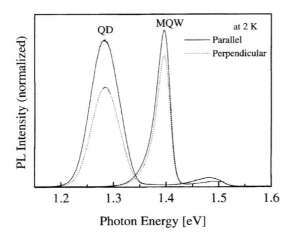

**Fig. 1.48.** PRPL spectra of a 20-periodic QD sample grown on a (311)B substrate (after Lee et al. [36])

perpendicular to the QD arrays ([011] direction). Clear polarization dependence was seen in the spectrum from the QD region, but not in the GaAs-related spectrum at 1.49 eV. The degree of polarization of PL, defined as $(I_\parallel - I_\perp)/(I_\parallel + I_\perp)$ is 12.5% (where $I_\parallel$ ($I_\perp$) is the PL peak intensity whose polarization is parallel (perpendicular) to [011]). Polarization dependence is also observed in QW PL spectra, because polarization of optical matrix elements depends on the substrate orientation [39]. The degree of polarization of PL for the QD array sample is about twice that of the QW sample (6%). The PRPL spectra of QD arrays are similar to those for QWR structures, suggesting the existence of carrier coupling in 1D systems originating from the short distance between self-aligned QDs.

## 1.4   Real-Time Monitoring of Self-Organized Quantum Dot Formation

The growth and optical characterization of self-assembled QDs have been widely studied, mostly in the In(Ga)As/GaAs and InP/InGaP systems by MBE and MOVPE. Problems awaiting solutions include non-uniform size distribution and uncertainties of QD height, which have led to difficulty in comparing the observed quantization effect with the theoretical one. The precise control of growth conditions and determination of the optimum growth parameters are crucial to achieving well-defined QD structures. In situ monitoring is an important tool for ensuring reproducibility and feedback on the structure formation. Reflection high-energy electron diffraction (RHEED) is used routinely for this in MBE. The diffraction pattern observed is also used for different monitoring purposes, e.g., deoxidization of the substrate surface, temperature calibration, and thickness control via monolayer oscillations in island growth mode.

On the other hand, problems remain in reproducible QD fabrication by means of MOVPE due to the lack of in situ characterization methods. Considerable attention has recently been directed to developing optical in situ monitoring techniques applicable to MOVPE, such as reflectance difference spectroscopy (RDS, also termed RAS, reflectance anisotropy spectroscopy), surface photo absorption (SPA), and ellipsometry. In situ RDS and ellipsometry have been successfully used to monitor layer thickness in monolayer order, bulk composition, and surface reconstruction during III–V growth.

In this section, real-time monitoring techniques for MBE and MOVPE are summarized. Growth rate evaluation in monolayer order and determination of QD formation onset by in situ monitoring techniques are presented.

### 1.4.1  Reflection High-Energy Electron Diffraction in Molecular Beam Epitaxy

Reflection high-energy electron diffraction (RHEED) is a highly surface sensitive ultra-high-vacuum technique used to monitor growth in MBE systems. It provides information on morphology and surface reconstruction [40].

The conditions for constructive interference of the elastically scattered electrons may be inferred, using the Ewald construction in the reciprocal lattice. In the case where the electron beam essentially interacts with a 2D atomic net, the reciprocal lattice is composed of rods in reciprocal space in a direction normal to the real surface. Figure 1.49 shows the Ewald sphere and reciprocal lattice rods for a simple square net. It should be noted that

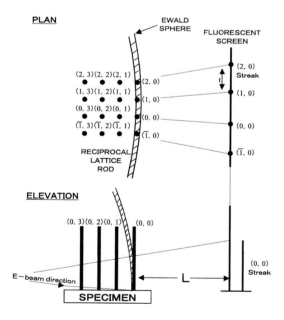

**Fig. 1.49.** The Ewald sphere and reciprocal lattice rods for a simple square net

the reciprocal lattice rods have finite thickness due to lattice imperfections and thermal vibrations, and that the Ewald sphere also has finite thickness, due, in this case, to electron energy spread and to beam convergence. The radius of the Ewald sphere is very much larger than the separation of the rods. This can be verified from a simple calculation of electron wavelength: if the surface lattice net has a lattice constant $a$ of 5.65 Å (unreconstructed GaAs along the [100] direction), then the distance between adjacent rods in the reciprocal space ($=2\pi/a$) will be 1.1 Å. Electrons have a wavelength $\lambda$ related to the potential difference $V$ through which they have been accelerated by the equation

$$\lambda \sim \sqrt{\frac{150}{V(1+10^{-6}V)}} \ [\text{Å}].$$

From this equation, $\lambda$ is 0.12~0.06 Å when $V$ is 10~50 kV.

Using this, it follows that the radius of the Ewald sphere is 36.3 Å. As a result, the intersection of the sphere and rods occurs some way along their length, resulting in a streaked rather than a spotty diffraction pattern, examples of which are shown in Fig. 1.50. It is easy to show that if the distance between the crystal surface and the screen is $L$, and if the separation of the streaks is $t$, then the periodicity in the surface is given by

$$a = \lambda L/t \ [\text{Å}].$$

Real semiconductor surfaces are more complex and their detailed structure must be inferred from diffraction patterns taken at different azimuths. The polar surfaces of GaAs such as the (100) and (111) planes have an excess of Ga or As atoms, depending on growth conditions or postgrowth treatment. The different reconstructions of this surface are related to its variable stoichiometry, and several attempts have been made to determine the effective As coverage as a function of the particular reconstruction. It is generally agreed that both the (2×4) and c(4×4) structures being As rich, with the latter having the higher As coverage, are stable on the GaAs (100) surface. For the (2×4) structure, a comparison between theoretical and experimental results gives strong evidence for the presence of asymmetric As dimers

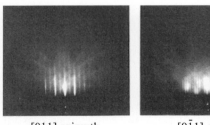

[011] azimuth          [0$\bar{1}$1] azimuth

**Fig. 1.50.** Observed RHEED patterns of GaAs (100)–(2 × 4) surface reconstruction (after Neave et al. [41])

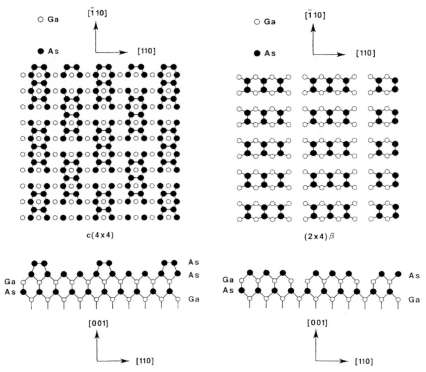

**Fig. 1.51.** Schematic crystal arrangement of surface reconstruction on GaAs (100) surface

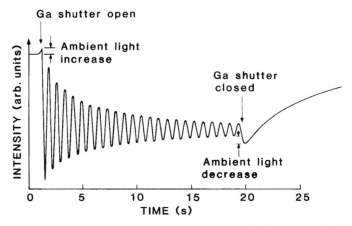

**Fig. 1.52.** Intensity oscillations of the specular beam in the RHEED pattern from a GaAs (100)–(2×4) reconstructed surface, [011] azimuth (after Neave et al. [41])

in the surface layer (Fig. 1.51). An energy minimization approach for the c(4×4) surface based on the assumption that this surface is stoichiometrically As-terminated suggests an equal number of symmetric and asymmetric As dimers. It is immediately clear from the three low-index azimuth patterns that the surface has a c(4×4) structure and this symmetry is maintained over a wide range of As exposure. These results relating surface structure and composition have been confirmed with some excellent mass spectroscopy and Auger studies. In the case of growth of GaAs films on GaAs (100) substrates by MBE, there are oscillations in the intensity of the RHEED pattern [41]. A typical example is shown in Fig. 1.52. The period of the oscillation corresponds exactly to the growth of a single monolayer, i.e., one complete layer of Ga, plus one complete layer of As. It is obvious, therefore, that the oscillations provide an absolute measurement of growth rate.

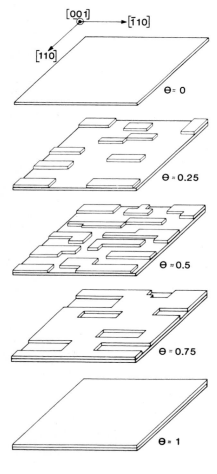

**Fig. 1.53.** Real-space representation of the formation of a single complete layer (after Neave et al. [41])

Figure 1.53 shows a real-space representation of the formation of a single complete layer, which illustrates how the oscillation in the intensity of the specular beam occurs [50]. There is a maximum in reflectivity for the initial and final smooth surfaces and a minimum for the intermediate stage when the growing layer is approximately half complete. In early stages, one layer is likely to be almost complete before the next layer starts so the reflectivity increases as the surface again becomes smooth on the atomic scale, but with subsequent surface roughening as the next layer develops. This repetitive process causes the oscillations in reflectivity gradually to be damped as the surface becomes statistically distributed over several incomplete atomic levels.

In principle, this combination of MBE and RHEED enables the response of a growing surface to a perturbation to be monitored easily and continuously to provide information on morphology and structure. Upon transformation of an initially 2D ordered surface with monolayer-high islands into a corrugated structure, the RHEED pattern changes from streaky to spotty. Figure 1.54 shows the RHEED patterns observed during InAs growth. Since streak patterns can be seen in Fig. 1.54a, the surface is flat before InAs growth. When 1 ML of InAs was grown, the streak patterns remained, i.e., 1 ML of InAs grew two-dimensionally on GaAs. However, when 2 ML of InAs

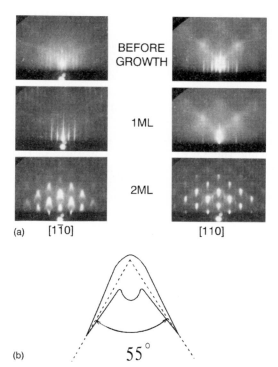

**Fig. 1.54.** (a) Observed RHEED patterns during InAs growth. (b) Two typical oblique streaks starting from the same reciprocal lattice point (after Nabetani et al. [42])

were grown, the RHEED patterns changed from streaky to spotty, indicating that InAs grows three-dimensionally. Moreover, Nabetani et al. reported observed oblique streaks in the RHEED pattern with the electron beam incident along the [0$\bar{1}$1] direction [42]. Since the angle between two streaks starting from same reciprocal lattice point is about 55°, as shown in Fig. 1.54b, the InAs 3D structures have (113)A facets. On the other hand, no specific facets can be observed with the incident electron beam along the [110] direction. The pattern indicates merely that InAs grows three-dimensionally. Lee et al. reported that the bounding facets of MBE-grown InAs/GaAs QDs are of the {136} family [43].

### 1.4.2 Optical in situ Measurement in Metal-Organic Vapor-Phase Epitaxy

**In situ Monitoring in Monolayer Order.** In metal organic vapor-phase epitaxy (MOVPE) under gas-phase conditions, electron-based in situ techniques cannot be used because of beam scattering in the ambient. To overcome this difficulty, MOVPE has been studied by techniques such as RDS, SPA, and spectroscopic ellipsometry (SE).

RDS and ellipsometry coupled to the MOVPE process have shown a high potential for the in situ monitoring of many relevant growth parameters such as substrate temperature, surface reconstruction, growth rate, composition of ternary compounds, and surface morphology.

RDS is a useful technique to measure the optical anisotropy induced by the reconstruction of semiconductor surfaces. Typical RD (reflectance difference) spectra for the four primary reconstructions (4×2), (2×4), and c(4×4) are shown in Fig. 1.55 [44]. The features in these spectra originate from electronic transitions between energy levels of the local atomic structures, and can be uniquely related to specific surface dimers. The 1.9 eV feature of the (4×2) surface is due to transitions between bonding Ga dimer orbitals and empty Ga lone-pair states, while the 2.6 eV and 4.2 eV features of the (2×4) surface are due to transitions between filled As lone-pair states and the unoccupied As antibonding dimer orbitals, and transitions between bonding and antibonding As dimer orbitals, respectively. The positive 2.6 eV feature that is inverted in the c(4×4) spectrum can be explained by the difference in As dimer orientation: [$\bar{1}$10] and [110], respectively. In RDS the d(4×4) surface is distinguished from c(4×4) by the peak at 4.0 eV.

Spectroscopic difference ellipsometry (SDE), measuring the surface-induced optical anisotropy $\Delta\varepsilon=\varepsilon_{\bar{1}10}-\varepsilon_{110}$ of the surface layer, shows similar features. Figure 1.56 compares the RD and SE spectra for the (2×4), (4×2), and c(4×4) surface reconstructions. For the (2×4) As terminated reconstruction, the imaginary part of $\Delta\varepsilon$ is positive around 2.6 eV [45]. On the other hand, $\Delta\varepsilon$ around 2.6 eV for c(4×4) is negative, indicating the c(4×4) surface reconstruction contains additional As dimers rotated by 90° on the top of those of the (2×4) reconstruction. Similar to the response of the c(4×4)

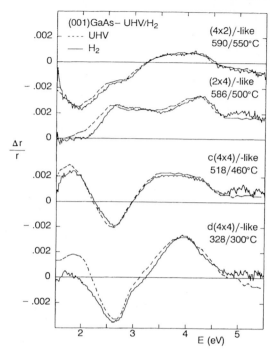

**Fig. 1.55.** RD spectra for the four primary reconstructions (4×2), (2×4), and c(4×4) (after Kamiya et al. [44])

reconstruction, the signal of the Ga-terminated (4×2) reconstruction is negative in agreement with the 90° rotation of the Ga dimers with respect to the As dimers of the (2×4) reconstruction.

As described above, surface dimers obviously have oscillator-like anisotropic absorption resonances with characteristic resonance energies, i.e., about 2.6 eV for As dimers and about 2.0 eV for the Ga dimers. These can be utilized to sensitively monitor the surface status through the spectral shape or in very rapid transport situations by taking one-wavelength transients. Taking the above results and step density change during 1 ML of growth into account, the monolayer oscillations in RDS and ellipsometry signals are predicted as follows [46–48].

In MOVPE, c(4×4) reconstruction is stable on the (100) GaAs surface (Fig. 1.57a). During growth, the (2×4)-like dimer constellation related to a smooth surface covers the centers of islands (Fig. 1.57b). On the other hand, the Ga dimer constellation is related to the (111)A step at island edges. The $(n×6)$-like ($n$=1 or 4) structure, the mixture of As and Ga dimers, is apparently more likely to exist close to steps.

The Bruggemann effective medium approximation (EMA) has been shown to describe these effects accurately and is used here to describe oscillations in the ellipsometry response through island formation [49]. The effective complex dielectric functions of films, obtained by EMA theory, are connected with

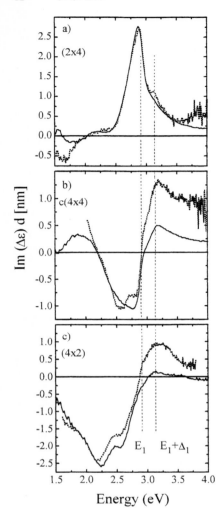

**Fig. 1.56.** Imaginary part of the surface-induced optical anisotropy $\Delta\varepsilon = \varepsilon_{\bar{1}10} - \varepsilon_{110}$ of (**a**) the GaAs (100)–(2×4) surface reconstruction; (**b**) the GaAs (100)–c(4×4) surface reconstruction; and (**c**) the GaAs (100)–(4×2) surface reconstruction. The *solid line* represents the RDS and the *dashed line* the SDE data (after Wassermeier et al. [45])

complex dielectric functions of the GaAs surface with a mixture of reconstructions and ambient. Accordingly, the effective complex dielectric function $\varepsilon$ of the GaAs topmost layer during growth is obtained from the following equations:

$$f_a \frac{\varepsilon_a - \varepsilon}{\varepsilon_a + \varepsilon} + \sum_{k=1}^{n} f_k \frac{\varepsilon_k - \varepsilon}{\varepsilon_k + \varepsilon} = 0$$

with

$$f_a + \sum_{k=1}^{n} f_k = 1\,,$$

## (a) Smooth surface (1 monolayer growth completes)

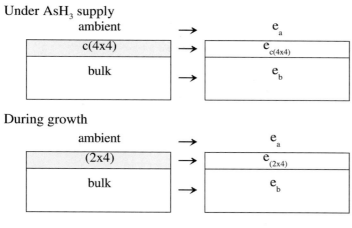

## (b) Rough surface (half monolayer growth)
Surface reconstruction model

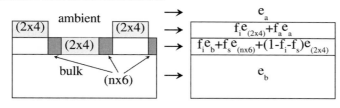

Ga-C bond model

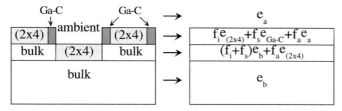

**Fig. 1.57.** Analytical models for Bruggemann EMA: (**a**) model of perfectly smooth layer; (**b**) model of layer with surface roughness with step-related $(n \times 6)$-like area fraction and rough surface with Ga–C bond at step edges (after Lee et al. [47, 48])

where $n$ is the number of distinct constituent media in the mixture, $f_k$ (for example, $f_{n \times 6}$ and $f_{2 \times 4}$) is the volume fraction of the $k$th component, and $\Delta \varepsilon_k$ is the dielectric function of the $k$th component. The values $f_a$ and $\varepsilon_a$ correspond to the volume fraction and dielectric function of the ambient. During growth, the surface is assumed to oscillate between more $(2 \times 4)$-like As dimers and more $(n \times 6)$-like As and Ga dimers. Thus, $\varepsilon$ oscillates following the above equations. Smooth and rough surface models and pseudodielectric functions

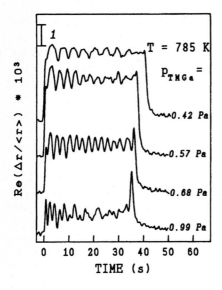

**Fig. 1.58.** RD oscillations during GaAs growth on (100) GaAs substrate (after Reinhardt et al. [50])

derived from these models during 1 ML of growth are shown in Fig. 1.57a,b. The contribution of Ga–C bond absorption has also been suggested as the origin of oscillation. Organometallic groups are likely present at island edges since lone Ga pairs at edges prevent As incorporation (Fig. 1.57b) [48].

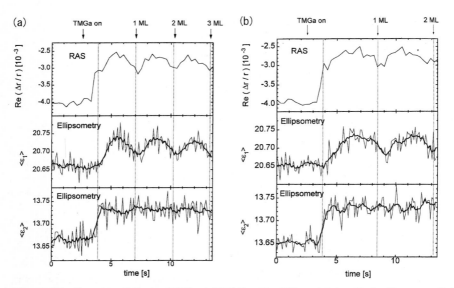

**Fig. 1.59.** Growth of GaAs (100) at 500°C with different trimethylgallium partial pressures of (**a**) 0.33 and (**b**) 0.22 Pa. RDS oscillations compared to the respective transients in $\varepsilon_1$ and $\varepsilon_2$ (after Zettler et et al. [51])

An example of RD oscillations during the GaAs growth on the (100) GaAs substrate is given in Fig. 1.58 [50]. A fast rise in the 2.65 eV RDS signal is seen when trimethylgallium (TMG) is supplied. This rise is followed by a number of well-resolved oscillations. The time period of the oscillation decreases with increasing partial pressure of TMG. A linear dependence on TMG partial pressure and an excellent agreement between growth-rate data derived from the oscillation period and those derived from the layer thickness measurements indicates that the time period corresponds to 1 ML of GaAs. After TMG is switched off, the surface returns to the c(4×4) state within seconds.

Using ellipsometry signals, Zettler et al. reported GaAs growth oscillation with monolayer periodicity [51]. Figure 1.59 shows RDS and SE transients for the different growth rates measured with a photon energy of 2.65 eV at the As dimer-related resonance energy. Comparing the results of the RDS and ellipsometry experiments, good agreement was obtained. In the ellipsometry signal, however, the signal-to-noise ratio was still low. It is very difficult to recognize the oscillations without a guide (thick line in Fig. 1.59).

On the other hand, Lee et al. observed clear oscillations using laser as a probe light (Fig. 1.60). An immediate decrease in $\varepsilon_1$ and an increase in $\varepsilon_2$ are seen when triethylgallium (TEG) is switched on, followed by a number of well-resolved oscillations. Similar oscillations have been observed when trimethylindium (TMI) was used in InAs growth (Fig. 1.61). The time period

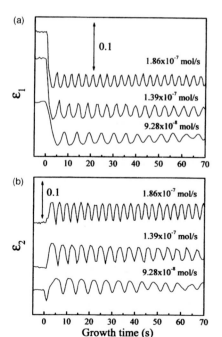

**Fig. 1.60.** Transients in (**a**) real ($\varepsilon_1$) and (**b**) imaginary ($\varepsilon_2$) parts of pseudodielectric functions during GaAs growth at 500°C (after Lee et al. [48])

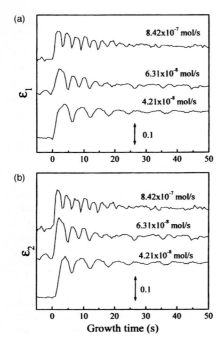

**Fig. 1.61.** Transients in (**a**) real ($\varepsilon_1$) and (**b**) imaginary ($\varepsilon_2$) parts of pseudodielectric functions during InAs growth at 400°C (after Lee et al. [48])

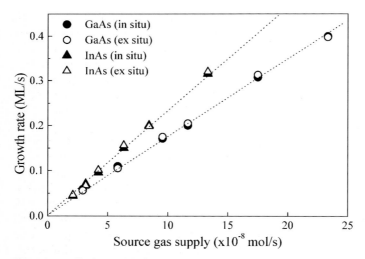

**Fig. 1.62.** GaAs and InAs growth-rate dependence on source gas supply evaluated from the growth oscillation and from the postgrowth thickness determination (after Lee et al. [48])

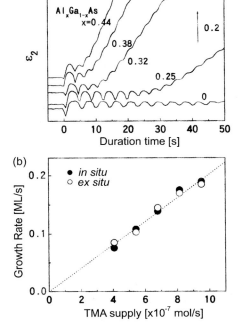

**Fig. 1.63.** (a) Transients in imaginary ($\varepsilon_2$) part of pseudodielectric functions during AlGaAs growth at 520°C. (b) AlAs growth dependence on source gas supply evaluated from the AlGaAs growth oscillation and from the post-growth thickness determination (after Lee et al. [52])

of oscillations decreases with increasing source gas supply independent of the material, indicating that the time period corresponds to 1 ML of growth. The growth rate determined from the layer thickness after growth (ex situ) and that obtained assuming that oscillations describe the completion of a monolayer of GaAs and InAs (in situ) show excellent agreement (Fig. 1.62). Thus, ellipsometry in MOVPE growth has a potential equivalent to RHEED for MBE growth in characterizing the surface before growth and calibrating growth on a monolayer scale.

Using ellipsometry monolayer oscillation, precise alloy composition is obtained directly from the ellipsometry signal. Figure 1.63a shows transients in $\varepsilon_2$ during AlGaAs growth at 520°C [52]. The Al composition was changed from 0% (GaAs) to 44% by changing the trimethylaluminum (TMA) flow rate under constant TEG flow rate. The oscillation period decreases with increasing TMA flow, and Al composition is directly obtained by measuring the increase in oscillation frequency. AlAs growth deduced from oscillation periods agrees well with that from ex situ measurement by X-ray diffraction (Fig. 1.63(b)).

**Real-time Monitoring of Quantum Dot Formation Onset.** In a highly lattice-mismatched system, QD structures are self-formed due to the transition of the growth mode from 2D to 3D at a certain layer thickness. Aspnes

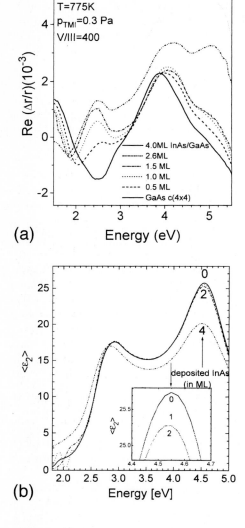

**Fig. 1.64.** (a) Reflectance anisotropy and (b) ellipsometry spectra as measured after the deposition of well-defined amounts of InAs on GaAs (100) (after Steimetz et al. [54])

reported detailed spectral investigations by RDS and SE of the initial steps of QD formation and a comparison with effective medium calculations [53]. The optical spectra gained for a stepwise deposition of InAs (in 0.5 ML steps) on GaAs are given in Fig. 1.64 [54]. Significant change in sign in the RDS signal at about 2.6 V (the As dimer-related energy on GaAs) is clearly observed even for sub-monolayer coverage. An instantaneous reconstruction change from an As-rich c(4×4) GaAs surface to a (1×3) InAs-reconstructed surface was reported from MBE growth, where simultaneous RHEED measurements were performed [55]. With an increasing amount of InAs on the surface, the anisotropy increases and the structure in RDS redshifts to a more

InAs-bulk-like As dimer energy. When the total coverage exceeds the critical layer thickness of about 1.8 ML, the anisotropy is increased in the whole spectral range due to anisotropic light scattering originating from the new 3D morphology. This effect is pronounced for the high energy side, because light with a shorter wavelength is scattered more strongly.

The SE spectra taken simultaneously during the same experiment change only weakly for a total InAs deposition up to 2 ML, but as the inset in Fig. 1.64b demonstrates, the effective dielectric function is strongly reduced for higher coverages. As observed in RDS, the increased surface roughness for the 3D surface morphology is responsible for the strong decrease in $\varepsilon_2$ for InAs coverage beyond 2 ML.

The spectra taken at certain coverages (Fig. 1.64a,b) represent the different stages at which the surface morphology inhibits growth, but time-resolved measurements at well-selected energies provide a real-time view of the whole process. Therefore, time-resolved measurements at 2.6 eV in RDS and reflectance (As dimer-related energy), and at 4.8 eV in SE (sensitive to surface roughness) were performed during the one-step deposition of 4 ML of InAs. From the ellipsometry, reflectance, and RDS signals, QD formation onset and the influence of the growth rate on both the QD formation process and the resulting island size can be investigated. The results are grouped in Fig. 1.65 [55]. Starting the deposition, using ellipsometry signals, the first approximately 2 ML grow two dimensionally, causing only a weak decrease in $\varepsilon_2$, in good agreement with calculated $\varepsilon_2$ for layer-by-layer deposition. When the growth mode transition takes place, a strong decrease in $\varepsilon_2$ is found. For all growth rates, the final level in $\varepsilon_2$ is the same, indicating that the final level of roughness at the surface is independent from the growth rate. However, analyzing the intensity of the reflected light, significant differences for different growth rates were observed. The magnitude of light scattering increases significantly with decreasing growth rates. Obviously, the formation of large islands is enhanced by low growth rates. This should be related to the fact that the growth front is generally much rougher for higher growth rates – the incoming atoms simply do not have enough time to find their positions in the lattice. After the deposition of 2 ML of InAs with a rather high growth rate, a surface with a large number of monoatomic steps and kinks is present, which offers nucleation centers for the following 3D growth of islands. Therefore, a higher density of smaller QDs can be achieved with higher deposition rates. No differences were found in the $\varepsilon_2$ data due to the island size larger than the wavelength. Quantitative information is taken from the ellipsometry data only as long as the nanoclusters formed are much smaller than the wavelength of light, i.e., as long as light scattering is neglected.

At the growth initiation, the RDS signal changes from negative values (GaAs c(4×4)-level) to positive ones. The maximum RDS level reached after more than 2 ML of InAs deposition compares well to the respective level for strained InAs. After the transition to 3D growth has already started, the

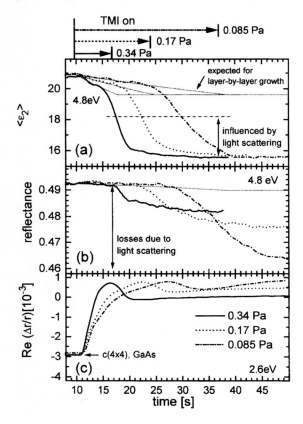

**Fig. 1.65.** RDS and ellipsometry transients taken during the continuous deposition of 4 ML InAs on GaAs (100) (after Steimetz et al. [55])

upper InAs layer relaxes and causes the 2.6 eV RD level to bounce back to values characteristic for only a single InAs monolayer (Fig. 1.65c).

The onset of intensity losses in the reflectivity data is a hint that large islands start to form at the surface – this situation is undesirable for QD fabrication. Thus, between the growth-mode transition (2D to 3D transition sensed by a strong reduction in $\varepsilon_2$) and the formation of large clusters, the deposition should be stopped in order to avoid coalescence of the islands. The optical measurements estimate that the smallest and fewest clusters were formed for the highest chosen deposition rate.

Lee et al. also investigated QD formation onset by ellipsometry [56]. Figure 1.66 shows the trajectory on the $\Delta$–$\Psi$ plane with increasing layer thickness of InGaAs in the QD formation growth and in the layer-by-layer growth on GaAs (100) substrates at 495°C. In the figure, the $In_{0.3}Ga_{0.7}As$ layer-by-layer growth trajectory shows continuous change, mainly with an increase in $\Psi$ as the layer thickness increases. In the wider $\Delta$–$\Psi$ plane, the trajectory spirals, indicating a continuous increase in layer thickness. Observed and simulation curves are in good agreement in the $In_{0.3}Ga_{0.7}As$ case. The deviation of the

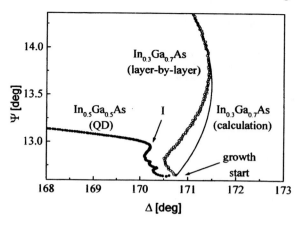

Fig. 1.66. $\Delta$–$\Psi$ trajectories with increasing film thickness, measured for InGaAs film deposition on GaAs substrates at 495°C and analogous calculation for $In_{0.3}Ga_{0.7}As$ layer-by-layer growth (after Lee et al. [56])

observed growth curve from the simulation curve at the initial stages of InGaAs growth is due to the island structure or strain effect caused by a large lattice mismatch. The $In_{0.5}Ga_{0.5}As$ trajectory growth, on the other hand, does not spiral, but inflects at an early growth stage (I in Fig. 1.66). The drastic change in surface morphology from 2D to 3D at a certain layer thickness in the S–K growth mode reflects the occurrence of inflection points in the ellipsometric trajectory.

To obtain intuitive comprehension, the pseudodielectric function $\varepsilon_2$, derived from the ellipsometric signal as a function of growth time, was plotted. Figure 1.67a shows a transient in $\varepsilon_2$ during InGaAs growth at 510°C [47, 48]. The composition of In changed from 0% (GaAs) to 50% by changing the TMI flow rate with a fixed TEG flow rate. The oscillation period decreases with increasing TMI flow, and In composition is directly obtained by measuring the rate of increase. The In composition deduced from oscillation periods and from ex situ measurement by X-ray diffraction show good agreement (Fig. 1.67b). The quantity $\varepsilon_2$ increases rapidly when the $\Delta$–$\Psi$ trajectory passes through the inflection point. The rapid increase in $\varepsilon_2$ is considered to be due to rapidly increased surface roughness because of QD formation. Ellipsometry data and ex situ AFM images show good agreement. The rapid increase in $\varepsilon_2$ (arrows in Fig. 1.67a) is caused by probe light scattering because of increased surface roughness coming from QD formation. Ex situ AFM m easurement of $In_{0.5}Ga_{0.5}As$ sample surfaces whose growth was stopped before and after the inflection point (indicated by arrows $\alpha$ and $\beta$ in Fig. 1.67a) clearly proved that deviation from the spiral trajectory comes from the growth-mode transition (Fig. 1.67c). Quantum dot structures were observed only on the surface after the $\Delta$–$\Psi$ trajectory passes through the inflection points. Thus, the rapid increase in $\varepsilon_2$ originates from the increase in surface roughness caused by morphological surface change.

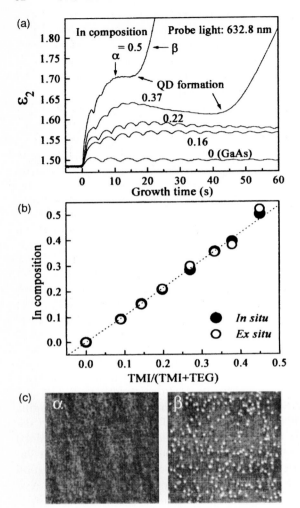

**Fig. 1.67.** Transients in (**a**) imaginary ($\varepsilon_2$) part of pseudodielectric functions during InGaAs growth at different TMI flow rates; (**b**) In composition dependence on source gas evaluated from growth oscillation and from the postgrowth determination; (**c**) AFM images of the $In_{0.5}Ga_{0.5}As$ layers corresponding to the surface of samples for which growth was stopped at $\alpha$ and $\beta$ in (**a**). Image area is $1\,\mu m \times 1\,\mu m$ (after Lee et al. [48])

For the origin of the increase in $\varepsilon_2$, Mie scattering is suggested to dominate when the island size becomes larger than $\lambda/10$, where $\lambda$ is the wavelength of the probe light [57]. Taking the size of the islands (350 Å in diameter and 100 Å in height, in Fig. 1.67c), probe light wavelength of 6328 Å, and refractive index of InGaAs ($n\sim4$) into account, an increase in $\varepsilon_2$ by scattering is plausible. The monotonic increase in $\varepsilon_2$ observed after the inflection indicates a continuous increase in surface roughness. The increase in QD density and/or the increase in size with thicker $In_{0.5}Ga_{0.5}As$ layers enhances surface roughening. The linear increase in $\varepsilon_2$ with the increase in growth time indicates an increase in the number of QDs rather than an increase in size because scattering efficiency is linearly proportional to the number of dots and the fourth power of size ($(R/\lambda)^4$, where $R$ is the radius of particle) [57].

Figure 1.67 also shows the critical point of In composition for QD formation. In the experiment, no inflection point is seen, and ellipsometry oscillation remains for several tens of periods when In composition is below 0.3, implying it is the critical In composition for QD formation.

The critical layer thickness for QD formation is directly calculated by dividing the oscillation period into the time period between the growth onset and the inflection point. The critical In composition for QD formation is estimated at 0.32 and the critical layer thickness is 14.2 ML. From Fig. 1.67a, the critical layer thickness for QD formation is 5.3 (11.1) ML when In composition is 0.5 (0.37).

These results indicate that ellipsometric oscillation is a practical means for accurate in situ determination of growth and gas-flow ratio in MOVPE, which is important for precise control of lattice matching, alloy composition, QW thickness, and QD formation.

As described above, the evolution of uncovered InAs islands after growth has been studied in situ. However, a very important step in QD device fabrication is the overgrowth of the islands by a GaAs or AlGaAs cap layer. Figure 1.68 shows a typical ellipsometry trajectory during the InGaAs QD and GaAs capping layer growth [56]. When the capping layer growth starts,

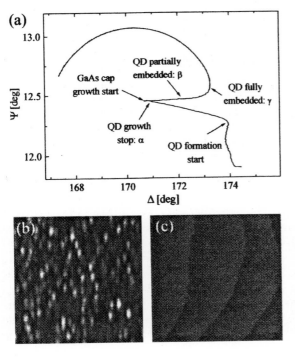

**Fig. 1.68.** (a) $\Delta$–$\Psi$ trajectory of GaAs capping layer growth following the $In_{0.5}Ga_{0.5}As$ QD formation; 1 μm×1 μm surface AFM images of $In_{0.5}Ga_{0.5}As$ QD samples (**b**) at point β and (**c**) at γ in (**a**) (after Lee et al. [56])

$\Delta$ increased but $\Psi$ scarcely changed, indicating surface smoothing starts in capping layer growth. The completion of QD embedding is known from the ellipsometric signal at the point where the trajectory begins to spiral. The surface AFM images of the samples whose GaAs capping layer growths were stopped at points $\beta$ and $\gamma$ are shown in Fig. 1.68b,c. As seen in the figures, AFM results show good agreement with the ellipsometric signal. The image shows the surface morphology of QDs partially embedded at the point $\beta$, and a flat and smooth surface was observed at the point $\gamma$ where the trajectory begins to spiral. AFM images also proved that there is no intermixing in this case. However, in most cases of In(Ga)As QD growth, temperature-dependent In segregation was observed during cap layer growth and the postgrowth annealing. The suppression of an intermixing of the InAs QDs with the cap layer is one of the remaining challenges for InAs QD-based laser production, where elevated temperatures for the growth of the mirror layers containing aluminum are necessary.

Steimetz et al. also investigated the overgrowth of a GaAs cap layer on InAs islands by using optical in situ measurements [58]. Figure 1.69 shows the RDS transient measurements for InAs QD/GaAs cap layer at 500°C and 475°C. The starting GaAs surfaces for both temperatures showed typical RDS spectra for a c(4×4)-reconstructed surface. With InAs deposition, the reconstruction changed immediately from c(4×4) via (1×3) toward (2×4) for InAs. This fact is responsible for the strong signal change in the RDS transient

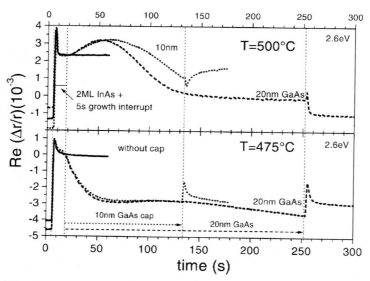

**Fig. 1.69.** Time-resolved RDS measurement at 2.6 eV during InAs QD deposition and cap layer growth for different growth temperatures and total cap layer thicknesses (after Steimetz et al. [58])

at 2.6 eV. When the growth-mode transition takes place, the RDS signal is reduced again, caused by a thickness reduction of the 2D InAs wetting layer. The islands themselves do not have any influence on the RDS signal as long as they are small and isotropic. The overgrowth of GaAs on the InAs islands leads to an As-rich c(4×4) reconstructed surface again, causing a reduction of the RDS signal back to the starting level.

For the low-temperature growth at 475°C, a rapid reduction of the RDS signal toward the GaAs level is observed, indicating a rapid covering of the 3D surface. However, for the higher growth at 500°C, a strong increase in the RDS transient is found before the RDS signal returns to the GaAs c(4×4) level. Since RDS at 2.6 eV is predominantly sensitive to the As coverage of the surface and the thickness of the growing InGaAs, this increase might be caused by more In-rich conditions during the GaAs growth due to In segregation. Further, the RDS transient again showed an increase when the deposition was stopped after 10 nm GaAs growth, indicating that the surface, which still contains In, is now rearranging, leading again to a (1×3)-like surface reconstruction. It is known from various investigations that larger islands are formed at higher growth temperatures. Therefore, covering the surface and flattening it again might be more difficult at elevated temperatures.

# References

1.  A. Scherer, H.G. Craighead: Appl. Phys. Lett. **49**, 1284 (1986)
2.  Y. Sugiyama, Y. Sakuma, S. Muto, N. Yokoyama: Jpn. J. Appl. Phys. **34**, 4384 (1995)
3.  T. Sekiguchi, Y. Sakuma, Y. Awano, N. Yokoyama: J. Appl. Phys. **83**, 4944 (1998)
4.  T. Fukui, S. Ando, Y. Tokura, T. Toriyama: Appl. Phys. Lett. **58**, 2018 (1991)
5.  K. Kumakura, K. Nakakoshi, J. Motohisa, T. Fukui, H. Hasegawa: Jpn. J. Appl. Phys. **34**, 4387 (1995)
6.  Y. Nagamune, M. Nishioka, S. Tsukamoto, Y. Arakawa: Appl. Phys. Lett. **64**, 2495 (1994)
7.  M. Grundmann, D. Bimberg: Phys. Rev. B **55**, 4054 (1997)
8.  R. Heitz, T.R. Ramachandran, A. Kalbuge, Q. Xie, I. Mukhametzhanov, P. Chen, A. Madhukar: Phys. Rev. Lett. **78**, 4071 (1997)
9.  K. Georgsson, N. Carlsson, L. Samuelson, W. Seifert, L.R. Wallenberg: Appl. Phys. Lett. **68**, 2971 (1998)
10. J. Tulkki, A. Heinamaki: Phys. Rev. B **52**, 8239 (1995)
11. M. Sopanen, H. Lipsanen, J. Ahopelto: Appl. Phys. Lett. **66**, 2364 (1995)
12. M. Sopanen, H. Lipsanen, J. Tulkki, J. Ahopelto: Physica E **2**, 19 (1998)
13. D. Leonard, M. Krishnamurthy, C.M. Reaves, S.P. Denbaas, P.M. Petroff: Appl. Phys. Lett. **63**, 3203 (1993)
14. J.M. Moison, F. Houzay, R. Bhate, L. Leprince, E. Andre, O. Vatel: Appl. Phys. Lett. **64**, 196 (1994)
15. Y. Nabetani, T. Ishikawa, S. Noda, A. Sasaki: J. Appl. Phys. **76**, 347 (1994)
16. H. Lee, R.L. Webb, W. Yang, P.C. Sercel: Appl. Phys. Lett. **72**, 812 (1998)

17. H. Saito, K. Nishi, S. Sugou: Appl. Phys. Lett. **74**, 1224 (1999)
18. S.P. Guo, H. Ohno, A. Shen, F. Matsukura, Y. Ohno: Appl. Phys. Lett. **70**, 2738 (1997)
19. D. Lacombe, A. Ponchet, S. Frechengues, V. Drouot, N. Bertu, B. Lambert, A. LeCorre: Appl. Phys. Lett. **74**, 1680 (1999)
20. J.-S. Lee, K. Nishi, Y. Masumoto: J. Cryst. Growth **221**, 586 (2000)
21. E. Pehlke, N. Moll, A. Kley, M. Scheffler: Appl. Phys. A **65**, 525 (1997)
22. D.A. Woolf, D.I. Westwood, R.H. Williams: Appl. Phys. Lett. **62**, 1370 (1993)
23. J.-S. Lee, H. Isshiki, T. Sugano, Y. Aoyagi: J. Cryst. Growth **183**, 53 (1998)
24. K. Yang, L.J. Schowalter, T. Thundat: Appl. Phys. Lett. **64**, 1641 (1994)
25. V.A. Schukin, N.N. Ledentsov, P.S. Kop'ev, D. Bimberg: Phys. Rev. Lett. **75**, 2968 (1995)
26. Y.H. Xie, S.B. Samavedam, M. Bulsara, T.A. Langdo, E.A. Fitzgerald: Appl. Phys. Lett. **71**, 3567 (1997)
27. D.S.L. Mui, D. Leonald. L.A. Coldren, P.M. Petroff: Appl. Phys. Lett. **66**, 1620 (1995)
28. D.E. Jeppesen, M.S. Miller, D. Hessman, B. Kowalski, I. Maximov, L. Samuelson: Appl. Phys. Lett. **68**, 2228 (1996)
29. G. Jin, J.L. Liu, S.G. Thomas, Y.H. Luo, K.L. Wang, B.Y. Nguyen: Appl. Phys. Lett. **75**, 2752 (1999)
30. M. Kitamura, M. Nishioka, J. Oshinowo, Y. Arakawa: Appl. Phys. Lett. **66**, 3663 (1995)
31. J.-S. Lee, S. Sugou, Y. Masumoto: J. Cryst. Growth **205**, 467 (1999)
32. R. Nötzel, J. Temmyo, H. Kumada, T. Furuta, T. Tamamura: Appl. Phys. Lett. **65**, 457 (1994)
33. Q. Xie, A. Madhukar, P. Chen, N. Kobayashi: Phys. Rev. Lett. **75**, 2542 (1995)
34. Y. Nakata, Y. Sugiyama, T. Futatsugi, N. Yokoyama: J. Cryst. Growth **175–176**, 713 (1997)
35. K. Mukai, N. Ohtuka, H. Shoji, M. Sugawara: Appl. Surf. Sci. **112**, 102 (1997)
36. J.-S. Lee, M. Sugisaki, H.-W. Ren, S. Sugou, Y. Masumoto: J. Cryst. Growth **200**, 77 (1999)
37. J.-S. Lee, M. Sugisaki, H.-W. Ren, S. Sugou, Y. Masumoto: Physica E **7**, 303 (2000)
38. R. Nötzel, J. Temmyo, T. Tamamura: Jpn. J. Appl. Phys. 2, Lett. **33**, L275 (1994)
39. A.A. Yamaguchi, K. Nishi, A. Usui: Jpn. J. Appl. Phys. 2, Lett. **33**, L912 (1994)
40. P.K. Larsen: Phys. Rev. B **27**, 4966 (1983)
41. J.H. Neave, B.A. Joyce, P.J. Dobson, N. Norton: Appl. Phys. A **31**, 1 (1983)
42. Y. Nabetani, T. Ishikawa, S. Noda, A. Sasaki: J. Appl. Phys. **76**, 347 (1994)
43. H. Lee, R.L. Webb, W. Yang, P.C. Sercel: Appl. Phys. Lett. **72**, 812 (1998)
44. I. Kamiya, D.E. Aspnes, H. Tanaka, L.T. Florez, E. Colas, J.P. Harbison, R. Bhat: Phys. Rev. Lett. **68**, 627 (1992)
45. M. Wassermeier, J. Behrend, J.-T. Zettler, K. Stahrenberg, K.H. Ploog: Appl. Surf. Sci. **107**, 48 (1996)
46. J.-S. Lee, S. Sugou, Y. Masumoto: Jpn. J. Appl. Phys. 2, Lett. **38**, L614 (1999)
47. J.-S. Lee, S. Sugou, Y. Masumoto: J. Cryst. Growth **209**, 614 (2000)
48. J.-S. Lee, S. Sugou, Y. Masumoto: J. Appl. Phys. **88**, 196 (2000)
49. D.E. Aspnes: Am. J. Phys. **50**, 704 (1982)

50. F. Reinhardt, J. Jonsson, M. Zorn, W. Richter, K. Ploska, J. Rumberg, P. Kurpas: J. Vac. Sci. Technol. B **12**, 2541 (1994)
51. J.-T. Zettler, T. Wethkamp, M. Zorn, M. Pristovsek, C. Meyne, K. Ploska, W. Richter: Appl. Phys. Lett. **67**, 3783 (1995)
52. J.-S. Lee, Y. Masumoto: J. Cryst. Growth **221**, 111 (2000)
53. D.E. Aspnes: Appl. Surf. Sci. **107**, 203 (1996)
54. E. Steimetz, F. Schienle, J.-T. Zettler, W. Richter: J. Cryst. Growth **170**, 208 (1997)
55. E. Steimetz, J.-T. Zettler, F. Schienle, T. Trepk, T. Wethkamp, W. Richter, I. Sieber: Appl. Surf. Sci. **107**, 203 (1996)
56. J.-S. Lee, S. Sugou, H.-W. Ren, Y. Masumoto: J. Vac. Sci. Technol. B **17**, 1341 (1999)
57. C.F. Bohren, D.B. Huffman: *Absorption and Scattering of Light by Small Particles* (John Wiley and Sons, New York 1983) p. 136
58. E. Steimetz, T. Haberl, J.-T. Zettler, W. Richter: J. Cryst. Growth **195**, 530 (1998)

# 2 Excitonic Structures and Optical Properties of Quantum Dots

Toshihide Takagahara

## 2.1 Introduction

The spatial confinement of electrons and holes along all three dimensions leads to a discrete energy level structure with sharp optical absorption lines. In this sense a semiconductor quantum dot (QD) can be regarded as an artificial solid-state atom. The concentration of the oscillator strength to sharp exciton transitions makes QDs very attractive for electro-optic and nonlinear optical applications. To be more specific, QDs are considered as promising elements for implementing the coherent control of the quantum state, which is an essential function to achieve quantum information processing and quantum computation [1]. Recently, very fine structures were observed in exciton photoluminescence (PL) from GaAs quantum wells (QWs) [2–4]. These sharp lines are interpreted as luminescence from localized excitons at island structures in the QW. These island structures can be regarded as zero-dimensional QDs. The excitonic wave function has been manipulated by controlling the optical phase of the pulse sequence through timing and polarization [5]. In this wave function manipulation, important features are the sharp spectral lines, which indicate a long coherence time, and well-defined polarization characteristics. In view of this recent progress, in this section we discuss the fundamental physics of QDs, focusing on their excitonic optical properties.

In general, the valence band in semiconductors has a p-like character and splits into the fourfold degenerate ($j=3/2$) band and the doubly degenerate ($j=1/2$) band due to the spin-orbit interaction. The energetic order of these bands is dependent on the sign of the spin-orbit interaction. In most semiconductors the topmost valence band is the $j=3/2$ band, whereas in a few semiconductors, e.g., CuCl, the $j=1/2$ band is the highest valence band. The excitonic and biexcitonic energy level structures in the latter case are discussed in detail in Chap. 10. The excitonic structures for the former case, e.g., GaAs and InAs QDs, will be discussed in this chapter. In addition to this topic, we will discuss several important aspects that appear most pronounced in QDs, namely, the dielectric confinement effect, the electron–hole exchange interaction, the nonlocality of the radiation–matter interaction, and the optical properties of an array of QDs.

## 2.2    Quantum and Dielectric Confinement Effect

In a quantum dot the electronic states are quantized and the energy levels become discretized. In calculating the excitonic energy spectra, the boundary conditions on the exciton wave function play an important role. In actual samples of quantum dots, e.g., semiconductor nanocrystals embedded in a glass matrix, the energy gap difference between the semiconductor and the surrounding medium is rather large, and at the same time the bonding characters of both materials are considerably different. Thus, the potential discontinuity for the electron and the hole between the two materials is determined not only by the energy gap difference, but also by the surface potential. Our knowledge about these is still poor. Another important problem in calculating the excitonic energy spectra is how many electron and hole sub-bands are to be included in the calculation. The answer to this problem is dependent on the quantum dot size, and the important parameter is the ratio of the dot radius to the exciton Bohr radius in the bulk material. This can be seen by noting that the sub-band energy difference of the electron or hole is given by

$$\frac{\hbar^2 D}{2m_e R^2} \quad \text{or} \quad \frac{\hbar^2 D}{2m_h R^2} \; , \tag{2.1}$$

where $R$ is the radius of a quantum dot, $m_{e(h)}$ the effective mass of the electron (hole), and $D$ is a numerical factor dependent on the shape of quantum dot and is typically about 10 for a spherical dot. On the other hand, the exciton binding energy in the bulk is given by

$$\frac{\hbar^2}{2\mu(a_B^*)^2} \; , \tag{2.2}$$

where $\mu$ is the electron–hole reduced mass and $a_B^*$ is the exciton Bohr radius. Thus when assuming $m_e \ll m_h$, the criterion for the higher hole sub-bands to be included is given by

$$\frac{D}{m_h R^2} \leq \frac{1}{m_e (a_B^*)^2} \; , \tag{2.3}$$

and is rewritten as

$$\left(\frac{R}{a_B^*}\right)^2 \geq \frac{m_e}{m_h} D \; . \tag{2.4}$$

In the strong confinement regime, i.e., $R/a_B^* \ll 1$, where the sub-band energy separations are much larger than the electron–hole Coulomb interaction of the order of the exciton binding energy, the exciton ground state is mainly composed of the lowest-energy sub-band states. As the quantum dot size is increased, the energy spacings of the quantized sub-bands become comparable to or smaller than the exciton binding energy. In this regime of the intermediate confinement, higher sub-band states are mixed into even the exciton ground state. At the same time, the excited states with respect to the electron–hole relative motion are also to be included in the exciton wave function. In the weak-confinement regime where the quantum dot size is much larger than the exciton Bohr radius, the quantized sub-bands are distributed almost continuously and the electron–hole binding energy is nearly the same as in the bulk material. Only the center-of-mass motion of the exciton is regarded as quantized in a large quantum dot. These properties are discussed in Chap. 10 in more detail.

Another interesting property that is manifested in low-dimensional structures is the dielectric confinement effect. Semiconductor nanocrystals are usually embedded within a material having a relatively small dielectric constant. The electric force lines emerging from charged particles within a semiconductor nanocrystal pass through the surrounding medium with a smaller dielectric constant than that of the nanocrystal. Thus the screening effect is reduced, and the Coulomb interaction between charged particles becomes enhanced, resulting in the enhancement of the exciton binding energy and the exciton oscillator strength. This dielectric confinement effect was first investigated by Keldysh [6] for a layered structure. Recently we studied this effect more thoroughly for the dielectric quantum well structure and clarified the enhancement of the exciton binding energy and the exciton oscillator strength [7]. A similar enhancement effect has been discussed for the quantum wire structure [8,9]. The dielectric confinement effect is expected to appear more pronounced in the quantum dot structure than in quantum well and quantum wire structures because of more probable penetration of the electric force lines into the surrounding medium having a smaller dielectric constant. The critical size that characterizes the reduction in the dimensionality is of the order of the electron–hole separation, namely, the exciton Bohr radius. In a quantum dot whose size is of the same order as the exciton Bohr radius, an electron and a hole composing an exciton feel the dielectric boundary and their motions are strongly affected. On the other hand, when the quantum dot size is much larger than this length scale, the situation is similar to that in the bulk medium, and the dielectric confinement effect would not be effective.

As a consequence of the quantum confinement effect and the dielectric confinement effect, the electron–hole exchange interaction is also enhanced, which yields the energy difference between the spin-singlet and spin-triplet excitons. As is well known, the long-range part of this interaction in the bulk material gives rise to the longitudinal–transverse (LT) splitting energy of excitons. In semiconductor quantum dots the translational symmetry is totally lost, and the long-range part of the electron–hole exchange interaction is expected to be highly dependent on the size and shape of the quantum dot.

We will discuss the general aspects of the dielectric confinement effect. In the effective-mass approximation the relevant Hamiltonian for an electron–hole pair in a spherical quantum dot with radius $R$ was derived by Brus [10] and is given as

$$H = -\frac{\hbar^2}{2m_e}\nabla_e^2 - \frac{\hbar^2}{2m_h}\nabla_h^2 - \frac{e^2}{\epsilon_1|r_e - r_h|} + \frac{e^2}{2R}\sum_{n=0}^{\infty}\alpha_n\left[\left(\frac{r_e}{R}\right)^{2n} + \left(\frac{r_h}{R}\right)^{2n}\right]$$

$$-\frac{e^2}{R}\sum_{n=0}^{\infty}\alpha_n\left(\frac{r_e r_h}{R^2}\right)^n P_n(\cos\Theta_{eh}) , \tag{2.5}$$

where $r_e$ and $r_h$ denote the coordinates of an electron and a hole, respectively, $m_e$ and $m_h$ are their effective masses, $P_n$ is the Legendre polynomial of the $n$th order, and $\Theta_{eh}$ is the angle between $r_e$ and $r_h$. Here the degeneracy of the valence band is not taken into account. The dielectric constants of the semiconductor material and the surrounding medium are denoted by $\epsilon_1$ and $\epsilon_2$, respectively, and $\alpha_n$ is defined by

$$\alpha_n = \frac{(n+1)(\epsilon-1)}{\epsilon_1(n\epsilon+n+1)} , \tag{2.6}$$

with $\epsilon=\epsilon_1/\epsilon_2$. The first two terms in (2.5) represent the kinetic energy, and the third term is the direct Coulomb interaction between an electron and a hole. The last two terms are usually called the surface polarization energy, which arises from the difference in the dielectric constant between the quantum dot and the surrounding medium. The former is the self-energy of an electron and a hole due to their own image charges, whereas the latter is the mutual interaction energy between an electron and a hole via image charges. The expression of the dielectric confinement energy for an exciton is dependent on the shape of the quantum dot [11,12].

In calculating the excitonic energy spectra we prefer to restrict our calculations within the model of the surrounding medium with an infinite potential barrier, although there are a few theoretical attempts to incorporate the finiteness of the potential barrier height [13,14]. Within the model of an infinite potential barrier, the most suitable basis set for the one-particle wave functions that vanish at the surface of a spherical quantum dot is given as

$$\phi_{lmj}(r,\theta,\varphi) = j_l\left(k_j^l\frac{r}{R}\right)Y_{lm}(\theta,\varphi) , \tag{2.7}$$

where $j_l$ is the $l$th-order spherical Bessel function, $k_j^l$ is its $j$th zero, and $Y_{lm}$ is one of the spherical harmonics. The exciton wave function can be constructed from a linear combination of products of $\phi_{lmj}$ for the electron and the hole. At the same time, we must take into account the fact that the total angular momentum of an electron–hole pair is a good quantum number because of the spherical symmetry of the Hamiltonian in (2.5). The angular momenta of the electron, the hole, and the pair are denoted by $l_e$, $l_h$, and $L = l_e + l_h$, respectively. The angular part of the exciton wave function with a total angular momentum $L$ and a magnetic quantum number $M$ can be written as

$$\sum_{m_e}\langle l_e, m_e, l_h, m_h|L, M\rangle Y_{l_e,m_e}(\Omega_e)Y_{l_h,m_h}(\Omega_h) , \tag{2.8}$$

using the Clebsch–Gordan coefficient, where $\Omega$ denotes the angular v. As for the radial partwave function, we have to include not only the sub-band state but also the higher sub-band states and the excited concerning the electron–hole relative motion. Thus a general exciton function can be written as

$$\Phi_{LM}(r_{\mathrm{e}}, r_{\mathrm{h}}) = C_\Phi\, e^{-\alpha r_{\mathrm{eh}}} \sum_n \sum_{l_{\mathrm{e}},l_{\mathrm{h}},m_{\mathrm{e}}} \langle l_{\mathrm{e}}, m_{\mathrm{e}}, l_{\mathrm{h}}, m_{\mathrm{h}} | L, M \rangle$$

$$\times\, Y_{l_{\mathrm{e}},m_{\mathrm{e}}}(\Omega_{\mathrm{e}}) Y_{l_{\mathrm{h}},m_{\mathrm{h}}}(\Omega_{\mathrm{h}})\, r_{\mathrm{eh}}^n\, R_{l_{\mathrm{e}},l_{\mathrm{h}}}^{(n)}(r_{\mathrm{e}}, r_{\mathrm{h}})\ , \tag{2.9}$$

where $r_{\mathrm{eh}} = |r_{\mathrm{e}} - r_{\mathrm{h}}|$ and $R_{l_{\mathrm{e}},l_{\mathrm{h}}}^{(n)}$ is a radial function that vanishes at the nanocrystal surface.

It is easy to show that the optical transition is allowed only to the $L=0$ exciton states. Namely, the transition matrix element from the ground state to an exciton state given by (2.9) is calculated as

$$\langle X | P | 0 \rangle = \sqrt{2} p_{\mathrm{cv}}^0 \sum_{r_{\mathrm{s}}} \Phi_{LM}(r_{\mathrm{s}}, r_{\mathrm{s}})\ , \tag{2.10}$$

with

$$|X\rangle = \sum_{r_{\mathrm{e}},r_{\mathrm{h}}} \Phi_{LM}(r_{\mathrm{e}}, r_{\mathrm{h}}) \frac{1}{\sqrt{2}} \left( a_{\mathrm{cre}\alpha}^\dagger a_{\mathrm{vrh}\alpha} + a_{\mathrm{cre}\beta}^\dagger a_{\mathrm{vrh}\beta} \right) |0\rangle\ , \tag{2.11}$$

$$P = \sum_{r_{\mathrm{s}}} p_{\mathrm{cv}}^0 \left( a_{\mathrm{cre}\alpha}^\dagger a_{\mathrm{vrh}\alpha} + a_{\mathrm{cre}\beta}^\dagger a_{\mathrm{vrh}\beta} \right)\ , \tag{2.12}$$

where $\alpha(\beta)$ denotes the spin up(down) state and $p_{\mathrm{cv}}^0$ is the momentum matrix element between the valence-band top and the conduction-band bottom at the $\Gamma$-point. Then we have

$$\sum_{r_{\mathrm{s}}} \Phi_{LM}(r_{\mathrm{s}}, r_{\mathrm{s}}) \propto \sum_{l_{\mathrm{h}},m_{\mathrm{e}}} \int d\Omega\, \langle l_{\mathrm{e}}, m_{\mathrm{e}}, l_{\mathrm{h}}, m_{\mathrm{h}} | L, M \rangle\, Y_{l_{\mathrm{e}},m_{\mathrm{e}}}(\Omega) Y_{l_{\mathrm{h}},m_{\mathrm{h}}}(\Omega)$$

$$= \sum_{l_{\mathrm{h}},m_{\mathrm{e}}} \langle l_{\mathrm{e}}, m_{\mathrm{e}}, l_{\mathrm{h}}, m_{\mathrm{h}} | L, M \rangle\, (-1)^{m_{\mathrm{h}}} \delta_{l_{\mathrm{e}},l_{\mathrm{h}}} \delta_{m_{\mathrm{e}},-m_{\mathrm{h}}}$$

$$= \sum_{m_{\mathrm{e}}} \langle l_{\mathrm{e}}, m_{\mathrm{e}}, l_{\mathrm{e}}, -m_{\mathrm{e}} | L, 0 \rangle\, (-1)^{m_{\mathrm{e}}}\ . \tag{2.13}$$

From the property of the Clebsch–Gordan coefficient, i.e., $\langle l, m, l, -m | 0, 0 \rangle = (-1)^{l-m}/\sqrt{2l+1}$, we see that

$$\sum_{m_{\mathrm{e}}} \langle l_{\mathrm{e}}, m_{\mathrm{e}}, l_{\mathrm{e}}, -m_{\mathrm{e}} | L, 0 \rangle\, (-1)^{m_{\mathrm{e}}}$$

$$\propto \sum_{m_{\mathrm{e}}} \langle l_{\mathrm{e}}, m_{\mathrm{e}}, l_{\mathrm{e}}, -m_{\mathrm{e}} | L, 0 \rangle \langle l_{\mathrm{e}}, m_{\mathrm{e}}, l_{\mathrm{e}}, -m_{\mathrm{e}} | 0, 0 \rangle \propto \delta_{L,0}\ , \tag{2.14}$$

where the orthogonality relation of the Clebsch–Gordan coefficient is used. Thus the optically active exciton state should have the total angular momentum $L=0$.

The $L=0$ wave function is given as

$$\Phi_{00}(r_{\mathrm{e}}, r_{\mathrm{h}}) = C_\Phi \, \mathrm{e}^{-\alpha r_{\mathrm{eh}}} \sum_n \sum_{l_{\mathrm{e}}, m_{\mathrm{e}}} \langle l_{\mathrm{e}}, m_{\mathrm{e}}, l_{\mathrm{e}}, -m_{\mathrm{e}} | 0, 0 \rangle$$

$$\times Y_{l_{\mathrm{e}}, m_{\mathrm{e}}}(\Omega_{\mathrm{e}}) Y_{l_{\mathrm{e}}, -m_{\mathrm{e}}}(\Omega_{\mathrm{h}}) \, r_{\mathrm{eh}}^n \, R_{l_{\mathrm{e}}, l_{\mathrm{e}}}^{(n)}(r_{\mathrm{e}}, r_{\mathrm{h}}) \,. \tag{2.15}$$

Again using the property of the Clebsch–Gordan coefficient mentioned above, we have

$$\sum_m \langle l, m, l, -m | 0, 0 \rangle \, Y_{l, m}(\Omega_{\mathrm{e}}) Y_{l, -m}(\Omega_{\mathrm{h}})$$

$$\propto \sum_m Y_{l, m}(\Omega_{\mathrm{e}}) Y_{l, m}^*(\Omega_{\mathrm{h}}) \propto P_l(\cos\Theta_{\mathrm{eh}}) \,. \tag{2.16}$$

Thus the optically active exciton state with $L=0$ can be written as

$$\Phi_{00}(r_{\mathrm{e}}, r_{\mathrm{h}}) = C_\Phi \, \mathrm{e}^{-\alpha r_{\mathrm{eh}}} \Big\{ \sum_n r_{\mathrm{eh}}^n R_{00}^{(n)} + P_1(\cos\Theta_{\mathrm{eh}}) \sum_n r_{\mathrm{eh}}^n R_{11}^{(n)}$$

$$+ P_2(\cos\Theta_{\mathrm{eh}}) \sum_n r_{\mathrm{eh}}^n R_{22}^{(n)} + \cdots \Big\} \,. \tag{2.17}$$

More general exciton wave functions for $L \neq 0$ are discussed in Chap. 10.

Now we clarify the salient features of the dielectric confinement effect in the strong confinement regime. As mentioned before, the dielectric confinement effect appears most pronounced in the smaller quantum dot because of the larger opportunity for the electric force lines between an electron and a hole to penetrate through the surrounding medium, which has a relatively smaller dielectric constant. Thus the excitonic energy spectra in the strong confinement regime are expected to be strongly modified by this dielectric confinement effect. Here the effect on the exciton ground state is examined in detail.

In the extremely strong confinement regime, the envelope function of the exciton ground state can be given by

$$\Phi_{00}(r_{\mathrm{e}}, r_{\mathrm{h}}) = C_\Phi \, \mathrm{e}^{-\alpha r_{\mathrm{eh}}} j_0\left(k_1^0 \frac{r_{\mathrm{e}}}{R}\right) j_0\left(k_1^0 \frac{r_{\mathrm{h}}}{R}\right) \,. \tag{2.18}$$

Since $r_{\mathrm{eh}}$ is very small, the exponential can be expanded as

$$\Phi_{00}(r_{\mathrm{e}}, r_{\mathrm{h}}) = C_\Phi \, (1 - \alpha r_{\mathrm{eh}}) \, j_0\left(k_1^0 \frac{r_{\mathrm{e}}}{R}\right) j_0\left(k_1^0 \frac{r_{\mathrm{h}}}{R}\right) \,, \tag{2.19}$$

and this expression facilitates the analytical treatment. The spherical coordinates are more convenient than the Hylleraas coordinates [15] for dealing with the surface polarization energy. A key relation in the spherical coordinates is given as

$$
\begin{aligned}
r_{\mathrm{eh}} &= \frac{1}{r_{\mathrm{eh}}} r_{\mathrm{eh}}^2 \\
&= \sum_{l=0}^{\infty} \left\{ \frac{r_<^l}{r_>^{l+1}} (r_>^2 + r_<^2) - 2 \frac{r_<^l}{r_>^{l-1}} \frac{l}{2l-1} - 2 \frac{r_<^{l+2}}{r_>^{l+1}} \frac{l+1}{2l+3} \right\} P_l(\cos \Theta_{\mathrm{eh}}) \\
&= \sum_{l=0}^{\infty} \mathcal{R}_l(r_{\mathrm{e}}, r_{\mathrm{h}}) \, P_l(\cos \Theta_{\mathrm{eh}}) \,, \quad (2.20)
\end{aligned}
$$

where $r_> = \max(r_{\mathrm{e}}, r_{\mathrm{h}})$, $r_< = \min(r_{\mathrm{e}}, r_{\mathrm{h}})$ and $\mathcal{R}_l(r_{\mathrm{e}}, r_{\mathrm{h}})$ is introduced for abbreviation. Minimizing the exciton energy with respect to $\bar{\alpha} = \alpha a_{\mathrm{B}}^*$ ($a_{\mathrm{B}}^*$: the exciton Bohr radius in the bulk) in the limit of $R \longrightarrow 0$, we find the optimum value of $\bar{\alpha}$ as

$$
\bar{\alpha} = -\frac{8}{\pi^5} \left[ 4\pi \left( \frac{\pi^4}{16} - I_1 I_2 \right) + (4I_4 - 2I_5) \frac{\pi^2}{4} + I_1 (2I_3 - \pi^2 \alpha_0) \right] \,, \quad (2.21)
$$

with

$$
I_1 = \int_0^\pi \mathrm{d}x_{\mathrm{e}} \int_0^\pi \mathrm{d}x_{\mathrm{h}} \, \sin^2 x_{\mathrm{e}} \, \sin^2 x_{\mathrm{h}} \, \mathcal{R}_0(x_{\mathrm{e}}, x_{\mathrm{h}}) \,, \quad (2.22)
$$

$$
I_2 = \int_0^\pi \mathrm{d}x_{\mathrm{e}} \int_0^\pi \mathrm{d}x_{\mathrm{h}} \, \frac{1}{x_>} \, \sin^2 x_{\mathrm{e}} \, \sin^2 x_{\mathrm{h}} \,, \quad (2.23)
$$

$$
I_3 = \pi \int_0^\pi \mathrm{d}x \, \sin^2 x \sum_{n=0}^\infty \alpha_n \left( \frac{x}{\pi} \right)^{2n} \,, \quad (2.24)
$$

$$
I_4 = \int_0^\pi \mathrm{d}x_{\mathrm{e}} \int_0^\pi \mathrm{d}x_{\mathrm{h}} \, \sin^2 x_{\mathrm{e}} \, \sin^2 x_{\mathrm{h}} \sum_{n=0}^\infty \alpha_n \left( \frac{x_{\mathrm{e}} x_{\mathrm{h}}}{\pi^2} \right)^n \frac{\mathcal{R}_n(x_{\mathrm{e}} x_{\mathrm{h}})}{2n+1} \,, \quad (2.25)
$$

$$
\begin{aligned}
I_5 = &\int_0^\pi \mathrm{d}x_{\mathrm{e}} \int_0^\pi \mathrm{d}x_{\mathrm{h}} \, \sin^2 x_{\mathrm{e}} \, \sin^2 x_{\mathrm{h}} \, \mathcal{R}_0(x_{\mathrm{e}}, x_{\mathrm{h}}) \\
&\times \sum_{n=0}^\infty \alpha_n \left\{ \left( \frac{x_{\mathrm{e}}}{\pi} \right)^{2n} + \left( \frac{x_{\mathrm{h}}}{\pi} \right)^{2n} \right\} \,, \quad (2.26)
\end{aligned}
$$

where $\{\alpha_n\}$ ($n = 0, 1, 2, \cdots$) are defined in (2.6). The first term within the square brackets of (2.21) corresponds to the result in the absence of the dielectric confinement effect and the second and third terms arise from the surface polarization energy. The energy of the exciton ground state in units of the effective Rydberg ($Ry^*$) is calculated as

$$
\begin{aligned}
E &= \frac{\pi^2}{\bar{R}^2} + \frac{-8\pi I_2 + 4I_3 - 2\pi^2 \alpha_0}{\pi^2} \frac{1}{\bar{R}} - \bar{\alpha}^2 + \mathcal{O}(\bar{R}) \\
&= \frac{\pi^2}{\bar{R}^2} + \frac{A_1}{\bar{R}} + A_0 + \mathcal{O}(\bar{R}) \,, \quad (2.27)
\end{aligned}
$$

where $\bar{R} = R/a_{\mathrm{B}}^*$. This is a general formula for the energy of the exciton ground state in the strong confinement regime that includes the dielectric confinement effect. In the second term proportional to $1/\bar{R}$, the first term in the numerator comes from the electron–hole direct Coulomb interaction, whereas the second and third terms arise from the surface polarization energy. When we set $\epsilon_1/\epsilon_2 = 1$, i.e., $\alpha_n = 0$ for all $n$, the above formula reproduces the previous result [16], namely

$$\bar{\alpha} \cong 0.498, \qquad E \cong \frac{\pi^2}{\bar{R}^2} - \frac{3.572}{\bar{R}} - 0.248 . \tag{2.28}$$

The optimum value of $\bar{\alpha}$ and the coefficients $A_0$ and $A_1$ in (2.27) are plotted in Figs. 2.1 and 2.2, respectively, as a function of the dielectric constant ratio $\epsilon_1/\epsilon_2$. It is seen that in the strong confinement regime, the dielectric confinement has a significant effect on the exciton energy and cannot be treated as a minor perturbation. In the case of a CdS or CdSe nanocrystal embedded in silicate glasses, the dielectric-constant ratio $\epsilon_1/\epsilon_2$ is estimated to be about 3~4 and the dielectric confinement effect would be significant.

The exciton binding energy is known to increase in low-dimensional systems because the spatial overlap between an electron and a hole is increased due to the quantum confinement effect. Thus we expect that the exciton binding energy is most strongly enhanced in the zero-dimensional system. At the same time, the dielectric confinement effect is also enhanced in the low-dimensional system due to the penetration of the electric force lines into the surrounding medium with a relatively small dielectric constant, resulting in the enhancement of the exciton binding energy. These features are discussed in this section. First of all, we have to note the peculiarity of the

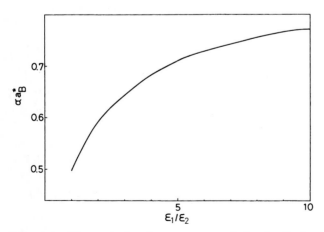

**Fig. 2.1.** The variational parameter $\alpha a_{\mathrm{B}}^*$ in the limit of $R \longrightarrow 0$ is plotted as a function of the dielectric-constant ratio $\epsilon_1/\epsilon_2$. In this limit the electron-to-hole mass ratio is irrelevant to the results (from T. Takagahara [11])

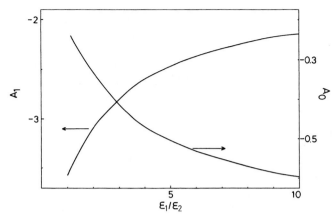

**Fig. 2.2.** The coefficients $A_0$ and $A_1$ of the lowest exciton energy as given in (2.27) are plotted as a function of the dielectric-constant ratio $\epsilon_1/\epsilon_2$. As in Fig. 2.1, the electron-to-hole mass ratio is irrelevant to the results (from T. Takagahara [11])

zero-dimensional system in defining the exciton binding energy. The exciton binding energy is usually defined by the energy difference between the exciton state and the continuum state. The continuum exciton state in a true sense is associated with the sub-band states, which are entirely extended over the surrounding medium. These continuum exciton states are absent when a model of infinite potential barrier for the surrounding medium is employed, as in this section. Thus the definition of the exciton binding energy in the zero-dimensional system is rather indefinite. However, restricting the argument to the lowest-energy exciton, we can reasonably define the exciton binding energy in reference to the sum of energies of the lowest electron and hole subbands, namely,

$$B_X = \langle \phi_{001}(r_e)|H_e|\phi_{001}(r_e)\rangle + \langle \phi_{001}(r_h)|H_h|\phi_{001}(r_h)\rangle - E_{\min} , \qquad (2.29)$$

where $\phi_{001}$ is the lowest-energy sub-band state defined in (2.7), $E_{\min}$ the lowest exciton energy, and the single-particle Hamiltonian $H_i(i{=}e, h)$ contains the kinetic energy and the self-energy part of the surface polarization energy, i.e.,

$$H_i = -\frac{\hbar^2 \nabla_i^2}{2m_i} + \frac{e^2}{2R} \sum_{n=0}^{\infty} \alpha_n \left(\frac{r_i}{R}\right)^{2n} . \qquad (2.30)$$

Then the exciton binding energy in effective Rydberg units $(Ry^*)$ is calculated as

$$B_X = \left(\frac{\pi}{R}\right)^2 + \frac{4}{R} \sum_{n=0}^{\infty} \frac{\epsilon_1 \alpha_n}{\pi^{2n+1}} \int_0^{\pi} \mathrm{d}x\, x^{2n} \sin^2 x - E_{\min} , \qquad (2.31)$$

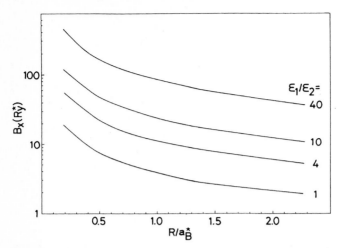

**Fig. 2.3.** The binding energy of the lowest exciton state is plotted as a function of the quantum dot radius for a few values of the dielectric constant ratio $\epsilon_1/\epsilon_2$ with a fixed value of the electron-to-hole mass ratio, i.e., $m_e/m_h = 0.2$ (from T. Takagahara [11])

where $\bar{R} = R/a_B^*$. This binding energy is proportional to $1/\bar{R}$ in the limit of $R \longrightarrow 0$ because the kinetic energy term proportional to $1/\bar{R}^2$ is subtracted. In Fig. 2.3 the size dependence of $B_X$ is shown for a fixed value of the electron-to-hole mass ratio. The enhancement of the exciton binding energy in the smaller crystallites is due to the increased spatial overlap between an electron and a hole. At the same time, the exciton binding energy increases strongly with an increase in the dielectric constant ratio $\epsilon_1/\epsilon_2$.

Now the oscillator strength of the lowest-energy exciton state will be examined with regard to its dependence on the quantum dot size and the dielectric constant ratio. As mentioned before, the matrix element of the momentum operator is calculated as

$$\langle \Phi | P | 0 \rangle = \sqrt{2}\, p_{cv}^0 \sum_{r_s} \Phi_{00}(r_s, r_s)$$

$$= \sqrt{2}\, p_{cv}^0 C_\Phi \sum_{r_s} \left\{ R_{00}^{(0)}(r_s, r_s) + R_{11}^{(0)}(r_s, r_s) + \cdots \right\}, \qquad (2.32)$$

where the notation in (2.17) is used. The oscillator strength of the excitonic transition is defined by

$$f_X = \frac{1}{m_0 \hbar \omega_X} |\langle \Phi | P | 0 \rangle|^2 , \qquad (2.33)$$

and is calculated as

$$f_0 \left| C_\Phi \sum_{r_s} \left\{ R_{00}^{(0)}(r_s, r_s) + R_{11}^{(0)}(r_s, r_s) + \cdots \right\} \right|^2 , \qquad (2.34)$$

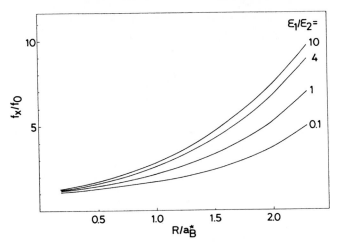

**Fig. 2.4.** The normalized oscillator strength of the lowest exciton state is plotted as a function of the quantum dot radius for a few values of the dielectric constant ratio $\epsilon_1/\epsilon_2$ with a fixed value of the electron-to-hole mass ratio, i.e., $m_\mathrm{e}/m_\mathrm{h}=0.2$ (from T. Takagahara [11])

with

$$f_0 = \frac{2}{m_0 \hbar \omega_X} |p_\mathrm{cv}^0|^2 \, , \tag{2.35}$$

where $m_0$ is the free-electron mass. In the following, the excitonic transition energy $\hbar \omega_X$ is assumed as a constant independent of the crystallite size. Then $f_0$ has a meaning of the oscillator strength of the band-to-band transition at the $\Gamma$-point. In Fig. 2.4 the normalized oscillator strength $f_X/f_0$ is shown as a function of the quantum dot radius for a few values of the dielectric constant ratio $\epsilon_1/\epsilon_2$ with a fixed value of the electron-to-hole mass ratio, i.e., $m_\mathrm{e}/m_\mathrm{h}=0.2$. In the limit of $R \longrightarrow 0$, the exciton oscillator strength $f_X$ approaches $f_0$. Since $f_0$ has a meaning of the oscillator strength of the atomic transition, it is reasonable that the limit of $R \longrightarrow 0$ corresponds to the case of a single atom. The magnitude of the oscillator strength is determined by two factors concerning the electron–hole relative motion and the exciton center-of-mass motion. The former is related to the parameter $\alpha$ in (2.17), whereas the latter factor comes from the summation over $r_\mathrm{s}$ in (2.34) and is roughly proportional to the number of unit cells in a quantum dot. As a result, the exciton oscillator strength increases with increasing crystallite size. This increasing trend is stronger for a larger value of the dielectric constant ratio. This is a manifestation of the dielectric confinement effect via the increased electron–hole Coulomb binding.

## 2.3    Nonlocal Response Theory of Radiative Decay Rate of Excitons in Quantum Dots: Size Dependence and Temperature Dependence

The radiative decay rate of excitons in semiconductor nanocrystals was predicted to increase in proportion to the volume of the nanocrystal [17,18] and was called "mesoscopic enhancement." This property was observed in the size-dependent decay characteristics of luminescence from CuCl nanocrystals [19,20]. However, the increase of the radiative decay rate with increasing crystallite size does not continue indefinitely, but saturates at a certain level. The physical mechanism of this saturation of the radiative decay rate and the relevant factors determining the crystallite size at which the saturation occurs have not been clarified comprehensively until recently [21], although an argument was given for CdS nanocrystals [22].

There are two possible mechanisms that could explain the saturation of the radiative decay rate. One is the breakdown of the coherent addition of atomic transition dipoles throughout a nanocrystal due to the finiteness of the wavelength of light. We will call this mechanism the retardation effect. The other mechanism is the breakdown of the mesoscopic enhancement due to the homogeneous broadening effect, which induces the overlap of the excited excitonic states within the homogeneous linewidth of the lowest exciton state.

In the limit that the wavelength of light having an energy of the relevant excitonic transition is much larger than the nanocrystal size, the radiative decay rate is proportional to the oscillator strength of the exciton. In this limit, the oscillator strength is, in turn, proportional to the volume of the nanocrystal, because the transition dipole moment of each atom is added coherently throughout the nanocrystal. However, when the crystallite size becomes larger than the wavelength of light, the coherent addition of atomic dipoles becomes incomplete, resulting in the saturation of the radiative decay rate. This saturation occurs at the nanocrystal size comparable with the wavelength of light.

On the other hand, when the energy spacings of excitonic levels become smaller than the homogeneous linewidth of the lowest excitonic state, several excited excitonic states overlap. In this situation, even under the resonant excitation of the lowest exciton state, the excited excitonic states are simultaneously excited and contribute to the radiative decay process. In general, the excited excitonic states have smaller radiative decay rates than the lowest exciton state. Thus the overlap of excited excitonic levels within the homogeneous linewidth of the lowest excitonic state induces the thermal population of these excited states and leads to the saturation of the radiative decay rate.

### 2.3.1    Formulation

In order to estimate the size-dependent radiative decay rate of excitons without employing the long-wavelength approximation, we must solve Maxwell's equations making use of the nonlocal polarizability tensor, which incorporates the excitonic energy spectra in a nanocrystal. The formal aspect of this kind of nonlocal treatment was discussed previously [23]. However, since some improvement concerning the dielectric boundary condition is necessary, we will outline this formalism.

The optical response of a system will be discussed in the Coulomb gauge. The Maxwell electric field $\boldsymbol{E}(\boldsymbol{r}, t)$ is represented as

$$\boldsymbol{E}(\boldsymbol{r}, t) = -\frac{1}{c}\frac{\partial \boldsymbol{A}}{\partial t} , \tag{2.36}$$

where $\boldsymbol{A}$ is the vector potential. The electron–electron interaction which produces the longitudinal electric field is included in the material Hamiltonian. The radiation–matter interaction up to the first order in the vector potential $\boldsymbol{A}$ can be written as

$$H' = -\sum_{\omega} \int \mathrm{d}^3r \, \tilde{p}(\boldsymbol{r}, \omega) \left[ \boldsymbol{E}_{\mathrm{s}}^{(-)}(\boldsymbol{r}, \omega)\mathrm{e}^{-\mathrm{i}\omega t} - \boldsymbol{E}_{\mathrm{s}}^{(+)}(\boldsymbol{r}, \omega)\mathrm{e}^{\mathrm{i}\omega t} \right] , \tag{2.37}$$

with

$$\tilde{p}(\boldsymbol{r}, \omega) = \frac{\mathrm{i}e}{2m_0\omega}\sum_{\mathrm{s}}[\delta(\boldsymbol{r} - \boldsymbol{r}_s)\boldsymbol{p} + \boldsymbol{p}\delta(\boldsymbol{r} - \boldsymbol{r}_s)] , \tag{2.38}$$

where $\boldsymbol{p} = (-\mathrm{i}\hbar)\nabla$ and the summation over relevant frequency components is carried out, and $\boldsymbol{E}_{\mathrm{s}}^{(\pm)}$ is the positive or negative frequency component of the source electric field. Then the polarization density induced by $\boldsymbol{E}_{\mathrm{s}}^{(-)}$ can be given by

$$P_i^{(-)}(\boldsymbol{r}, \omega) = \int \mathrm{d}^3r' \, \alpha_{ij}(\boldsymbol{r}, \boldsymbol{r}'; \omega)E_{\mathrm{s}j}^{(-)}(\boldsymbol{r}', \omega) , \tag{2.39}$$

with

$$\alpha_{ij}(\boldsymbol{r}, \boldsymbol{r}'; \omega) = -\sum_{\lambda} \left[ \frac{\langle 0|\tilde{p}_i(\boldsymbol{r}, \omega)|\lambda\rangle\langle\lambda|\tilde{p}_j(\boldsymbol{r}', \omega)|0\rangle}{E_\lambda - \hbar\omega - \mathrm{i}\hbar\gamma_\lambda} \right.$$
$$\left. + \frac{\langle 0|\tilde{p}_j(\boldsymbol{r}', \omega)|\lambda\rangle\langle\lambda|\tilde{p}_i(\boldsymbol{r}, \omega)|0\rangle}{E_\lambda + \hbar\omega + \mathrm{i}\hbar\gamma_\lambda} \right] , \tag{2.40}$$

where $\alpha_{ij}$ is the polarizability tensor, the summation over the excitonic states denoted by $\lambda$ is carried out, and $E_\lambda(\gamma_\lambda)$ denotes the energy (homogeneous linewidth) of the $\lambda$ exciton state.

When the material system is polarized, a test charge at the position $\boldsymbol{r}$ feels a local electric field induced by all the dipoles namely,

$$\int \mathrm{d}^3r' \, \mathrm{grad\ div}\frac{1}{|\boldsymbol{r} - \boldsymbol{r}'|}\boldsymbol{P}(\boldsymbol{r}', \omega)$$
$$= \int \mathrm{d}^3r'\frac{3(\boldsymbol{r} - \boldsymbol{r}')^t(\boldsymbol{r} - \boldsymbol{r}') - |\boldsymbol{r} - \boldsymbol{r}'|^2\hat{1}}{|\boldsymbol{r} - \boldsymbol{r}'|^5}\boldsymbol{P}(\boldsymbol{r}', \omega) , \tag{2.41}$$

where grad and div are operating on the coordinates of $\boldsymbol{r}$, and a dyadic form is used on the right hand side. This field is composed of the depolarization field arising from the surface polarization charges and the Lorentz cavity field. The depolarization field is given by

$$\boldsymbol{E}_{\text{dep}}(\boldsymbol{r}, \omega) = \text{grad div} \int d^3 r' \frac{1}{|\boldsymbol{r} - \boldsymbol{r}'|} \boldsymbol{P}(\boldsymbol{r}', \omega) . \tag{2.42}$$

The proof of this relation is given in Appendix A. The macroscopic or Maxwell electric field is composed of the external source field and the depolarization field. Thus the source electric field is given by

$$\boldsymbol{E}_{\text{s}}(\boldsymbol{r}, \omega) = \boldsymbol{E}(\boldsymbol{r}, \omega) - \boldsymbol{E}_{\text{dep}}(\boldsymbol{r}, \omega) . \tag{2.43}$$

So far, the background dielectric constant has not been taken into account. In actual systems, however, when the excitonic transitions are considered, the dielectric screening by non-resonant transitions is inevitably present and should be included to describe the excitonic properties adequately. This effect can be included in terms of the background dielectric constant $\epsilon_{\text{b}}$. Then the expression of the depolarization field should be modified appropriately. First of all, since the excitonic dipoles are embedded in a background material with the dielectric constant $\epsilon_{\text{b}}$, the depolarization field is different from that in (2.42) where the vacuum is assumed as the background. Secondly, the polarization dipoles of the background material induced by the source field give rise to a depolarization field. The former contribution is derived in Appendix B, taking account of the dielectric boundary condition. The latter contribution is given by

$$\text{grad div} \int d^3 r' \frac{\chi_{\text{b}}}{|\boldsymbol{r} - \boldsymbol{r}'|} \boldsymbol{E}_{\text{s}}(\boldsymbol{r}', \omega) , \tag{2.44}$$

with $\chi_{\text{b}} = (\epsilon_{\text{b}} - 1)/4\pi$. It is to be noted that the depolarization field due to the polarization dipoles of the background material induced by the excitonic polarization is included in the former contribution. Consequently, the depolarization field is given by

$$\begin{aligned}
\boldsymbol{E}_{\text{dep}}(\boldsymbol{r}, \omega) &= \frac{1}{\epsilon_{\text{b}}} \text{grad div} \int d^3 r' \frac{1}{|\boldsymbol{r} - \boldsymbol{r}'|} \boldsymbol{P}(\boldsymbol{r}', \omega) \\
&+ \text{grad div} \int d^3 r' \frac{\chi_{\text{b}}}{|\boldsymbol{r} - \boldsymbol{r}'|} \boldsymbol{E}_{\text{s}}(\boldsymbol{r}', \omega) \\
&- \frac{1}{R} \text{grad} \int d^3 r' \sum_{l=1}^{\infty} \frac{\alpha_l \, r^l (r')^{l-1}}{R^{2l}} [l(\boldsymbol{n}_{r'} \cdot \boldsymbol{P}(\boldsymbol{r}', \omega)) P_l(\cos\theta_{rr'}) \\
&- \{(\boldsymbol{n}_{r'} \cdot \boldsymbol{n}_r)(\boldsymbol{n}_{r'} \cdot \boldsymbol{P}(\boldsymbol{r}', \omega)) - (\boldsymbol{n}_r \cdot \boldsymbol{P}(\boldsymbol{r}', \omega))\} P_l'(\cos\theta_{rr'})],
\end{aligned} \tag{2.45}$$

with

$$\alpha_l = \frac{(l+1)(\epsilon_{\text{b}} - 1)}{\epsilon_{\text{b}}(l\epsilon_{\text{b}} + l + 1)} , \tag{2.46}$$

where $P_l$ and $P'_l$ are the Legendre polynomial and its first derivative, $\boldsymbol{n}_r(\boldsymbol{n}_{r'})$ is the unit vector in the direction of $\boldsymbol{r}(\boldsymbol{r}')$, and $\theta_{rr'}$ the angle between $\boldsymbol{r}$ and $\boldsymbol{r}'$.

The Maxwell equation for the electric field within the material system is written as

$$(\text{rot rot} - k_{\mathrm{b}}^2)\boldsymbol{E}(\boldsymbol{r}, \omega) = 4\pi k_0^2 \boldsymbol{P}(\boldsymbol{r}, \omega) , \tag{2.47}$$

with

$$\boldsymbol{P}(\boldsymbol{r}, \omega) = \int d^3r' \, \alpha(\boldsymbol{r}, \boldsymbol{r}'; \omega)\boldsymbol{E}_{\mathrm{s}}(\boldsymbol{r}', \omega) , \tag{2.48}$$

where $k_{\mathrm{b}}^2 = \epsilon_{\mathrm{b}} k_0^2$ and $k_0^2 = \omega^2/c^2$. The general solution of (2.47) is given as

$$\boldsymbol{E}(\boldsymbol{r}, \omega) = \boldsymbol{E}_0(\boldsymbol{r}, \omega) + k_0^2 \int d^3r' \frac{\exp(ik_{\mathrm{b}}|\boldsymbol{r} - \boldsymbol{r}'|)}{|\boldsymbol{r} - \boldsymbol{r}'|} \boldsymbol{P}(\boldsymbol{r}', \omega)$$
$$+ \frac{1}{\epsilon_{\mathrm{b}}} \text{grad div} \int d^3r' \frac{\exp(ik_{\mathrm{b}}|\boldsymbol{r} - \boldsymbol{r}'|)}{|\boldsymbol{r} - \boldsymbol{r}'|} \boldsymbol{P}(\boldsymbol{r}', \omega) . \tag{2.49}$$

Here $\boldsymbol{E}_0(\boldsymbol{r}, \omega)$ is a field satisfying a homogeneous equation

$$(\text{rot rot} - k_{\mathrm{b}}^2)\boldsymbol{E}_0(\boldsymbol{r}, \omega) = 0 , \tag{2.50}$$

and can be written as

$$\boldsymbol{E}_0(\boldsymbol{r}, \omega) = \sum_{lm} \left[ a_{lm}^< \boldsymbol{M}_{lm}^{(0)}(k_{\mathrm{b}}r) + b_{lm}^< \boldsymbol{N}_{lm}^{(0)}(k_{\mathrm{b}}r) \right] , \tag{2.51}$$

where $\boldsymbol{M}_{lm}$ and $\boldsymbol{N}_{lm}$ are the transverse vector spherical harmonics given in Appendix C, and $a_{lm}^<$ and $b_{lm}^<$ are to be determined from the external (incident) field through the Maxwell boundary conditions.

Denoting the excitonic wavefunction by $\Psi_\lambda(\boldsymbol{r}_{\mathrm{e}}, \boldsymbol{r}_{\mathrm{h}})$, we introduce $\rho_\lambda^*(\boldsymbol{r})$ defined by

$$\langle \lambda | \tilde{p}(\boldsymbol{r}, \omega) | 0 \rangle \equiv \rho_\lambda^*(\boldsymbol{r}) = \frac{e}{m_0\omega} \boldsymbol{P}^{cv} \sum_s \delta(\boldsymbol{r} - \boldsymbol{r}_s)\Psi_\lambda^*(\boldsymbol{r}_s, \boldsymbol{r}_s) , \tag{2.52}$$

$$\langle 0 | \tilde{p}(\boldsymbol{r}, \omega) | \lambda \rangle \equiv -\rho_\lambda(\boldsymbol{r}) = -\frac{e}{m_0\omega} \boldsymbol{P}^{cv} \sum_s \delta(\boldsymbol{r} - \boldsymbol{r}_s)\Psi_\lambda(\boldsymbol{r}_s, \boldsymbol{r}_s) , \tag{2.53}$$

where $\boldsymbol{P}^{cv} = \hbar\langle c|\nabla|v\rangle = -\hbar\langle v|\nabla|c\rangle$ is the matrix element between the Bloch functions at the conduction band bottom and at the valence band top and is assumed to be real. Then taking into account only the resonant term in (2.40), we obtain from (2.39)

$$\boldsymbol{P}(\boldsymbol{r}, \omega) = \sum_\lambda g_\lambda(\omega)\boldsymbol{\rho}_\lambda(\boldsymbol{r}) \int d^3r' \rho_\lambda^*(\boldsymbol{r}') \cdot \boldsymbol{E}_{\mathrm{s}}(\boldsymbol{r}', \omega)$$
$$\equiv \sum_\lambda g_\lambda(\omega)\boldsymbol{\rho}_\lambda(\boldsymbol{r})F_\lambda , \tag{2.54}$$

with $g_\lambda(\omega) = 1/(E_\lambda - \hbar\omega - i\hbar\gamma_\lambda)$, where the superscript $(-)$ is dropped since only the negative frequency part will be considered in the following, and $F_\lambda$ is defined by the integral in the middle term.

In the case of CuCl, since the exciton Bohr radius is rather small ($\sim 7$ Å) and the exciton binding energy is fairly large ($\sim 200$ meV), we calculate the exciton wavefunctions in the center-of-mass confinement regime with the electron–hole relative motion fixed in the lowest state (1s). More explicitly, the exciton wavefunction can be written as

$$\Psi_{\bar{l}\bar{m}j}(r_{\rm e}, r_{\rm h}) = C j_{\bar{l}}\left(k_j^{\bar{l}} r_M\right) Y_{\bar{l}\bar{m}}(\Omega_M) \phi_{1s}(r_{\rm e} - r_{\rm h}) , \qquad (2.55)$$

where $C$ is the normalization constant, $r_M(\Omega_M)$ denotes the radial(angular) part of the center-of-mass coordinate of the exciton, and $\phi_{1s}$ is the ground state of the electron–hole relative motion. In the long-wavelength limit, the optical transition is allowed only for $\bar{l}=0$. However, in the nonlocal response regime where the spatial extent of the exciton wavefunction or the size of the quantum dot is comparable to or larger than the wavelength of light corresponding to the exciton energy, the above selection rule is relaxed and the $\bar{l}\neq 0$ states can have a finite spontaneous emission rate.

Expanding $\boldsymbol{P}^{cv} Y_{\bar{l}\bar{m}}$ in terms of the mutually orthogonal angular harmonics defined in Appendix C as

$$\boldsymbol{P}^{cv} Y_{\bar{l}\bar{m}}(\Omega) = \sum_{lm} \left[ a_{lm}(\bar{l}, \bar{m}) \boldsymbol{P}_{lm}(\Omega) \right.$$
$$\left. + b_{lm}(\bar{l}, \bar{m}) \boldsymbol{B}_{lm}(\Omega) + c_{lm}(\bar{l}, \bar{m}) \boldsymbol{C}_{lm}(\Omega) \right] , \qquad (2.56)$$

and using the formulas

$$\text{grad div}(F(r)\boldsymbol{P}_{lm}(\Omega)) = \frac{\rm d}{\rm dr}\left[ \frac{\rm dF(r)}{\rm dr} + \frac{2}{r}F(r) \right] \boldsymbol{P}_{lm}(\Omega)$$
$$+ \frac{\sqrt{l(l+1)}}{r}\left[ \frac{\rm dF(r)}{\rm dr} + \frac{2}{r}F(r) \right] \boldsymbol{B}_{lm}(\Omega), \quad (2.57)$$

$$\text{grad div}(F(r)\boldsymbol{B}_{lm}(\Omega)) = -\sqrt{l(l+1)}\frac{\rm d}{\rm dr}\left[ \frac{F(r)}{r} \right] \boldsymbol{P}_{lm}(\Omega)$$
$$- \frac{l(l+1)}{r^2}F(r)\boldsymbol{B}_{lm}(\Omega) , \qquad (2.58)$$

$$\text{grad div}(F(r)\boldsymbol{C}_{lm}(\Omega)) = 0 , \qquad (2.59)$$

we can express the Maxwell field $\boldsymbol{E}(\boldsymbol{r}, \omega)$ as

$$\boldsymbol{E}(\boldsymbol{r}, \omega) = \sum_{lm} \left[ a_{lm}^< \boldsymbol{M}_{lm}^{(0)}(k_{\rm b}r) + b_{lm}^< \boldsymbol{N}_{lm}^{(0)}(k_{\rm b}r) \right]$$
$$+ \frac{4\pi i k_{\rm b} k_0^2 e}{m_0 \omega} \sum_\lambda g_\lambda(\omega) F_\lambda \sum_{lm} [R_{lm}^\lambda(r) \boldsymbol{P}_{lm}(\Omega)$$
$$+ S_{lm}^\lambda(r) \boldsymbol{B}_{lm}(\Omega) + T_{lm}^\lambda(r) \boldsymbol{C}_{lm}(\Omega)] , \qquad (2.60)$$

with

$$R_{lm}^{\lambda}(r) = a_{lm}(\bar{l}, \bar{m}) \left\{ A_{\lambda}(r) + \frac{1}{k_{\mathrm{b}}^2} \frac{\mathrm{d}}{\mathrm{d}r} \left[ \frac{\mathrm{d}A_{\lambda}(r)}{\mathrm{d}r} + \frac{2}{r} A_{\lambda}(r) \right] \right\}$$

$$- b_{lm}(\bar{l}, \bar{m}) \frac{\sqrt{l(l+1)}}{k_{\mathrm{b}}^2} \frac{\mathrm{d}}{\mathrm{d}r} \left( \frac{A_{\lambda}(r)}{r} \right) , \tag{2.61}$$

$$S_{lm}^{\lambda}(r) = a_{lm}(\bar{l}, \bar{m}) \frac{\sqrt{l(l+1)}}{k_{\mathrm{b}}^2 r} \left[ \frac{\mathrm{d}A_{\lambda}(r)}{\mathrm{d}r} + \frac{2}{r} A_{\lambda}(r) \right]$$

$$+ b_{lm}(\bar{l}, \bar{m}) \left( 1 - \frac{l(l+1)}{k_{\mathrm{b}}^2 r^2} \right) A_{\lambda}(r) , \tag{2.62}$$

$$T_{lm}^{\lambda}(r) = c_{lm}(\bar{l}, \bar{m}) A_{\lambda}(r) , \tag{2.63}$$

$$A_{\lambda}(r) = \int_0^R \mathrm{d}r' \, (r')^2 h_{\bar{l}}^{(1)}(k_{\mathrm{b}} r_>) j_{\bar{l}}(k_{\mathrm{b}} r_<) \Phi_{\lambda}(r') , \tag{2.64}$$

where $r_> = \max(r, r')$, $r_< = \min(r, r')$, and $\Phi_{\lambda}(\lambda = (\bar{l}, \bar{m}, j)$ in (2.55)) is defined by $\Psi_{\lambda}(\boldsymbol{r}, \boldsymbol{r}) = Y_{\bar{l}\bar{m}}(\Omega) \Phi_{\lambda}(r)$.

By using the formulas

$$\mathrm{grad} \, \mathrm{div} \int \mathrm{d}^3 r' \frac{1}{|\boldsymbol{r} - \boldsymbol{r}'|} \boldsymbol{M}_{lm}^{(0)}(k_{\mathrm{b}} r') = 0 , \tag{2.65}$$

$$\mathrm{grad} \, \mathrm{div} \int \mathrm{d}^3 r' \frac{1}{|\boldsymbol{r} - \boldsymbol{r}'|} \boldsymbol{N}_{lm}^{(0)}(k_{\mathrm{b}} r')$$

$$= \Pi_l(r) \boldsymbol{P}_{lm}(\Omega) + \beta_l(r) \boldsymbol{B}_{lm}(\Omega) , \tag{2.66}$$

with

$$\Pi_l(r) = -\frac{4\pi l^2 (l+1)}{(2l+1) k_{\mathrm{b}}} \frac{r^{l-1}}{R^l} j_l(k_{\mathrm{b}} R) , \tag{2.67}$$

and

$$\beta_l(r) = -\frac{4\pi [l(l+1)]^{3/2}}{(2l+1) k_{\mathrm{b}}} \frac{r^{l-1}}{R^l} j_l(k_{\mathrm{b}} R) , \tag{2.68}$$

the depolarization field can be calculated as

$$\boldsymbol{E}_{\mathrm{dep}}(\boldsymbol{r}, \omega) = \chi_{\mathrm{b}} \sum_{lm} b_{lm}^< \left[ \Pi_l(r) \boldsymbol{P}_{lm}(\Omega) + \beta_l(r) \boldsymbol{B}_{lm}(\Omega) \right]$$

$$+ \chi_{\mathrm{b}} \frac{4\pi \mathrm{i} k_{\mathrm{b}} k_0^2 e}{m_0 \omega} \mathrm{grad} \, \mathrm{div} \int \mathrm{d}^3 r' \frac{1}{|\boldsymbol{r} - \boldsymbol{r}'|}$$

$$\times \sum_{\lambda} g_{\lambda}(\omega) F_{\lambda} \left\{ \frac{m_0 \omega}{4\pi \mathrm{i} k_{\mathrm{b}} k_0^2 e \epsilon_{\mathrm{b}} \chi_{\mathrm{b}}} \boldsymbol{\rho}_{\lambda}(\boldsymbol{r}') \right.$$

$$+ \sum_{lm} \left[ R_{lm}^{\lambda}(r') \boldsymbol{P}_{lm}(\Omega') + S_{lm}^{\lambda}(r') \boldsymbol{B}_{lm}(\Omega') \right.$$

$$\left. + T_{lm}^{\lambda}(r') \boldsymbol{C}_{lm}(\Omega') \right] \right\} + \frac{e}{m_0 \omega} \sum_{\lambda} g_{\lambda}(\omega) F_{\lambda}$$

$$\times \sum_{lm} \left[ \gamma_{lm}^{\lambda}(r) \boldsymbol{P}_{lm}(\Omega) + \delta_{lm}^{\lambda}(r) \boldsymbol{B}_{lm}(\Omega) \right] , \tag{2.69}$$

with

$$\gamma^\lambda_{lm}(r) = -\left[la_{lm}(\bar{l},\bar{m}) + \sqrt{l(l+1)}b_{lm}(\bar{l},\bar{m})\right]\frac{l\xi_l}{R^3}\left(\frac{r}{R}\right)^{l-1}, \tag{2.70}$$

$$\delta^\lambda_{lm}(r) = -\left[la_{lm}(\bar{l},\bar{m}) + \sqrt{l(l+1)}b_{lm}(\bar{l},\bar{m})\right]\frac{\sqrt{l(l+1)}\xi_l}{R^3}\left(\frac{r}{R}\right)^{l-1}, \tag{2.71}$$

and

$$\xi_l = \frac{4\pi\alpha_l}{2l+1}\int_0^R \mathrm{d}r'\,(r')^2\Phi_\lambda(r')\left(\frac{r'}{R}\right)^{l-1}. \tag{2.72}$$

Then the source field is given by $\boldsymbol{E}_s(\boldsymbol{r},\omega){=}\boldsymbol{E}(\boldsymbol{r},\omega){-}\boldsymbol{E}_{dep}(\boldsymbol{r},\omega)$, and the equation for $F_\lambda{=}\int \mathrm{d}^3r\boldsymbol{\rho}^*_\lambda(\boldsymbol{r})\cdot\boldsymbol{E}_s(\boldsymbol{r},\omega)$ is obtained as

$$F_\lambda = \sum_{lm}\left[a^<_{lm}D^\lambda_{lm} + b^<_{lm}\left(E^\lambda_{lm} - \chi_{\mathrm{b}}H^\lambda_{lm}\right)\right]$$
$$+ \sum_\nu (C_{\lambda\nu} - \chi_{\mathrm{b}}D_{\lambda\nu})g_\nu(\omega)F_\nu, \tag{2.73}$$

with

$$D^\lambda_{lm} = \int \mathrm{d}^3r\,\boldsymbol{\rho}^*_\lambda(\boldsymbol{r})\cdot\boldsymbol{M}^{(0)}_{lm}(k_{\mathrm{b}}r), \tag{2.74}$$

$$E^\lambda_{lm} = \int \mathrm{d}^3r\,\boldsymbol{\rho}^*_\lambda(\boldsymbol{r})\cdot\boldsymbol{N}^{(0)}_{lm}(k_{\mathrm{b}}r), \tag{2.75}$$

$$H^\lambda_{lm} = \int \mathrm{d}^3r\,\boldsymbol{\rho}^*_\lambda(\boldsymbol{r})\cdot\left(\Pi_l(r)\boldsymbol{P}_{lm}(\Omega) + \beta_l(r)\boldsymbol{B}_{lm}(\Omega)\right), \tag{2.76}$$

$$C_{\lambda\nu} = \frac{4\pi\mathrm{i}k_{\mathrm{b}}k_0^2 e}{m_0\omega}\int \mathrm{d}^3r\,\boldsymbol{\rho}^*_\lambda(\boldsymbol{r})\times\sum_{lm}\left[\left(R^\nu_{lm}(r) - \frac{\gamma^\nu_{lm}(r)}{4\pi\mathrm{i}k_{\mathrm{b}}k_0^2}\right)\boldsymbol{P}_{lm}(\Omega)\right.$$
$$\left. + \left(S^\nu_{lm}(r) - \frac{\delta^\nu_{lm}(r)}{4\pi\mathrm{i}k_{\mathrm{b}}k_0^2}\right)\boldsymbol{B}_{lm}(\Omega) + T^\nu_{lm}(r)\boldsymbol{C}_{lm}(\Omega)\right], \tag{2.77}$$

and

$$D_{\lambda\nu} = \frac{4\pi\mathrm{i}k_{\mathrm{b}}k_0^2 e}{m_0\omega}\int \mathrm{d}^3r\,\boldsymbol{\rho}^*_\lambda(\boldsymbol{r})\cdot\mathrm{grad}\,\mathrm{div}\int \mathrm{d}^3r'\frac{1}{|\boldsymbol{r}-\boldsymbol{r}'|}$$
$$\left\{\frac{m_0\omega}{4\pi\mathrm{i}k_{\mathrm{b}}k_0^2 e\epsilon_{\mathrm{b}}\chi_{\mathrm{b}}}\rho_\nu(\boldsymbol{r}') + \sum_{lm}\left[R^\nu_{lm}(r')\boldsymbol{P}_{lm}(\Omega')\right.\right.$$
$$\left.\left. + S^\nu_{lm}(r')\boldsymbol{B}_{lm}(\Omega') + T^\nu_{lm}(r')\boldsymbol{C}_{lm}(\Omega')\right]\right\}. \tag{2.78}$$

Equation (2.73) is a linear equation for $\{F_\lambda(\lambda = 1, 2, \cdots)\}$ and can easily be solved in terms of $\{a^<_{lm},\, b^<_{lm}\}$.

Now we consider the Maxwell boundary condition at the surface of a quantum dot. The Maxwell electric field outside the quantum dot is written as

$$\boldsymbol{E}^{>}(\boldsymbol{r},\omega) = \boldsymbol{E}_0 e^{ik_0 z} + \sum_{lm} \left[ a_{lm}^{>} \boldsymbol{M}_{lm}^{(1)}(k_0 r) + b_{lm}^{>} \boldsymbol{N}_{lm}^{(1)}(k_0 r) \right] , \qquad (2.79)$$

where the first term is the external (incident) field, and $\boldsymbol{M}^{(1)}$ and $\boldsymbol{N}^{(1)}$ are the vector spherical harmonics associated with the spherical Hankel function of the first kind $h_l^{(1)}(k_0 r)$. The Maxwell boundary conditions are given by the continuity of tangential components of the electric field and the magnetic field at the boundary. In the case of a spherical body, the boundary conditions can be derived easily by expanding the fields in terms of the mutually orthogonal angular harmonics $\boldsymbol{P}_{lm}(\Omega)$, $\boldsymbol{B}_{lm}(\Omega)$, and $\boldsymbol{C}_{lm}(\Omega)$, since the tangential components are given by $\boldsymbol{B}_{lm}(\Omega)$ and $\boldsymbol{C}_{lm}(\Omega)$. The whole procedure amounts to the extension of the Mie scattering problem to the case of nonlocal media.

The boundary condition yields a set of linear equations concerning $a_{lm}^{>}$, $b_{lm}^{>}$, $a_{lm}^{<}$ and $b_{lm}^{<}$. Then the zeros with respect to $\omega$ of the determinant of these coefficients give rise to peaks in the absorption and extinction spectra. The radiative decay rate can be estimated as twice the imaginary part of these zeros.

## 2.3.2   Size Dependence of Excitonic Radiative Decay Rate

The calculated radiative decay rates for CuCl nanocrystals are given in Fig 2.5 for radii of 30 Å and 100 Å, respectively. In the small-size regime, the long-wavelength approximation is applicable, and the radiative decay rate is very small for the exciton states having nonzero angular momenta of the center-of-mass motion. On the other hand, in the large-size regime, these exciton states have a finite probability of spontaneous emission. As can be seen in Fig. 2.5, the exciton state with an index of $(l, m, j)=(1, 0, 1)$ has a fairly large radiative decay rate.

The radiative decay rate of the exciton ground state is plotted in Fig. 2.6 as a function of the quantum dot radius. In the range of radii up to several hundred angstroms, the radiative decay rate increases almost in proportion to the volume of the quantum dot. This is a well-known behavior of the mesoscopic enhancement. However, the radiative decay rate assumes the maximum value around $R{\sim}800$ Å and then decreases. This feature can be understood from the fact that the radial distribution of the electromagnetic field intensity within the quantum dot is determined mainly by $|j_0(k_b r)|^2$, and its spatial extent is about $2\pi/k_b{\simeq}1600$ Å. When the diameter of a quantum dot is of the same order as $2\pi/k_b$, the atomic dipole moments are coherently added throughout the quantum dot, resulting in a large radiative decay rate. When the diameter exceeds this length, there exists a spatial region where the electromagnetic field is out-of-phase with atomic dipole moments, and

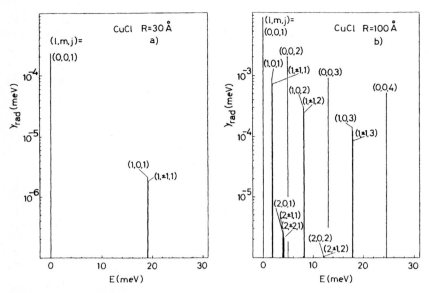

**Fig. 2.5.** The radiative decay rates of excitons in a CuCl nanocrystal are plotted at their energy positions. The exciton states are specified by the angular momentum $(l, m)$ of the center-of-mass motion and the radial quantum number $j$. The radius is 30 Å in (**a**) and 100 Å in (**b**)

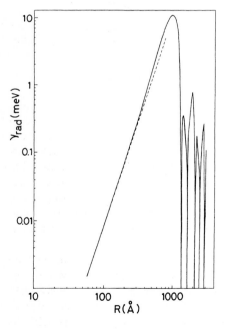

**Fig. 2.6.** The radiative decay rate of the exciton ground state in a CuCl nanocrystal is plotted as a function of the radius $R$. The *dotted line* represents the $R^3$ dependence (from T. Takagahara [21])

the radiative decay rate tends to decrease. In the larger-size region, sharp structures appear that can be identified as the morphology-dependent resonances [24,25]. It is important to note that for radii up to several hundred angstroms, the radiative decay rate continues to increase in proportion to the volume of the quantum dot and does not show any saturation, in contrast to the experimental results [20]. Thus the retardation or nonlocal effect cannot explain the experiment.

### 2.3.3   Effect of Homogeneous Broadening on Excitonic Radiative Decay Rate

Now we discuss the breakdown of the mesoscopic enhancement in the presence of homogeneous broadening. As the quantum dot size is increased, the energy spacings among excitonic levels become smaller, and several levels overlap within the homogeneous linewidth. In this situation, even under the resonant excitation of the lowest exciton state, the excited excitonic states are simultaneously excited and contribute to the radiative decay process. The redistribution of population among excitonic levels is, in general, induced by multiphonon processes. Here a phenomenological approach is employed, assuming rapid redistribution among excitonic levels. We consider that the observed radiative decay rate is an average of the radiative decay rate of excitonic states within the homogeneous linewidth of the lowest exciton state. Namely, we define as

$$\gamma_{\mathrm{rad}} = \frac{\sum_i \gamma_i g(E_i) \mathrm{e}^{-E_i/k_{\mathrm{B}}T}}{\sum_i g(E_i) \mathrm{e}^{-E_i/k_{\mathrm{B}}T}} \ , \tag{2.80}$$

with

$$g(E_i) = \frac{1}{(E_i - E_0)^2 + \Gamma_{\mathrm{h}}^2(T,R)} \ , \tag{2.81}$$

where $E_i$ and $\gamma_i$ are the energy and the radiative decay rate of the $i$th excitonic level, respectively. $\Gamma_{\mathrm{h}}$ is the size- and temperature-dependent homogeneous linewidth of the lowest excitonic state, and the experimental values [26] are employed in the calculation. In the summation in (2.80), the excitonic levels up to 30 meV from the lowest exciton level are included. It is important to note that the excitonic states that are optically forbidden in the long-wavelength limit have a finite radiative decay rate in the present treatment where the retardation effect is taken into account.

The calculated radiative decay rates are plotted in Fig. 2.7 as a function of the quantum dot radius for several values of temperature. At low temperatures, the proportionality to the volume of the quantum dot is clearly seen. However, with increasing temperature this proportionality breaks down at smaller values of the radius. This feature reflects the thermal excitation to excited excitonic states. At 80 K, the size dependence begins to deviate from

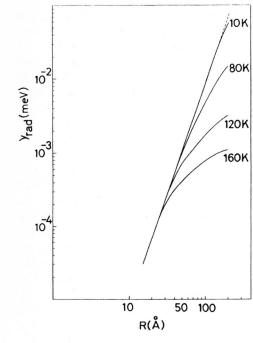

**Fig. 2.7.** The excitonic radiative decay rate in a CuCl nanocrystal is plotted as a function of the radius for several values of temperature. The *dotted line* represents the $R^3$ dependence (from T. Takagahara [21])

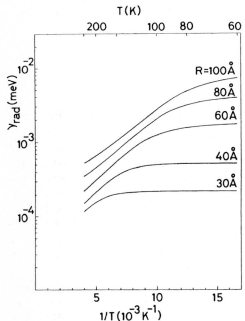

**Fig. 2.8.** The excitonic radiative decay rate in a CuCl nanocrystal is plotted as a function of the inverse temperature for several values of the radius $R$. The *upper abscissa* represents the temperature (from T. Takagahara [21])

the $R^3$ proportionality around $R \simeq 50$ Å, in good agreement with the experiment [20], which was carried out at 77 K. In Fig. 2.8, the radiative decay rates are plotted as a function of the inverse temperature for several values of the quantum dot radius. They are almost constant at low temperatures, whereas they decrease almost exponentially at higher temperatures. The temperature at which the radiative decay rate begins to decrease exponentially shifts to lower values for larger quantum dots. This feature can be explained by considering that thermal population of the excited excitonic states occurs more readily for larger quantum dots because of the smaller energy spacings among excitonic states.

Combining these results, we can conclude that the observed size dependence of the radiative decay rate of excitons in CuCl nanocrystals cannot be explained by the electromagnetic retardation effect, but should be interpreted in terms of the thermal excitation of excited excitonic states within the homogeneous linewidth of the lowest excitonic state. Although the present arguments are developed only for zero-dimensional structures, the latter mechanism is surely the most important and general one that saturates the mesoscopic enhancement of the excitonic radiative decay rate and determines the effective exciton coherence length in low-dimensional structures [27,28].

Recently, the temperature dependence of the photoluminescence lifetime of InGaAs quantum disks has been measured, and the dimensional crossover between two-dimensional and zero-dimensional behaviors has been clearly identified by changing the size of lateral confinement, which can be interpreted satisfactorily by the above theoretical model [29,30].

## 2.4  Electron–Hole Exchange Interaction in Degenerate Valence Band Structures

Very fine structures were observed in exciton photoluminescence (PL) from GaAs quantum wells (QWs) [2–4]. These sharp lines are interpreted as luminescence from localized excitons at island structures in QWs. These island structures can be regarded as zero-dimensional quantum dots (QDs). Furthermore, in the photoluminescence excitation spectra, doublet structures with mutually orthogonal linear polarizations were discovered. The direction of linear polarization is closely related to the geometrical shape of the QW islands. As observed by scanning tunneling microscopy (STM) [4], QW islands are elongated along the [1$\bar{1}$0] direction. Correspondingly, the exciton states are polarized along the [1$\bar{1}$0] or [110] direction. The splitting energy of each doublet state is about several tens of μeV and is of the same order of magnitude as the exciton longitudinal–transverse (LT) splitting energy in bulk GaAs. From these experimental results, we can deduce that these splittings arise from the electron–hole exchange interaction, especially its long-range part, which is highly dependent on the geometrical shape of quantum dots. Recently the important role of the electron–hole exchange interaction in exciton fine structures has been discussed for compound semiconductor quantum dots, both experimentally [31–34] and theoretically [35–37].

### 2.4.1    Formulation

We now formulate a theory of excitons in a quantum dot, taking into account the multiband structure of the valence band and the electron–hole exchange interaction for the case of direct-gap semiconductors at the $\Gamma$-point. For the valence band, we take into account four states belonging to the $\Gamma_8$ symmetry ($J=3/2$) and employ the full Luttinger Hamiltonian [38] without making the spherical approximation. We neglect the spin–orbit split $J=1/2$ valence band because the splitting energy between the $J=3/2$ and the $J=1/2$ valence bands is much larger than the typical energy level spacings discussed here. Then we denote two conduction-band Bloch functions by $|c\tau\rangle$, where $\tau$ indicates the spin-up ($\alpha$) or the spin-down ($\beta$) state, and similarly four valence-band Bloch functions by $|v\sigma\rangle$, where $\sigma$ stands for the $z$-component of the angular momentum ($3/2$, $1/2$, $-1/2$, $-3/2$). Thus the exciton state is composed of eight combinations of the conduction band and the valence band, namely

$$|X\rangle = \sum_{\tau,\sigma} \sum_{r_e,r_h} f_{\tau\sigma}(r_e,r_h) a^\dagger_{c\tau r_e} a_{v\sigma r_h} |0\rangle \ , \tag{2.82}$$

where the Wannier function representation of the annihilation and creation operators is used, and $f_{\tau\sigma}$ is the envelope function. The eigenvalue equation for $f_{\tau\sigma}$ is given as

$$\sum_{\tau'\sigma'r'_e r'_h} H(c\tau r_e, v\sigma r_h; c\tau' r'_e, v\sigma' r'_h) f_{\tau'\sigma'}(r'_e, r'_h) = E\, f_{\tau\sigma}(r_e, r_h) \ , \tag{2.83}$$

with

$$
\begin{aligned}
&H\left(c\tau r_e, v\sigma r_h; c\tau' r'_e, v\sigma' r'_h\right) \\
&= \delta_{\sigma\sigma'}\delta_{r_h r'_h}\left[\epsilon_{c\tau,c\tau'}(-i\nabla_e) + \delta_{\tau\tau'} v_e(r_e)\right] \\
&\quad - \delta_{\tau\tau'}\delta_{r_e r'_e}\left[\epsilon_{v\sigma',v\sigma}(i\nabla_h) + \delta_{\sigma\sigma'} v_h(r_h)\right] \\
&\quad + V(c\tau r_e, v\sigma' r'_h; c\tau' r'_e, v\sigma r_h) - V(c\tau r_e, v\sigma' r'_h; v\sigma r_h, c\tau' r'_e) \ , \tag{2.84}
\end{aligned}
$$

where $\epsilon_{c\tau,c\tau'}(\epsilon_{v\sigma',v\sigma})$ is the band energy of the conduction (valence) band electron, $v_e(v_h)$ is the confinement potential for the conduction (valence) band electron, and $V(i,j;k,l)$ is the Coulomb or exchange matrix element defined by

$$V(i,j;k,l) = \int \mathrm{d}^3 r \int \mathrm{d}^3 r' \ \phi_i^*(r)\phi_j^*(r')\frac{e^2}{\epsilon|r-r'|}\phi_k(r')\phi_l(r) \ , \tag{2.85}$$

in terms of the Wannier functions $\{\phi_i\}$ and the dielectric constant $\epsilon$. The quantity $\epsilon_{c\tau,c\tau'}$ is simply approximated as

$$\epsilon_{c\tau,c\tau'}(k) = -\frac{\hbar^2}{2m_e}k^2\delta_{\tau\tau'} \ , \tag{2.86}$$

with the effective mass $m_e$, and $\epsilon_{v\sigma',v\sigma}(k)$ is the Luttinger Hamiltonian given by

$$
-\frac{\hbar^2}{m_0}\Bigg[(\gamma_1+\frac{5}{2}\gamma_2)\frac{k^2}{2}-\gamma_2(k_x^2 J_x^2+k_y^2 J_y^2+k_z^2 J_z^2)
$$
$$
-\gamma_3 k_y k_z(J_y J_z+J_z J_y)-\gamma_3 k_z k_x(J_z J_x+J_x J_z)
$$
$$
-\gamma_3 k_x k_y(J_x J_y+J_y J_x)\Bigg]
\tag{2.87}
$$

in terms of the Luttinger parameters $\gamma_1, \gamma_2$, and $\gamma_3$, the free electron mass $m_0$, and the angular momentum operators $J_x, J_y$, and $J_z$ for $J=3/2$. The Coulomb matrix element can be approximated due to the localized nature of the Wannier functions as

$$
V(c\tau r_e, v\sigma' r_h'; v\sigma r_h, c\tau' r_e') \approx \delta_{r_e r_e'}\delta_{r_h r_h'}\delta_{\tau\tau'}\delta_{\sigma\sigma'}\frac{e^2}{\epsilon|r_e-r_h|} .
\tag{2.88}
$$

The exchange term can also be approximated as

$$
V\,(c\tau r_e, v\sigma' r_h'; c\tau' r_e', v\sigma r_h)
$$
$$
\approx \delta_{r_e r_h}\delta_{r_e' r_h'}[\delta_{r_e r_e'}V(c\tau r_e, v\sigma' r_e; c\tau' r_e, v\sigma r_e)
$$
$$
+(1-\delta_{r_e r_e'})V(c\tau r_e, v\sigma' r_e'; c\tau' r_e', v\sigma r_e)] .
\tag{2.89}
$$

The first (second) term in the parentheses is usually called the short-range (long-range) exchange term. The long-range term can be decomposed into the multipole expansion, and the lowest-order term is retained as

$$
V(c\tau r_e, v\sigma' r_e'; c\tau' r_e', v\sigma r_e) \approx e^2 \mu_{c\tau,v\sigma}\frac{[1-3n\cdot^t n]}{|r_e-r_e'|^3}\mu_{v\sigma',c\tau'} ,
\tag{2.90}
$$

with

$$
n=\frac{r_e-r_e'}{|r_e-r_e'|} ,
\tag{2.91}
$$

$$
\mu_{c\tau,v\sigma}=\int \mathrm{d}^3 r\, \phi^*_{c\tau R}(r)(r-R)\phi_{v\sigma R}(r) ,
\tag{2.92}
$$

where $\phi_{c\tau(v\sigma)R}(r)$ is a Wannier function localized at the site $R$. Hereafter the vector symbols will be dropped because there is no fear of confusion. Thus the long-range exchange term is described as the dipole–dipole interaction between the exciton transition dipole moments. In the calculation of the matrix elements of this long-range exchange term, the singular nature in (2.90) can be avoided by a mathematical artifice [11,39]. As a result, we have

$$
\sum_{r_e\neq r_e'} f^*_{\tau\sigma}(r_e,r_e)f_{\tau'\sigma'}(r_e',r_e')\mu_{c\tau,v\sigma}\frac{[1-3n\cdot^t n]}{|r_e-r_e'|^3}\mu_{v\sigma',c\tau'}
$$
$$
=-\frac{4\pi}{3}(\mu_{c\tau,v\sigma}\cdot\mu_{v\sigma',c\tau'})\int \mathrm{d}^3 r f^*_{\tau\sigma}(r,r)f_{\tau'\sigma'}(r,r)
$$
$$
+\int \mathrm{d}^3 r\, \mathrm{div}_r(f^*_{\tau\sigma}(r,r)\mu_{c\tau,v\sigma})\mathrm{div}_r\left[\int \mathrm{d}^3 r'\frac{\mu_{v\sigma',c\tau'}}{|r-r'|}f_{\tau'\sigma'}(r',r')\right]. \tag{2.93}
$$

It can be shown that this long-range exchange term vanishes for the $s$-like envelope functions [11]. Thus the admixture of the higher angular-momentum states is necessary to have a finite magnitude of the long-range exchange energy, which would be induced by the anisotropic shape of quantum dots or by the strained environment around the quantum dots [40]. On the other hand, the matrix element of the short-range exchange term is simply given as

$$V(c\tau r_0, v\sigma' r_0 \; ; c\tau' r_0, v\sigma r_0) \int d^3r \; f^*_{\tau\sigma}(r,r) f_{\tau'\sigma'}(r,r) \; . \tag{2.94}$$

Now we proceed to more detailed calculation of these matrix elements. The angular momentum and spin parts of the $J=3/2$ valence band Bloch functions can be written as

$$\left|\frac{3}{2}\frac{3}{2}\right\rangle = |v_1\rangle = Y_{11}\alpha \; ,$$

$$\left|\frac{3}{2}\frac{1}{2}\right\rangle = |v_2\rangle = \sqrt{\frac{2}{3}}Y_{10}\alpha + \sqrt{\frac{1}{3}}Y_{11}\beta \; ,$$

$$\left|\frac{3}{2}-\frac{1}{2}\right\rangle = |v_3\rangle = \sqrt{\frac{1}{3}}Y_{1-1}\alpha + \sqrt{\frac{2}{3}}Y_{10}\beta \; ,$$

$$\left|\frac{3}{2}-\frac{3}{2}\right\rangle = |v_4\rangle = Y_{1-1}\beta \; , \tag{2.95}$$

where $\alpha(\beta)$ denotes the up(down) spin state. Then we can identify the nonzero matrix elements of the short-range exchange interaction at the same site $r_0$ as

$$V(c\alpha r_0, v_1 r_0; c\alpha r_0, v_1 r_0) = V(c\beta r_0, v_4 r_0; c\beta r_0, v_4 r_0) = E^S_X \; ,$$

$$V(c\alpha r_0, v_2 r_0; c\alpha r_0, v_2 r_0) = V(c\beta r_0, v_3 r_0; c\beta r_0, v_3 r_0) = \frac{2}{3}E^S_X \; ,$$

$$V(c\alpha r_0, v_3 r_0; c\alpha r_0, v_3 r_0) = V(c\beta r_0, v_2 r_0; c\beta r_0, v_2 r_0) = \frac{1}{3}E^S_X \; ,$$

$$V(c\alpha r_0, v_2 r_0; c\beta r_0, v_1 r_0) = V(c\beta r_0, v_1 r_0; c\alpha r_0, v_2 r_0) = \sqrt{\frac{1}{3}}E^S_X \; ,$$

$$V(c\alpha r_0, v_3 r_0; c\beta r_0, v_2 r_0) = V(c\beta r_0, v_2 r_0; c\alpha r_0, v_3 r_0) = \frac{2}{3}E^S_X \; ,$$

$$V(c\alpha r_0, v_4 r_0; c\beta r_0, v_3 r_0) = V(c\beta r_0, v_3 r_0; c\alpha r_0, v_4 r_0) = \sqrt{\frac{1}{3}}E^S_X \; , \tag{2.96}$$

with

$$E^S_X = \frac{4\pi e^2}{3} \int dr \; r^2 \int dr' \; (r')^2 \; \phi_s(r)\phi_p(r')\frac{r_<}{r_>^2}\phi_s(r')\phi_p(r) \; , \tag{2.97}$$

where $r_< = \min.(r, r')$, $r_> = \max.(r, r')$, and $\phi_s (\phi_p)$ denotes the radial part of the s(p)-like Wannier function at a site.

As mentioned before, the exciton is a linear combination state with respect to the electron–hole pair index $\tau\sigma$ in (2.82), namely $\alpha v_1$, $\alpha v_2$, $\alpha v_3$, $\alpha v_4$, $\beta v_1$, $\beta v_2$, $\beta v_3$, and $\beta v_4$ ($v_1 \sim v_4$ are defined in (2.95)). The short-range exchange term in this 8×8 matrix representation is given as

$$
H_{\text{exch}}^{SR} = \begin{pmatrix}
|\alpha v_1\rangle & |\alpha v_2\rangle & |\alpha v_3\rangle & |\alpha v_4\rangle & |\beta v_1\rangle & |\beta v_2\rangle & |\beta v_3\rangle & |\beta v_4\rangle \\
1 & 0 & 0 & 0 & 0 & \sqrt{\tfrac{1}{3}} & 0 & 0 \\
0 & \tfrac{2}{3} & 0 & 0 & 0 & 0 & \tfrac{2}{3} & 0 \\
0 & 0 & \tfrac{1}{3} & 0 & 0 & 0 & 0 & \sqrt{\tfrac{1}{3}} \\
0 & 0 & 0 & 0 & 0 & 0 & 0 & 0 \\
0 & 0 & 0 & 0 & 0 & 0 & 0 & 0 \\
\sqrt{\tfrac{1}{3}} & 0 & 0 & 0 & 0 & \tfrac{1}{3} & 0 & 0 \\
0 & \tfrac{2}{3} & 0 & 0 & 0 & 0 & \tfrac{2}{3} & 0 \\
0 & 0 & \sqrt{\tfrac{1}{3}} & 0 & 0 & 0 & 0 & 1
\end{pmatrix} \cdot E_X^S , \quad (2.98)
$$

where the first line in the matrix is inserted to indicate the Bloch function bases, and each matrix element is a block matrix having the dimension of the number of the envelope function bases.

The matrix elements of the interband transition in (2.92) that appear in the long-range exchange term are given as

$$\langle c\alpha | \mathbf{r} | v_1 \rangle = {}^t(-1, -i, 0)\mu , \qquad \langle c\alpha | \mathbf{r} | v_2 \rangle = {}^t\left(0, 0, \frac{2}{\sqrt{3}}\right)\mu ,$$

$$\langle c\alpha | \mathbf{r} | v_3 \rangle = {}^t\left(\frac{1}{\sqrt{3}}, \frac{-i}{\sqrt{3}}, 0\right)\mu , \qquad \langle c\alpha | \mathbf{r} | v_4 \rangle = {}^t(0, 0, 0) ,$$

$$\langle c\beta | \mathbf{r} | v_1 \rangle = {}^t(0, 0, 0) , \qquad \langle c\beta | \mathbf{r} | v_2 \rangle = {}^t\left(\frac{-1}{\sqrt{3}}, \frac{-i}{\sqrt{3}}, 0\right)\mu ,$$

$$\langle c\beta | \mathbf{r} | v_3 \rangle = {}^t\left(0, 0, \frac{2}{\sqrt{3}}\right)\mu , \qquad \langle c\beta | \mathbf{r} | v_4 \rangle = {}^t(1, -i, 0)\mu \qquad (2.99)$$

with

$$\mu = \sqrt{\frac{2\pi}{3}} \int_0^\infty \mathrm{d}r \; r^2 \phi_s(r) \, r \, \phi_p(r) , \qquad (2.100)$$

where $\phi_s$ and $\phi_p$ have the same meaning as in (2.97). Then the factor $(\mu_{c\tau, v\sigma} \cdot \mu_{v\sigma', c\tau'})$ in (2.93) can be written in the 8×8 matrix representation as

$$(\mu_{c\tau,v\sigma} \cdot \mu_{v\sigma',c\tau'}) =$$

$$
\begin{array}{c}
\begin{array}{cccccccc}
|\alpha v_1\rangle & |\alpha v_2\rangle & |\alpha v_3\rangle & |\alpha v_4\rangle & |\beta v_1\rangle & |\beta v_2\rangle & |v_3\rangle & |\beta v_4\rangle
\end{array} \\
\left(
\begin{array}{cccccccc}
1 & 0 & 0 & 0 & 0 & \sqrt{\tfrac{1}{3}} & 0 & 0 \\
0 & \tfrac{2}{3} & 0 & 0 & 0 & 0 & \tfrac{2}{3} & 0 \\
0 & 0 & \tfrac{1}{3} & 0 & 0 & 0 & 0 & \sqrt{\tfrac{1}{3}} \\
0 & 0 & 0 & 0 & 0 & 0 & 0 & 0 \\
0 & 0 & 0 & 0 & 0 & 0 & 0 & 0 \\
\sqrt{\tfrac{1}{3}} & 0 & 0 & 0 & 0 & \tfrac{1}{3} & 0 & 0 \\
0 & \tfrac{2}{3} & 0 & 0 & 0 & 0 & \tfrac{2}{3} & 0 \\
0 & 0 & \sqrt{\tfrac{1}{3}} & 0 & 0 & 0 & 0 & 1
\end{array}
\right) \cdot 2\mu^2 .
\end{array}
\tag{2.101}
$$

As can be guessed from (2.99), the second term of the long-range exchange energy in (2.93) has a form given by

$$
H_{\mathrm{exch}}^{LR(2)} =
\begin{array}{c}
\begin{array}{cccccccc}
|\alpha v_1\rangle & |\alpha v_2\rangle & |\alpha v_3\rangle & |\alpha v_4\rangle & |\beta v_1\rangle & |\beta v_2\rangle & |\beta v_3\rangle & |\beta v_4\rangle
\end{array} \\
\left(
\begin{array}{cccccccc}
* & * & * & 0 & 0 & * & * & * \\
* & * & * & 0 & 0 & * & * & * \\
* & * & * & 0 & 0 & * & * & * \\
0 & 0 & 0 & 0 & 0 & 0 & 0 & 0 \\
0 & 0 & 0 & 0 & 0 & 0 & 0 & 0 \\
* & * & * & 0 & 0 & * & * & * \\
* & * & * & 0 & 0 & * & * & * \\
* & * & * & 0 & 0 & * & * & *
\end{array}
\right) ,
\end{array}
\tag{2.102}
$$

where $*$ indicates a nonzero block matrix whose dimension is given by the number of the envelope function bases. Similar matrix representations of the exciton exchange interaction in QWs were discussed by Maialle et al. [41].

Here we consider localized excitons at quantum dot-like islands in QW structures that are formed by the lateral fluctuations of the QW thickness. In these island-like structures, the confinement in the direction of the crystal growth is strong, whereas the confinement in the lateral direction is rather weak. Furthermore, the island structures were found to be elongated along the $[1\bar{1}0]$ direction [4]. Thus these island structures can be modeled by an anisotropic quantum disk. In order to facilitate the calculation, the lateral confinement potential in the $x$- and $y$- directions is assumed to be Gaussian as

$$v_{e(h)}(r) = v_{e(h)}^0 \exp\left[-\left(\frac{x}{a}\right)^2 - \left(\frac{y}{b}\right)^2\right],
\tag{2.103}$$

where the lateral size parameters $a$ and $b$ can be fixed, in principle, from the measurement of morphology by, e.g., STM, but the actual size would be different from the morphological size because of the presence of the surface depletion layer, and so on. Thus the parameters $a$ and $b$ are left as adjustable parameters. The exciton envelope function in such an anisotropic quantum disk can be approximated as

$$f(r_e, r_h) = \sum_{l_e, l_h, m_e, m_h} C(l_e, l_h, m_e, m_h) \left(\frac{x_e}{a}\right)^{l_e} \left(\frac{x_h}{a}\right)^{l_h} \left(\frac{y_e}{b}\right)^{m_e} \left(\frac{y_h}{b}\right)^{m_h}$$
$$\cdot \exp\left\{ -\frac{1}{2}\left[ \left(\frac{x_e}{a}\right)^2 + \left(\frac{x_h}{a}\right)^2 + \left(\frac{y_e}{b}\right)^2 + \left(\frac{y_h}{b}\right)^2 \right]\right.$$
$$\left. - \alpha_x(x_e - x_h)^2 - \alpha_y(y_e - y_h)^2 \right\} \varphi_0(z_e) \left[\varphi_0(z_h) \text{ or } \varphi_1(z_h)\right]$$

$$(2.104)$$

with

$$\varphi_0(z) = \sqrt{\frac{2}{L_z}} \cos\left(\frac{\pi z}{L_z}\right) ,$$
$$\varphi_1(z) = \sqrt{\frac{2}{L_z}} \sin\left(\frac{2\pi z}{L_z}\right) ,$$

$$(2.105)$$

where $C(l_e, l_h, m_e, m_h)$ is the expansion coefficient, $L_z$ the QW thickness, the factor $1/2$ in the exponent is attached to make the probability distribution $|f(r_e, r_h)|^2$ to follow the functional form of the confining potential, and $\alpha_x$ and $\alpha_y$ indicate the degree of the electron–hole correlation, and are determined variationally. Concerning the envelope function in the $z$-direction, the lowest two functions are taken into account for the hole to guarantee sufficient accuracy. The electron–hole relative motion within the exciton state is not much different from that in the bulk because the lateral confinement is rather weak. As a result, the parameters $\alpha_x$ and $\alpha_y$ are weakly dependent on the lateral size. Because of the inversion symmetry of the confining potential, the parity is a good quantum number, and the wavefunction can be classified in terms of the combination of parities of $x^{l_e+l_h}$, $y^{m_e+m_h}$, and $\varphi_0$ or $\varphi_1$. In actual calculations, terms up to the sixth power are included, namely, $0 \leq l_e + l_h$, $m_e + m_h \leq 6$ to ensure the convergence of the calculation. The potential depth for the lateral motion of the electron and the hole can be guessed from the splitting energies of the exciton states due to the monolayer fluctuation of the QW thickness. The value of $|v_e^0 - v_h^0|$ is typically about 10 meV for a nominal QW thickness of about 3 nm [5]. Of course, even if $v_e^0 - v_h^0$ is fixed, each value of $v_e^0$ and $v_h^0$ cannot be determined uniquely. Here we employ $v_e^0 = -6$ meV and $v_h^0 = 3$ meV in most parts of this section, referring to the experimental results on the exciton energy spectra and assuming $|v_e^0| : |v_h^0| = 2 : 1$, but we will later examine several cases of the potential depth.

It is important to examine the size dependence of the exchange energy. For that purpose, we first note that

$$f(r_e, r_h) \propto 1/L^3 \text{(quantum dot)} , \quad 1/L^2 \text{(quantum disk)} , \qquad (2.106)$$

where $L$ is the characteristic size of a quantum dot or the lateral size of a quantum disk with a fixed disk height. This scaling comes simply from the normalization

$$\int d^3r_e \int d^3r_h \ |f(r_e, r_h)|^2 = 1 \ . \tag{2.107}$$

Then the short-range exchange term in (2.94) scales as

$$L^3 \frac{1}{L^3} \frac{1}{L^3} \propto \frac{1}{L^3} (\text{quantum dot}) \ , \quad L^2 \frac{1}{L^2} \frac{1}{L^2} \propto \frac{1}{L^2} (\text{quantum disk}) \ , \tag{2.108}$$

where the first factor $L^3$ or $L^2$ comes from the volume integral. The first term of the long-range exchange energy in (2.93) has the same size dependence. In the second term of the long-range exchange energy, the operator div and $1/|r{-}r'|$ give rise to a factor of $1/L$, and the integral in (2.93) scales as

$$L^3 \frac{1}{L} \frac{1}{L^3} \frac{1}{L} \ L^3 \frac{1}{L} \frac{1}{L^3} \propto \frac{1}{L^3} (\text{quantum dot}) \ ,$$

$$L^2 \frac{1}{L} \frac{1}{L^2} \frac{1}{L} \ L^2 \frac{1}{L} \frac{1}{L^2} \propto \frac{1}{L^3} (\text{quantum disk}) \ . \tag{2.109}$$

Thus the exchange energy in quantum dots scales as $1/L^3$ for both the short-range and the long-range parts. On the other hand, in quantum disks, the short-range and the long-range parts of the exchange energy depend on the size differently as $1/L^2$ and $1/L^3$, respectively. Of course, this size dependence holds asymptotically only in the strong confinement regime. In general, details of the envelope function are substantially dependent on the size through, for example, the variational parameters $\alpha_x$ and $\alpha_y$, leading to possible deviation from the rough scaling laws estimated above.

It is important to note that the long-range part of the electron–hole exchange interaction is not absent for the odd-parity exciton states. For these exciton states,

$$\int d^3r \ f(r, r) = 0 \ , \tag{2.110}$$

and the oscillator strength vanishes. However, the integrals in (2.93) do not vanish, in general.

### 2.4.2   Exciton Doublet Structures

Now we discuss the schematic exciton energy level structures in an anisotropic quantum disk as shown in Fig. 2.9. The conduction band levels are doubly degenerate due to the spin degree of freedom. The valence band levels are also doubly degenerate owing to Kramer's theorem in the absence of a magnetic field, even when valence band mixing is taken into account. Thus each exciton state is fourfold degenerate in the absence of the electron–hole exchange interaction (left column). When the short-range exchange term is included

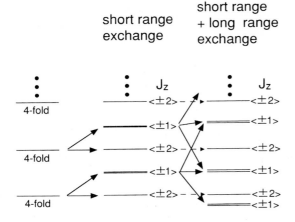

short range
exchange

short range
+ long range
exchange

**Fig. 2.9.** Exciton energy level structures in an anisotropic quantum dot are shown schematically for the cases with no exchange interaction (*left column*), with only short-range exchange interaction (*middle column*), and with both short-range and long-range exchange interactions (*right column*) (from T. Takagahara [37])

(middle column), the quadruplet is split into two doublets. The first doublet consists of almost degenerate levels, which are mainly composed of states having the $z$-component of the total angular momentum $J_z$ of 2 and $-2$, namely $|\tau\sigma\rangle = |\alpha v_4\rangle$ or $|\beta v_1\rangle$ in (2.82) and is denoted as $<\pm 2>$ in Fig. 2.9. Another doublet is mainly composed of the states having the $J_z$-component of $-1$, 0, and 1, and is denoted as $<\pm 1>$. This doublet is split slightly by the short-range exchange term. When the long-range exchange term is added (right column), the levels denoted by $<\pm 2>$ remain almost the same because exchange interactions are absent for the $J_z = \pm 2$ states. On the other hand, the levels denoted by $<\pm 1>$ are affected rather strongly due to mixing among themselves. Namely, the doublet splitting increases and the lowest doublet is pushed down by the exchange interaction with excited exciton states. This level restructuring is depicted by solid arrows in Fig. 2.9. As a result, the exciton ground state can become an optically active state depending on the magnitude of the long-range exchange energy.

This schematic understanding can be confirmed by numerical calculations in which we employ a quantum disk model to simulate a localized exciton in a QW island. In the following, material parameters of GaAs [42] are employed (see Appendix D for details). The chosen size parameters in (2.104) are $a$=20 nm, $b$=15 nm, and $L_z$=3 nm. The exciton level structure is shown in Fig. 2.10 for the case in which only the short-range exchange energy is included. The exciton ground state is optically dark and the first optically active exciton doublet lies about 0.3 meV above the exciton ground state. The splitting energy of the first optically active exciton doublet is much smaller than the observed splitting. Also for the higher-lying exciton doublets, the splitting energy is very small ($<1$ μeV). On the other hand, when both the short-range and the long-range exchange terms are included as shown in Fig. 2.11, the exciton ground state is optically active and its doublet splitting is 29 μeV, in agreement with experiments [4]. The optically dark exciton states are mainly

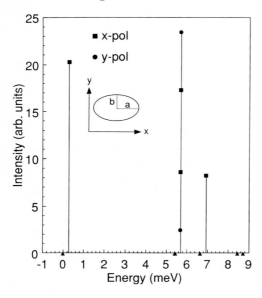

**Fig. 2.10.** Exciton energy level structures are shown for a GaAs elliptical quantum disk with the size parameters $a$=20 nm, $b$=15 nm, and $L_z$(disk height)=3 nm, in which only the short-range exchange term is taken into account. The $x(y)$-direction of the polarization is along the semimajor (semiminor) axis of the potential ellipse. For the first and the fourth doublets, the energy position and the intensity are almost degenerate. The *triangles* indicate the energy positions of the even-parity dark states (from T. Takagahara [37])

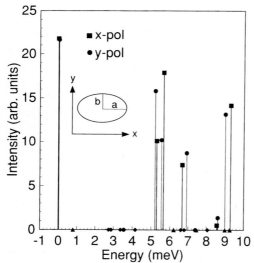

**Fig. 2.11.** Exciton energy level structures are shown for a GaAs elliptical quantum disk with the size parameters $a$=20 nm, $b$=15 nm, and $L_z$(disk height)=3 nm, in which both the short-range and the long-range exchange terms are taken into account. The meaning of the $x(y)$-direction and the *triangles* are the same as in Fig. 2.10. The *diamonds* indicate the energy positions of the odd-parity exciton states (from T. Takagahara [37])

composed of the $J_z$=±2 components, and its first doublet lies about 0.8 meV above the exciton ground state. Thus the long-range exchange term is important to understand the experimental results. So far we have considered the even-parity exciton states, which include both the optically active and the dark states. In addition to these, there are odd-parity exciton states, all of which are optically dark. They are plotted by diamonds in Fig. 2.11 to show their energy positions. They do not contribute to the luminescence, but play

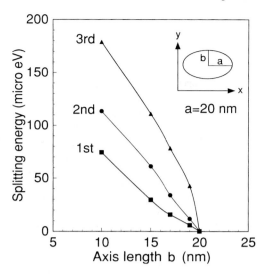

**Fig. 2.12.** The variation of the doublet splitting energy in a GaAs elliptical quantum disk is shown for the lowest three optically active exciton doublets when the semimajor axis length $a$ is fixed at 20 nm and the semiminor axis length $b$ is varied around 20 nm with a fixed value of $L_z=3$ nm (from T. Takagahara [37])

a role in the intermediate states of relaxation processes, as will be discussed in Chap. 9.

Furthermore, the importance of the long-range exchange interaction can be seen in the dependence of the doublet splitting energy on the aspect ratio of the confinement potential ellipse. The variation of the doublet splitting energy is shown in Fig. 2.12 for the lowest three optically active exciton doublets when the semimajor axis length $a$ is fixed at 20 nm and the semiminor axis length $b$ is varied around 20 nm with a fixed value of $L_z=3$ nm. When the potential ellipse approaches a circular shape, the doublet splitting energy decreases to zero. This demonstrates the sensitivity of the long-range exchange interaction to the quantum disk shape and its significance in determining the doublet splitting.

We have examined the dependence of the doublet splitting energy on the size of a quantum disk with a fixed aspect ratio $a/b$. The results are shown in Fig. 2.13 for $a/b=4/3$. The size dependence of the doublet splitting energy for the lowest doublet is rather weak, suggesting that the exciton wavefunction is localized in the interior of the confinement potential and is not strongly affected by the potential size. On the other hand, for the higher-lying exciton doublets, the splitting energy shows a dependence of $1/a^n (n=1.3\sim1.5)$, as depicted by solid lines in Fig. 2.13. A simple argument based on (2.108) and (2.109) indicates the $1/a^2$ dependence for the short-range exchange term and the $1/a^2$ or $1/a^3$ dependence for the long-range exchange term. The calculated size dependence deviates from these simple predictions. As mentioned before, this deviation might be ascribed to the size dependence of details of the exciton wavefunctions.

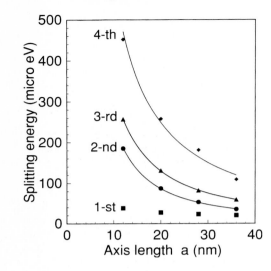

**Fig. 2.13.** The doublet splitting energy in a GaAs elliptical quantum disk is plotted for the lowest four exciton doublets as a function of the semimajor axis length $a$ with a fixed aspect ratio $a/b=4/3$ and a fixed value of $L_z=3$ nm. The *solid lines* show a dependence of $1/a^n(n=1.3\sim1.5)$ (from T. Takagahara [37])

So far we have employed the values of $v_e^0=-6$ meV and $v_h^0=3$ meV for the lateral confinement potential in (2.103). We examined, however, the dependence of the doublet splitting energy on the potential depth. The splitting energy of the ground exciton doublet is 27 µeV for $(v_e^0, v_h^0)=(-4$ meV, 2 meV), 29 µeV for $(v_e^0, v_h^0)=(-6$ meV, 3 meV), 34 µeV for $(v_e^0, v_h^0)=(-9$ meV, 4 meV) and 46 µeV for $(v_e^0, v_h^0)=(-15$ meV, 10 meV). The trend of these numerical results can be understood as follows. When the lateral confinement is strengthened, both the electron and the hole wavefunctions shrink spatially and the overlap between them increases, resulting in the increase of both the short-range and the long-range exchange interactions and the enhancement of the doublet splitting energy.

### 2.4.3   Polarization Characteristics of Exciton Doublets

The polarization characteristics of exciton doublets are quite interesting and reveal an important physical aspect. For the lowest doublet, the lower exciton state is polarized along the $x$-direction, namely along the semimajor axis of the potential ellipse. On the other hand, the upper exciton state is polarized along the $y$-direction (the semiminor axis). However, for the second and third doublets the relation between the energetic order and the polarization direction is reversed. According to our calculation, the polarization direction of the lower exciton state is $x, y, y,$ and $x$ for the lowest four exciton doublets, in complete agreement with experimental results [4]. These features can be understood in view of the spatial distribution of exciton polarization.

The transition dipole moment of an exciton state is given by

$$P_i = \sum_r \sum_{\tau,\sigma} \langle c\tau|\mu_i|v\sigma\rangle \, f_{\tau\sigma}^*(r,r) \,, \quad (i=x,y,z) \tag{2.111}$$

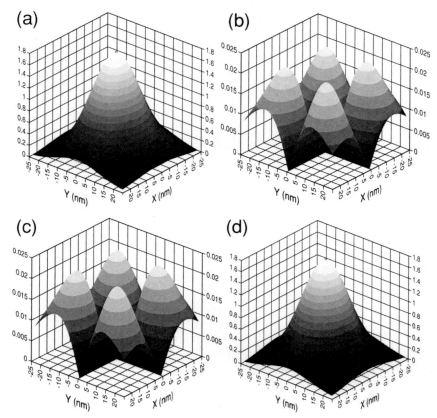

**Fig. 2.14.** The polarization distribution function is plotted for the first exciton doublet states in a GaAs elliptical quantum disk with the size parameters $a=20$ nm, $b=15$ nm, and $L_z$ (disk height)$=3$ nm: (**a**) $x$-polarization distribution for the $x$-polarized exciton state (1 state); (**b**) $y$-polarization distribution for the 1 state; (**c**) $x$-polarization distribution for the $y$-polarized exciton state (1′ state); (**d**) $y$-polarization distribution for the 1′ state. The weight ($z$-axis) is plotted in the arbitrary units, but its scale is common for all the cases (from T. Takagahara [37])

where $\mu_i$ is the dipole moment operator in the $i$-direction. In the following we interpret $\sum_{\tau,\sigma}\langle c\tau|\mu_i|v\sigma\rangle f^*_{\tau\sigma}(r,r)$ as the polarization density in the $i$-direction at position $r$. But this quantity is, in general, complex and we consider the absolute value to understand the qualitative features. The polarization distribution calculated in this way is shown in Fig. 2.14 for the first exciton doublet in a confinement potential with $(a,b)=(20$ nm, $15$ nm). For the lower exciton state (1 state), the distribution function of the $x$-polarization is single-peaked (Fig. 2.14a), corresponding to a uniform distribution of the polarization direction. On the other hand, the distribution function of the $y$-polarization (Fig. 2.14b) is four-peaked, indicating a staggered distribution of the polar-

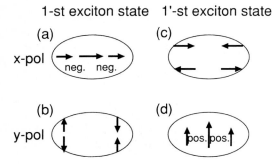

**Fig. 2.15.** Schematic representation of the polarization distribution in Fig. 2.14 for the first exciton doublet: (**a**) $x$-polarization distribution for the 1 state; (**b**) $y$-polarization distribution for the 1 state; (**c**) $x$-polarization distribution for the 1′ state; (**d**) $y$-polarization distribution for the 1′ state (from T. Takagahara [37])

ization direction. For the upper exciton state of the first doublet (1′ state), the spatial distribution functions of the $x$-and $y$-polarization (Fig. 2.14c,d) are interchanged relative to the case of the lower exciton state. These features of polarization distribution are schematically depicted in Fig. 2.15a–d.

Now we consider the dipole–dipole interaction originating from the long-range exchange interaction. The dipole–dipole interaction energy changes its sign according to the polarization configuration, as shown in Fig. 2.16. Then we see that for the $x$-polarized (1) exciton state the negative contribution is significant, whereas for the $y$-polarized (1′) exciton state the positive contribution is substantial. For the staggered configuration of polarizations like in Fig. 2.15b and c, the positive and negative dipole–dipole interactions largely cancel each other, and the resulting contribution is rather small compared to that from Fig. 2.15a and d. As a consequence, the $x$-polarized exciton state has a lower energy than the $y$-polarized state.

The same argument can be extended to the higher-lying doublet states. For the $n$-th doublet, the exciton levels are named $n$ and $n′$, corresponding to the $x$- and $y$-polarized states, respectively. In the excited states, the number of nodes of the exciton wavefunction increases and correspondingly, the polarization distribution becomes more staggered and complicated. For the second and third doublets, there are two nodes in the $x$-direction, whereas for the fourth doublet there are two nodes in the $y$-direction. These polarization

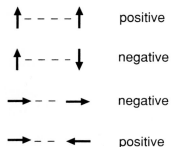

**Fig. 2.16.** The sign of the dipole–dipole interaction energy for various configurations of polarization direction

2-nd exciton state    2'-nd exciton state

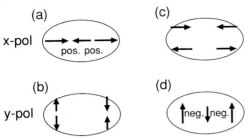

**Fig. 2.17.** Schematic representation of the polarization distribution for the second exciton doublet: (**a**) $x$-polarization distribution for the 2 state; (**b**) $y$-polarization distribution for the 2 state; (**c**) $x$-polarization distribution for the 2′ state; (**d**) $y$-polarization distribution for the 2′ state (from T. Takagahara [37])

3-rd exciton state    3'-rd exciton state

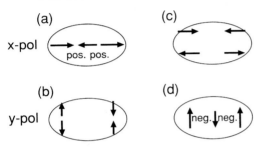

**Fig. 2.18.** Schematic representation of the polarization distribution for the third exciton doublet: (**a**) $x$-polarization distribution for the 3 state; (**b**) $y$-polarization distribution for the 3 state; (**c**) $x$-polarization distribution for the 3′ state; (**d**) $y$-polarization distribution for the 3′ state (from T. Takagahara [37])

4-th exciton state    4'-th exciton state

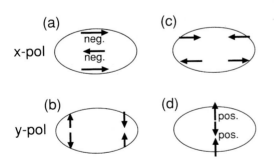

**Fig. 2.19.** Schematic representation of the polarization distribution for the fourth exciton doublet: (a) $x$-polarization distribution for the 4 state; (b) $y$-polarization distribution for the 4 state; (c) $x$-polarization distribution for the 4′ state; (d) $y$-polarization distribution for the 4′ state (from T. Takagahara [37])

distribution functions are schematically depicted in Figs. 2.17, 2.18, and 2.19. The sign of the dipole–dipole interaction energy deduced from Fig. 2.16 is also given in the figures. Then we see that, for the second and third doublets, the $y$-polarized state (2′ or 3′ state) is energetically lower than the $x$-polarized state (2 or 3 state). On the other hand, for the fourth doublet, the energy of the $x$-polarized state (4 state) is lower than that of the $y$-polarized state (4′ state). Thus the key concepts to understanding the energetic order of the orthogonally polarized exciton doublet states are the node configuration of the polarization distribution and the dipole–dipole interaction originating from the long-range exchange term.

We have examined the dependence on the lateral size of the quantum disk of the energetic order of the orthogonally polarized exciton doublet states. For quantum disks with a fixed aspect ratio, namely $a/b=4/3$, we examined the cases of $(a, b)=$(12 nm, 9 nm), (20 nm, 15 nm), (28 nm, 21 nm), and (36 nm, 27 nm) with a fixed disk height of $L_z=3$ nm and found that the polarization direction of the lower exciton state of the lowest four exciton doublets is commonly $x$, $y$, $y$, and $x$. This reflects the same node configuration of the exciton polarization for these quantum disks having common values of the aspect ratio.

We also examined cases with different aspect ratios, namely the cases of $(a, b)=$(30 nm, 10 nm) and (40 nm, 10 nm). The polarization direction of the lower exciton state of the lowest four exciton doublets is $x$, $y$, $y$, and $y$ for $(a, b)=$(30 nm, 10 nm) and $x$, $x$, $y$, and $y$ for $(a, b)=$(40 nm, 10 nm), respectively. These features can be understood by inspecting the polarization distribution functions, although they are not shown. Thus the polarization characteristics of the exciton doublet states are highly dependent on the aspect ratio of the lateral confinement potential.

## 2.5    Enhancement of Excitonic Optical Nonlinearity in Quantum Dot Arrays

We now discuss advantages of a quantum dot array for enhancing the excitonic optical nonlinearity, especially its figures of merit. One of the most important mechanisms for enhancing optical nonlinearity is the quantum confinement effect, which leads to a discrete level structure and causes the concentration of oscillator strength to the lowest energy transitions. Since the density of electronic states becomes sharper by reducing the dimensionality of structures, a sharp excitonic feature would be realized in low-dimensional materials. In this respect, the zero-dimensional structures, namely quantum dots, seem to be most favorable for enhancing the optical nonlinearity [17,18,43]. This enhancement comes from the high oscillator strength of the excitonic state, which increases almost proportionally to the volume of the quantum dot. Correspondingly, the optical nonlinearity increases with the quantum dot size as long as the quantum confinement effect is effectively working on the exciton center-of-mass motion. This type of enhancement was called mesoscopic enhancement.

However, when the quantum dot size is increased further, the energy spacing among excitonic levels is reduced, and several excited excitonic states are overlapped within the homogeneous linewidth of the lowest excitonic state. In this situation, even under resonant excitation of the lowest exciton state, the excited excitonic states are simultaneously excited and contribute to the

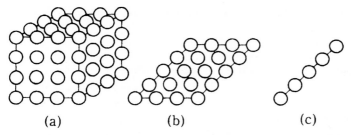

**Fig. 2.20.** Schematic illustration of the quantum dot array; (**a**) a three-dimensional structure, (**b**) a planar structure, and (**c**) a one-dimensional structure

linear and nonlinear optical processes. In general, the excited excitonic states have smaller oscillator strengths than the lowest exciton state. Thus the overlap of excited excitonic levels within the homogeneous linewidth of the lowest excitonic state induces the thermal population of these excited states and leads to the saturation of the exciton oscillator strength [21].

In this section, we propose to use a regular array of quantum dots as shown in Fig. 2.20 in order to overcome these limitations and to further enhance the optical nonlinearity. These kinds of structures are already realized or synthesized in the fields of semiconductors and organic materials. An example of the one-dimensional structure is realized in a J-aggregate of dye molecules [44]. Two-dimensional structures are synthesized by the crystal growth method using compound semiconductors [45,46] and are realized in Langmuir–Blodgett films of various organic materials [47]. Examples of the three-dimensional structure are now numerous. The most beautiful examples are Si cluster solids [48] and $C_{60}$ crystals [49,50]. In addition to these, various molecular crystals and semiconductor clusters formed in zeolite materials [51,52] are also examples of three-dimensional structures.

In the quantum dot array, the excitonic state is not localized within each quantum dot, but a Frenkel-type exciton can propagate through the dot array via exciton transfer due to the dipole–dipole interaction. Then the exciton coherence extends over many quantum dots, and the Frenkel-type exciton gathers all the oscillator strength of the quantum dots within the range of the exciton coherence length. Thus the concentration of oscillator strength occurs in two hierarchical stages. In a quantum dot, the quantum confinement effect concentrates the oscillator strength of constituent atoms to the lowest-energy excitonic state. Then in the quantum dot array, the Frenkel exciton effect concentrates the oscillator strength of many quantum dots to a single excitonic state. As a consequence, the Frenkel exciton obtains a large oscillator strength and the optical nonlinearity is greatly enhanced [53].

### 2.5.1   Exciton Band Structure in Quantum Dot Arrays

The most fundamental quantity to characterize the exciton coherence among many quantum dots is the exciton transfer energy between two quantum dots. When the two quantum dots are far apart, this transfer energy is primarily determined by the dipole–dipole interaction. On the other hand, when the distance between the two quantum dots is comparable to the quantum dot size, the exciton transfer energy is composed of many components of multipolar interaction. To calculate the transfer energy exactly including these multipolar interactions, we consider two spherical quantum dots and solve the electrostatic problem employing bispherical coordinates. To obtain a universal curve of the exciton transfer energy as a function of the distance between two quantum dots, we scale the transfer energy as

$$T / \left( \frac{r_{\mathrm{cv}}}{a_{\mathrm{B}}^*} \right)^2 Ry^* \,, \tag{2.112}$$

where $r_{\mathrm{cv}}$ is the length of the transition dipole moment between the conduction and the valence bands, and $a_{\mathrm{B}}^*(Ry^*)$ the bulk exciton Bohr radius (binding energy) of the quantum dot material. This scaled transfer energy is plotted in Fig. 2.21 as a function of the ratio of the distance to the quantum dot diameter, i.e., $D/2R$. The $D^{-3}$ dependence of the transfer energy due to the dipole–dipole interaction is recovered in the long-distance region

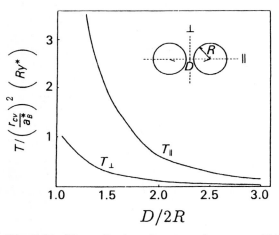

**Fig. 2.21.** Normalized exciton transfer energy $T$ between two spherical quantum dots is plotted as a function of the ratio of the interparticle distance to the quantum dot diameter, where $r_{\mathrm{cv}}$ is the length of the transition dipole moment between the conduction and the valence bands and $a_{\mathrm{B}}^*(Ry^*)$ is the exciton Bohr radius (binding energy) in the bulk material of the quantum dot. $T_\parallel (T_\perp)$ corresponds to the exciton polarization parallel (perpendicular) to the direction through the centers of two quantum dots (from T. Takagahara [53])

of about $D/2R{\geq}5{\sim}6$. Thus the transfer energy shown in Fig. 2.21 appears to deviate strongly from this dependence. It also should be noted that the transfer energy is sensitive to the polarization direction of the exciton. The transfer energy for the case in which the exciton polarization is parallel to the line through the centers of two quantum dots is several times larger than that for the case of perpendicular configuration. This is due to the depolarization effect.

Making use of this transfer energy, the exciton band structure can be calculated for any structure of the quantum dot array. The exciton dispersion curve is essentially given by the Fourier transform of the exciton transfer energy, namely

$$E(k) = \sum_n e^{ikR_n} T(R_n) , \qquad (2.113)$$

where $T(R_n)$ is the exciton transfer energy between two quantum dots separated by a vector $R_n$. As typical examples of the quantum dot array, we consider a planar array of quantum dots in the square lattice structure and stacks of a few such unit layers. In Fig. 2.22, the exciton band structures are shown for the single-, double-, and triple-layer cases where $D/2R=1.5$ and the interlayer spacing is the same as the lattice constant of the square lattice in the unit layer. The $x$- and $y$-axes are taken along the square lattice, and the $z$-axis is chosen perpendicular to the unit layer. The wavevector $k$ is taken along the $x$-axis in Fig. 2.22.

In the single-layer case, the $y$-polarized branch is located in the lowest energy due to the gain of the dipolar interaction between quantum dots. The $x$-polarized branch has a larger dispersion than the $z$-polarized branch, again due to the dipolar interaction. In the double-layer case, there appear six exciton branches because each unit layer accompanies three exciton states corresponding to the $x$-, $y$- and $z$-polarizations. The $x$- and $z$-polarized exciton states are coupled through interlayer coupling, showing typical level anticrossing behaviors on the dispersion curves. On the other hand, the $y$-polarized exciton states do not couple with the $x$- and $z$-polarized exciton states and lie in the lowest-energy region in Fig. 2.22b. However, they are mixed with each other through interlayer coupling and show the level anti-crossing behavior. In the triple-layer case, there appear nine exciton branches, and they are classified into six polarization-mixed exciton branches between the $x$- and $z$-polarizations and three $y$-polarized exciton branches. In this case also, the $y$-polarized exciton branches are lying in the lowest energy region.

As a limit where the number of stacked layers goes to infinity, we consider the three-dimensional simple cubic lattice. We show the calculated exciton band structure in Fig. 2.23, where the lattice parameter is again fixed as $D/2R=1.5$. As can be expected from the anisotropy of the transfer energy, the dispersion curve is highly dependent on the directions of the $k$-vector and the polarization vector of the exciton. In the case of $k\|[100]$, the $y$- and $z$-polarized branches are degenerate, and the $x$-polarized branch has a higher

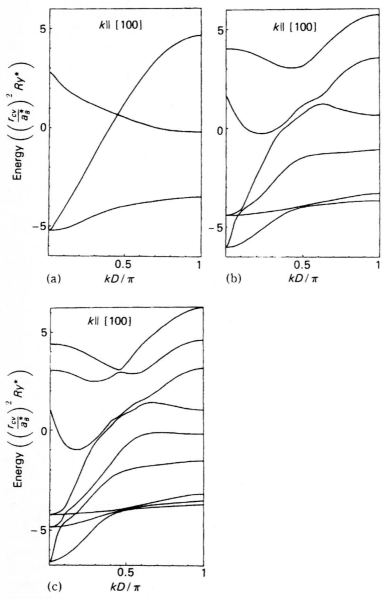

**Fig. 2.22.** Exciton band structures are plotted for a planar quantum dot array having the square lattice structure and for stacks of a few unit layers: (**a**) single-layer case; (**b**) double-layer case; and (**c**) triple-layer case. The wavevector $k$ is along the direction of [100]. The ratio of the lattice constant to the quantum dot diameter, i.e., $D/2R$, is 1.5

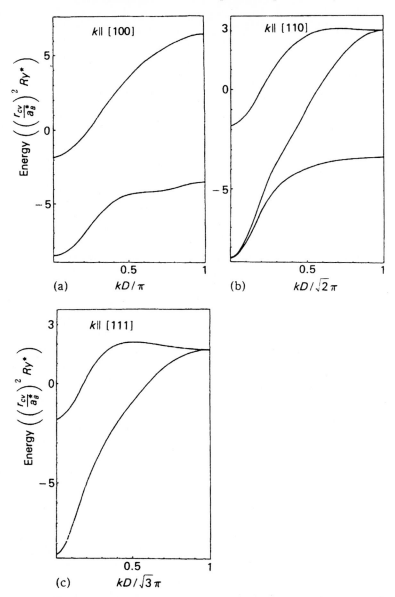

**Fig. 2.23.** Exciton band structures in a three-dimensional quantum dot array having the simple cubic structure are plotted for three directions of the wavevector: (a) $k\|[100]$; (b) $k\|[110]$; and (c) $k\|[111]$. The ratio of the lattice constant to the quantum dot diameter, i.e., $D/2R$ is 1.5 (from T. Takagahara [53])

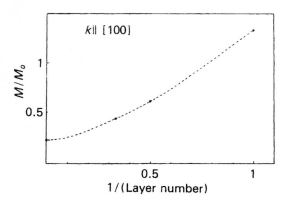

**Fig. 2.24.** The translational mass at the $\Gamma$-point of the lowest-branch exciton state in Fig. 2.22 is plotted as a function of the inverse layer number. The point of 1/(layer number)=0 corresponds to the case of bulk material

energy. In the case of $k\|[110]$, the $x$- and $y$-polarized branches are mixed together, while the $z$-polarized branch is isolated below them. For the case of $k\|[111]$, the higher branch is a longitudinal mode and the lower branch is doubly degenerate, having a transverse polarization.

The lowest-energy branch of the exciton plays a major role in determining the linear and nonlinear optical properties. One of the most important parameters in determining the magnitude of the optical nonlinearity is the exciton translational mass at the $\Gamma$-point. The translational mass of the lowest exciton state estimated from the curvature of the dispersion curve is plotted in Fig. 2.24 as a function of the inverse layer number, where $M_0$ is defined by

$$M_0 = \mu \left[ \frac{(a_B^*)^2}{r_{cv} D} \right]^2 , \tag{2.114}$$

where $\mu$ is the electron–hole reduced mass. It is clear that the exciton translational mass decreases as the layer number is increased. This is because the number of available sites for exciton transfer is increased and the exciton transfer becomes easier. In the case of materials having a small exciton Bohr radius $a_B^*$, $M_0$ can be made rather small compared to $\mu$ by choosing a large value of $D$. Consequently $M/\mu$ can be rather small because $M/M_0$ is of the order of unity, as seen in Fig. 2.24. This feature arises from the fact that the exciton transfer occurs through the electromagnetic dipolar and higher multipolar interactions, but not through the wave-function overlap. Thus there is no intrinsic restriction on the relative magnitude between the exciton translational mass and the constituent carrier masses. From these arguments, we see that the quantum dot array can yield a very small translational mass of the Frenkel-type exciton. This property is favorable for enhancing the excitonic optical nonlinearity.

### 2.5.2   Excitonic Optical Nonlinearity of Quantum Dot Arrays

The third-order nonlinear susceptibility $\chi^{(3)}$ is calculated by the third-order perturbation theory with summation over the excitonic states [17]. In this calculation, an important quantity is the oscillator strength of the excitonic transition. This oscillator strength is determined by the exciton coherence volume, which is closely related to the homogeneous linewidth of the excitonic transition [27,28]. The homogeneous linewidth reflects the energy fluctuation of the excitonic state and represents the wave-vector fluctuation according to the exciton band structure. This wave-vector fluctuation restricts the exciton coherent motion within a range of the order of the inverse wave-vector fluctuation. In other words, the oscillator strength of the $k = 0$ exciton is redistributed among all states within the homogeneous linewidth.

The redistributed oscillator strength per one exciton state can be written in terms of the exciton coherence length $l_c$ as [28]

$$f_X = f_0(l_c)^3|F(0)|^2 ,\tag{2.115}$$

where $f_0$ is the oscillator strength of the band-to-band transition at the $\Gamma$-point, and $F(0)$ is the value at the origin of the envelope function concerning the electron–hole relative motion. The exciton coherence length in the three-dimensional system is given as

$$l_c = \left(\frac{3\pi^2}{\sqrt{2}}\right)^{1/3}\frac{\hbar}{\sqrt{M\hbar\Gamma_h}} ,\tag{2.116}$$

where $M$ is the exciton translational mass and $\hbar\Gamma_h$ is the homogeneous linewidth of the excitonic transition. In terms of this exciton coherence length, the maximum value of the imaginary part of $\chi^{(3)}$ at the exciton resonance for the quantum dot array is given as

$$\mathrm{Im}\chi^{(3)}|_{\max} = \frac{e^4}{(2m_0\omega_0)^2}\frac{f_0^2}{\hbar\gamma_\|\gamma_\perp^2}l_c^3\left(\frac{v_d|F(0)|^2}{D^3}\right)^2 ,\tag{2.117}$$

where the contribution from the biexcitonic states is neglected, assuming that the off-resonance of biexcitonic states relative to twice the exciton energy is rather large. Here $v_d$ is the volume of a quantum dot, $m_0$ the free electron mass, $\hbar\omega_0$ the exciton energy and $\gamma_\|(\gamma_\perp)$ is the longitudinal (transverse) relaxation constant of the excitonic transition ($\gamma_\perp = \Gamma_h$). Thus a longer coherence length leads to a larger value of $\chi^{(3)}$.

The most fundamental quantities describing the efficiency of the optical nonlinearity are the figures of merit. The figures of merit are usually defined by the changes in the absorption coefficient $\alpha$ and the refractive index n induced by one electron–hole pair (exciton) in a unit volume, which are denoted by $\sigma$ and $\eta$, respectively, and given as $\alpha = \alpha_0 + \sigma N$ and $n = n_0 + \eta N$, where $\alpha_0(n_0)$ is the absorption coefficient (refractive index) under no excitation, and $N$ is the number density of electron–hole pairs (excitons). The figure of merit $\sigma$ for the absorptive optical nonlinearity is related to the saturation density $N_s$ of the absorption coefficient and to the exciton coherence length as [28]

$$1/N_s = \sigma/\alpha_0 = 2(l_c)^3 .\tag{2.118}$$

This result indicates that an exciton has a coherent range of the order of $l_c$ with respect to the center-of-mass motion, and the creation of another exciton within that range is inhibited. Thus, in order to enhance the figures of merit, it is primarily required to increase the exciton coherence length. The enhancement of the exciton coherence length is in turn achieved by reducing the exciton translational mass, as can be seen from (2.116) or, in other words, by increasing the exciton transfer energy.

In order to more clearly see the effect of array formation on the enhancement of optical nonlinearity, we examine the ratio of the $\sigma$-value of the quantum dot array to the $\sigma$-value of an isolated dot. This quantity can be shown to be proportional to $(l_c/D)^3$. Thus there are two ways to increase this ratio: one is a decrease of $D$, the other is an increase of $l_c$. In order to reduce $D$ while keeping the dot size within the regime of mesoscopic enhancement, it is desirable to employ materials having a small exciton Bohr radius. On the other hand, in order to increase $l_c$ or to increase the exciton transfer energy, it is favorable to use materials having a large oscillator strength. Combining these requirements, we see that CuCl is one of the most favorable materials for quantum dot arrays.

The excitonic optical nonlinearity of quantum dot array is estimated for typical materials of GaAs, CdS, and CuCl, assuming a three-dimensional simple cubic structure. The results are shown in Table 2.1. The quantum dot radius is chosen to be a few to ten times the bulk exciton Bohr radius. In the estimation of the optical nonlinearity, the homogeneous linewidth $\hbar\Gamma_h$ is chosen to be compatible with known data. The radiative lifetime $\tau_R$ of the exciton is one of the factors that determine the response speed of the optical nonlinearity. For CuCl, the transfer energy is rather large and the figure of

**Table 2.1.** Nonlinear optical properties are estimated for quantum dot arrays of GaAs, CdS, and CuCl having the simple cubic structure. $R$ is the radius of the quantum dot, $D$ the lattice constant of the quantum dot array, $T$ the exciton transfer energy, $\Gamma_h$ the homogeneous linewidth of the excitonic transition, $\mathrm{Im}\chi^{(3)}$ the peak value of the imaginary part of the third-order nonlinear susceptibility, $\tau_R$ the radiative lifetime of the exciton, $\sigma$ the figure of merit for the absorptive optical nonlinearity, $\sigma/\sigma(\mathrm{dot})$ the enhancement factor of $\sigma$ due to the array formation, $\Delta R$ the tolerance limit for the fluctuation of the quantum dot radius, and $\Delta D$ the tolerance limit for the fluctuation of the lattice constant (from T. Takagahara [53])

| Medium | R (Å) | D (Å) | T (meV) | $\Gamma_h$ (meV) | $\tau_R$ (ps) | $\|\mathrm{Im}\chi^{(3)}\|$ (esu) | $\sigma$ (cm$^2$) | $\dfrac{\sigma}{\sigma(\mathrm{dot})}$ | $\Delta R$ (Å) | $\Delta D$ (Å) |
|---|---|---|---|---|---|---|---|---|---|---|
| GaAs | 220 | 500 | 0.05 | 1 | 550 | $3 \cdot 10^{-2}$ | $1 \cdot 10^{-12}$ | 2 | 40 | 40 |
| CdS | 150 | 400 | 1 | 10 | 90 | $3 \cdot 10^{-4}$ | $3 \cdot 10^{-13}$ | 4 | 30 | 60 |
| CuCl | 80 | 200 | 5 | 2 | 140 | $7 \cdot 10^{-3}$ | $4 \cdot 10^{-10}$ | 120 | 17 | 10 |

merit $\sigma$ is $4 \cdot 10^{-10}$ cm$^2$, which is several orders of magnitude larger than the usually known values ($\sim 10^{-14}$ cm$^2$) [54]. The enhancement factor of $\sigma$ relative to that of a single dot is more than 100. On the other hand, for GaAs and CdS, the exciton transfer energy is rather small or the homogeneous linewidth is rather large. As a consequence, the $\sigma$-value is not much enhanced compared to that of a single dot. Thus, it is confirmed that CuCl is one of the most favorable materials for the quantum dot array. Recent measurements of single-dot spectroscopy revealed very small values of the homogeneous linewidth at low temperatures ($\sim$ several K) [2–4]. This narrow linewidth encourages us to observe the proposed effect in quantum dot arrays at low temperatures.

### 2.5.3 Tolerance Limits for the Fluctuation of Structure Parameters of the Quantum Dot Array

Now we discuss the conditions for the enhancement of optical nonlinearity in quantum dot arrays. So far, we have assumed implicitly that the quantum dot size and the lattice constant are completely uniform. This uniformity is primarily required to achieve a long exciton coherence length and to obtain a large optical nonlinearity. However, in actual samples, the fluctuation of the quantum dot size and the lattice constant cannot be avoided. Thus it is very important to estimate the tolerance limits for the fluctuation of these structural parameters.

A criterion for the allowable size fluctuation can be derived as follows. If the fluctuation of the exciton energy due to the size fluctuation is smaller than the homogeneous linewidth of the excitonic transition, the coherent cooperation of many quantum dots can be achieved. Namely, the condition determines the tolerance limits for the size fluctuation $\Delta R$ is given by

$$\frac{\hbar^2 \pi^2}{2M} \left| \frac{1}{R^2} - \frac{1}{(R \pm \Delta R)^2} \right| < \Gamma_h . \tag{2.119}$$

In a similar way, the tolerance limits for the fluctuation of the lattice constant can be derived. The fluctuation of the lattice constant induces a fluctuation of the exciton transfer energy, which leads to the fluctuation of the exciton energy. To simplify the analysis, we use an analytical expression of the exciton energy for the case of a simple cubic lattice with the nearest-neighbor transfer given as

$$\begin{aligned} E(k_x, k_y, k_z) &= 2T(\cos k_x D + \cos k_y D + \cos k_z D) \\ &\cong 6T - TD^2(k_x^2 + k_y^2 + k_z^2) . \end{aligned} \tag{2.120}$$

We see that the exciton energy fluctuation is roughly given by six times the transfer energy fluctuation. Then the allowable fluctuation $\Delta D$ of the lattice constant is determined by the condition that the fluctuation of the exciton energy is smaller than the homogeneous linewidth of the excitonic transition, namely

$$6\Delta T < \Gamma_{\rm h} \ . \tag{2.121}$$

The tolerance limits for $\Delta R$ and $\Delta D$ estimated from these criteria are given in Table 2.1. Fortunately, these values are about ten or twenty percent of the nominal values, and these tolerance limits do not seem to be very difficult to satisfy at the present stage of various fabrication techniques.

### 2.5.4    The Polariton Effect and Photonic Band Structures

We proposed the quantum dot array for enhancing excitonic optical nonlinearity, especially its figures of merit. The most important aspect of the quantum dot array for enhancing optical nonlinearity is the enlarged exciton coherence length due to exciton transfer among quantum dots via the electromagnetic dipolar and higher-multipolar interactions. The important requisites for the material of the quantum dot array to enhance the figures of merit of the optical nonlinearity were shown to be large oscillator strength of the excitonic transition and small exciton Bohr radius.

In the arguments in Sect. 2.5.3, we did not take into account the polariton effect, which is one of the factors determining the exciton coherence length. The polariton effect becomes important when the quantum dot size and the lattice constant of the quantum dot array are comparable to or larger than the optical wavelength of the relevant excitonic transition. In the case of a single quantum dot, the polariton effect on the excitonic radiative decay rate was discussed in Sect. 2.3, incorporating the nonlocality of the electromagnetic interaction. The observed size dependence of the radiative decay rate for CuCl microcrystallites was explained quantitatively well by the theory in Sect. 2.3.2. According to the theory, the saturation of the excitonic radiative decay rate with increasing quantum dot size, namely the breakdown of the mesoscopic enhancement, is mainly determined by the homogeneous broadening of the excitonic transition rather than the polariton effect. In other words, the excitonic coherence length determined by the homogeneous linewidth is actually much shorter than the optical wavelength. In the case of the quantum dot array, a similar situation prevails. The excitonic coherence length estimated from the homogeneous linewidth given in Table 2.1 and from the exciton translational mass in Fig. 2.23 is several hundreds of angstroms, or about one order of magnitude smaller than the relevant optical wavelength. Thus, within the range of the structural parameters in Table 2.1, the magnitude of the optical nonlinearity is mainly determined by the homogeneous linewidth, and the polariton effect would not be significant. However, the polariton effect in the quantum dot array is quite interesting in relation to the fundamental aspects of the radiation–matter interaction in the presence of photonic band gaps [55–58] and is left for future study.

## 2.6   Summary

The excitonic energy levels and their oscillator strengths are discussed for spherical quantum dots embedded in dielectric materials clarifying the dielectric confinement effect, which appears most pronounced in zero-dimensional systems. In the weak-confinement regime, the exciton oscillator strength is proportional to the volume of a quantum dot. This is due to the formation of a large transition dipole moment as a result of the coherent superposition of the atomic dipole moments within a quantum dot. But this proportionality does not persist over the quantum dot size exceeding the wavelength of a photon corresponding to the exciton transition energy. This intriguing property of the size- and temperature-dependent radiative decay rate of excitons in quantum dots is clarified on the basis of the nonlocal response theory. Exciton fine structures arising from the electron–hole exchange interaction are discussed, taking into account the degenerate valence band structures, and quantitative comparison with recent results of single dot spectroscopy is made. The prerequisite to enhance the excitonic optical nonlinearity is a large transition dipole moment. This can be achieved by formation of a regular array of quantum dots. Each quantum dot corresponds to an atom with a large oscillator strength. Through dipole–dipole coupling among quantum dots, the coherent superposition of the dipole moments of quantum dots can be accomplished, in analogy to the Frenkel-type exciton.

## Appendix A. Expression of Depolarization Field

When a polarization field $\boldsymbol{P}(\boldsymbol{r})$ is present within a material system, the electric field on a test charge at the position $\boldsymbol{r}$ due to all the polarization dipoles is given by

$$\int \mathrm{d}^3 r' \, \mathrm{grad} \, \mathrm{div} \frac{1}{|\boldsymbol{r} - \boldsymbol{r}'|} \boldsymbol{P}(\boldsymbol{r}')$$
$$= \sum_{\mathrm{s}} \frac{3(\boldsymbol{r} - \boldsymbol{r}_{\mathrm{s}})^t (\boldsymbol{r} - \boldsymbol{r}_{\mathrm{s}}) - |\boldsymbol{r} - \boldsymbol{r}_{\mathrm{s}}|^2 \hat{1}}{|\boldsymbol{r} - \boldsymbol{r}_{\mathrm{s}}|^5} \boldsymbol{P}(\boldsymbol{r}_{\mathrm{s}}) \, , \tag{2.122}$$

where operators grad and div are acting on the coordinates of $\boldsymbol{r}$ and a dyadic form is used on the right hand side. Dividing the sum or the integral in (2.122) into two components: one over the dipoles within a small sphere around the point $\boldsymbol{r}$ and the other containing the residual dipoles, we have the local field at the position $\boldsymbol{r}$ given by

$$\boldsymbol{E}_{\mathrm{local}}(\boldsymbol{r}) = \boldsymbol{E}_{\mathrm{s}}(\boldsymbol{r}) + \int_{\mathrm{c}} \mathrm{d}^3 r' \, \mathrm{grad} \, \mathrm{div} \frac{1}{|\boldsymbol{r} - \boldsymbol{r}'|} \boldsymbol{P}(\boldsymbol{r}') \, , \tag{2.123}$$

where $\boldsymbol{E}_{\mathbf{s}}(\boldsymbol{r})$ is the source (external) field and the subscript c on the integral symbol means that the integration is carried out excluding the small sphere around the point $r$. This is exact because the contribution from dipoles within that sphere is vanishing. Physically, the second term on the right hand side of (2.123) is composed of the depolarization field arising from the polarization charges on the outer surface of the material system and the Lorentz cavity field due to the polarization charges on the surface of the cavity. Then making use of a general formula for an arbitrary vector field $Q(r)$ given as [59]

$$\frac{4\pi}{3} Q(r) + \text{grad div} \int_c d^3r' \frac{1}{|r - r'|} Q(r')$$

$$= \int_c d^3r' \text{ grad div} \frac{1}{|r - r'|} Q(r') \,, \tag{2.124}$$

and noticing that $4\pi P(r)/3$ is the Lorentz cavity field $E_{\text{Lorentz}}(r)$, we have

$$E_{\text{local}}(r) = E_s(r) + E_{\text{Lorentz}}(r) + \text{grad div} \int_c d^3r' \frac{1}{|r - r'|} P(r') \,, \tag{2.125}$$

and identify the depolarization field as

$$E_{\text{dep}}(r) = \text{grad div} \int_c d^3r' \frac{1}{|r - r'|} P(r') \,. \tag{2.126}$$

The integrand in (2.126) is integrable around the point $r$, and that contribution can be made arbitrarily small by choosing a very small radius of the sphere. Thus we obtain

$$E_{\text{dep}}(r) = \text{grad div} \int d^3r' \frac{1}{|r - r'|} P(r') \,, \tag{2.127}$$

where the integration can be carried out over all the relevant material systems.

## Appendix B. Depolarization Field in the Presence of a Background Dielectric Constant

We consider a spherical semiconductor nanocrystal having a background dielectric constant $\epsilon_b$ that is embedded in a vacuum. When a charge $q_0$ is fixed at the position $r_0$, the potential energy for a test charge $q$ at the position $r$ is given by [60,10]

$$V(r, r_0) = \frac{q q_0}{\epsilon_b |r - r_0|} + \frac{q q_0}{R} \sum_{l=0}^{\infty} \alpha_l \left( \frac{r r_0}{R^2} \right)^l P_l(\cos \theta_{r r_0}) \,, \tag{2.128}$$

with

$$\alpha_l = \frac{(l + 1)(\epsilon_b - 1)}{\epsilon_b(l \epsilon_b + l + 1)} \,, \tag{2.129}$$

where both $r$ and $r_0$ are assumed to be within the nanocrystal, $r = |r|$, $r_0 = |r_0|$, $P_l$ is a Legendre polynomial, $\theta_{rr_0}$ is the angle between $r$ and $r_0$, and the second term represents the contribution from the image charge due to the dielectric boundary. Considering a dipole composed of a charge $q_0$ at the position $r_0$ and a charge $-q_0$ at the position $r_0-d$, we can calculate the potential energy for a test charge $q$ at the position $r$ by

$$V(r, r_0) - V(r, r_0 - d) . \tag{2.130}$$

Since $|d|$ can be assumed to be much smaller than $r$ and $r_0$, we can reduce a typical term in (2.130) as

$$\left(\frac{rr_0}{R^2}\right)^l P_l(\cos\theta_{rr_0}) - \left(\frac{r|r_0 - d|}{R^2}\right)^l P_l(\cos\theta_{rr_0-d})$$

$$\sim \frac{r^l r_0^{l-1}}{R^{2l}} \{ l(n_0 \cdot d) P_l(\cos\theta_{rr_0})$$

$$- [(n_0 \cdot n_r)(n_0 \cdot d) - (n_r \cdot d)] P_l'(\cos\theta_{rr_0}) \} , \tag{2.131}$$

where $n_0 = r_0/|r_0|$, $n_r = r/|r|$, and $P_l'$ is the first derivative of the Legendre polynomial. Then the electric field at the position $r$ due to a polarization dipole $p_d = q_0 d$ at the position $r_0$ is given by

$$E(r) = -\mathrm{grad}\phi(r) , \tag{2.132}$$

with

$$\phi(r) = -\frac{1}{\epsilon_b}\nabla_r \frac{1}{|r - r_0|} \cdot p_d + \sum_{l=1}^{\infty} \frac{\alpha_l r^l r_0^{l-1}}{R^{2l+1}} \{ l(n_0 \cdot p_d) P_l(\cos\theta_{rr_0})$$

$$- \{[(n_0 \cdot n_r)(n_0 \cdot p_d) - (n_r \cdot p_d)] P_l'(\cos\theta_{rr_0}) \} , \tag{2.133}$$

where the second term is the potential induced by the presence of the background dielectric constant and the $l=0$ term vanishes automatically.

## Appendix C. Vector Spherical Harmonics

The vector Helmholtz equation given by

$$[\nabla^2 + k^2]\Psi(r) = 0 \tag{2.134}$$

has three linearly independent solutions. They are usually denoted by $L_{lm}(kr)$, $M_{lm}(kr)$, and $N_{lm}(kr)$ as given in (2.51) [61], where the radial part can generally include spherical Hankel functions as well as spherical Bessel functions. These $L$, $M$, and $N$ fields are generally not orthogonal to each other. On the other hand, concerning the angular part, there are mutually orthogonal harmonic bases [62]. In spherical coordinates, they are defined by

$$\boldsymbol{P}_{lm}(\Omega) = {}^{t}(1,0,0)Y_{lm}(\Omega) , \tag{2.135}$$

$$\boldsymbol{B}_{lm}(\Omega) = \frac{1}{\sqrt{l(l+1)}} {}^{t}\left(0, \frac{\partial}{\partial\theta}, \frac{1}{\sin\theta}\frac{\partial}{\partial\varphi}\right) Y_{lm}(\Omega) , \tag{2.136}$$

$$\boldsymbol{C}_{lm}(\Omega) = \frac{1}{\sqrt{l(l+1)}} {}^{t}(0, \frac{1}{\sin\theta}\frac{\partial}{\partial\varphi}, -\frac{\partial}{\partial\theta})Y_{lm}(\Omega) , \tag{2.137}$$

where $\boldsymbol{P}_{lm}$ is defined for all $l(\geq 0)$, whereas $\boldsymbol{B}_{m}$ and $\boldsymbol{C}_{lm}$ are defined for $l \geq 1$. They are orthonormalized concerning the angular integral, namely

$$\int d\Omega \boldsymbol{P}_{lm}^{*}(\Omega) \cdot \boldsymbol{P}_{l'm'}(\Omega) = \int d\Omega \boldsymbol{B}_{lm}^{*}(\Omega) \cdot \boldsymbol{B}_{l'm'}(\Omega)$$

$$= \int d\Omega \boldsymbol{C}_{lm}^{*}(\Omega) \cdot \boldsymbol{C}_{l'm'}(\Omega) = \delta_{ll'}\delta_{mm'} , \tag{2.138}$$

$$\int d\Omega \boldsymbol{P}_{lm}^{*}(\Omega) \cdot \boldsymbol{B}_{l'm'}(\Omega) = \int d\Omega \boldsymbol{B}_{lm}^{*}(\Omega) \cdot \boldsymbol{C}_{l'm'}(\Omega)$$

$$= \int d\Omega \boldsymbol{C}_{lm}^{*}(\Omega) \cdot \boldsymbol{P}_{l'm'}(\Omega) = 0 . \tag{2.139}$$

In terms of these angular harmonics, $\boldsymbol{L}$, $\boldsymbol{M}$, and $\boldsymbol{N}$ fields can be written as

$$\boldsymbol{L}_{lm}(kr) = z_l'(kr)\boldsymbol{P}_{lm}(\Omega) + \frac{\sqrt{l(l+1)}}{kr}z_l(kr)\boldsymbol{B}_{lm}(\Omega) , \tag{2.140}$$

$$\boldsymbol{M}_{lm}(kr) = \sqrt{l(l+1)}z_l(kr)\boldsymbol{C}_{lm}(\Omega) , \tag{2.141}$$

$$\boldsymbol{N}_{lm}(kr) = \frac{l(l+1)}{kr}z_l(kr)\boldsymbol{P}_{lm}(\Omega) + \frac{\sqrt{l(l+1)}}{kr}\frac{d(krz_l(kr))}{d(kr)}\boldsymbol{B}_{lm}(\Omega) , \tag{2.142}$$

where $z_l(kr)$ is a radial function. In the text, the superscript $(0)$ or $(1)$ is attached to the $\boldsymbol{L}$, $\boldsymbol{M}$, and $\boldsymbol{N}$ fields corresponding to $z_l(kr)=j_l(kr)$ (spherical Bessel function) or $h_l^{(1)}(kr)$ (spherical Hankel function of the first kind).

## Appendix D. Parameters Related to the Electron–Hole Exchange Energies

Here we discuss the determination of parameters $E_X^S$ in (2.97) and $\mu$ in (2.100). These quantities can be fixed from the singlet–triplet splitting energy $\Delta_{\mathrm{ST}}$ and the longitudinal–transverse splitting energy $\Delta_{\mathrm{LT}}$ in bulk materials. In bulk materials, the exciton state can be written as

$$|X\rangle_K = \frac{1}{\sqrt{N}} \sum_{\tau,\sigma} \sum_{n,m} e^{iKR_m} F_{\tau\sigma}(R_n) a_{c\tau R_n+R_m}^{\dagger} a_{v\sigma R_m} |0\rangle , \tag{2.143}$$

where $K$ is the center-of-mass wavevector, $F$ the envelope function describing the electron–hole relative motion and the Wannier function representation is used. The electron–hole exchange energy is given by

$$\sum_{\tau\sigma\tau'\sigma'} F_{\tau\sigma}(0)F_{\tau'\sigma'}(0)V(c\tau R_0, v\sigma' R_0; c\tau' R_0, v\sigma R_0)$$

$$+ \sum_{\tau\sigma\tau'\sigma'} F_{\tau\sigma}(0)F_{\tau'\sigma'}(0)$$

$$\times \sum_{m'\neq m} e^{-iK(R_m - R_{m'})}V(c\tau R_m, v\sigma' R_{m'}; c\tau' R_{m'}, v\sigma R_m) , \quad (2.144)$$

where the first(second) term is the short(long)-range exchange term. By the multipole expansion, the second term can be rewritten as

$$-\frac{4\pi}{3v_0} \sum_{\tau\sigma\tau'\sigma'} F_{\tau\sigma}(0)F_{\tau'\sigma'}(0)\,{}^t\boldsymbol{\mu}_{c\tau,v\sigma} [\hat{1} - 3\hat{K}^t\hat{K}]\,\boldsymbol{\mu}_{v\sigma',c\tau'} , \quad (2.145)$$

where $\hat{K}$ is a unit vector in the direction of $K$. In GaAs the top valence band is the fourfold degenerate $\Gamma_8(J=3/2)$ band. Carrying out a straightforward calculation in the $8{\times}8$ matrix form and assuming the ground state envelope function as

$$F(r) = \sqrt{\frac{v_0}{\pi a_B^{*3}}}\, e^{-r/a_B^*} , \quad (2.146)$$

where $a_B^*$ is the exciton Bohr radius and $v_0$ is the volume of a unit cell, we find the energy level structure as shown in Fig. 2.25 with

$$E_S = \frac{v_0}{\pi a_B^{*3}} E_X^S \quad \text{and} \quad E_L = \frac{4e^2\mu^2}{3a_B^{*3}} . \quad (2.147)$$

The lowest state is the fivefold degenerate $J=2$ dark exciton state, the next lowest state is the doubly degenerate $J=1$ transverse exciton state, and the uppermost state is the $J=1$ longitudinal exciton state. Then we can identify as

$$\Delta_{ST} = \frac{4v_0}{3\pi a_B^{*3}} E_X^S , \quad \Delta_{LT} = \frac{32e^2\mu^2}{3a_B^{*3}} . \quad (2.148)$$

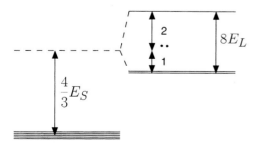

Fig. 2.25. Schematic level structure of the exciton ground state in bulk materials like GaAs having the four-fold degenerate ($\Gamma_8$) valence band. The definitions of $E_S$ and $E_L$ are given in (2.147)

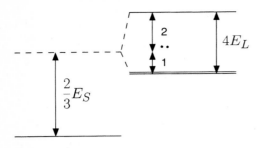

**Fig. 2.26.** Schematic level structure of the exciton ground state in bulk materials like CuCl having the doubly degenerate ($\Gamma_7$) valence band. The definitions of $E_S$ and $E_L$ are given in (2.147)

The parameters $E_X^S$ and $\mu$ can be fixed from the experimental values of $\Delta_{ST}$ and $\Delta_{LT}$ [42]. It is to be noted that the subtle problem of screening of the electron–hole exchange interaction is circumvented in the above formulation because the effect of screening is implicitly included in the parameters $E_X^S$ and $\mu$.

For the case where the top valence band is the doubly degenerate $\Gamma_7(j=1/2)$ band as in, e.g., CuCl, we find the excitonic energy level structure as shown in Fig. 2.26. The lowest state is the dark ($J=0$) exciton state, and the $J=1$ exciton states split into a doubly degenerate transverse exciton state and a longitudinal exciton state. Using the same notation as in (2.148), we find

$$\Delta_{ST} = \frac{2v_0}{3\pi a_B^{*3}} E_X^S , \quad \Delta_{LT} = \frac{16e^2\mu^2}{3a_B^{*3}} . \tag{2.149}$$

Thus it is to be noted that the relations for estimating $E_X^S$ and $\mu$ from the experimental values of $\Delta_{ST}$ and $\Delta_{LT}$ are different for the $\Gamma_7$ and $\Gamma_8$ valence bands.

# References

1. S. Lloyd: Science **261**, 1569 (1993); A. Barenco, D. Deutsch, A. Ekert, R. Jozsa: Phys. Rev. Lett. **74**, 4083 (1995)
2. H.F. Hess, E. Betzig, T.D. Harris, L.N. Pfeiffer, K.W. West: Science **264**, 1740 (1994)
3. K. Brunner, G. Abstreiter, G. Böhm, G. Tränkle, G. Weimann: Phys. Rev. Lett. **73**, 1138 (1994)
4. D. Gammon, E.S. Snow, B.V. Shanabrook, D.S. Katzer, D. Park: Science **273**, 87 (1996); Phys. Rev. Lett. **76**, 3005 (1996)
5. N.H. Bonadeo, J. Erland, D. Gammon, D. Park, D.S. Katzer, D.G. Steel: Science **282**, 1473 (1998)
6. L.V. Keldysh: Pis'ma Zh. Eksp. Teor. Fiz. **29**, 716 (1979) [JETP Lett. **29**, 658 (1979)]; Superlatt. Microstruct. **4**, 637 (1988)
7. M. Kumagai, T. Takagahara: Phys. Rev. B **40**, 12359 (1989)
8. L. Banyai, I. Galbraith, C. Ell, H. Haug: Phys. Rev. B **36**, 6099 (1987)
9. T. Ogawa, T. Takagahara: Phys. Rev. B **44**, 8138 (1991)

10. L.E. Brus: J. Chem. Phys. **80**, 4403 (1984)
11. T. Takagahara: Phys. Rev. B **47**, 4569 (1993)
12. P.G. Bolcatto, C.R. Proetto: Phys. Rev. B **59**, 12487 (1999)
13. Y. Kayanuma, H. Momiji: Phys. Rev. B **41**, 10261 (1990)
14. D.B. Tran Thoai, Y.Z. Hu, S.W. Koch: Phys. Rev. B **42**, 11261 (1990)
15. E.A. Hylleraas: Z. Phys. **54**, 347 (1929)
16. Y. Kayanuma: Phys. Rev. B **38**, 9797 (1988)
17. T. Takagahara: Phys. Rev. B **36**, 9293 (1987)
18. E. Hanamura: Phys. Rev. B **37**, 1273 (1988)
19. A. Nakamura, H. Yamada, T. Tokizaki: Phys. Rev. B **40**, 8585 (1989)
20. T. Itoh, T. Ikehara, Y. Iwabuchi: J. Lumin. **45**, 29 (1990)
21. T. Takagahara: Phys. Rev. B **47**, 16639 (1993)
22. J.R. Kuklinski, S. Mukamel: Chem. Phys. Lett. **189**, 119 (1992)
23. K. Cho: Prog. Theor. Phys. Suppl. **106**, 225 (1991)
24. R. Fuchs, K.L. Kliewer: J. Opt. Soc. Am. **58**, 319 (1968)
25. S.C. Hill, R.E. Benner: J. Opt. Soc. Am. B **3**, 1509 (1986)
26. T. Itoh, M. Furumiya: J. Lumin. **48 & 49**, 704 (1991)
27. J. Feldmann, G. Peter, E.O. Göbel, P. Dawson, K. Moore, C. Foxon, R.J. Elliott: Phys. Rev. Lett. **59**, 2337 (1987)
28. T. Takagahara: Solid State Commun. **78**, 279 (1991)
29. H. Gotoh, H. Ando, T. Takagahara: J. Appl. Phys. **81**, 1785 (1997)
30. H. Gotoh, H. Ando, T. Takagahara, H. Kamada, A. Chavez-Pirson, J. Temmyo: Jpn. J. Appl. Phys. Part 1 **36**, 4204 (1997)
31. M. Bayer, A. Kuther, A. Forchel, A. Gorbunov, V.B. Timofeev, F. Schäfer, J.P. Reithmaier, T.L. Reinecke, S.N. Walck: Phys. Rev. Lett. **82**, 1748 (1999)
32. V.D. Kulakovskii, G. Bacher, R. Weigand, T. Kümmell, A. Forchel, E. Borovitskaya, K. Leonardi, D. Hommel: Phys. Rev. Lett. **82**, 1780 (1999)
33. M. Bayer, O. Stern, A. Kuther, A. Forchel: Phys. Rev. B **61**, 7273 (2000)
34. L. Besombes, K. Kheng, D. Martrou: Phys. Rev. Lett. **85**, 425 (2000)
35. S.V. Gupalov, E.L. Ivchenko, A.V. Kavokin: JETP Lett. **86**, 388 (1998); E.L. Ivchenko: Phys. Status Solidi A **164**, 487 (1997)
36. I. Kang, F.W. Wise: J. Opt. Soc. Am. B **14**, 1632 (1997)
37. T. Takagahara: Phys. Rev. B **62**, 16840 (2000)
38. J.M. Luttinger: Phys. Rev. **102**, 1030 (1956)
39. T. Takagahara: Phys. Rev. B **60**, 2638 (1999)
40. M. Sugisaki, H.W. Ren, S.V. Nair, K. Nishi, S. Sugou, T. Okuno, Y. Masumoto: Phys. Rev. B **59**, R5300 (1999)
41. M.Z. Maialle, E.A. de Andrada e Silva, L.J. Sham: Phys. Rev. B **47**, 15776 (1993)
42. *Physics of Group IV Elements and III-V Compounds, Landolt-Börnstein, Vol. 17a*, ed. by O. Madelung, M. Schulz, H. Weiss (Springer, Berlin 1982)
43. T. Takagahara: Phys. Rev. B **39**, 10206 (1989)
44. Y. Wang: J. Opt. Soc. Am. B **8**, 981 (1991)
45. R. Nötzel, J. Temmyo, T. Tamamura: Nature (London) **369**, 131 (1994); R. Nötzel, J. Temmyo, H. Kamada, T. Furuta, T. Tamamura: Appl. Phys. Lett. **65**, 457 (1996)
46. D. Bimberg, M. Grundmann, N.N. Ledentsov: *Quantum Dot Heterostructures*, (John Wiley & Sons, New York 1999) and references therein
47. *Langmuir–Blodgett Films*, ed. by G.G. Roberts (Plenum, New York 1990)

48. K. Furukawa, M. Fujino, N. Matsumoto: Appl. Phys. Lett. **60**, 2744 (1992)
49. L. Pintschovius, B. Renker, F. Gompf, R. Heid, S.L. Chaplot, M. Haluska, H. Kuzmany: Phys. Rev. Lett. **69**, 2662 (1992)
50. S. Huant, J.B. Robert, G. Chouteau, P. Bernier, C. Fabre, A. Rassat: Phys. Rev. Lett. **69**, 2666 (1992)
51. Y. Wang, N. Herron, W. Mahler, A. Suna: J. Opt. Soc. Am. B **6**, 808 (1989)
52. G.D. Stucky, J.E. MacDougall: Science **247**, 669 (1990)
53. T. Takagahara: Surf. Sci. **267**, 310 (1992); in *Nonlinear Optics, Fundamentals, Materials and Devices*, ed. by S. Miyata (Elsevier, Amsterdam 1992), p.85
54. D.S. Chemla, D.A.B. Miller, P.W. Smith, A.C. Gossard, W. Wiegmann: IEEE J. Quantum Electron. **QE-20**, 265 (1984)
55. See e.g., the feature issue, J. Opt. Soc. Am. B **10**, 279 (1993)
56. J.D. Joannopoulos, R.D. Meade, J. N. Winn: *Photonic Crystals* (Princeton Univ. Press, Princeton 1995)
57. J.M. Gerard, B. Sermage, B. Gayral, B. Legrand, E. Costard, V. Thierry-Mieg: Phys. Rev. Lett. **81**, 1110 (1998)
58. M. Bayer, T. Gutbrod, A. Forchel, T.L. Reinecke, P.A. Knipp, R. Werner, J.P. Reithmaier: Phys. Rev. Lett. **83**, 5374 (1999)
59. M. Born, E. Wolf: *Principles of Optics*, 5th edn. (Pergamon, New York 1975)
60. C.J.F. Böttcher: *Theory of Electric Polarization, Vol. 1* (Elsevier, Amsterdam 1973)
61. J.A. Stratton: *Electromagnetic Theory*, (McGraw-Hill, New York 1941)
62. P.M. Morse, H. Feshbach: *Methods of Theoretical Physics* (McGraw-Hill, New York 1953)

# 3 Electron–Phonon Interactions in Semiconductor Quantum Dots

Toshihide Takagahara

## 3.1 Introduction

Semiconductor nanocrystals of a size comparable to or smaller than the exciton Bohr radius in the bulk material are attracting much attention from the fundamental physics viewpoint and from interest in their application to functional devices [1–3]. In particular, their novel optical properties because of discrete electronic energy levels have been investigated extensively [4,5]. In semiconductor nanocrystals, not only the electronic energy levels but also the lattice vibrational modes become discrete due to three-dimensional confinement. The consequences of the latter feature, namely phonon confinement, have been studied extensively. The longitudinal optical (LO) phonons in semiconductor nanocrystals were observed by resonance Raman scattering [6–8], and the size dependence of the electron–LO-phonon coupling strength was discussed [7,9,10]. Also, the size-quantized acoustic phonon modes were observed by low-frequency Raman scattering [11] and were discussed extensively [12].

In addition to these studies, excitonic dephasing in various semiconductor nanocrystals has been measured in detail as a function of the nanocrystal size and the temperature. In CuCl nanocrystals, the homogeneous linewidth of the excitonic transition was measured from the luminescence linewidth under size-selective excitation [13] and by spectral hole burning [14]. In nanocrystals of II–VI compound semiconductors, the excitonic dephasing constant was measured by spectral hole burning [15–18] and by four-wave mixing [19–21]. The commonly observed temperature-linear behavior of the excitonic dephasing rate suggested the importance of the electron–phonon interaction with acoustic phonon modes, although the relevant temperature range is dependent on the nanocrystal size and on the material. These earlier studies were renewed by the recent advent of single-dot spectroscopy. The exciton fine structures have been revealed with unprecedented accuracy, and the origin of linewidth of sharp exciton lines is extensively investigated. This subject is discussed in detail in Chaps. 8 and 9.

In this chapter, we derive the electron–phonon interactions with acoustic and polar optical phonons in semiconductor nanocrystals and clarify the size dependence of the coupling strength for typical coupling mechanisms, i.e., deformation potential coupling, piezoelectric coupling, and Fröhlich coupling.

As an application of the theory, we formulate the electron–phonon interaction in Si nanocrystals for acoustic phonon modes to calculate the luminescence Stokes shift, and the Huang–Rhys factor. As a result, we find that the luminescence onset energy can be explained mainly in terms of the excitonic exchange splitting.

## 3.2    Energy Spectra of Acoustic Phonon Modes in Spherical Nanocrystals

A quantum dot is usually embedded in another material with different elastic constants. The boundary conditions on the lattice vibrations are highly dependent on the interface structures. There are two typical cases of the stress-free boundary condition and the smooth contact condition. The former case corresponds to an isolated quantum dot, whereas the latter case is exemplified by epitaxially grown semiconductor quantum dots. As long as the nanocrystal size is not too small, its acoustic properties can be described in terms of the elastic vibration of a homogeneous particle. In the following, the shape of a nanocrystal is assumed to be spherical and the anisotropy of the elastic constants is neglected for simplicity of the arguments. Then the vibrations of an elastically isotropic sphere can be described by

$$\rho \frac{\partial^2}{\partial t^2} \boldsymbol{u} = (\lambda + \mu) \, \text{grad div} \, \boldsymbol{u} + \mu \nabla^2 \boldsymbol{u} \,, \tag{3.1}$$

where $\boldsymbol{u}$ is the lattice displacement vector, $\rho$ the mass density, and $\lambda$ and $\mu$ are the Lame's constants [22]. A similar equation holds for the surrounding medium. The eigenmodes of the lattice vibration of the whole system should be determined under an appropriate boundary condition. In the case of spherical geometry, the stress on a plane perpendicular to a radial unit vector $\boldsymbol{e}_{\rm r}$ is given by

$$\sigma \boldsymbol{e}_{\rm r} = \frac{1}{r} \left( \lambda \boldsymbol{r} \, \text{div} \boldsymbol{u} + \mu \nabla (\boldsymbol{r} \cdot \boldsymbol{u}) - \mu \boldsymbol{u} + \mu r \frac{\partial}{\partial r} \boldsymbol{u} \right) \,, \tag{3.2}$$

where $\sigma$ is the stress tensor. When the time dependence is assumed as

$$\boldsymbol{u}(r, t) = \boldsymbol{u}(r) \, {\rm e}^{-{\rm i}\omega t} \,, \tag{3.3}$$

we have

$$\left( \nabla^2 + \frac{\rho \, \omega^2}{\mu} \right) \, \text{rot} \, \boldsymbol{u} = 0 \,, \tag{3.4}$$

$$\left( \nabla^2 + \frac{\rho \, \omega^2}{\lambda + 2\mu} \right) \, \text{div} \, \boldsymbol{u} = 0 \,. \tag{3.5}$$

These are the vector and scalar Helmholtz equations.

## 3.2.1   The Case of the Stress-Free Boundary Condition

The eigenmodes under the stress-free boundary condition were first studied by Lamb [23,22]. There are two kinds of eigenmodes, namely torsional modes and spheroidal modes. Under spherical geometry, each mode can be expanded in terms of vector spherical harmonics [24]. These harmonics for an angular momentum $l$ and its $z$-component $m$ are defined as

$$L_{lm}(kr) = \frac{1}{k}\nabla\Psi_{lm}(kr) ,$$

$$M_{lm}(kr) = \mathrm{rot}[r\Psi_{lm}(kr)] = \frac{1}{k}\mathrm{rot}N_{lm}(kr) ,$$

$$N_{lm}(kr) = \frac{1}{k}\mathrm{rot}M_{lm}(kr) , \tag{3.6}$$

with $\Psi_{lm}(kr) = j_l(kr)Y_{lm}(\Omega)$, where $j_l(kr)$ is an $l$-th order spherical Bessel function and $Y_{lm}$ is a spherical harmonic. These vector fields satisfy the vector Helmholtz equation

$$(\nabla^2 + k^2)L_{lm} = (\nabla^2 + k^2)M_{lm} = (\nabla^2 + k^2)N_{lm} = 0 . \tag{3.7}$$

In the following, the phonon displacement vector will be considered only for the component corresponding to the angular momentum $l$ and its $z$-component $m$, unless otherwise stated. In the case of the torsional modes, the displacement vector, apart from a normalization factor, is written as

$$u(r) = R\,M_{lm}(kr) = R\,j_l(kr)\begin{pmatrix} 0 \\ \frac{1}{\sin\theta}\frac{\partial}{\partial\varphi} \\ -\frac{\partial}{\partial\theta} \end{pmatrix} Y_{lm}(\Omega) , \tag{3.8}$$

where $R$ is the radius of a spherical nanocrystal and is included since $M_{lm}$ is defined as a dimensionless quantity. Here the vector components are given in spherical coordinates and $M_{lm}$ vanishes identically for $l=0$. This mode is purely transversal since $\mathrm{div}\,u=0$. The eigenfrequency $\omega$ is related to $k$ by $\mu k^2 = \rho\omega^2$. Here $r \cdot u = 0$ and

$$\sigma e_{\mathrm{r}} = \frac{\mu}{r}(-u + r\frac{\partial}{\partial r}u) . \tag{3.9}$$

Then the stress-free boundary condition leads to an equation that determines the size-quantized $k$-values, namely,

$$kRj_l'(kR) - j_l(kR) = 0 . \tag{3.10}$$

It is interesting to note that this equation is independent of the elastic constants of the material and the $k$-value is determined only by the radius.

In the case of spheroidal modes, the displacement vector is written as

$$u(r) = p_{lm}R\,L_{lm}(hr) + q_{lm}R\,N_{lm}(kr) , \tag{3.11}$$

where $R$ is included because $\boldsymbol{L}_{lm}$ and $\boldsymbol{N}_{lm}$ fields are dimensionless, and the eigenfrequency $\omega$ is related to $h$ and $k$ by $\rho\omega^2=(\lambda+2\mu)h^2=\mu k^2$. This mode has a mixed character of the longitudinal and transversal modes. The condition that the stress vanishes on the quantum dot surface $(r=R)$ yields a set of linear equations

$$\begin{pmatrix} \alpha_{lm} & \beta_{lm} \\ \gamma_{lm} & \delta_{lm} \end{pmatrix} \begin{pmatrix} p_{lm} \\ q_{lm} \end{pmatrix} = 0 \ , \tag{3.12}$$

with

$$\alpha_{lm} = -\sigma^2 h R j_l(hR) + 2(l+2)j_{l+1}(hR) \ , \tag{3.13}$$

$$\beta_{lm} = lkR j_l(kR) - 2l(l+2)j_{l+1}(kR) \ , \tag{3.14}$$

$$\gamma_{lm} = -\sigma^2 h R j_l(hR) + 2(l-1)j_{l-1}(hR) \ , \tag{3.15}$$

$$\delta_{lm} = (l+1)[2(l-1)j_{l-1}(kR) - kR j_l(kR)] \ , \tag{3.16}$$

where $\sigma^2=(\lambda+2\mu)/\mu$, and $h$ and $k$ are determined from $\alpha_{lm}\delta_{lm} - \beta_{lm}\gamma_{lm}=0$. In the case of $l=m=0$, $\boldsymbol{N}_{lm}=0$ and the above equation should be replaced by

$$-\sigma^2 h R j_0(hR) + 4j_1(hR) = 0 \ . \tag{3.17}$$

Now, the typical phonon energy spectra of a CdSe nanocrystal with 20 Å radius are shown in Fig. 3.1a and b for the torsional and spheroidal modes, respectively. The material constants employed are [25]

$$\lambda + 2\mu = c_{11} = 7.49 \times 10^{11} \text{ dyn cm}^{-2} \ ,$$
$$\mu = c_{44} = 1.315 \times 10^{11} \text{ dyn cm}^{-2} \ , \text{ and}$$
$$\rho = 5.81 \text{ g cm}^{-3} \ . \tag{3.18}$$

These size-quantized acoustic phonon modes were observed by low-frequency Raman scattering [11].

In the calculation of the electron–phonon interaction, the second-quantized form of the phonon displacement becomes necessary. Here the procedure of the second quantization is described. The torsional and spheroidal modes will be denoted by $\boldsymbol{T}_{lmj}(r)$ and $\boldsymbol{S}_{lmj}(r)$, respectively, where $l$ and $m$ are the angular momentum indices as above and $j$ denotes the radial quantum number. They are normalized within a nanocrystal as

$$\int \mathrm{d}^3 r \ {}^t\boldsymbol{T}_{lmj}(r) \cdot \boldsymbol{T}_{lmj}(r) = \int \mathrm{d}^3 r \ {}^t\boldsymbol{S}_{lmj}(r) \cdot \boldsymbol{S}_{lmj}(r) = 1 \ . \tag{3.19}$$

The displacement vector can be generally expanded as

$$\boldsymbol{u}(r) = \sum_{lmj} \tau_{lmj}(a_{lmj} + (-1)^m a^\dagger_{l-mj})\boldsymbol{T}_{lmj}(r)$$
$$+ \sum_{lmj} \sigma_{lmj}(b_{lmj} + (-1)^m b^\dagger_{l-mj})\boldsymbol{S}_{lmj}(r) \ , \tag{3.20}$$

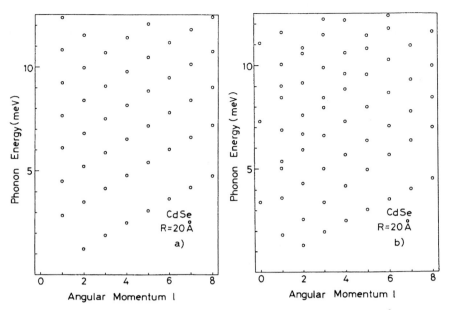

**Fig. 3.1.** The acoustic phonon energies of a CdSe nanocrystal with 20 Å radius are plotted as a function of the angular momentum $l$ for (**a**) torsional modes and (**b**) spheroidal modes (from T. Takagahara [12])

where $a_{lmj}$ and $b_{lmj}$ are the boson annihilation operators of the torsional and spheroidal phonon modes, respectively, and the factor $(-1)^m$ appears due to the property, i.e., $Y_{lm}^* = (-1)^m Y_{l-m}$. The coefficients $\tau_{lmj}$ and $\sigma_{lmj}$ are now fixed. Starting from the Lagrangian density given as

$$\mathcal{L} = \frac{1}{2}\rho(\dot{\boldsymbol{u}})^2 - \frac{1}{2}(\lambda + 2\mu)(\mathrm{div}\boldsymbol{u})^2 - \frac{1}{2}\mu(\mathrm{rot}\boldsymbol{u})^2 \ , \qquad (3.21)$$

and postulating the commutation relation between the displacement vector $\boldsymbol{u}$ and the canonically conjugate momentum vector $\rho\dot{\boldsymbol{u}}$, we find that $\tau_{lmj}$ and $\sigma_{lmj}$ must satisfy

$$2\rho\omega_{lmj}^T \tau_{lmj}^2 = \hbar \quad \text{and} \quad 2\rho\omega_{lmj}^S \sigma_{lmj}^2 = \hbar \ , \qquad (3.22)$$

respectively, where $\omega_{lmj}^T (\omega_{lmj}^S)$ is the eigenfrequency of the torsional (spheroidal) phonon mode. The total phonon Hamiltonian and the second-quantized form of the acoustic phonon displacement can then be written as

$$H = \sum_{lmj} \hbar\omega_{lmj}^T \left( a_{lmj}^\dagger a_{lmj} + \frac{1}{2} \right) + \sum_{lmj} \hbar\omega_{lmj}^S \left( b_{lmj}^\dagger b_{lmj} + \frac{1}{2} \right) \ , \qquad (3.23)$$

and

$$\boldsymbol{u}(r) = \sum_{lmj} \sqrt{\frac{\hbar}{2\rho\omega_{lmj}^T}} \left[ a_{lmj} + (-1)^m a_{l-mj}^\dagger \right] \boldsymbol{T}_{lmj}(r)$$

$$+ \sum_{lmj} \sqrt{\frac{\hbar}{2\rho\omega_{lmj}^S}} \left[ b_{lmj} + (-1)^m b_{l-mj}^\dagger \right] \boldsymbol{S}_{lmj}(r) . \qquad (3.24)$$

### 3.2.2    The Case of Smooth Contact Between a Quantum Dot and the Surrounding Medium

This case was studied in order to discuss the matrix effect on the acoustic lattice vibrations of a nanocrystal [26,27]. But these studies seem to contain erroneous points and will be corrected in this section.

In the case of the torsional modes, the lattice displacement is given as

$$\boldsymbol{u}(r) = \begin{cases} c_< R\, j_\ell(k_< r)\, \boldsymbol{C}_{lm}(\Omega) \cdots r < R\,, \\[2mm] c_> R\, z_\ell(k_> r)\, \boldsymbol{C}_{lm}(\Omega) \cdots r > R\,, \end{cases} \qquad (3.25)$$

where $\rho_{<(>)}\omega^2 = \mu_{<(>)}\, k_{<(>)}^2$, the subscript $<(>)$ is attached to the quantity in the quantum dot (surrounding medium), $\boldsymbol{C}_{lm}(\Omega)$ is defnded in Appendix B, and $z_\ell$ represents symbolically the spherical Bessel function $j_\ell$ or the spherical Neumann function $n_\ell$. The continuity of $\boldsymbol{u}$ and $\sigma e_r$ at the quantum dot surface yields conditions as

$$c_< j_\ell(k_< R) = c_> z_\ell(k_> R) \,, \qquad (3.26)$$

$$c_< \mu_< \left[ k_< R\, j_\ell'(k_< R) - j_\ell(k_< R) \right] = c_> \mu_> \left[ k_> R\, z_\ell'(k_> R) - z_\ell(k_> R) \right] . \qquad (3.27)$$

Then the eigenfrequency is determined from a secular equation:

$$\begin{vmatrix} j_\ell(k_< R) & , & z_\ell(k_> R) \\[2mm] \mu_< \left[ k_< R\, j_\ell'(k_< R) - j_\ell(k_< R) \right] & , & \mu_> \left[ k_> R\, z_\ell'(k_> R) - z_\ell(k_> R) \right] \end{vmatrix} = 0 . \qquad (3.28)$$

We can put the case of the spheroidal modes in the same way as (3.11):

$$\boldsymbol{u}(r) = \begin{cases} a_< R\, \boldsymbol{L}_{lm}^{(j)}(h_< r) + b_< R\, \boldsymbol{N}_{lm}^{(j)}(k_< r) \cdots r < R\,, \\[2mm] a_> R\, \boldsymbol{L}_{lm}^{(z)}(h_> r) + b_> R\, \boldsymbol{N}_{lm}^{(z')}(k_> r) \cdots r > R\,, \end{cases} \qquad (3.29)$$

where $\rho_{<(>)}\omega^2 = (\lambda_{<(>)} + 2\mu_{<(>)})h_{<(>)}^2 = \mu_{<(>)}k_{<(>)}^2$, and the superscript $j$ or $z(z')$ on $\boldsymbol{L}_{lm}$ and $\boldsymbol{N}_{lm}$ fields indicates the spherical Bessel function $j_\ell$ for $r < R$

and $j_\ell$ or $n_\ell$ for $r>R$. Using the decomposition of $\boldsymbol{L}_{lm}$ and $\boldsymbol{N}_{lm}$ in terms of $\boldsymbol{P}_{lm}$ and $\boldsymbol{B}_{lm}$ fields given in Appendix B, $\boldsymbol{u}$ can be written as

$$
\boldsymbol{u}(r) = 
\begin{cases}
R\left[a_< j'_\ell(h_< r) + b_< \frac{\ell(\ell+1)}{k_< r} j_\ell(k_< r)\right]\boldsymbol{P}_{lm}(\Omega) \\
\quad +\sqrt{\ell(\ell+1)}R\left(a_< \frac{j_\ell(h_< r)}{h_< r} + b_< \frac{[k_< r j_\ell(k_< r)]'}{k_< r}\right)\boldsymbol{B}_{lm}(\Omega) \cdots r < R\,, \\
R\left[a_> z_\ell^{(L)'}(h_> r) + b_> \frac{\ell(\ell+1)}{k_> r} z_\ell^{(N)}(k_> r)\right]\boldsymbol{P}_{lm}(\Omega) \\
\quad +\sqrt{\ell(\ell+1)}R\left(a_> \frac{z_\ell^{(L)}(h_> r)}{h_> r} + b_> \frac{[k_> r z_\ell^{(N)}(k_> r)]'}{k_> r}\right)\boldsymbol{B}_{lm}(\Omega) \cdots r > R\,,
\end{cases}
$$

$$(3.30)$$

where $z_\ell^{(L)}(z_\ell^{(N)})$ represents the $j_\ell$ or $n_\ell$ function associated with the $\boldsymbol{L}_{lm}(\boldsymbol{N}_{lm})$ field outside the quantum dot and the prime indicates, e.g., $(zj_\ell(z))'=d/dz(zj_\ell(z))$. Then the continuity condition of $\boldsymbol{u}$ at the surface $(r=R)$ can be written as

$$
a_< j'_\ell(h_< R) + b_< \frac{\ell(\ell+1)}{k_< R} j_\ell(k_< R) = a_> z_\ell^{(L)'}(h_> R)
$$
$$
+ b_> \frac{\ell(\ell+1)}{k_> R} z_\ell^{(N)}(k_> R)\,,
$$

$$(3.31)$$

$$
a_< \frac{j_\ell(h_< R)}{h_< R} + b_< \frac{(k_< R\, j_\ell(k_< R))'}{k_< R} = a_> \frac{z_\ell^{(L)}(h_> R)}{h_> R}
$$
$$
+ b_> \frac{(k_> R\, z_\ell^{(N)}(k_> R))'}{k_> R}\,.
$$

$$(3.32)$$

In the calculation of the stress tensor in (3.2) the following expressions are necessary:

$$
\mathrm{div}\,\boldsymbol{u}(r) = 
\begin{cases}
-a_< h_< R\, j_\ell(h_< r)Y_{lm}(\Omega) \quad \cdots r < R\,, \\
-a_> h_> R\, z_\ell^{(L)}(h_> r)Y_{lm}(\Omega) \cdots r > R\,,
\end{cases}
$$

$$(3.33)$$

$$
\boldsymbol{r}\cdot\boldsymbol{u} = 
\begin{cases}
R\left[a_< r\, j'_\ell(h_< r) + b_< \frac{\ell(\ell+1)}{k_<} j_\ell(k_< r)\right]Y_{lm}(\Omega) \quad \cdots r < R\,, \\
R\left[a_> r\, z_\ell^{(L)'}(h_> r) + b_> \frac{\ell(\ell+1)}{k_>} z_\ell^{(N)}(k_> r)\right]Y_{lm}(\Omega) \cdots r > R\,.
\end{cases}
$$

$$(3.34)$$

Then the continuity condition of the stress at $r=R$ can be written as

$$
[-\lambda_< h_< R\, j_\ell(h_< R) + 2\mu_< G_1(h_< R; j_\ell)]a_< + 2\mu_< \ell(\ell+1)G_2\,(k_< R; j_\ell)\,b_<
$$
$$
= \left[-\lambda_> h_> R\, z_\ell^{(L)}(h_> R) + 2\mu_> G_1\left(h_> R; z_\ell^{(L)}\right)\right]a_>
$$
$$
+ 2\mu_> \ell(\ell+1)G_2\left(k_> R; z_\ell^{(N)}\right)b_>\,,
$$

$$(3.35)$$

$$2\mu_< G_2(h_< R; j_\ell)a_< + \mu_< G_3(k_< R; j_\ell)b_<$$
$$= 2\mu_> G_2\left(h_> R; z_\ell^{(L)}\right)a_> + \mu_> G_3\left(k_> R; z_\ell^{(N)}\right)b_> \ , \tag{3.36}$$

with

$$G_1(x; z_\ell) = 2z_{\ell+1}(x) - \left[x - \frac{\ell(\ell-1)}{x}\right]z_\ell(x) \ , \tag{3.37}$$

$$G_2(x; z_\ell) = \frac{\ell-1}{x}z_\ell(x) - z_{\ell+1}(x) \ , \tag{3.38}$$

$$G_3(x; z_\ell) = 2z_{\ell+1}(x) - \left[x - \frac{2(\ell^2-1)}{x}\right]z_\ell(x) \ , \tag{3.39}$$

where $z_\ell$ represents the $j_\ell$ or $n_\ell$ function. Four equations (3.31), (3.32), (3.35) and (3.36) form a set of linear relations among $a_<$, $b_<$, $a_>$, and $b_>$. The eigenfrequency $\omega$ is determined by a 4×4 secular equation given by

$$\begin{vmatrix} a_{11} & a_{12} & a_{13} & a_{14} \\ a_{21} & a_{22} & a_{23} & a_{24} \\ a_{31} & a_{32} & a_{33} & a_{34} \\ a_{41} & a_{42} & a_{43} & a_{44} \end{vmatrix} = 0 \ , \tag{3.40}$$

with

$$a_{11} = j_\ell'(h_< R) \ , \quad a_{12} = \frac{\ell(\ell+1)}{k_< R}j_\ell(k_< R) \ , \quad a_{13} = z_\ell^{(L)'}(h_> R) \ ,$$

$$a_{14} = \frac{\ell(\ell+1)}{k_> R}z_\ell^{(N)}(k_> R) \ ,$$

$$a_{21} = \frac{j_\ell(h_< R)}{h_< R} \ , \quad a_{22} = \frac{[k_< R\, j_\ell(k_< R)]'}{k_< R} \ , a_{23} = \frac{z_\ell^{(L)}(h_> R)}{h_> R} \ ,$$

$$a_{24} = \frac{\left[k_> R\, z_\ell^{(N)}(k_> R)\right]'}{k_> R} \ ,$$

$$a_{31} = -\lambda_< h_< R\, j_\ell(h_< R) + 2\mu_< G_1(h_< R; j_\ell) \ ,$$

$$a_{32} = 2\mu_< \ell(\ell+1)G_2(k_< R; j_\ell) \ ,$$

$$a_{33} = -\lambda_> h_> R\, z_\ell^{(L)}(h_> R) + 2\mu_> G_1\left(h_> R; z_\ell^{(L)}\right) \ ,$$

$$a_{34} = 2\mu_> \ell(\ell+1)G_2\left(k_> R; z_\ell^{(N)}\right) \ ,$$

$$a_{41} = 2\mu_< G_2(h_< R; j_\ell) \ , \quad a_{42} = \mu_< G_3(k_< R; j_\ell) \ ,$$

$$a_{43} = 2\mu_> G_2\left(h_> R; z_\ell^{(L)}\right) \ , \quad a_{44} = \mu_> G_3\left(k_> R; z_\ell^{(N)}\right) \ . \tag{3.41}$$

It should be noted that there are four cases corresponding to the combination of $z_\ell^{(L)} = j_\ell$ or $n_\ell$ and $z_\ell^{(N)} = j_\ell$ or $n_\ell$. The second-quantization of the phonon modes proceeds in the same way as in the case of the stress-free boundary condition. The lattice displacement $\boldsymbol{u}$ in the surrounding medium is extended and behaves asymptotically as $1/r$ when $r \longrightarrow \infty$. This behavior corresponds to a plane-wave eigenmode in the bulk medium. In some cases, however, this kind of extended phonon modes would not be realized due to the presence of various inhomogeneities at the interface between a quantum dot and the surrounding medium. Also, the effect of neighboring dots would not be negligible. Thus the nature of acoustic phonon modes of a nanocrystal would be affected significantly by the surroundings.

Now let us survey some recent experimental reports. For CdSe and CdS nanocrystals embedded in glasses, the low-frequency Raman scattering was studied [28,29]. The size dependence of energies of quantized acoustic phonon modes was discussed in comparison of various boundary conditions. On the other hand, for CdS nanocrystals embedded in a glass matrix, the results of low-frequency Raman scattering and spin-flip Raman scattering were interpreted in terms of the interaction of the size-quantized exciton states with nonconfined acoustic phonon modes [30] since the sound velocities in a glass matrix and the crystalline CdS are similar. For CuCl nanocrystals embedded in a glass matrix and in a NaCl crystal, confined acoustic phonons were observed in the hole-burning spectra, which are discussed in detail in Chap. 5. Very recently, low-frequency Raman spectroscopy has been carried out on Si nanocrystals embedded in a $SiO_2$ matrix. It was demonstrated that the stress-free boundary condition is appropriate for describing the confined acoustic phonon modes and that the matrix effects are negligible [31]. Thus the confinement of the acoustic phonon modes is highly dependent on the combination of the quantum dot material and the embedding medium.

## 3.3 Derivation of the Electron–Acoustic-Phonon Interactions

The electron–phonon interaction with acoustic phonon modes arises mainly through deformation-potential coupling and piezoelectric coupling. In this section their expressions will be derived.

When the crystal lattice deforms, the electronic energy structure changes in proportion to the strain tensor and the proportionality constant is called a deformation potential. Although the detailed form of the deformation potential coupling is dependent on the crystal symmetry [32], the most dominant term can be described by $E_d \mathrm{div} \boldsymbol{u}$, where $E_d$ is a deformation potential constant. Hereafter only this term will be taken into account. Within this simplification, the torsional modes do not contribute to this coupling because of their transversal character. In the case of a spheroidal mode, only the longitudinal component contributes to the deformation potential coupling, and we obtain

$$
\begin{aligned}
E_d \mathrm{div} \boldsymbol{S}_{lmj} &= E_d \mathrm{div} \left[ p_{lmj} \boldsymbol{L}_{lm}(h_j r) + q_{lmj} \boldsymbol{N}_{lm}(k_j r) \right] \\
&= -E_d h_j p_{lmj} j_l(h_j r) Y_{lm}(\Omega) \ .
\end{aligned}
\tag{3.42}
$$

This interaction is of the form of a contact coupling between an electron and the lattice displacement; the relevant Hamiltonian is given by (3.42) only by substituting the electron coordinates. The exciton–phonon interaction Hamiltonian is composed of the interaction terms of the conduction band electron and of the valence band hole. The deformation potentials in a nanocrystal might not be very much different from those in bulk materials as long as the nanocrystal is not in the regime of small clusters.

In polar semiconductors, the lattice strain produces the lattice polarization, and this polarization interacts with an electron. The lattice polarization in Cartesian coordinates is given as

$$\boldsymbol{P}_{\mathrm{PZ}}(r) = {}^{t}(e_{15}e_{xz}, e_{15}e_{yz}, e_{31}(e_{xx} + e_{yy}) + e_{33}e_{zz}) , \tag{3.43}$$

for the wurtzite structure, and as

$$\boldsymbol{P}_{\mathrm{PZ}}(r) = e_{14} \, {}^{t}(e_{yz}, e_{zx}, e_{xy}) , \tag{3.44}$$

for the zincblende structure, respectively, where $e_{ij}$ is the strain tensor defined by

$$e_{xx} = \frac{\partial u_x}{\partial x} , \quad e_{xy} = \frac{\partial u_x}{\partial y} + \frac{\partial u_y}{\partial x} \tag{3.45}$$

and by their cyclic permutations and $e_{15}, e_{31}, e_{33}$, and $e_{14}$ are the piezoelectric constants [33]. The explicit expressions of these tensor components in spherical coordinates are given in Appendix A. The electron–lattice interaction is given by the potential energy of the lattice polarization in the electric field induced by an electron, and is represented as

$$-\frac{e}{\epsilon_\infty} \int \mathrm{d}^3 r \nabla_r \left( \frac{1}{|\boldsymbol{r} - \boldsymbol{r}_\mathrm{e}|} \right) \cdot \boldsymbol{P}_{\mathrm{PZ}}(r) , \tag{3.46}$$

where $\epsilon_\infty$ is the high-frequency dielectric constant. The integral is over a quantum dot (the whole space) in case of Sect. 3.2.1 (3.2.2). In the calculation of this integral, it is convenient to employ the expression

$$
\nabla_r \left( \frac{1}{|\boldsymbol{r} - \boldsymbol{r}_\mathrm{e}|} \right) = \sum_{lm} \frac{4\pi}{2l + 1} \times \left[ \frac{\mathrm{d}}{\mathrm{d}r} \left( \frac{r_<^l}{r_>^{l+1}} \right) \boldsymbol{P}_{lm}^*(\Omega) Y_{lm}(\Omega_\mathrm{e}) \right.
$$
$$
\left. + \frac{\sqrt{l(l+1)}}{r} \frac{r_<^l}{r_>^{l+1}} \boldsymbol{B}_{lm}^*(\Omega) Y_{lm}(\Omega_\mathrm{e}) \right] , \tag{3.47}
$$

where $r_> = \max(r, r_\mathrm{e})$, $r_< = \min(r, r_\mathrm{e})$, and $\boldsymbol{P}_{lm}$ and $\boldsymbol{B}_{lm}$ are the vector spherical harmonics as given in Appendix B.

Substituting (3.47) into (3.46) and carrying out the integration, we obtain the electron–phonon interaction Hamiltonian as

$$\sum_{l_\mathrm{e}, m_\mathrm{e}, l, m, j} \mathcal{P}_{l_\mathrm{e}, m_\mathrm{e}, l, m, j}(r_\mathrm{e}) Y_{l_\mathrm{e}, m_\mathrm{e}}(\Omega_\mathrm{e}) \left( b_{l,m,j} + (-1)^m b_{l,-m,j}^\dagger \right) , \tag{3.48}$$

and

$$\sum_j \sigma_{ij} \, n_j = 2\rho\beta_{\mathrm{T}}^2 \begin{pmatrix} \epsilon_{rr} \\ \epsilon_{\theta r} \\ \epsilon_{\varphi r} \end{pmatrix}$$

$$= 2\rho\beta_{\mathrm{T}}^2 \, r \frac{\mathrm{d}}{\mathrm{d}r} \left( \frac{g(r)}{r} \right) \, \sqrt{\ell(\ell+1)} \, \boldsymbol{C}_{lm}(\Omega) \, , \tag{3.82}$$

where $g(r) = c_< j_\ell(Q_< r)$ or $c_> z_\ell(Q_> r)$, and the expressions in Appendix A are used. Then the condition of (3.74) leads to

$$c_< \, \rho_< \beta_{\mathrm{T}<}^2 \, \frac{\mathrm{d}}{\mathrm{d}r} \left( \frac{j_\ell(Q_< r)}{r} \right) \Bigg|_{r=R} = c_> \, \rho_> \beta_{\mathrm{T}>}^2 \, \frac{\mathrm{d}}{\mathrm{d}r} \left( \frac{z_\ell(Q_> r)}{r} \right) \Bigg|_{r=R} . \tag{3.83}$$

Finally, the secular equation

$$\begin{vmatrix} j_\ell(Q_< R) & , & z_\ell(Q_> R) \\ \rho_< \beta_{\mathrm{T}<}^2 \, \frac{\mathrm{d}}{\mathrm{d}r} \left( \frac{j_\ell(Q_< r)}{r} \right) \Big|_{r=R} & , & \rho_> \beta_{\mathrm{T}>}^2 \, \frac{\mathrm{d}}{\mathrm{d}r} \left( \frac{z_\ell(Q_> r)}{r} \right) \Big|_{r=R} \end{vmatrix} = 0 \tag{3.84}$$

determines the eigenfrequency $\omega$.

In the case of a mixed mode of the longitudinal and transverse characters, we can assume the displacement vector $\boldsymbol{u}$ within the quantum dot as

$$\boldsymbol{u}_<(r) = \nabla\Psi_<(r) + c_{1<} R\boldsymbol{N}_{lm}(Q_< r) \, , \tag{3.85}$$

where $\Psi_<(r)$ has the dimension of area. Now taking into account (3.76) and noting that div $\boldsymbol{u}_< = \nabla^2\Psi_<(r)$, we can postulate as

$$\Psi_<(r) = \left[ c_{2<} \, R^2 \, j_\ell(q_< r) + c_{3<} \, R^2 \left( \frac{r}{R} \right)^\ell \right] Y_{lm}(\Omega) \, , \tag{3.86}$$

$$\phi_<(r) = \frac{4\pi\alpha_<}{\epsilon_{\infty<}} \left[ c_{2<} \, R^2 \, j_\ell(q_< r) + c_{4<} \, R^2 \left( \frac{r}{R} \right)^\ell \right] Y_{lm}(\Omega) \, , \tag{3.87}$$

where $q_<$ is defined by the quantity in (3.76) for the quantum dot material, and we find that the condition

$$c_{3<} \, \rho_< \, (\omega^2 - \omega_{\mathrm{T}<}^2) = \frac{4\pi\alpha_<^2}{\epsilon_{\infty<}} c_{4<} \tag{3.88}$$

is necessary to satisfy (3.67). In the same way, the phonon displacement vector outside the quantum dot is given by

$$\boldsymbol{u}_>(r) = \nabla\Psi_>(r) + c_{1>} R\boldsymbol{N}_{lm}(Q_> r) \, , \tag{3.89}$$

with

$$\Psi_>(r) = \left[ c_{2>} \, R^2 \, z_\ell(q_> r) + c_{3>} \, R^2 \left( \frac{R}{r} \right)^{\ell+1} \right] Y_{lm}(\Omega) \, , \tag{3.90}$$

$$\phi_>(r) = \frac{4\pi\alpha_>}{\epsilon_{\infty>}} \left[ c_{2>} \ R^2 \ z_\ell(q_>r) + c_{4>} \ R^2 \left(\frac{R}{r}\right)^{\ell+1} \right] Y_{lm}(\Omega) , \qquad (3.91)$$

$$c_{3>} \ \rho_> \ \left(\omega^2 - \omega_{T>}^2\right) = \frac{4\pi\alpha_>^2}{\epsilon_{\infty>}} c_{4>} , \qquad (3.92)$$

where $z_\ell(z)$ represents symbolically the $j_\ell(z)$ or $n_\ell(z)$ function. For the sake of later use we write down the expression of $\boldsymbol{u}$:

$$\begin{aligned}
\boldsymbol{u}_<(r) = \ & c_{1<}R\bigg\{ \frac{\ell(\ell+1)}{Q_<r} j_\ell(Q_<r)\boldsymbol{P}_{lm}(\Omega) \\
& + \frac{\sqrt{\ell(\ell+1)}}{Q_<r} [Q_<rj_\ell(Q_<r)]' \, \boldsymbol{B}_{lm}(\Omega) \bigg\} \\
& + R^2 \left[ c_{2<}q_< j_\ell'(q_<r) + c_{3<}\ell\frac{r^{\ell-1}}{R^\ell} \right] \boldsymbol{P}_{lm}(\Omega) \\
& + R^2 \left[ c_{2<} \ j_\ell(q_<r) + c_{3<} \left(\frac{r}{R}\right)^\ell \right] \frac{\sqrt{\ell(\ell+1)}}{r} \boldsymbol{B}_{lm}(\Omega) , \qquad (3.93) \\
\boldsymbol{u}_>(r) = \ & c_{1>}R\bigg\{ \frac{\ell(\ell+1)}{Q_>r} z_\ell^{(N)}(Q_>r)\boldsymbol{P}_{lm}(\Omega) \\
& + \frac{\sqrt{\ell(\ell+1)}}{Q_>r} \left[Q_>rz_\ell^{(N)}(Q_>r)\right]' \boldsymbol{B}_{lm}(\Omega) \bigg\} \\
& + R^2 \left[ c_{2>}q_> z_\ell'(q_>r) - c_{3>}(\ell+1)\frac{R^{\ell+1}}{r^{\ell+2}} \right] \boldsymbol{P}_{lm}(\Omega) \\
& + R^2 \left[ c_{2>}z_\ell(q_>r) + c_{3>} \left(\frac{R}{r}\right)^{\ell+1} \right] \frac{\sqrt{\ell(\ell+1)}}{r} \boldsymbol{B}_{lm}(\Omega) , \qquad (3.94)
\end{aligned}$$

where $\boldsymbol{P}_{lm}$ and $\boldsymbol{B}_{lm}$ are defined in Appendix B, and $z_\ell^{(N)}=j_\ell$ or $n_\ell$, but is not always the same as $z_\ell$ in the third and fourth lines of (3.94). Then the continuity of $\boldsymbol{u}$ and $\phi$ at the boundary $(r=R)$ yields the conditions:

$$\begin{aligned}
& c_{1<}\frac{\ell(\ell+1)}{Q_<} j_\ell(Q_<R) + R^2 \left[ c_{2<}q_< j_\ell'(q_<R) + c_{3<}\frac{\ell}{R} \right] \\
& = c_{1>}\frac{\ell(\ell+1)}{Q_>} z_\ell^{(N)}(Q_>R) + R^2 \left[ c_{2>}q_> z_\ell'(q_>R) - c_{3>}\frac{\ell+1}{R} \right] , \qquad (3.95)
\end{aligned}$$

$$\begin{aligned}
& \frac{c_{1<}}{Q_<} [Q_<Rj_\ell(Q_<R)]' + R [c_{2<}j_\ell(q_<R) + c_{3<}] \\
& = \frac{c_{1>}}{Q_>} \left[Q_>Rz_\ell^{(N)}(Q_>R)\right]' + R [c_{2>}z_\ell(q_>R) + c_{3>}] , \qquad (3.96)
\end{aligned}$$

$$\frac{\alpha_<}{\epsilon_{\infty<}} [c_{2<}j_\ell(q_<R) + c_{4<}] = \frac{\alpha_>}{\epsilon_{\infty>}} [c_{2>}z_\ell(q_>R) + c_{4>}] . \qquad (3.97)$$

The continuity condition (3.73) in which the component of $\boldsymbol{D}$ proportional to $\boldsymbol{P}_{lm}$ is relevant leads to

$$\alpha_< \left[ c_{1<} \frac{\ell(\ell+1)}{Q_<} j_\ell(Q_< R) + R\ell(c_{3<} - c_{4<}) \right]$$
$$= \alpha_> \left[ c_{1>} \frac{\ell(\ell+1)}{Q_>} z_\ell^{(N)}(Q_> R) - R(\ell+1)(c_{3>} - c_{4>}) \right] . \tag{3.98}$$

Now the continuity condition (3.74) of the stress will be discussed in detail. The stress tensor is given by (3.63) and the vector $\boldsymbol{n}$ is the radial unit vector. Then, in spherical coordinates, we have

$$\sigma \boldsymbol{n} = \rho \left( \beta_{\mathrm{L}}^2 - 2\beta_{\mathrm{T}}^2 \right) \mathrm{div} \boldsymbol{u} \begin{pmatrix} 1 \\ 0 \\ 0 \end{pmatrix} + 2\rho \, \beta_{\mathrm{T}}^2 \begin{pmatrix} \epsilon_{rr} \\ \epsilon_{\theta r} \\ \epsilon_{\varphi r} \end{pmatrix} . \tag{3.99}$$

From Appendix A we see that

$$\epsilon_{rr} = \frac{\partial u_r}{\partial r} , \quad \epsilon_{\theta r} = \frac{1}{2} \left[ \frac{1}{r} \frac{\partial u_r}{\partial \theta} + r \frac{\partial}{\partial r} \left( \frac{u_\theta}{r} \right) \right] ,$$
$$\epsilon_{\varphi r} = \frac{1}{2} \left[ \frac{1}{r \sin\theta} \frac{\partial u_r}{\partial \varphi} + r \frac{\partial}{\partial r} \left( \frac{u_\varphi}{r} \right) \right] . \tag{3.100}$$

The spherical components of $\boldsymbol{u}$ in (3.93) and (3.94) can be written in the form

$$u_r = f(r) Y_{lm}(\Omega) , \quad u_\theta = g(r) \frac{\partial}{\partial \theta} Y_{lm}(\Omega) , \quad u_\varphi = g(r) \frac{1}{\sin\theta} \frac{\partial}{\partial \varphi} Y_{lm}(\Omega) . \tag{3.101}$$

Thus we find

$$\begin{pmatrix} \epsilon_{rr} \\ \epsilon_{\theta r} \\ \epsilon_{\varphi r} \end{pmatrix} = \frac{\mathrm{d}f(r)}{\mathrm{d}r} \boldsymbol{P}_{lm}(\Omega) + \frac{\sqrt{\ell(\ell+1)}}{2} \left[ \frac{f(r)}{r} + r \frac{\mathrm{d}}{\mathrm{d}r} \left( \frac{g(r)}{r} \right) \right] \boldsymbol{B}_{lm}(\Omega) . \tag{3.102}$$

The explicit expressions of $f_<(r)(f_>(r))$ and $g_<(r)(g_>(r))$ for the region of $r<R$ $(r>R)$ are given as

$$f_<(r) = c_{1<} R \frac{\ell(\ell+1)}{Q_< r} j_\ell(Q_< r) + R^2 \left[ c_{2<} q_< j_\ell'(q_< r) + c_{3<} \ell \frac{r^{\ell-1}}{R^\ell} \right] , \tag{3.103}$$

$$g_<(r) = c_{1<} R \frac{1}{Q_< r} \left[ Q_< r j_\ell(Q_< r) \right]' + \frac{R^2}{r} \left[ c_{2<} j_\ell(q_< r) + c_{3<} \left( \frac{r}{R} \right)^\ell \right] , \tag{3.104}$$

$$f_>(r) = c_{1>} R \frac{\ell(\ell+1)}{Q_> r} z_\ell^{(N)}(Q_> r) + R^2 \left[ c_{2>} q_> z_\ell'(q_> r) - c_{3>} (\ell+1) \frac{R^{\ell+1}}{r^{\ell+2}} \right],$$

(3.105)

$$g_>(r) = c_{1>} R \frac{1}{Q_> r} \left[ Q_> r z_\ell^{(N)}(Q_> r) \right]' + \frac{R^2}{r} \left[ c_{2>} z_\ell(q_> r) + c_{3>} \left( \frac{R}{r} \right)^{\ell+1} \right].$$

(3.106)

Then condition (3.74) leads to

$$-\rho_< \left( \beta_{L<}^2 - 2\beta_{T<}^2 \right) c_{2<} (q_< R)^2 j_\ell(q_< R) + 2\rho_< \beta_{T<}^2 \left. \frac{df_<(r)}{dr} \right|_{r=R}$$

$$= -\rho_> \left( \beta_{L>}^2 - 2\beta_{T>}^2 \right) c_{2>} (q_> R)^2 z_\ell(q_> R) + 2\rho_> \beta_{T>}^2 \left. \frac{df_>(r)}{dr} \right|_{r=R},$$

(3.107)

$$\rho_< \beta_{T<}^2 \left[ \frac{f_<(R)}{R} + r \frac{d}{dr} \left( \frac{g_<(r)}{r} \right) \Big|_{r=R} \right]$$

$$= \rho_> \beta_{T>}^2 \left[ \frac{f_>(R)}{R} + r \frac{d}{dr} \left( \frac{g_>(r)}{r} \right) \Big|_{r=R} \right].$$

(3.108)

Finally, we obtain six conditions: (3.95)–(3.98), (3.107), and (3.108) for the six parameters, $c_{1<}$, $c_{2<}$, $c_{3<}$, $c_{1>}$, $c_{2>}$, and $c_{3>}$; the secular equation for this $6\times6$ matrix determines the eigenvalues.

From the Hermiticity of the equations of motion (3.64) and (3.65), we can prove the orthogonality of eigenfunctions:

$$\int d^3 r \, \rho(r) \boldsymbol{u}_\lambda^*(r) \cdot \boldsymbol{u}_\mu(r) = \Omega_0 \, \delta_{\lambda \mu},$$

(3.109)

where $\lambda$ and $\mu$ represent symbolically the index of the radial eigenstate and the angular momentum indices $\ell$ and $m$, and we can confirm the completeness of the eigenfunctions:

$$\sum_\lambda \rho(r) \boldsymbol{u}_\lambda^*(r) \cdot \boldsymbol{u}_\lambda(r') = \Omega_0 \, \mathbf{1} \, \delta^{(3)}(r - r'),$$

(3.110)

where $\mathbf{1}$ is a unit matrix. The quantization of the polar-optical vibrations proceeds as follows. Since we postulated the time dependence as in (3.66), we have

$$\boldsymbol{u}(r,t) = \sum_\lambda \left[ \tilde{c}_\lambda \, \boldsymbol{u}_\lambda(r) \, e^{-i\omega_\lambda t} + c.c. \right],$$

(3.111)

$$\phi(r,t) = \sum_\lambda \left[ \tilde{c}_\lambda \, \phi_\lambda(r) \, e^{-i\omega_\lambda t} + c.c. \right],$$

(3.112)

where the summation over the eigenfunctions $(\boldsymbol{u}_\lambda, \phi_\lambda)$ is carried out. The variable $\boldsymbol{\Pi}$ canonically conjugate to $\boldsymbol{u}$ is given by

$$\boldsymbol{\Pi} = \frac{\partial \mathcal{L}}{\partial \dot{\boldsymbol{u}}} = \rho \dot{\boldsymbol{u}} = \sum_\lambda \left[ \tilde{c}_\lambda \, (-\mathrm{i}) \omega_\lambda \, \rho(r) \, \boldsymbol{u}_\lambda(r) \, \mathrm{e}^{-\mathrm{i}\omega_\lambda t} + \mathrm{c.c.} \right] . \tag{3.113}$$

Introducing the annihilation operator $b_\lambda$ and the creation operator $b_\lambda^\dagger$ corresponding to

$$\mathrm{e}^{-\mathrm{i}\omega_\lambda t} \longrightarrow b_\lambda , \qquad \mathrm{e}^{\mathrm{i}\omega_\lambda t} \longrightarrow b_\lambda^\dagger , \tag{3.114}$$

we have

$$\boldsymbol{u}(r,t) = \sum_\lambda \left[ c_\lambda \, \boldsymbol{u}_\lambda(r) b_\lambda + c_\lambda^* \, \boldsymbol{u}_\lambda^*(r) b_\lambda^\dagger \right] , \tag{3.115}$$

$$\phi(r,t) = \sum_\lambda \left[ c_\lambda \, \phi_\lambda(r) b_\lambda + c_\lambda^* \, \phi_\lambda^*(r) b_\lambda^\dagger \right] , \tag{3.116}$$

$$\boldsymbol{\Pi}(r) = \sum_\lambda \left[ c_\lambda \, (-\mathrm{i}) \omega_\lambda \, \rho(r) \, \boldsymbol{u}_\lambda(r) \, b_\lambda + c_\lambda^* \, \mathrm{i}\omega_\lambda \, \rho(r) \, \boldsymbol{u}_\lambda^*(r) \, b_\lambda^\dagger \right] . \tag{3.117}$$

Now by postulating the commutation relation

$$[\boldsymbol{u}(r), \boldsymbol{\Pi}(r')] = \mathrm{i}\hbar \, \mathbf{1} \, \delta^{(3)}(r - r') , \tag{3.118}$$

we can fix $c_\lambda$ as

$$c_\lambda = \sqrt{\frac{\hbar}{2\omega_\lambda \Omega_0}} , \tag{3.119}$$

and we have

$$\boldsymbol{u}(r) = \sum_\lambda \sqrt{\frac{\hbar}{2\omega_\lambda \Omega_0}} \left[ \boldsymbol{u}_\lambda(r) b_\lambda + \boldsymbol{u}_\lambda^*(r) b_\lambda^\dagger \right] , \tag{3.120}$$

$$\phi(r) = \sum_\lambda \sqrt{\frac{\hbar}{2\omega_\lambda \Omega_0}} \left[ \phi_\lambda(r) b_\lambda + \phi_\lambda^*(r) b_\lambda^\dagger \right] . \tag{3.121}$$

The electron–phonon interaction is derived from the potential energy for a single electron induced by the polar-optical vibrations and is given by

$$H_{\mathrm{ep}} = -e\phi(r) = -e \sum_\lambda \sqrt{\frac{\hbar}{2\omega_\lambda \Omega_0}} \left[ \phi_\lambda(r) b_\lambda + \phi_\lambda^*(r) b_\lambda^\dagger \right] . \tag{3.122}$$

Alternatively, this interaction can be derived from the potential energy for the lattice polarization within the longitudinal electric field induced by a single electron at $r_{\mathrm{e}}$, namely,

$$-e \int \mathrm{d}^3 r \, \nabla \frac{1}{|\boldsymbol{r} - \boldsymbol{r}_{\mathrm{e}}|} \, \boldsymbol{P}(r) , \tag{3.123}$$

with

$$\boldsymbol{P} = \alpha \boldsymbol{u} + \frac{\epsilon_\infty - 1}{4\pi} \boldsymbol{E} \ . \tag{3.124}$$

Using the relation

$$\mathrm{div}\boldsymbol{P} = \frac{1}{4\pi} \nabla^2 \phi \ , \tag{3.125}$$

we can reduce (3.123) as

$$\frac{e}{4\pi} \int \mathrm{d}^3 r \frac{1}{|r - r_\mathrm{e}|} \nabla^2 \phi(r) = \frac{e}{4\pi} \int \mathrm{d}^3 r \nabla^2 \frac{1}{|r - r_\mathrm{e}|} \phi(r) = -e\phi(r_\mathrm{e}) \ . \tag{3.126}$$

The most general formulation has been given in the above. Usually a simple boundary condition, i.e., $\boldsymbol{u}=0$ at the interface is imposed, and an expression of the electron-phonon interaction is derived [34]. It is, however, to be noted that this simple boundary condition is not compatible with the continuity condition (3.74) concerning the stress because $\sigma\boldsymbol{n}$ does not vanish at the interface inside the quantum dot, whereas $\sigma\boldsymbol{n}=0$ in the surrounding medium. Based on these formulations, the selection rules of Raman scattering and hyper-Raman scattering were discussed [35–37].

A more simplified formulation was presented in [7]. In this formulation, the lattice displacement vector is represented by

$$\boldsymbol{u}(r) \propto \nabla[j_l(k_j^l r) Y_{lm}(\Omega)] \ , \tag{3.127}$$

where $k_j^l$ is fixed by $j_l(k_j^l R)=0$. Then the Fröhlich interaction in (3.122) can be written as

$$\sqrt{\frac{4\pi e^2 \hbar \omega_\mathrm{LO}}{R} \left(\frac{1}{\epsilon_\infty} - \frac{1}{\epsilon_0}\right)} \frac{1}{k_j^l R |j_{l+1}(k_j^l R)|} j_l\left(k_j^l r_\mathrm{e}\right) Y_{lm}\left(\Omega_\mathrm{e}\right) \ , \quad (l \neq 0) \tag{3.128}$$

and

$$\sqrt{\frac{4\pi e^2 \hbar \omega_\mathrm{LO}}{R} \left(\frac{1}{\epsilon_\infty} - \frac{1}{\epsilon_0}\right)} j_0\left(k_j^0 r_\mathrm{e}\right) Y_{00}\left(\Omega_\mathrm{e}\right) \ , \quad (l = 0) \tag{3.129}$$

where $\hbar\omega_\mathrm{LO}$ is the LO phonon energy and the spatial dispersion of the mode is neglected. Thus the coupling strength is proportional to $1/\sqrt{R}$. This size dependence can be confirmed also in (3.122) for the most general formulation.

The intriguing and controversial problem concerning the exciton–polar-optical-phonon interaction has been the unexpectedly large value of the Huang–Rhys factor $S$ observed for various quantum dots, which was estimated from the intensity ratio among the LO phonon sidebands in the photoluminescence excitation spectra and in the Raman spectra. The theoretically estimated Huang–Rhys factor $S$ for the polar optical phonons is very small in the strong-confinement regime for excitons [38]. This small value can be traced back to the expression of the matrix element of the exciton–phonon coupling, which is proportional to

$$\int d^3 r_e \int d^3 r_h \, F_f^* (r_e, r_h) \, [\phi(r_e) - \phi(r_h)] \, F_i(r_e, r_h) \,, \tag{3.130}$$

where $\phi(r)$ is given in (3.122), and $F_f (F_i)$ is the envelope function of the final (initial) exciton state. When the envelope functions of the electron and the hole are similar to each other, cancellation occurs in (3.130) and leads to a small value of $S$. Thus in the strong-confinement regime, which is realized in quantum dots of materials having a large exciton Bohr radius in the bulk, a small value of $S$ is expected. However, the experimentally observed values are about 0.5–1.0 for, e.g., CdSe nanocrystals doped in a glass matrix [7,39]. This discrepancy can be reconciled by taking into account the deformation of the envelope functions by an extra charge within and/or outside a quantum dot [10]. Recently, for an InAs/GaAs quantum dot, the same discrepancy was explained in terms of the spatial separation of the electron and the hole wavefunctions due to the piezoelectric potential induced by the shear strains [40,41].

## 3.5   A Formal Theory on the Exciton–Phonon System Within the Franck–Condon Approximation

Now that the electron–phonon interactions in semiconductor nanocrystals are derived, we can calculate various quantities, e.g., the excitonic dephasing constant. The excitonic dephasing rate is the decay rate of the excitonic polarization and, in general, consists of the pure (adiabatic) dephasing constant and a half of the longitudinal decay constant. The former part arises from the fluctuation of the excitonic energy due to the virtual emission and absorption of phonons without change of the excitonic state, whereas the latter part comes from the phonon-assisted transitions of the relevant excitonic state to other excitonic states. When specified to a two-level model consisting of the electronic ground state $|g\rangle$ and the lowest excitonic state $|e\rangle$, the diagonal part of the relevant Hamiltonian in the Franck–Condon approximation can be written as

$$H = |g\rangle \sum_j \hbar\omega_j b_j^\dagger b_j \langle g| + |e\rangle \left\{ E_0 + \sum_j \left[ \hbar\omega_j b_j^\dagger b_j + \gamma_j \left( b_j + b_j^\dagger \right) \right] \right\} \langle e|, \tag{3.131}$$

where $b(b^\dagger)$ is the annihilation (creation) operator of phonons, the subscript $j$ is the phonon mode index, $\omega_j$ the mode frequency, and $E_0$ is the adiabatic electronic excitation energy in which the lattice relaxation energy is not included. The coupling constant $\gamma_j$ is the matrix element of the electron-phonon interaction Hamiltonian taken between the same exciton wavefunctions. The term proportional to $\gamma_j$ in (3.131) represents the shift of the origin of the lattice vibration in the excitonic state relative to the ground state and can be interpreted also as representing the adiabatic fluctuation of the excitonic energy level.

The homogeneous linewidth can be estimated from the linewidth of the absorption spectrum. The absorption spectrum of the above system is proportional to [42]

$$Re\left\{ \int_0^\infty dt \exp\left[ i\frac{(E - E_0 + E_{LR})}{\hbar}t - \frac{\gamma_\parallel}{2}t - S + S_+(t) + S_-(t) \right] \right\}, \quad (3.132)$$

with

$$S_\pm(t) = \sum_j \left(\frac{\gamma_j}{\hbar\omega_j}\right)^2 e^{\mp i\omega_j t} \left[ n(\hbar\omega_j) + \frac{1}{2} \pm \frac{1}{2} \right], \quad (3.133)$$

$$S = S_+(0) + S_-(0), \quad (3.134)$$

and

$$E_{LR} = \sum_j \frac{\gamma_j^2}{\hbar\omega_j}, \quad (3.135)$$

where $\gamma_\parallel$ is the longitudinal relaxation rate of the excitonic state, $n(\hbar\omega)$ the phonon occupation number, and $E_{LR}$ is usually called the lattice relaxation energy. Expanding the exponential function in (3.132) in powers of $S_+(t)$ and $S_-(t)$, we can in principle obtain the multitude of phonon sidebands. When only a single phonon mode with a frequency $\omega_0$ is considered, we obtain

$$I(E) \propto \sum_{l_+, l_-} \frac{(S_+)^{l_+}(S_-)^{l_-}}{l_+! \, l_-!} \delta\left[ E - E_0 + E_{LR} - (l_+ - l_-)\hbar\omega_0 \right], \quad (3.136)$$

where $S_+ = S_+(0)$ and $S_- = S_-(0)$. When the summation is carried out fixing $l_+ - l_- = p$, we have

$$I(E) \propto \left(\frac{S_+}{S_-}\right)^{p/2} I_p(2\sqrt{S_+ S_-}) \, \delta(E - E_0 + E_{LR} - p\hbar\omega_0), \quad (3.137)$$

where $I_p$ is a modified Bessel function of the first kind. In the limit of $T \longrightarrow 0$ K, $S_-(t) = 0$, and the phonon-assisted absorption spectrum is present only in the Stokes side:

$$I(E) \propto e^{-S_+} \sum_p \frac{(S_+)^p}{p!} \, \delta(E - E_0 + E_{LR} - p\hbar\omega_0). \quad (3.138)$$

This is the Poissonian distribution.

When the low-energy phonon modes are distributed quasi-continuously, the above approach to calculate the absorption spectrum is not realistic. Also when there are several excitonic states and exciton–phonon coupling among these states is present, an analytical expression like that in (3.132) cannot be obtained. In this case, the Green function formalism is much more effective in discussing the homogeneous linewidth of excitonic states as presented in Chap. 9.

The Huang–Rhys factor $S$ can be estimated from the relative intensity ratio between the phonon sidebands in the Raman spectra. This feature will be discussed within the above simple model with only a single phonon mode. Let us denote the eigenstates of the ground state and the exciton state by $|gm\rangle$ and $|en\rangle$, respectively, where $m$ and $n$ indicate the phonon number states. Then the transition probability from $|g0\rangle$ to $|gn\rangle$ is given by

$$\left| \sum_m \frac{\langle gn|H_R|em\rangle \, \langle em|H_R|g0\rangle}{\hbar\omega - E_0 - m\hbar\omega_0 - i\hbar\gamma_{em}} \right|^2 , \tag{3.139}$$

where $\hbar\omega$ is the incident light energy, $H_R$ the electron-radiation interaction and $\gamma_{em}$ is the population decay rate of the $|em\rangle$ state. In the Born–Oppenheimer approximation, we can approximate as

$$\langle em|H_R|g0\rangle \cong \langle e|H_R|g\rangle \, \langle em|g0\rangle , \tag{3.140}$$

where the last factor is the overlap integral of the vibrational states. Then we can simplify (3.139) as

$$|\langle e|H_R|g\rangle|^2 \left| \sum_m \frac{\langle gn|em\rangle \, \langle em|g0\rangle}{\hbar\omega - E_0 - m\hbar\omega_0 - i\hbar\gamma_{em}} \right|^2 . \tag{3.141}$$

The vibrational overlap integral can be calculated by noting

$$|en\rangle = \exp[\Delta(b - b^\dagger)] \, |gn\rangle , \tag{3.142}$$

where $\Delta = \gamma/\hbar\omega$ for the case of a single phonon mode in (3.131). Typical results are

$$\langle em|g0\rangle = e^{-\Delta^2/2} \frac{\Delta^m}{\sqrt{m!}} , \tag{3.143}$$

$$\langle em|g1\rangle = e^{-\Delta^2/2} \frac{\Delta^m}{\sqrt{m!}} \left( \frac{m}{\Delta} - \Delta \right) , \tag{3.144}$$

$$\langle em|g2\rangle = \frac{e^{-\Delta^2/2}}{\sqrt{2}\Delta^2} \frac{\Delta^m}{\sqrt{m!}} \left[ \left( m - \Delta^2 \right)^2 - m \right] . \tag{3.145}$$

In resonant Raman scattering, i.e., $\hbar\omega \sim E_0$, the intermediate state with $m=0$ contributes dominantly, and we find

$$I_{2\text{phonon}}/I_{1\text{phonon}} = \Delta^2/2 = S_+(0)/2 . \tag{3.146}$$

Thus we can estimate the Huang–Rhys factor $S_+$ from the intensity ratio between the phonon sidebands at low temperatures.

## 3.6    Luminescence Stokes Shift and Huang–Rhys Factor

Now as an application of the theory in Sect. 3.5, we discuss an intriguing phenomenon caused by the exciton–acoustic-phonon interaction, namely, the intrinsic Stokes shift between edges of the absorption and luminescence spectra. The term "intrinsic" is meant to exclude the luminescence associated with defects and impurities. There was some controversy on the origin of the luminescence onset energy in Si nanocrystals [43,44]. In conjunction with this problem, the luminescence Stokes shift and the excitonic exchange splitting were discussed. In order to clarify these features and to search for a quantitative interpretation of the observed onset energy, we calculated the excitonic energy spectra in Si nanocrystals, focusing on the strong enhancement of the exchange splitting, employing the multi-band effective mass theory, and assuming a spherical nanocrystal shape to simplify the analysis but to retain the essential physics [45].

Here, we formulate the exciton–phonon interaction in Si nanocrystals for acoustic phonon modes to calculate the Stokes shift. Since Si is a non-polar material, piezoelectric coupling is absent, and only the deformation-potential coupling is effective. The explicit form of the electron–acoustic-phonon interaction in bulk Si was derived by Laude et al. [46]. At the valence band top at the $\Gamma$-point, the interaction can be written as

$$
\begin{aligned}
H_{\text{strain}}^{v} = {} & -a_1(\epsilon_{xx} + \epsilon_{yy} + \epsilon_{zz}) - 3b_1\left[\left(L_x^2 - \frac{1}{3}L^2\right)\epsilon_{xx} + \text{c. p.}\right] \\
& -\sqrt{3}d_1[(L_xL_y + L_yL_x)\epsilon_{xy} + \text{c. p.}] \\
& -3b_2[(L_x\sigma_x - \frac{1}{3}L\cdot\sigma)\epsilon_{xx} + \text{c. p.}] \\
& -\sqrt{3}d_2[(L_x\sigma_y + L_y\sigma_x)\epsilon_{xy} + \text{c. p.}]\,,
\end{aligned}
\tag{3.147}
$$

where $L_i(i = x, y, z)$ are the angular momentum operators in the manifold of $L=1$, $\sigma_i(i = x, y, z)$ the spin operators, $a_1, b_1, b_2, d_1, d_2$ deformation potential constants, c. p. means the cyclic permutation, and the last two terms describe the stress-dependent spin–orbit interaction. On the other hand, at the bottom of the conduction band valley on the $\Delta$-line, the interaction Hamiltonian is written as

$$
H_{\text{strain}}^{c} = \boldsymbol{n} \cdot \left\{ \mathcal{E}_1(\epsilon_{xx} + \epsilon_{yy} + \epsilon_{zz})\mathbf{1} + \mathcal{E}_2\left[\boldsymbol{\epsilon} - \frac{1}{3}(\epsilon_{xx} + \epsilon_{yy} + \epsilon_{zz})\mathbf{1}\right] \right\} \cdot \boldsymbol{n}\,,
\tag{3.148}
$$

where $\mathcal{E}_1$ and $\mathcal{E}_2$ are deformation potential constants, $\boldsymbol{\epsilon}$ the strain tensor, $\mathbf{1}$ the unit tensor, $\boldsymbol{n}=\boldsymbol{k}_\Delta/|\boldsymbol{k}_\Delta|$, and $\boldsymbol{k}_\Delta$ is the wavevector at the valley bottom. Although these interactions are derived for a spatially homogeneous strain field, they can be extended to the case of nanocrystals, as long as the wavevector dependence of deformation potentials is reasonably weak around the valence band top and the conduction valley bottom. Then the matrix element of the exciton–phonon interaction can be calculated by integrating (3.147) and (3.148) multiplied by exciton envelope functions in nanocrystals. In the following calculation, the inter-valley electron–phonon interaction is not taken into account for simplicity.

The acoustic phonon modes in Si nanocrystals can be classified as torsional modes and spheroidal modes under assumptions of the stress-free boundary condition and isotropic elastic constants. The mode spectra for a Si nanocrystal with 20 Å radius are shown in Fig. 3.2. The material constants employed are [47]

$$\lambda + 2\mu = c_{11} = 16.772 \times 10^{11} \text{ dyn cm}^{-2} ,$$
$$\mu = c_{44} = 8.036 \times 10^{11} \text{ dyn cm}^{-2} , \text{ and} \qquad (3.149)$$
$$\rho = 2.329 \text{ g cm}^{-3} .$$

The mode spectra are rather sparse as compared with those in Fig. 3.1 for a CdSe nanocrystal with the same radius. This is due to the larger elastic constants and the smaller mass density for Si compared with CdSe.

The relevant exciton–phonon interaction Hamiltonian can be written in the form of (3.131). Then the luminescence Stokes shift $\delta_{st}$, which is twice the lattice relaxation energy in (3.135), and the dimensionless coupling strength

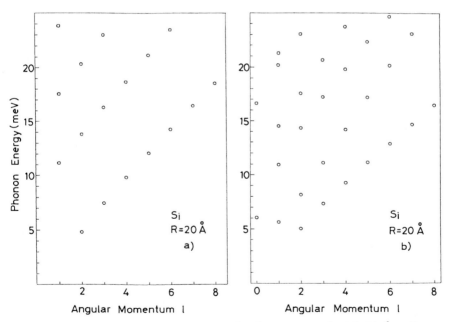

**Fig. 3.2.** The acoustic phonon energies of a Si nanocrystal with 20 Å radius are plotted as a function of the angular momentum for (**a**) torsional modes and (**b**) spheroidal modes (from T. Takagahara [12])

$\tilde{S}$, which is a sum of the Huang–Rhys factors, are given in the Franck–Condon approximation as

$$\delta_{st} = 2\sum_j \gamma_j^2/\hbar\omega_j \quad \text{and} \quad \tilde{S} = \sum_j \gamma_j^2/(\hbar\omega_j)^2 \, , \tag{3.150}$$

where the summation is taken over relevant acoustic phonon modes. The above $\tilde{S}$ is equal to $S$ in (3.134) at $T$=0 K and will be called simply the Huang–Rhys factor. The contribution from the torsional modes turns out to be two or three orders of magnitude smaller than that from the spheroidal modes. Through a similar analysis to that in Sect. 3.3, we see that $\delta_{st}(\tilde{S})$ is inversely proportional to $R^3(R^2)$ in the small-size regime.

The maximum value of the Huang–Rhys factor within the manifold of the $S_{3/2}$ exciton levels was calculated to vary from 0.7 to 0.3 for nanocrystal radii from 14 to 20 Å [45], indicating the weak-coupling regime of the exciton–acoustic-phonon interaction. However, the luminescence Stokes shift is not negligible and is of the order of several meV in the above size range. The Stokes shift is dependent on the excitonic levels. In the estimation of the onset energy of photoluminescence, the photon-absorbing state is identified with the excitonic state in the $S_{3/2}$ manifold having the largest oscillator strength, while the luminescent state is identified with that having the lowest energy when the Stokes shift is included. The onset energy estimated in this way is plotted in Fig. 3.3 by a solid line as a function of the nanocrystal radius (upper abscissa) or as a function of the exciton absorption energy (lower ab-

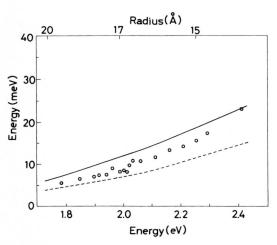

**Fig. 3.3.** The theoretical excitonic exchange splitting and the result including the Stokes shift are plotted by a *dashed line* and a *solid line*, respectively, as a function of the excitonic transition energy. The experimental data of [43] are exhibited by *circles*. On the *upper abscissa*, the nanocrystal radius corresponding to the exciton energy in the *lower abscissa* is shown (from T. Takagahara [45])

scissa). The calculated excitonic exchange splitting [45] is plotted by a dashed line. Thus the exchange splitting contributes to the onset energy more than the Stokes shift. The quantitative agreement between the experiment (open dots) and the theory (solid line) can be regarded as satisfactory, although the fluctuation of the exciton energies due to inhomogeneities in the nanocrystal size and shape is not taken into account. An important conclusion is that the observed onset energy of photoluminescence can be explained mainly in terms of excitonic exchange splitting. Very recently, this conclusion was confirmed by measuring separately the light-absorbing and light-emitting states in Si nanocrystals by magneto-optical spectral hole-burning spectroscopy [48].

## 3.7   Summary

The most general formulation of the electron–phonon interactions in semiconductor nanocrystals has been presented for both the acoustic phonon modes and the polar optical phonon modes. The size dependence of the coupling strength is clarified for typical coupling mechanisms. As an application of this theory, the Huang–Rhys factor and the Stokes shift of photoluminescence due to acoustic phonon modes in Si nanocrystals are calculated and their size dependencies are clarified. It is found that the observed onset energy of photoluminescence from Si nanocrystals can be interpreted mainly in terms of excitonic exchange splitting.

## Appendix A. Strain Tensor Components in General Orthogonal Curvilinear Coordinates

In this Appendix, we derive the strain tensor components in a general orthogonal curvilinear coordinate system and the transformation formulas for the strain tensor components between different coordinate systems. First of all, it is instructive to see the situation in the usual Cartesian coordinates. Let us consider two points denoted by $P(x, y, z)$ and $Q(x+lr, y+mr, z+nr)$, where $r$ is the distance $\overline{PQ}$, and $l$, $m$, and $n$ are the direction cosines of the vector $\boldsymbol{PQ}$. We assume that by an elastic deformation the point $P$ moves to a new point $P'(x + u, y + v, z + w)$. Then the point $Q$ moves to a new point $Q'(X, Y, Z)$ given by

$$X = x + lr + u + lr\,\frac{\partial u}{\partial x} + mr\frac{\partial u}{\partial y} + nr\frac{\partial u}{\partial z}\,, \tag{3.151}$$

$$Y = y + mr + v + lr\,\frac{\partial v}{\partial x} + mr\frac{\partial v}{\partial y} + nr\frac{\partial v}{\partial z}\,, \tag{3.152}$$

$$Z = z + nr + w + lr\,\frac{\partial w}{\partial x} + mr\frac{\partial w}{\partial y} + nr\frac{\partial w}{\partial z}\,, \tag{3.153}$$

and the distance $\overline{P'Q'}$ is calculated as

$$\overline{P'Q'} \cong r\left[1 + l^2 e_{xx} + m^2 e_{yy} + n^2 e_{zz} + lm e_{xy} + mn e_{yz} + ln e_{xz}\right], \quad (3.154)$$

with

$$e_{xx} = \frac{\partial u}{\partial x}, \quad e_{xy} = \frac{\partial u}{\partial y} + \frac{\partial v}{\partial x}, \quad (3.155)$$

and their cyclic permutations.

Now we extend the argument to general orthogonal curvilinear coordinates denoted by $(\xi_1, \xi_2, \xi_3)$. In this system, the orthogonal unit vectors are defined as

$$e_i = \frac{1}{h_i}{}^t\left(\frac{\partial x}{\partial \xi_i}, \frac{\partial y}{\partial \xi_i}, \frac{\partial z}{\partial \xi_i}\right), \quad (3.156)$$

$$\text{with} \quad h_i = \sqrt{\left(\frac{\partial x}{\partial \xi_i}\right)^2 + \left(\frac{\partial y}{\partial \xi_i}\right)^2 + \left(\frac{\partial z}{\partial \xi_i}\right)^2} \quad (i = 1, 2, 3),$$

and the distance between two points denoted by $P(\xi_1, \xi_2, \xi_3)$ and $Q(\xi_1 + \delta\xi_1, \xi_2 + \delta\xi_2, \xi_3 + \delta\xi_3)$ is given by

$$\overline{PQ} = \sqrt{h_1^2(\delta\xi_1)^2 + h_2^2(\delta\xi_2)^2 + h_3^2(\delta\xi_3)^2}, \quad (3.157)$$

where, of course, $\delta\xi_i$ $(i=1,2,3)$ is assumed to be very small. Let us assume that the point $P$ moves to a new point $P'(\xi_1 + \Delta_1, \xi_2 + \Delta_2, \xi_3 + \Delta_3)$. Accordingly, the point $Q$ moves to a new point $Q'(X_1, X_2, X_3)$ given as

$$X_i = \xi_i + \delta\xi_i + \Delta_i + \frac{\partial \Delta_i}{\partial \xi_1}\delta\xi_1 + \frac{\partial \Delta_i}{\partial \xi_2}\delta\xi_2 + \frac{\partial \Delta_i}{\partial \xi_3}\delta\xi_3 \ (i = 1, 2, 3). \quad (3.158)$$

The distance $\overline{P'Q'}$ can be calculated by (3.157) using $h_i$ $(i=1,2,3)$ at the point $P'$, namely

$$\overline{P'Q'}^2 \cong \sum_{i=1}^{3}\left(h_i + \Delta_1\frac{\partial h_i}{\partial \xi_1} + \Delta_2\frac{\partial h_i}{\partial \xi_2} + \Delta_3\frac{\partial h_i}{\partial \xi_3}\right)^2$$

$$\times \left[\delta\xi_i + \frac{\partial \Delta_i}{\partial \xi_1}\delta\xi_1 + \frac{\partial \Delta_i}{\partial \xi_2}\delta\xi_2 + \frac{\partial \Delta_i}{\partial \xi_3}\delta\xi_3\right]^2. \quad (3.159)$$

Retaining the terms up to the first order in $\Delta_i$ $(i=1,2,3)$, we have

$$\overline{P'Q'} \cong \overline{PQ}\left[1 + l_1^2 e_{11} + l_2^2 e_{22} + l_3^2 e_{33} + l_1 l_2 e_{12} + l_2 l_3 e_{23} + l_3 l_1 e_{31}\right], \quad (3.160)$$

where $l_i$ is the direction cosine of the vector $\boldsymbol{PQ}$ in the system of $(e_1, e_2, e_3)$ at the point $P$ and is defined by $l_i = h_i \delta\xi_i / \overline{PQ}$. From the analogy to (3.154), we can define the strain tensor components in this curvilinear coordinates as

$$e_{11} = \frac{1}{h_1}\frac{\partial u_1}{\partial \xi_1} + \frac{u_2}{h_1 h_2}\frac{\partial h_1}{\partial \xi_2} + \frac{u_3}{h_1 h_3}\frac{\partial h_1}{\partial \xi_3} , \tag{3.161}$$

$$e_{22} = \frac{1}{h_2}\frac{\partial u_2}{\partial \xi_2} + \frac{u_1}{h_1 h_2}\frac{\partial h_2}{\partial \xi_1} + \frac{u_3}{h_2 h_3}\frac{\partial h_2}{\partial \xi_3} , \tag{3.162}$$

$$e_{33} = \frac{1}{h_3}\frac{\partial u_3}{\partial \xi_3} + \frac{u_1}{h_1 h_3}\frac{\partial h_3}{\partial \xi_1} + \frac{u_2}{h_2 h_3}\frac{\partial h_3}{\partial \xi_2} , \tag{3.163}$$

$$e_{12} = \frac{h_1}{h_2}\frac{\partial}{\partial \xi_2}\left(\frac{u_1}{h_1}\right) + \frac{h_2}{h_1}\frac{\partial}{\partial \xi_1}\left(\frac{u_2}{h_2}\right) , \tag{3.164}$$

$$e_{23} = \frac{h_2}{h_3}\frac{\partial}{\partial \xi_3}\left(\frac{u_2}{h_2}\right) + \frac{h_3}{h_2}\frac{\partial}{\partial \xi_2}\left(\frac{u_3}{h_3}\right) , \tag{3.165}$$

$$e_{31} = \frac{h_1}{h_3}\frac{\partial}{\partial \xi_3}\left(\frac{u_1}{h_1}\right) + \frac{h_3}{h_1}\frac{\partial}{\partial \xi_1}\left(\frac{u_3}{h_3}\right) , \tag{3.166}$$

where the displacement vector $u_i$ at the point $P$ can be identified with $h_i \Delta_i$ ($i=1,2,3$) as long as $\Delta_i$ is very small.

When specified to the spherical coordinates, namely

$$\xi_1 = r, \quad \xi_2 = \theta, \quad \xi_3 = \varphi, \quad h_1 = 1, \quad h_2 = r, \quad h_3 = r\sin\theta , \tag{3.167}$$

we have

$$e_{rr} = \frac{\partial u_r}{\partial r} , \tag{3.168}$$

$$e_{\theta\theta} = \frac{1}{r}\frac{\partial u_\theta}{\partial \theta} + \frac{u_r}{r} , \tag{3.169}$$

$$e_{\varphi\varphi} = \frac{1}{r\sin\theta}\frac{\partial u_\varphi}{\partial \varphi} + \frac{u_r}{r} + \frac{u_\theta}{r}\cot\theta , \tag{3.170}$$

$$e_{r\theta} = \frac{1}{r}\frac{\partial u_r}{\partial \theta} + r\frac{\partial}{\partial r}\left(\frac{u_\theta}{r}\right) , \tag{3.171}$$

$$e_{\theta\varphi} = \frac{1}{r\sin\theta}\frac{\partial u_\theta}{\partial \varphi} + \frac{\sin\theta}{r}\frac{\partial}{\partial \theta}\left(\frac{u_\varphi}{\sin\theta}\right) , \tag{3.172}$$

$$e_{\varphi r} = \frac{1}{r\sin\theta}\frac{\partial u_r}{\partial \varphi} + r\frac{\partial}{\partial r}\left(\frac{u_\varphi}{r}\right) . \tag{3.173}$$

In order to derive the transformation formulas for the strain tensor components between different orthogonal curvilinear coordinate systems, we need only to note that the distance $\overline{P'Q'}$ in (3.160) is the same for all the coordinate systems, namely

$$l_1^2 e_{11} + l_2^2 e_{22} + l_3^2 e_{33} + l_1 l_2 e_{12} + l_2 l_3 e_{23} + l_3 l_1 e_{31}$$

$$= l_1'^2 e_{11}' + l_2'^2 e_{22}' + l_3'^2 e_{33}' + l_1' l_2' e_{12}' + l_2' l_3' e_{23}' + l_3' l_1' e_{31}' , \tag{3.174}$$

where the primed quantities on the right side are associated with a coordinate system different from that on the left side. When the relation between the direction cosines in two coordinate systems is substituted, we obtain the desired transformation formulas for the strain tensor components. For example, the transformation formulas between Cartesian coordinates and spherical coordinates are given as

$$e_{xx} = \sin^2\theta\cos^2\varphi\, e_{rr} + \cos^2\theta\cos^2\varphi\, e_{\theta\theta} + \sin^2\varphi\, e_{\varphi\varphi} + \frac{1}{2}\sin 2\theta\cos^2\varphi\, e_{r\theta}$$
$$- \frac{1}{2}\cos\theta\sin 2\varphi\, e_{\theta\varphi} - \frac{1}{2}\sin\theta\sin 2\varphi\, e_{r\varphi}\,, \tag{3.175}$$

$$e_{yy} = \sin^2\theta\sin^2\varphi\, e_{rr} + \cos^2\theta\sin^2\varphi\, e_{\theta\theta} + \cos^2\varphi\, e_{\varphi\varphi} + \frac{1}{2}\sin 2\theta\sin^2\varphi\, e_{r\theta}$$
$$+ \frac{1}{2}\cos\theta\sin 2\varphi\, e_{\theta\varphi} + \frac{1}{2}\sin\theta\sin 2\varphi\, e_{r\varphi}\,, \tag{3.176}$$

$$e_{zz} = \cos^2\theta\, e_{rr} + \sin^2\theta\, e_{\theta\theta} - \frac{1}{2}\sin 2\theta\, e_{r\theta}\,, \tag{3.177}$$

$$e_{xy} = \sin^2\theta\sin 2\varphi\, e_{rr} + \cos^2\theta\sin 2\varphi\, e_{\theta\theta} - \sin 2\varphi\, e_{\varphi\varphi} + \frac{1}{2}\sin 2\theta\sin 2\varphi\, e_{r\theta}$$
$$+ \cos\theta\cos 2\varphi\, e_{\theta\varphi} + \sin\theta\cos 2\varphi\, e_{r\varphi}\,, \tag{3.178}$$

$$e_{yz} = \sin 2\theta\sin\varphi\, e_{rr} - \sin 2\theta\sin\varphi\, e_{\theta\theta} + \cos 2\theta\sin\varphi\, e_{r\theta} - \sin\theta\cos\varphi\, e_{\theta\varphi}$$
$$+ \cos\theta\cos\varphi\, e_{r\varphi}\,, \tag{3.179}$$

$$e_{xz} = \sin 2\theta\cos\varphi\, e_{rr} - \sin 2\theta\cos\varphi\, e_{\theta\theta} + \cos 2\theta\cos\varphi\, e_{r\theta} + \sin\theta\sin\varphi\, e_{\theta\varphi}$$
$$- \cos\theta\sin\varphi\, e_{r\varphi}\,. \tag{3.180}$$

## Appendix B. Vector Spherical Harmonics

The vector Helmholtz equation given by

$$[\nabla^2 + k^2]\boldsymbol{\Psi}(r) = 0 \tag{3.181}$$

has three linearly independent solutions. They are usually denoted by $\boldsymbol{L}_{lm}(kr)$, $\boldsymbol{M}_{lm}(kr)$, and $\boldsymbol{N}_{lm}(kr)$ as given in (3.6), although the radial part can generally include spherical Hankel functions as well as spherical Bessel functions. These $\boldsymbol{L}$, $\boldsymbol{M}$, and $\boldsymbol{N}$ fields are generally not orthogonal to each other. On the other hand, concerning the angular part, there are mutually orthogonal harmonic bases [24]. In spherical coordinates they are defined by

$$\boldsymbol{P}_{lm}(\Omega) = {}^t(1,0,0)Y_{lm}(\Omega)\,, \tag{3.182}$$

$$\boldsymbol{B}_{lm}(\Omega) = \frac{1}{\sqrt{l(l+1)}}\, {}^t\left(0\,,\ \frac{\partial}{\partial\theta}\,,\ \frac{1}{\sin\theta}\frac{\partial}{\partial\varphi}\right)Y_{lm}(\Omega)\,, \tag{3.183}$$

$$\boldsymbol{C}_{lm}(\Omega) = \frac{1}{\sqrt{l(l+1)}}\, {}^t\left(0\,,\ \frac{1}{\sin\theta}\frac{\partial}{\partial\varphi}\,,\ -\frac{\partial}{\partial\theta}\right)Y_{lm}(\Omega)\,, \tag{3.184}$$

where $\boldsymbol{P}_{lm}$ is defined for all $l(\geq 0)$, whereas $\boldsymbol{B}_{lm}$ and $\boldsymbol{C}_{lm}$ are defined for $l \geq 1$. They are orthonormalized concerning the angular integral, namely

$$
\int \mathrm{d}\Omega \boldsymbol{P}_{lm}^*(\Omega)\boldsymbol{P}_{l'm'}(\Omega) = \int \mathrm{d}\Omega \boldsymbol{B}_{lm}^*(\Omega)\boldsymbol{B}_{l'm'}(\Omega)
$$

$$
= \int \mathrm{d}\Omega \boldsymbol{C}_{lm}^*(\Omega)\boldsymbol{C}_{l'm'}(\Omega) = \delta_{ll'}\delta_{mm'} , \quad (3.185)
$$

$$
\int \mathrm{d}\Omega \boldsymbol{P}_{lm}^*(\Omega)\boldsymbol{B}_{l'm'}(\Omega) = \int \mathrm{d}\Omega \boldsymbol{B}_{lm}^*(\Omega)\boldsymbol{C}_{l'm'}(\Omega)
$$

$$
= \int \mathrm{d}\Omega \boldsymbol{C}_{lm}^*(\Omega)\boldsymbol{P}_{l'm'}(\Omega) = 0 . \quad (3.186)
$$

In terms of these angular harmonics, $\boldsymbol{L}$, $\boldsymbol{M}$, and $\boldsymbol{N}$ fields can be written as

$$
\boldsymbol{L}_{lm}(kr) = z_l'(kr)\boldsymbol{P}_{lm}(\Omega) + \frac{\sqrt{l(l+1)}}{kr}z_l(kr)\boldsymbol{B}_{lm}(\Omega) , \quad (3.187)
$$

$$
\boldsymbol{M}_{lm}(kr) = \sqrt{l(l+1)}z_l(kr)\boldsymbol{C}_{lm}(\Omega) , \quad (3.188)
$$

$$
\boldsymbol{N}_{lm}(kr) = \frac{l(l+1)}{kr}z_l(kr)\boldsymbol{P}_{lm}(\Omega) + \frac{\sqrt{l(l+1)}}{kr}\frac{\mathrm{d}\left[krz_l(kr)\right]}{\mathrm{d}(kr)}\boldsymbol{B}_{lm}(\Omega) ,
$$

$$
(3.189)
$$

where $z_l(kr)$ is a radial function.

# References

1. *Microcrystalline and Nanocrystalline Semiconductors*, M.R.S. Symp. Proc., Vol. 358, ed. by R.W. Collins, C.C. Tsai, M.Hirose, F.Koch, L. Brus (Materials Research Society, Pittsburgh 1995)
2. U. Woggon, *Optical Properties of Semiconductor Quantum Dots*, Springer Tracts in Modern Physics Vol. 136 (Springer-Verlag, Berlin 1997)
3. D. Bimberg, M. Grundmann, N. N. Ledentsov: *Quantum Dot Heterostructures* (John Wiley & Sons, New York 1999)
4. L.E. Brus: Appl. Phys. A **53**, 465 (1991); M.G. Bawendi, M.L. Steigerwald, L.E. Brus: Ann. Rev. Phys. Chem. **41**, 477 (1990)
5. C. Flytzanis, F. Hache, M.C. Klein, D. Ricard, P. Roussignol: *Progress in Optics, Vol. XXIX*, ed. by E. Wolf (Elsevier, Amsterdam 1991) p. 321
6. A.P. Alivisatos, T.D. Harris, P.J. Carroll, M.L. Steigerwald, L.E. Brus: J. Chem. Phys. **90**, 3463 (1989)
7. M.C. Klein, F. Hache, D. Ricard, C. Flytzanis: Phys. Rev. B **42**, 11123 (1990)
8. A. Tanaka, S. Onari, T. Arai: Phys. Rev. B **45**, 6587 (1992)
9. T. Takagahara: *Ultrafast Phenomena VI*, ed. by T. Yajima, K. Yoshihara, C.B. Harris, S. Shionoya: (Springer, Berlin 1988) p. 337
10. S. Nomura, T. Kobayashi: Phys. Rev. B **45**, 1305 (1992)
11. A. Tanaka, S. Onari, T. Arai: Phys. Rev. B **47**, 1237 (1993)

12. T. Takagahara: J. Lumin. **70**, 129 (1996)
13. T. Itoh, M. Furumiya: J. Lumin. **48–49**, 704 (1991)
14. T. Wamura, Y. Masumoto, T. Kawamura: Appl. Phys. Lett. **59**, 1758 (1991)
15. N. Peyghambarian, B. Fluegel, D. Hulin, A. Migus, M. Joffre, A. Antonetti, S.W. Koch, M. Lindberg: IEEE J. Quantum Electron. **25**, 2516 (1989)
16. P. Roussignol, D. Ricard, C. Flytzanis, N. Neuroth: Phys. Rev. Lett. **62**, 312 (1989)
17. M.G. Bawendi, W.L. Wilson, L. Rothberg, P.J. Carroll, T.M. Jedju, M.L. Steigerwald, L.E. Brus: Phys. Rev. Lett. **65**, 1623 (1990)
18. U. Woggon, S. Gaponenko, W. Langbein, A. Uhrig, C. Klingshirn: Phys. Rev. B **47**, 3684 (1993)
19. R.W. Schoenlein, D.M. Mittleman, J.J. Shiang, A.P. Alivisatos, C.V. Shank: Phys. Rev. Lett. **70**, 1014 (1993)
20. D.M. Mittleman, R.W. Schoenlein, J.J. Shiang, V.L. Colvin, A.P. Alivisatos, C.V. Shank: Phys. Rev. B **49**, 14435 (1994)
21. H. Shinojima, J. Yumoto, T. Takagahara, N. Uesugi: *Quantum Electronics and Laser Science Conf. Baltimore 1993*, QWH26
22. A.E.H. Love: *A Treatise on the Mathematical Theory of Elasticity* (Dover, New York 1944)
23. H. Lamb: Proc. Math. Soc. London **13**, 187 (1882)
24. P.M. Morse, H. Feshbach: *Methods of Theoretical Physics* (McGraw-Hill, New York 1953)
25. *Physics of II–VI and I–VII Compounds, Landolt-Börnstein Vol. 17b*, ed. by O. Madelung, M. Schulz, H. Weiss (Springer, Berlin 1982)
26. A. Tamura, K. Higeta, T. Ichinokawa: J. Phys. C **15**, 4975 (1982)
27. N.N. Ovsyuk, V.N. Novikov: Phys. Rev. B **53**, 3113 (1996)
28. L. Saviot, B. Champagnon, E. Duval, I.A. Kudriavtsev, A.I. Ekimov: J. Non-Cryst. Solids **197**, 238 (1996)
29. L. Saviot, B. Champagnon, E. Duval, A.I. Ekimov: Phys. Rev. B **57**, 341 (1998)
30. A.A. Sirenko, V.I. Belitsky, T. Ruf, M. Cardona, A.I. Ekimov, C. Trallero-Giner: Phys. Rev. B **58**, 2077 (1998)
31. X.L. Wu, Y.F. Mei, G.G. Siu, K.L. Wong, K. Moulding, M.J. Stokes, C.L. Fu, X.M. Bao: Phys. Rev. Lett. **86**, 3000 (2001)
32. G.L. Bir, G.E. Pikus: *Symmetry and Strain-Induced Effects in Semiconductors* (John Wiley & Sons, New York 1974)
33. W.G. Cady: *Piezoelectricity* (McGraw-Hill, New York 1946)
34. C. Trallero-Giner, F. Comas: Phil. Mag. B **70**, 583 (1994)
35. E. Menendez, C. Trallero-Giner, M. Cardona: Phys. Stat. Sol. B **199**, 81 (1997)
36. E. Menendez-Proupin, C. Trallero-Giner, A. Garcia-Cristobal: Phys. Rev. B **60**, 5513 (1999)
37. A.V. Fedorov, A.V. Baranov, K. Inoue: Phys. Rev. B **56**, 7491 (1997)
38. S. Schmitt-Rink, D.A.B. Miller, D.S. Chemla: Phys. Rev. B **35**, 8113 (1987)
39. D.J. Norris, Al.L. Efros, M. Rosen, M.G. Bawendi: Phys. Rev. B **53**, 16347 (1996)
40. R. Heitz, I. Mukhametzhanov, O. Stier, A. Madhukar, D. Bimberg: Phys. Rev. Lett. **83**, 4654 (1999)
41. O. Stier, M. Grundmann, D. Bimberg: Phys. Rev. B **59**, 5688 (1999)
42. S. Nakajima, Y. Toyozawa, R. Abe: *The Physics of Elementary Excitations* (Springer, Berlin 1980)

43. P.D.J. Calcott, K.J. Nash, L.T. Canham, M.J. Kane, D. Brumhead: J. Phys.: Condens. Matter **5**, L91 (1993)
44. E. Martin, C. Delerue, G. Allan, M. Lannoo: Phys. Rev. B **50**, 18258 (1994)
45. T. Takagahara, K. Takeda: Phys. Rev. B **53**, R4205 (1996)
46. L.D. Laude, F.H. Pollak, M. Cardona: Phys. Rev. B **3**, 2623 (1971)
47. *Physics of Group IV Elements and III–V Compounds, Landolt-Börnstein Vol. 17b*, ed. by O. Madelung, M. Schulz, H. Weiss (Springer, Berlin 1982)
48. J. Diener, D. Kovalev, H. Heckler, G. Polisski, F. Koch: Phys. Rev. B **63**, 073302 (2001)

# 4 Micro-Imaging
and Single Dot Spectroscopy
of Self-Assembled Quantum Dots

Mitsuru Sugisaki

## 4.1 Introduction

The last decade has witnessed considerable progress in the fabrication of low-dimensional semiconductor systems. In these systems, the carriers are confined in small regions formed by potential barriers. The spatial sizes of the confinement are comparable to the carrier de Broglie wavelengths or the exciton Bohr radii so that the quantization of their energy levels results. Experimentally, many novel features of the electronic and optical properties have been discovered in these systems [1,2]. Recent developments in crystal growth techniques have made semiconductors even more versatile. Following the success of semiconductor quantum wells, which permitted the study of two-dimensional (2D) optical and electronic effects of carriers due to lateral confinement, many investigations have been carried out in quasi-one-dimensional quantum wires and zero-dimensional (0D) quantum dots (QDs). Especially, in semiconductor QDs, since excitons are confined in all spatial directions on a length scale comparable to the exciton Bohr radius, many interesting phenomena are expected, such as a large blueshift of the band gap energy, discrete energy structure, $\delta$-function-like density of states, and high optical nonlinearity [2–4]. These characteristic properties of semiconductor QDs have attracted an enormous amount of interest for opto-electronic device applications. From the viewpoint of fundamental physics, understanding the nature of excitons in terms of the dimensionality is an important subject.

It is instructive to compare the optical properties between the bulk and low-dimensional semiconductors. Figure 4.1 shows how the density of states changes shape with the reduction of the dimension of the system. Since the excitons in a three-dimensional (3D, bulk) semiconductor system are thermally distributed at a finite temperature $T$, the photoluminescence (PL) band has a moderate width that is approximately proportional to the density of states

$$n_{3D} \propto \sqrt{E - E_0} \,, \tag{4.1}$$

and the Boltzmann distribution

$$f(E, T) = \exp\left(-\frac{E - E_0}{k_B T}\right) \,, \tag{4.2}$$

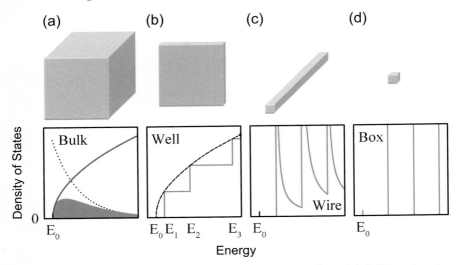

**Fig. 4.1.** Density of states in terms of system dimensionality: (**a**) A 3D system has the density of states of a parabolic shape. Since the PL spectrum (*shaded area of the spectrum*) is approximated by the product of the density of states (*solid curve*) and the thermal distribution function (*dotted curve*), a 3D system has a broad PL band; (**b**) The density of states of a 2D system is made up of a pile of step functions. The *dotted curve* shows the density of states of a 3D system for comparison; (**c**) The density of states in a 1D system diverges at $E_n$; (**d**) The PL spectra of 0D systems reflect the $\delta$-function shape of the density of states

where $E_0$ and $k_B$ are the exciton energy and the Boltzmann constant, respectively. Since the tail of the Boltzmann distribution extends toward the upper energy side with increasing temperature, the PL bandwidth becomes wider at high temperatures.

With the reduction of the dimensionality of systems, the density of state changes its form. A 2D system, namely a quantum well, has a thicknesses of a few tens of nanometers or less, while it has a macroscopic size (comparable to a 3D system) along the lateral directions. The density of states of a 2D system is given by

$$n_{2D} \propto \sum_n \Theta\left(E - E_n\right) , \tag{4.3}$$

with the step function,

$$\Theta(x) = \begin{cases} 1 \ (x \geq 0) \\ 0 \ (x < 0) \end{cases} . \tag{4.4}$$

That is, the density of state of a 2D system is made up of accumulated step functions, while that of a 3D system has a smooth parabolic shape.

The density of states of a one-dimensional (1D) system, namely a quantum wire, can be derived to be

$$n_{1D} \propto \sum_{E_n < E} \frac{1}{\sqrt{E - E_n}} \; . \tag{4.5}$$

As shown in Fig. 4.1c, the density of states diverges at $E_n$, which is a characteristic feature of 1D systems. Therefore, the quantum wire system is expected to be an excellent candidate for opto-electronic device applications because of their high quantum efficiency.

In the 0D systems on which we concentrate our main interest, reflecting the $\delta$-function-like density of states

$$n_{0D} \propto \sum_{E_n < E} \delta(E - E_n) \; , \tag{4.6}$$

the PL from a QD should show a very sharp line. Further, since the energy levels $E_n$ are separated from each other at intervals due to the quantization

$$E_n = \frac{\hbar^2 \pi^2}{2m} \left[ \left( \frac{n_x}{L_x} \right)^2 + \left( \frac{n_y}{L_y} \right)^2 + \left( \frac{n_z}{L_z} \right)^2 \right] \; , \tag{4.7}$$

the PL linewidth of 0D excitons is thus expected to be less sensitive to temperature and to be much sharper than that of a 3D exciton, even at room temperature. In (4.7), $L$, $m$, and $n$ are the size of the quantum box, the effective mass of the confined carrier, and the quantum number, respectively. In almost all cases, however, when we try to measure a PL spectrum of a sample containing QDs using a conventional detection setup (macro-PL), information of excitons in a single QD is obscured because of the fluctuations of the size and shape of QDs. Therefore, contrary to our expectations, a macro-PL spectrum of QDs unfortunately shows a much wider PL bandwidth than that of a 3D system, even at a cryogenic temperature. For instance, Fig. 4.2a shows the PL spectrum from QDs measured using a conventional system at 4 K. The PL bandwidth is about 20 times wider than that of bulk GaAs. In order to eliminate the ambiguity of the PL spectra due to the inhomogeneous distribution of the QDs, single dot spectroscopy, which we discuss in this chapter, is a powerful technique to overcome this difficulty and will help to clarify the nature of QD systems.

Figure 4.2b shows the micro-photoluminescence ($\mu$-PL) spectra from five individual QDs measured under weak excitation, as discussed in detail in Sect. 4.4. As expected, each PL line has an extremely sharp peak (more than one hundred times narrower than the macro-PL peak). These lines are very sharp with widths limited by the resolution of the spectrometer. Detailed experiments performed by Kamada et al. revealed that the homogeneous linewidth of a single InGaAs QD is as narrow as $\sim$30 $\mu$eV [5].

The shape of a macro-PL spectrum from QDs is generally considered to reflect the dot-size distribution because the macro-PL is composed of many

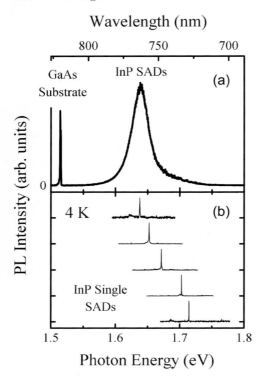

**Fig. 4.2.** Comparison between (**a**) macro- and (**b**) micro-PL spectra at 4 K. In the macro-PL spectrum, the *sharp* PL peak observed at 1.52 eV arises from $3D$ excitons from the GaAs substrate, while the *broad* PL band with a peak at 1.63 eV comes from the ensemble of $0D$ excitons from the InP self-assembled dots (SADs). The PL from the InP SADs shown in (**a**) is the sum of the sharp lines from the individual SADs. Some of the PL lines from single InP SADs are shown in (**b**)

μ-PL lines arising from individual QDs. Therefore, in the sample shown in Fig. 4.2, many QDs have PL peaks around 1.63 eV. The macro-PL intensity is very weak above 1.7 eV. However, there are some μ-PL lines even in this region, as shown in Fig. 4.2b, because a small number of QDs have a distribution even in this region.

There is no doubt that a μ-PL spectrum of a single QD contains rich information about the 0D system such as the electronic structure, the relaxation processes and kinetics of carriers, quantum statistical properties, and so on. However, since each QD shows a somewhat different character caused by its nonuniformity of size, shape, interface, etc., the measurements need to be carried out on several QDs in a sample to understand the general behavior of the PL characteristics.

In this chapter, in addition to the μ-PL measurements, the importance of the direct observation of μ-PL images is emphasized. The micro-images that show the integrated μ-PL intensities are useful for efficient measurements because many QDs can be observed at a time. They also help to isolate interesting exceptions, as will be seen in the following sections.

Here we survey the optical properties of individual QDs measured by means of macro- and micro-PL spectroscopy. The major portion of this chapter will be devoted to the optical properties of InP self-assembled dots (hereafter referred to as SADs) [6], because InP QDs are excellent candidates for opto-device applications in the spectral region from the visible red to the near-infrared (near-IR). Further, InP SADs exhibit rich physics as shown in the follow sections. The μ-PL images and spectra of individual SADs observed at various polarizations, excitation powers, and temperatures are compared. We will see that this information brings better understanding of the nature of 0D systems. The influence of the matrix on the optical properties of the QDs is discussed.

The organization of this chapter is as follows: Sect. 4.2 describes several methods to observe single QDs. Section 4.3 describes μ-PL images and spectra from individual InP SADs. The origin of a strong optical anisotropy in the SADs as high as that of quantum wires is discussed. Many carrier effects are presented in Sect. 4.4. The concept of efficiency in the state-filling effect is illustrated from the observation of μ-PL images. The exciton and biexciton states observed in the μ-PL spectra are also shown. The temperature dependence of the band gap energy and the thermal quenching behavior are discussed in Sect. 4.5. In Sect. 4.6, we describe one of the specific properties revealed by the observation of μ-PL images called fluorescence intermittency. Section 4.7 is devoted to the brief introduction of some other interesting phenomena, including external field effects. Finally, Sect. 4.8 summarizes the results obtained in these μ-PL measurements for future developments.

## 4.2   How to Get Access to a Single Quantum Dot

Various kinds of techniques have so far been developed for the growth of semiconductor nanocrystals, including diffusion-controlled phase decomposition (e.g., CuCl in glass), colloidal synthesis (e.g., CdSe in organic media), and self-assembly by the Stranski–Krastanow growth mode [4,7]. In addition to these methods, naturally formed quasi-0D excitons trapped at local potential minima are also widely studied, where the potential minima are provided by fluctuations of the thickness in quantum wells [8–15]. Further, the ternary and quaternary alloys are also known to form natural QDs due to fluctuations of the local concentration [16–20]. Among the numerous kinds of the QD systems that are actively studied, III–V and II–VI compound semiconductor QDs formed by Stranski–Krastanow growth, which are commonly referred to as SADs, are suitable systems for single-dot spectroscopy because of the following reasons:

− the QDs are grown in a single layer
− the dot density is controllable over a wide range by changing the growth conditions
− the photoluminescence (PL) efficiency is high

Let us estimate how low a density of QDs and how high a spatial resolution of equipment are required to observe a well spectrally resolved signal

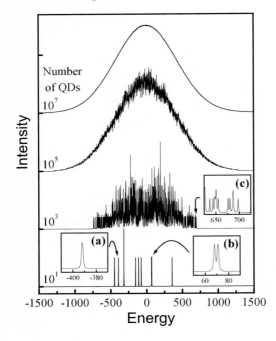

**Fig. 4.3.** Simulation of the absorption spectra of QDs. The number of QDs probed was varied from $10^7$ to 10. For the simulation, $\Gamma_i/\Gamma_h \sim 1000$ is assumed

from a single QD. Figure 4.3 shows a series of absorption spectra obtained by a simple simulation where various numbers of QDs are probed. In the simulation, a Gaussian size distribution and a Lorenzian line shape of each QD were assumed. Here, since we consider only the ground state of the exciton, the spectral shape is Gaussian when the number of the QDs is very high ($\geq 10^5$). The homogeneous line width $\Gamma_h$ of each line was set to be 1000 times smaller than the inhomogeneous line width $\Gamma_i$, which is typical in many III–V SADs. As the number of the QDs probed is decreased, the sharp line reflecting the energy state of each QD becomes recognizable. When the number of QDs probed becomes very small, each line is well-resolved (inset a of Fig. 4.3), while the energy states of some QDs may be very close together (inset b of Fig. 4.3). From the figure, we can see that the number of QDs should be limited to less than $\sim 10^3$ in order to observe single QDs over a wide energy range. Even if the number of QDs probed is as high as $10^3$, it is possible to observe some lines at both shoulders of the spectrum, as shown in inset c of Fig. 4.3, because the number of very large or small QDs is few, while the energy range that we can investigate is restricted in this case.

Now we consider the realistic situations of the experiments. When a conventional macroscopic setup for optical measurements is used, a laser beam can be focused to a diameter of about 100 μm. However, it is almost impossible to fabricate only one or a few QDs in this area at the present. In the case of III–V SADs, since the typical areal dot density is $10^7 \sim 10^{10}$ cm$^{-2}$, the

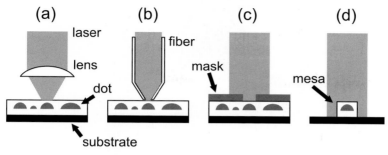

**Fig. 4.4.** Several techniques to get access to a single QD: (**a**) microscope; (**b**) SNOM; (**c**) a metal mask with an aperture; and (**d**) a QD in a mesa

number of SADs thus excited through the macroscopic setup is estimated to be about $10^3 \sim 10^6$. As can be seen in Fig. 4.3, the detection of well-resolved sharp lines is very difficult if a conventional optical method is used. We therefore need to make some effort to limit the number of QDs excited or detected. In order to select and probe a signal from a single QD, several techniques are usually adopted [21].

First, the simplest but very flexible method is to study PL through an objective lens of a microscope, which is known as the μ-PL measurement. As schematically drawn in Fig. 4.4a, the excitation laser beam is tightly focused and the sample PL is also detected through the same objective lens. Since the beam size is determined mainly by the wavelength of the laser, the typical spatial resolution of the μ-PL in the visible region is $\sim 1$ μm.

A comparable technique is to use a confocal microscope system as drawn in Fig. 4.5. Most of the μ-PL spectra and images presented in this chapter were successfully obtained using this technique. A single QD can be selected by a pinhole placed on the image plane, which determines the spatial resolution of the detection area. An advantage of the this technique is that, although a wide region of the sample is uniformly excited, the spatial resolution is better than the excitation through a microscope objective lens shown in Fig. 4.4a, which is an important point for image observations.

A μ-PL system referred to as far-field spectroscopy has low spectral resolution in comparison with near-field spectroscopy. However, in far-field spectroscopy, it is possible to observe several QDs simultaneously. Further, it is easy to introduce two spatially separated laser beams when the μ-PL setup is employed. It offers a great advantage for studying nonlinear effects by means of two-photon absorption and four-wave mixing, and such experiments are expected to be realized in the near future.

If the density of the QDs is higher than $10^8$ cm$^{-2}$ (more than one QD in 1 μm square), it is impossible to restrict the sample area to be excited or detected to a single QD by only using a microscope objective lens because of the diffraction limit of light. A scanning near-field optical microscopy (SNOM)

Sugisaki

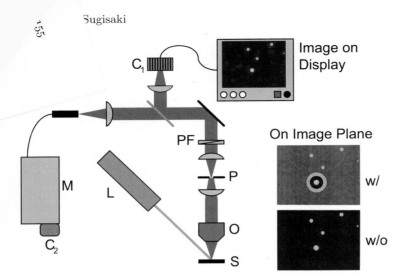

**Fig. 4.5.** Schematic drawing of a confocal μ-PL setup. An unfocused laser beam (L) is irradiated on the samples (S) set onto a cold-finger of a liquid He flow-type cryostat. The PL from the sample was collected using a microscope objective lens (O) with a numerical aperture of 0.42. Many bright PL spots from individual QDs are projected on the image plane. When the pinhole (P) is set on the image plane, the PL from the hatched area is masked, and consequently only the PL from a single QD at the center can be selected. A film polarizer and some band pass filters (PF) were set inside the microscope, and then the images were detected by a thermoelectrically cooled charge-coupled device (CCD) camera (C₁). The image shown on the display is the same as that on the image plane. In order to measure the μ-PL spectra, the sample PL is led to a single monochromator (M) through bundled optical fibers and then detected by a liquid-nitrogen-cooled CCD camera (C₂)

system, as shown in Fig. 4.4b, is a powerful and more sophisticated diagnostic tool for such samples containing high-density QDs [22–24]. In a SNOM system, the spatial resolution is determined by the aperture size on the fiber probe and is smaller than the diffraction limit of light. In some cases, both an optical fiber and a microscope objective lens are combined to increase the efficiency of the signal collection. Since a SNOM can access from a few microns to 0.1 μm range, it intermediates between μ-PL and a scanning tunneling microscope (SEM) or a transmission electron microscope (TEM). It is believed that SNOM has the potential to bridge the gap between these methods.

Processing steps on the sample surface are also widely used to reduce the number of QDs. As drawn in Fig. 4.4c, coating an opaque metal mask with small windows on the sample surface is a well-known technique [10,25]. The optical measurements are performed through the aperture, and consequently the spatial resolution is determined by the size of the window. The advantage of this technique is that it is less sensitive to small drifts, such as thermal drifts of the sample and small vibrations due to external origins, than the μ-PL measurement. Namely, in the case of the μ-PL measurements as drawn in Fig. 4.4a, the stability of the sample position is crucial and the drift should be less than the laser spot size. Otherwise, the effective spatial resolution becomes as large as such drifts. In the case of the sample covered with a metal mask, on the other hand, since the laser beam does not have to be focused on the sample surface, such drifts are negligible in comparison with the laser beam diameter. The main disadvantage of using a metal mask is that there is less freedom to select the excitation position.

Another popular processing technique is the fabrication of QDs in a mesa as shown in Fig. 4.4d, which is a similar method as that of a sample with a metal mask as mentioned above. After the sample is grown, most of the QDs and buffer layers are removed by chemical etching or ion milling leaving a small mesa [15,26–28]. This method has the same advantages and disadvantages as a metal mask shown in Fig. 4.4c. Some examples of μ-PL spectra using mesa samples are shown in Sect. 4.7.

In addition to the techniques shown in Fig. 4.4, a single QD structure realized by two-fold application of cleaved edge overgrowth is also interesting and promising [29]. This growth method is expected to be the one to obtain the ultimate low-density QD sample. Since the size, position, and shape of the QDs are controllable, it will be possible to characterize artificial molecules by fabricating a pair of such QDs.

Finally, cathodeluminescence has also been employed successfully to avoid inhomogeneous broadening in the spectroscopy of single QDs [30,31]. In this case, since the excitation energy is much higher than the band gap energy, the spatial resolution is limited by the diffusion length of the carriers. In this sense, combining this technique with a metal mask as mentioned previously is more effective.

## 4.3    Observation Energy Dependence and Optical Anisotropy

In this section, we compare the macro-PL spectra and μ-PL images. Figure 4.6a shows the macro-PL spectra of InP SADs measured at a cryogenic temperature. The samples were prepared by means of a metal organic vapor-phase epitaxy (MOVPE) [32,33]. The InP SADs are sandwiched by two $Ga_{0.5}In_{0.5}P$ layers grown on a GaAs substrate. The photo-excited carriers are generated in the $Ga_{0.5}In_{0.5}P$ matrix by a continuous wave (cw) Ar-ion laser. The PL bands coming from the excitons in the GaAs substrate and the $Ga_{0.5}In_{0.5}P$ matrix are observed at 1.52 eV and 1.87 eV, respectively. The broad PL band observed at 1.63 eV arises from the SADs. Since more than $10^4$ QDs are excited at the same time, the expected $\delta$-function-like character of the density of states of each SAD is hidden by size and shape fluctua-

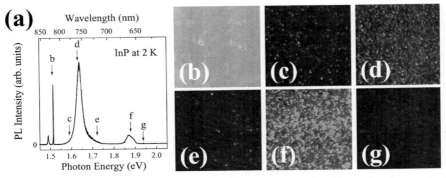

**Fig. 4.6.** (a) Macro-PL spectra of InP SADs at 2 K. The PL peaks observed at 1.87 eV, 1.63 eV and 1.52 eV come from the $Ga_{0.5}In_{0.5}P$ matrix, the InP SADs, and the GaAs substrate, respectively. (**b–g**) Micro-images observed at the various energies marked by the *arrows* in (**a**). The image area is 75 µm×77 µm. Each PL spot observed in (**c–e**) comes from a single InP SAD. Because of the formation of the natural superlattice domains, many PL spots are observed in the µ-PL image from the $Ga_{0.5}In_{0.5}P$ matrix shown in (**f**), while the bulk GaAs in (**b**) shows the PL from a whole surface. All images were taken at 6 K

tions. As discussed in Sect. 4.4 (see Fig. 4.13), the excitation power is weak enough to prevent the state-filling effect. Thus, the PL mainly comes from the radiative decay of the confined excitons in the ground state.

Figure 4.6b–g shows the micro-images of the sample PL observed at the various energies, as marked by arrows b–g in Fig. 4.6a, respectively. When the micro-image is taken at the PL peak from the free excitons of GaAs (1.52 eV), the PL from the whole surface of the sample is observed as shown in Fig. 4.6b. The images shown in Fig. 4.6c–e were taken in the vicinity of the PL band of the InP SADs. Each white spot shows the PL coming from a single InP SAD, which is also confirmed by the µ-PL measurement, as shown in Figs. 4.2 and 4.15. Namely, the expected $\delta$-function-like µ-PL spectrum reflecting the density of state of a 0D exciton is observed when one of the PL spots is selected.

The shape of macro-PL spectrum measured under weak excitation is expected to reflect the distribution of the density of QDs. The observed results show good correspondence to each other, i.e., the number of the PL spots observed at the PL peak of the SADs (Fig. 4.6d) is much higher than that observed at the lower- and higher-energy side of the PL band (Fig. 4.6c,e, respectively).

By observing the µ-PL image without a band-pass filter, the areal dot density is estimated to be about $10^8$ cm$^{-2}$. It should be mentioned that, in comparison with a sample without a cap layer for an atomic force microscope (AFM) observation, the dot density estimated by the µ-PL imaging is the same order of magnitude as that estimated using AFM.

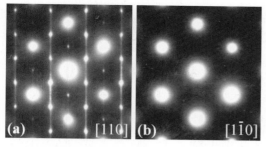

**Fig. 4.7.** Selective area transmission electron diffraction patterns along the **(a)** [110] and **(b)** [1$\bar{1}$0] zone-axes of the sample. Super spots at $(\frac{\bar{1}}{2}, \frac{1}{2}, \frac{1}{2})$ and $(\frac{1}{2}, \frac{\bar{1}}{2}, \frac{1}{2})$ and other equivalent positions are clearly observed along the [110] zone-axis. It indicates that the lattice periods are doublet in the [$\bar{1}$11] and [1$\bar{1}$1] directions of the Ga$_{0.5}$In$_{0.5}$P due to the presence of two variants of Cu–Pt$_B$-type ordering (from [32])

The interesting point about the InP SADs in the Ga$_{0.5}$In$_{0.5}$P matrix is that a number of small PL spots are observed at the PL band of the Ga$_{0.5}$In$_{0.5}$P matrix as well, as shown in Fig. 4.6f. As shown in Fig. 4.7, a selective-area transmission electron diffraction (TED) measurement clarified that it is due to Cu–Pt$_B$-type long-range ordering. The formation of the long-range ordered domains causes the strong optical anisotropy of the PL bands from the Ga$_{0.5}$In$_{0.5}$P matrix and the InP SADs, as discussed in Sect. 4.3.1. The PL was too weak to be detected when observed above 1.95 eV (e.g. Fig. 4.6g).

As mentioned in Sect. 4.2, the excitons trapped at local potential minima of the quantum wells due to interface roughness are well explained by considering them as quasi-0D systems. Lowisch et al. have pointed out the validity of this idea by observing μ-PL images at various energies [12]. Their results show the importance of a μ-PL image observation, which helps us characterize the real dimension of the systems.

### 4.3.1 Mechanism for Optical Anisotropy

Studying the polarization dependence is of great importance in low-dimensional systems because it directly provides information about the anisotropic electronic states inherent to each structure. In addition, the control of the polarization is a demand for opto-electronic applications. The thick and thin curves in Fig. 4.8a show the linear polarization dependence of the macro-PL spectra along the [110] and [1$\bar{1}$0] directions of the GaAs substrate, respectively [34]. Figure 4.8b shows the degree of linear polarization of the PL spectra shown in Fig. 4.8a, which is defined by

$$P = \frac{I_{[110]} - I_{[1\bar{1}0]}}{I_{[110]} + I_{[1\bar{1}0]}}, \tag{4.8}$$

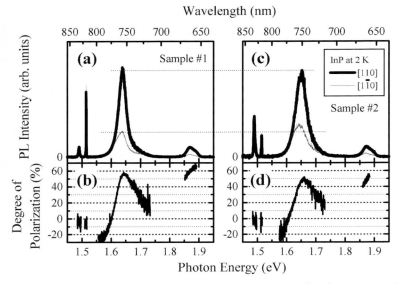

**Fig. 4.8.** (**a**) Macro-PL spectra at 2 K measured for the [110] (*thick curve*) and [1$\bar{1}$0] (*thin curve*) polarizations; (**b**) Degree of linear polarization of PL spectra shown in (a). The PL band from the GaAs observed at 1.5 eV is isotropic, while those from the InP SADs (1.63 eV) and Ga$_{0.5}$In$_{0.5}$P (1.87 eV) show strong optical anisotropy; (**c**) Polarization dependence of macro-PL spectra measured using another sample with a different degree of matrix ordering. Since the InP SADs layer of sample 2 was grown at the same temperature as that of sample 1, the average dot sizes are the same between the both samples; (**d**) Degree of polarization of the PL spectra shown in (**c**). Since the Ga$_{0.5}$In$_{0.5}$P matrix of sample 2 was grown at lower temperature than that of sample 1, the optical anisotropy of sample 2 is less than that of sample 1 (from [34])

where $I_{[110]}$ and $I_{[1\bar{1}0]}$ are the PL intensities observed for the [110] and [1$\bar{1}$0] polarizations, respectively. The PL peaks from the GaAs substrate observed around 1.5 eV do not show any optical anisotropy as expected for cubic ($T_d$) symmetry. The PL peaks from the Ga$_{0.5}$In$_{0.5}$P matrix and the InP SADs, however, show strong polarization dependence, although both bulk Ga$_{0.5}$In$_{0.5}$P and bulk InP have the same zinc-blende structure as GaAs. The polarization of the PL is not influenced by that of the excitation laser since the excitation photon energy is above the band gap of the Ga$_{0.5}$In$_{0.5}$P matrix, and thus the initial polarization is lost through interaction with phonons during carrier relaxation into the InP SADs. Therefore, the observed optical anisotropy reflects the symmetry of the luminescence states of the Ga$_{0.5}$In$_{0.5}$P matrix and the InP SADs. The peak intensity of the PL band observed for the [110] polarization is more than twice as strong as that observed for the [1$\bar{1}$0] polarization. The degrees of polarization at the PL peaks of the InP SADs and the Ga$_{0.5}$In$_{0.5}$P matrix are 55% and 59%, respectively.

**Table 4.1.** Growth temperature, size, aspect ratio, and optical properties of InP SADs in a $Ga_{0.5}In_{0.5}P$ matrix. For comparison, those of In(Ga)As SADs in the GaAs matrix are also summarized

| Sample | Size of SADs ($nm^3$) | Aspect Ratio | Degree of polarization in | |
|---|---|---|---|---|
| | | | matrix (%) | SADs (%) |
| InP (MOVPE #1)[a] | $25 \times 35 \times 7$ | 1.4 | 59 | 55 |
| InP (MOVPE #2)[a] | | | 48 | 44 |
| InP (MBE #1)[b] | $35 \times 45 \times 6$ | 1.3 | 83 | 44 |
| InP (MBE #2)[b] | $40 \times 48 \times 5$ | 1.2 | 30 | 13 |
| InGaAs[c] | $17 \times 33 \times 3$ | 1.9 | | 16 |
| InAs[d] | $13 \times 15 \times 3$ | 1.2 | | 32 |
| InAs[e] | | > 2 | | 13 |
| InAs[f] | $25 \times 30 \times 3$ | 1.2 | | 7 |
| InAs[g] | $16 \times 28 \times 4$ | 1.8 | | 9 |

[a] [34].   [b] [45].
[c] [35].   [d] [36].
[e] [37].   [f] [38].
[g] [39].

Optical anisotropy of In(Ga)As SADs has been reported by several authors [35–39] and attributed to the anisotropic shape of the SADs. As summarized in Table 4.1, the reported values of the degree of polarization of the In(Ga)As range from 32% to 7%, which are smaller than those of InP SADs, while the shape anisotropies (aspect ratios) of these samples are larger than those of the InP SADs. Further, the degree of linear polarization observed in the InP SADs is as high as that reported in several quantum wire systems, which are considered as the ultimate anisotropic QDs. This result implies that, in addition to the shape anisotropy, we have to take other effects into consideration in order to explain the strong optical anisotropy observed in the InP SADs.

Based on the idea that the optical anisotropy in InP SADs is induced by the anisotropic structure of the surrounding matrix, several samples having various degrees of $Cu–Pt_B$-type ordering have been prepared by changing the growth temperature of the $Ga_{0.5}In_{0.5}P$ matrix layer systematically [32,33]. The polarized PL spectra measured using another sample are shown in Fig. 4.8c. Since the InP SAD layers of the samples shown in Fig. 4.8a,c were all grown at 550°C, the average dot sizes of these samples estimated by AFM measurements were almost the same [32]. On the other hand, since the growth temperatures of the $Ga_{0.5}In_{0.5}P$ matrix of these samples were 660°C and 630°C, respectively, the degree of $Cu–Pt_B$-type ordering differs between the samples. This resulted in the degree of polarization (i.e., the optical anisotropy) of the sample shown in Fig. 4.8c being smaller than that shown in Fig. 4.8a.

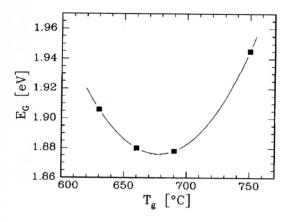

**Fig. 4.9.** PL peak energies of ordered $Ga_{0.5}In_{0.5}P$ grown at various temperatures (from [43])

As reported by several authors, the PL peak energy of the ordered $Ga_{0.5}In_{0.5}P$ strongly depends on the growth temperature and the V/III ratio, indicating the degree of formation of the $Cu–Pt_B$-type long-range ordered domains is governed by these parameters [40–43]. Scholz et al. have performed a systematic study of the band gap energy of $Ga_{0.5}In_{0.5}P$ over a wide growth temperature range [43]. As shown in Fig. 4.9, they found that the degree of ordering becomes maximum when a sample is grown at around 670°C, and that the degree of ordering is reduced on decreasing the growth temperature, which is consistent with the results shown in Fig. 4.8.

From a systematic study of the PL spectra of these samples, it was found that the degree of optical anisotropy in InP SADs is proportional to the degree of ordering of the $Ga_{0.5}In_{0.5}P$ matrix. Figure 4.10 plots the degree of polarization of InP SADs versus that of $Ga_{0.5}In_{0.5}P$ matrix. The most important

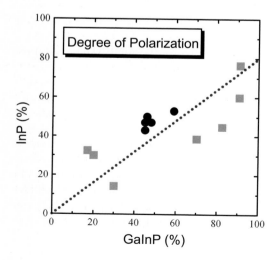

**Fig. 4.10.** Relation between the degree of polarization of the macro-PL peak from the InP SADs and that from the $Ga_{0.5}In_{0.5}P$ matrix. For comparison, the results obtained using the sample grown by a GS-MBE system are also shown by *squares*. The *dotted line* is a guide for the eyes. The degree of polarization in the InP SADs is proportional to that of the $Ga_{0.5}In_{0.5}P$ matrix (from [34])

observation is that the optical anisotropy of the InP SADs becomes strong when the InP SADs are confined in a highly ordered $Ga_{0.5}In_{0.5}P$ matrix. In Fig. 4.10, the degree of polarization of InP SADs embedded in a $Ga_{0.5}In_{0.5}P$ matrix grown by a GS-MBE system is also plotted for comparison. From these experimental results, we conclude that the most probable origin of the optical anisotropy in InP SADs is the anisotropic energy structure of the $Ga_{0.5}In_{0.5}P$ matrix.

It is important to note that although the optical anisotropy is observed in the samples grown by both MBE and MOVPE systems, the crystal structures and symmetry axes of the $Ga_{0.5}In_{0.5}P$ matrix differ between the samples obtained using these systems. In the case of the samples grown by an MBE system, the composition modulation planes with a period of a few tens of nanometers are formed along the $[1\bar{1}0]$ direction [44]. Thus the PL is polarized for the $[1\bar{1}0]$ direction [45]. On the other hand, in the case of an MOVPE system, the Cu–$Pt_B$-type ordered $Ga_{0.5}In_{0.5}P$ has $C_{3v}$ symmetry with the symmetry axis along the [111] crystal directions [32,46]. Since the optical measurements were performed using the samples grown on (100) surfaces, a two-fold symmetry polarized along the [110] direction is observed for the PL from the $Ga_{0.5}In_{0.5}P$ matrix. In both cases, the important point is that SADs embedded in an anisotropic matrix are subjected to an anisotropic strain that becomes stronger with the increase of the degree of asymmetry in the matrix, resulting in a strong optical anisotropy in the InP SADs. This explains the relation shown in Fig. 4.10.

### 4.3.2   Optical Anisotropy of Individual Quantum Dots

We will now take a close look at the μ-PL images. Figure 4.11a–d shows μ-PL images observed in the same region of the sample for various polarizations. The image shown in Fig. 4.11a was taken for the [110] polarization. Many SADs showing strong PL are clearly observed. The PL intensities of most of the SADs become weaker as the polarization of the observation is rotated (Fig. 4.11a–c), and then recover once again (Fig. 4.11c,d). In almost all cases, the PL from a single QD is polarized along the same direction, i.e., the μ-PL spots are strong in intensity when observed for the [110] polarization and become very weak for the $[1\bar{1}0]$ polarization (e.g., dots A and B). This is consistent with the macro-PL observations. The PL intensities from a few of the SADs, however, are almost isotropic (e.g., dot C).

The integrated PL intensities $I(\theta)$ of dots A, B, and C are plotted in Fig. 4.11e as a function of the polarization angle $\theta$ of the analyzer, where the polarization directions $\theta=0°$ and $90°$ are parallel to the [110] and $[1\bar{1}0]$ directions of the GaAs substrate, respectively. Dots A and B clearly exhibit a two-fold symmetry and are well fitted by

$$I(\theta) = X \sin^2 \theta + Y \cos^2 \theta \ , \tag{4.9}$$

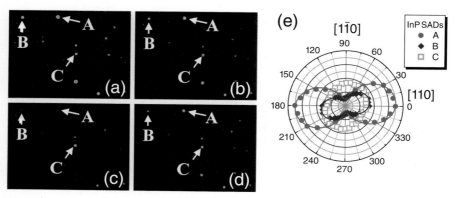

**Fig. 4.11.** (**a–d**) Micro-PL images observed for various polarizations: (**a**) 0°; (**b**) 45°; (**c**) 90°; and (**d**) 135°. The observed area is 40 μm×53 μm. (**e**) Polar plots of the integrated μ-PL intensities of dots A to C marked in the images (**a**) to (**d**), where 0° and 90° mean that the polarizations of observation are parallel to the [110] and [1\overline{1}0] directions of the GaAs substrate, respectively. The *solid curves* are using (4.9). The degrees of polarization of dots A and B are 75% and 55%, respectively (from [34])

where $X$ and $Y$ are constants. This suggests that the InP SADs have two-fold symmetric structures with symmetry axes along the [110] and [1\overline{1}0] directions. Notably, the degree of polarization of dot A is more than 70%, which is as high as that in 1D systems [47].

The $Ga_{0.5}In_{0.5}P$ containing Cu–Pt$_B$-type ordered domains within a disordered matrix can be regarded as a quasi-zero-dimensional system, as mentioned by DeLong et al. [48]. Recently, Kops et al. have reported a small diamagnetic shift and sharp μ-PL lines in bulk $Ga_{0.5}In_{0.5}P$, which supports DeLong's idea [17]. Further, it is known that the band gap energy strongly depends on the sample growth conditions and that the formation of the Cu–Pt$_B$-type ordered domains causes a reduction of the band gap energy (see Fig. 4.9). As shown in Fig. 4.8, the observed PL peak energy of the $Ga_{0.5}In_{0.5}P$ matrix (∼1.9 eV) is about 100 meV lower than that of the completely random $Ga_{0.5}In_{0.5}P$ alloy (∼2.01 eV at 5 K) [49], again indicating that the $Ga_{0.5}In_{0.5}P$ layers have a naturally ordered structure.

Cross-sectional TEM observations revealed that the sizes of ordered domains and the degree of ordering differ from site to site (see Fig. 4.12). The local fluctuations in the domain size and degree of ordering in the $Ga_{0.5}In_{0.5}P$ surrounding the InP SADs also induce variations in the strain applied to individual InP SADs. Therefore, it is concluded that the observed variations in the optical anisotropy of each InP SAD reflect the local fluctuation of the anisotropic strain from the $Ga_{0.5}In_{0.5}P$ matrix.

The important point found by this study is that the polarization of the QD is controllable by changing a symmetry of a matrix. This novel information

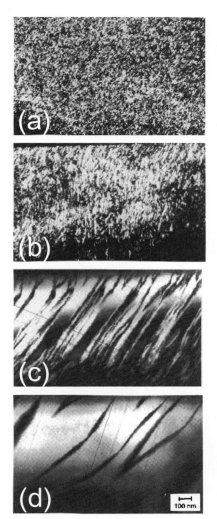

**Fig. 4.12.** Dark field TEM images of highly ordered $Ga_{0.5}In_{0.5}P$ layers grown at (**a**) 630; (**b**) 660; (**c**) 690; and (**d**) 750°C. The domain size and the degree of ordering strongly depend on the growth temperature. The lateral domain size varies systematically from (**a**) ~10 nm to (**d**) ~500 nm (from [43])

has potential to be applied to control the polarization of QD-based optoelectronic devices.

## 4.4  Many Carrier Effects

In 3D semiconductor systems, most of the radiative relaxation of the excitons occurs from the lowest $1s$ state. In 0D semiconductors, on the other hand, because of the finite degeneracy and discrete eigenstates, the lower lying states are occupied under moderate excitation power, and thus many confined excitons radiatively relax directly from the higher exciton states. Further, because of a theoretically predicted high nonlinearity of the 0D systems, multiexciton states are expected to be easily formed. It is thus interesting to study many carrier states in QDs.

In this section, the state-filling effect will be discussed first, followed by the many exciton states in a single QD.

### 4.4.1    State Filling Effects Studied by Micro-Imaging

Figure 4.13a shows macro-PL spectra observed under various excitation powers [50]. Under the weakest excitation, the bandwidth of the PL from the InP SAD was about 40 meV having the peak energy at 1.63 eV. With the increase of the excitation power density, the PL bandwidth becomes wider and the peak energy shifts to the higher-energy side due to the state-filling effect. A blueshift of the PL from the $Ga_{0.5}In_{0.5}P$ matrix is also observed. This provides additional evidence that the $Ga_{0.5}In_{0.5}P$ matrix forms the Cu–Pt$_B$-type natural superlattice domains, which causes a strong optical anisotropy in both the InP SADs and the $Ga_{0.5}In_{0.5}P$ matrix, as mentioned in Sect. 4.3.

The µ-PL images taken under various excitation power densities are shown in Fig. 4.13b–e. The images were observed at 1.82 eV as indicated by the arrow in Fig. 4.13a, which corresponds to the excited state energy of most of the SADs. Thus no PL from SADs was detected under very weak excitation. As marked by the dotted circle in Fig. 4.13b, when the excitation power

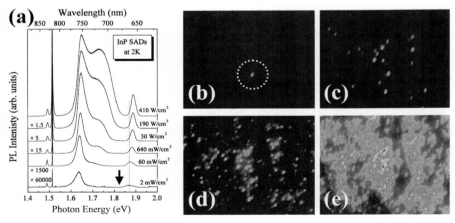

**Fig. 4.13.** (a) Excitation power dependence of the macro-PL spectra. The PL bandwidth of the InP SADs increases and the PL peak energy shifts to the higher energy side as the excitation power is increased because of the state-filling effect. (b–e) µ-PL images observed under various excitation powers at 1.82 eV as denoted by the *arrow* in (a). The observed area is 75 µm×95 µm. All images were taken using the same area of the sample. Under weak excitation only one SAD shows PL as marked by the *dotted circle* (b). With the increase of the excitation power (c–e), the number of the observed PL spots from single SADs is increased (from [50])

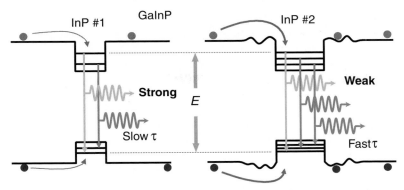

**Fig. 4.14.** Schematic drawing to explain the difference in the efficiency of the state-filling. Dot #1 has smooth interfaces and the radiative decay time $\tau$ of the lower laying exciton state is fast, while dot #2 has rough interfaces and fast $\tau$. In this case the efficiency of the state-filling of dot #1 is higher than that of dot #2. Further, since the second excited state of dot #1 and the third of dot #2 are observed simultaneously, the threshold of the detection differs from dot to dot (from [50])

reached $\sim$10 mW/cm$^2$, one PL spot from a single InP SAD was observed. This indicates that the lower exciton states of this SAD observed in Fig. 4.13b are filled at this excitation power density. With increasing excitation power (from Fig. 4.13c to Fig. 4.13e), the number of the PL spots observed also increased. This result indicates that state-filling easily occurs in some QDs, but is less efficient in other QDs.

As schematically drawn in Fig. 4.14, the variation of the efficiency in the state-filling effect can be explained as follows. The state-filling efficiency is mainly determined by two parameters. The first parameter is the relaxation rate of the lower-state excitons. If the excitons in the lower states relax efficiently, those in the higher states can successively relax to the lower state, and then radiatively annihilate in the lower state. In this case, the PL from the higher exciton states is weak, i.e., the state-filling does not occur efficiently.[1] The second parameter is the incoming number of excitons from the Ga$_{0.5}$In$_{0.5}$P matrix. If many carriers relax into the InP SADs from the Ga$_{0.5}$In$_{0.5}$P matrix, state-filling occurs easily. The efficiency of the incoming excitons from the Ga$_{0.5}$In$_{0.5}$P matrix depends on the quality of the interfaces between the matrix and the SADs. The SADs with rough interfaces have high potential barriers (InP #2 in Fig. 4.14), which prevent the relaxation of the carriers from the matrix to the SADs. Matsuda et al. have recently reported that the PL intensities from single InAs SADs under excitation below the absorption edge of the wetting layer are less dispersed than those measured under band-to-band excitation of the matrix [24]. This result supports the explanation that the incoming number of excitons to the QDs is influenced

---

[1] The PL decay time in InP SADs has been reported in [51–53].

by the interfaces between the QDs and the matrix. Thus, we can conclude the variation of efficiency in the state-filling effect is due to a competition between the inflow and outflow rates of the excitons in QDs. Moreover, we observed several excited states simultaneously even though the detection energy was fixed. For example, the observed PL comes from the second excited state of the exciton in one QD (InP #1 in Fig. 4.14) and from the third excited state in another QD (InP #2 in Fig. 4.14). The lower lying state is filled faster than the higher lying state, resulting in the low threshold of detection of the PL. The variation in the excitation threshold of the PL observation is due to these reasons.

## 4.4.2   Multiexciton States

In order to study the carrier–carrier interaction and the electronic structures of a 0D system, it is informative to observe the excitation power dependence of the μ-PL spectrum from a *single* QD. Figure 4.15 shows the μ-PL spectra of a single InP SAD observed over a wide range of excitation power density from 51 mW/cm$^2$ to 11 μW/cm$^2$. The PL peak observed at 1.88 eV comes from the Ga$_{0.5}$In$_{0.5}$P matrix. The PL peaks arising from the single InP SAD

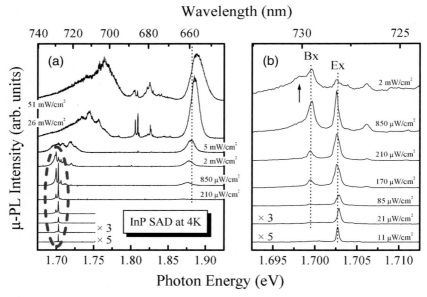

**Fig. 4.15.** Excitation power dependence of the μ-PL spectra from a single InP SAD at 4 K. The excitation power density was changed from 51 mW/cm$^2$ (*top curve of* (**a**)) to 11 μW/cm$^2$ (*bottom curves* of both (**a**) and (**b**)). (**a**) As the excitation power is decreased, the PL bandwidth from the InP SAD becomes narrower. (**b**) The magnified μ-PL spectra of the *dotted circle area* shown in (**a**). Ex and Bx stand for the PL bands from one- and two-exciton states, respectively (from [50])

are observed at energies below ∼1.85 eV. Under strong power excitation, the PL band from a single SAD is wide because of exciton–exciton interaction (the top curve of Fig. 4.15a). When the excitation power is reduced to 26 mW/cm$^2$, the PL peaks from the higher states of the confined excitons observed at ∼1.8 eV become sharp, while those observed below 1.77 eV are still broad. As the excitation power is decreased, the PL bandwidth becomes narrower and a few sharp PL lines remain. For example, the PL bands initially peaking at around 1.75 eV are as broad as ∼50 meV in width, which eventually become a single line with the linewidth of less than 1 meV accompanied by a redshift of the spectral weight by about 50 meV.

The magnified μ-PL spectra in the dotted circle region are shown in Fig. 4.15b. Only a single sharp PL line reflecting the density of states of the zero-dimensional exciton is observed at 1.7026 eV, as denoted by Ex (the bottom curve of Fig. 4.15b), where the estimated average number of excitons in an InP SAD is less than 0.01 under this excitation condition. Namely, we can conclude that the PL line comes from the confined excitons in the ground state.

Note that the μ-PL from a single InP SAD was observed by spatially selecting one of the bright PL spots observed in a μ-PL image (see Fig. 4.13). In almost all the samples presented here, sharp μ-PL lines from single InP SADs were observed at the lower-energy side of the Ga$_{0.5}$In$_{0.5}$P matrix by more than about 150 meV. The separations of the PL peak energies between the InP SADs and Ga$_{0.5}$In$_{0.5}$P matrix is consistent between the macro-PL and μ-PL observations. As mentioned in Sect. 4.3, the PL peak energy from the Ga$_{0.5}$In$_{0.5}$P matrix depends on the growth conditions because of the formation of Cu–Pt$_B$-type ordered domains [40–43]. The excitons trapped at the ordered domains of the Ga$_{0.5}$In$_{0.5}$P show quasi-0D behavior, such as sharp μ-PL lines and an anisotropic diamagnetic shift. In these samples, however, the μ-PL signal from Ga$_{0.5}$In$_{0.5}$P below 1.82 eV was too weak to be detected. Thus it is again obvious that each PL spot observed in Fig. 4.13b–e comes from the individual InP SAD, not from the local potential minimum in the Ga$_{0.5}$In$_{0.5}$P matrix.

In Fig. 4.15b, the PL line denoted by Bx grows superlinearly with increasing excitation power under moderate excitation powers. Its energy position is lower than the Ex line by about 3 meV. The μ-PL intensities of the Ex and Bx lines are plotted in Fig. 4.16 as a function of the excitation power density. The PL intensities of these lines increased with the excitation laser power below ∼1 mW/cm$^2$ and then decreased.

The master equation for the probability $f_N$ that $N$ excitons are created in a SAD may be written as

$$\frac{df_N}{dt} = \alpha(t)f_{N-1} - \alpha(t)f_N + (N+1)f_{N+1} - Nf_N , \tag{4.10}$$

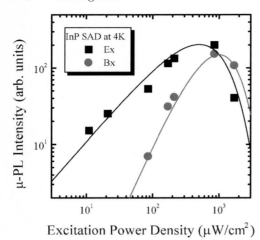

**Fig. 4.16.** Excitation power dependence of the Ex (*squares*) and Bx (*circles*) lines shown in Fig. 4.15. These results are well reproduced using (4.12) (from [50])

with

$$\sum_N f_N = 1 \ , \tag{4.11}$$

where $\alpha(t)$ is the exciton generation rate that is proportional to the excitation power $P$. The first and second terms of (4.10) reflect the photoexcitation and successive relaxation of the excitons to the $N-1$ and $N$ states, respectively, and the third and fourth terms represent the decay of the excitons from the $N+1$ and $N$ exciton states, respectively. Under cw excitation ($\alpha(t) = \alpha$), we obtain the Poisson distribution from (4.10) and (4.11),

$$f_N = \frac{\alpha^N}{N!} e^{-\alpha} \ . \tag{4.12}$$

As shown by the solid curves in Fig. 4.16, by using a single parameter $\alpha$, the excitation power dependence of both the Ex and Bx lines are well reproduced. It is thus concluded that the PL lines Ex and Bx arise from the radiative decay of the single excitons ($N = 1$) and biexcitons ($N = 2$), respectively.

In addition, some interesting phenomena in the μ-PL spectra have been found as shown in Fig. 4.15b. A shoulder, as indicated by the arrow, is observed just below the biexciton line at 1.698 eV. Recently, many exciton states have been theoretically studied (see Fig. 3 in [54]). The comparison of the experimental results with the theoretical model suggests that the shoulder arises from the three-exciton state. Further, the observed PL energies from single-excitons and biexcitons shift reproducibly by minute amounts with the excitation power. This phenomenon is observed in many InP SADs. The origin has not yet been clarified, but could be related to the spectral diffusion observed in various kind of QDs, as discussed in Sect. 4.6.

### 4.4.3   Biexciton Binding Energy

By examining several SADs, it was found that the PL peak energy of the confined exciton and the biexciton binding energy differ from dot to dot as shown in Fig. 4.17. The main peak of each spectrum comes from a single exciton state, while the small peak indicated by the arrow is due to the biexciton. In all the cases shown here, the μ-PL linewidth of the main peak is less than 300 μeV, which is limited by the resolution of the detectors. The dispersion of the PL energy is considered to reflect the fluctuations in the size and shape of each dot. As reported by Charbonneau et al., the binding energy of bulk InP is about 1 meV [55]. The binding energy of the biexciton in the InP SAD is, however, a few times larger than that of the bulk InP. This result indicates that the multiexciton state is stabilized due to spatial confinement in the SAD.

The multiexciton and charged exciton states are a matter of concern for many researchers studying low-dimensional semiconductors these days. Biexcitons have also been studied in In(Ga)As SADs. However, their formation and relaxation processes, especially in connection with the contribution of LO phonons, are still under discussion [5,56–58]. The charged excitons formed due to the supplement of excess carriers from doped barriers have recently been observed [59].

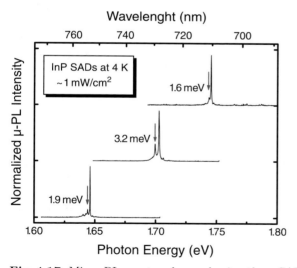

**Fig. 4.17.** Micro-PL spectra observed using three SADs. The main peak of each μ-PL spectrum arises from the confined exciton in the ground state, which corresponds to the Ex line in Fig. 4.15b. The peaks denoted by the *arrows* are the biexciton lines, which correspond to the Bx line in Fig. 4.15b. The binding energy of the biexciton differs from dot to dot, and is a few times larger than that of bulk InP. The numbers indicate the binding energies of the biexcitons (from [50])

More detailed investigations of this phenomenon using other materials will be an interesting subject for further study. Further, the observation of the excited states of biexcitons and charged excitons will be a challenging, but very intriguing topic.

## 4.5    Temperature Dependence

Micro-PL measurmets at various temperatures are important because they provide information about the electronic structures and the interaction between elementary excitations. In this section, we discuss the band gap energy shift and the thermal activation in individual InP SADs.

### 4.5.1    Band Gap Energy Shift

Figure 4.18a shows the μ-PL spectra measured under various temperatures ranging from 5 K to 80 K. The diameter of the observed area is about $\phi 100$ μm. Since each QD has different PL efficiency (as shown in Fig. 4.20a–f), each PL peak shows different intensity. Successive redshifts from the single

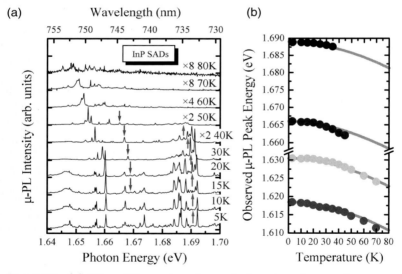

**Fig. 4.18.** (a) Micro-PL spectra observed at various temperatures ranging from 5 K to 80 K. The observed area size is $\sim\phi 100$ μm. Successive redshifts of the PL peaks from the QDs were clearly observed with increasing temperature. The PL lines disappear from the higher-energy side as the temperature is raised. Some peaks (e.g., ones marked by the *arrows*) that are not observed at low temperature appear when the temperature is raised (from [6]). (b) Temperature dependence of μ-PL peak energies of some individual InP SADs. The energy shifts are well fitted by (4.13)

QDs showing strong PL were clearly observed. The peak energies $E(T)$ of some PL lines are plotted in Fig. 4.18b as a function of the temperature $T$ and are fitted using an empirical function which is widely used in bulk materials

$$E(T) = E_0 + \frac{\alpha T^2}{T + \beta} , \tag{4.13}$$

where $E_0$ is the energy value at 0 K, and $\alpha$ and $\beta$ are constants that depend on the materials [60]. A good fit was obtained below 70 K by using the same values for $\alpha$ and $\beta$ as those in bulk InP, i.e., $\alpha = -4.91 \times 10^{-4}$ eV/K and $\beta = 327$ K [60,61].

The empirical expression (4.13) can be applied to many 0D systems. Figure 4.19 shows another example of the µ-PL spectra measured at various temperatures. Sharp PL lines come from single InAs SADs embedded in a GaAs matrix [28]. The peak energy shifts corresponding to the band gap energy shift are again well fitted by (4.13) with $\alpha = -3.3 \times 10^{-4}$ eV/K and $\beta = 248$ K.

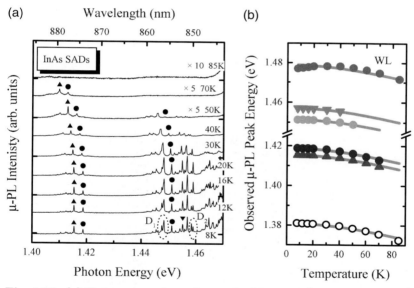

**Fig. 4.19.** (a) Temperature dependence of µ-PL spectra of InAs SADs embedded in GaAs. Successive redshifts of the PL peaks from the QDs were clearly observed with increasing temperature. As the temperature is raised from 8 K to 20 K, the PL intensities of the lower-energy side components of $D_1$ and $D_2$ decrease and those of the higher-energy side increase. The PL lines disappear from the higher-energy side as the temperature is raised further (from [28]). (b) The peak energy shifts of the sharp PL lines and wetting layer (WL) are well fitted by Varshni's expression $\Delta E(T) = -3.3 \times 10^{-4} T^2 / (T + 248)$

### 4.5.2 Thermal Activation

In Fig. 4.18a, when the sample temperature is raised from 5 K to 50 K, the PL peaks disappear successively from the higher-energy side of the spectra. Most of the PL lines are thermally quenched at 80 K. To our surprise, some PL lines as marked by the upward and downward pointing arrows with weak intensities at low temperatures become thermally active at about 30 K, and then become weak again at about 50 K.

Such unusual temperature dependence is observed in several systems [28,62]. Figure 4.19a shows another example of the "recovery" observed in the InAs SADs embedded in the GaAs matrix [28]. It was found that the higher-energy-side constituents of the doublets $D_1$ and $D_2$ increase with increasing temperature. Therefore, it seems that the recovery of the PL intensity is a general phenomenon in many quantum structures. In order to clarify the mechanism of this unusual behavior, temperature characteristics of individual SADs were investigated by means of $\mu$-PL image techniques.

### 4.5.3 Study of Thermal Activation by Micro-Photoluminescence Images

Micro-PL images of the same area of a sample were observed at various temperatures up to 100 K. Some of them are shown in Fig. 4.20a–f. At 4 K, many PL spots coming from single QDs are observed, as shown in Fig. 4.20a. Most of them become very weak in intensity due to thermal quenching at less than 20 K, but some remain (Fig. 4.20b). As the temperature is raised

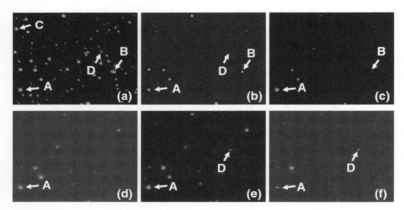

**Fig. 4.20.** Micro-PL images observed at (**a**) 4 K; (**b**) 16 K; (**c**) 40 K; (**d**) 60 K; (**e**) 73 K; and (**f**) 100 K. The observed area is 40 $\mu$m×53 $\mu$m. The PL from dot A is observed even at 100 K, while dots B and C, which show strong PL at 4 K, become inactive at 60 K and 16 K, respectively. It is interesting to note that the PL intensity of dot D is recovered at about 70 K, although its PL is very weak at 60 K

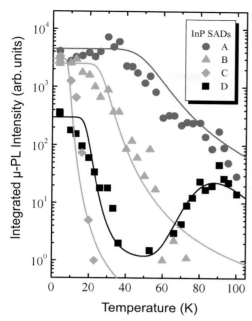

**Fig. 4.21.** Integrated μ-PL intensities of dots A to D shown in Fig. 4.20. The results are well reproduced (*solid curves*) by the simple model schematically drawn in Fig. 4.22. The estimated activation energies of dots A, B, and C are 39 meV, 26 meV, and 9 meV, respectively (from [6])

further, the number of the observed PL spots from the single SADs decrease (Fig. 4.20c–f). For example, the PL from dot A is observed even at 100 K, while dots B and C, which show strong PL at 4 K, become inactive at 60 K and 16 K, respectively. As an interesting phenomenon, it was found that the PL intensity of dot D is recovered at 73 K although its PL is very weak at 60 K (compare Fig. 4.20d and e).

The temperature dependence of the integrated μ-PL intensities from dots A to D is summarized in Fig. 4.21. The thermal quenching behaviors of dots A to C can be explained using a thermal activation model, as schematically drawn in Fig. 4.22a. The excitons relaxed from the $Ga_{0.5}In_{0.5}P$ matrix (state $|e\rangle$) annihilate radiatively at low temperature (state $|1\rangle$ to $|g\rangle$). As the temperature is raised, coupling with phonons is increased (state $|1\rangle$ to $|2\rangle$). Thus, some confined excitons relax through nonradiative process (state $|2\rangle$ to $|g\rangle$). In this case the rate equation is written as

$$\frac{dn}{dt} = I - n\gamma \exp\left(-\frac{E}{k_B T}\right) - \frac{n}{\tau} , \qquad (4.14)$$

where $E$ and $k_B$ stand for the thermal activation energy and the Boltzmann constant, respectively. Here $I$, $\tau$, and $\gamma$ are the relaxation rate of the excitons from $Ga_{0.5}In_{0.5}P$ matrix into the InP SAD, the exciton lifetime, and a constant indicating the efficiency of the thermal distribution, respectively.

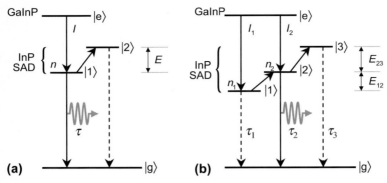

**Fig. 4.22.** Schematic energy levels of the InP SAD (states $|1\rangle$ to $|3\rangle$) and the $Ga_{0.5}In_{0.5}P$ matrix (state $|e\rangle$) under band-to-band excitation: (**a**) A thermal activation model. At low temperature, excitons in a SAD relax from state $|1\rangle$ to the ground state $|g\rangle$ by emitting photons. As the temperature is increased, since phonon population is enhanced, many excitons relax nonradiatively through state $|2\rangle$; (**b**) The excitons in states $|1\rangle$ and $|3\rangle$ relax through the nonradiative processes, while those in state $|2\rangle$ relax radiatively. When the temperature is increased, the excitons in state $|1\rangle$ are thermally populated to state $|2\rangle$, causing the μ-PL intensity to be recovered

By solving the rate equation (4.14) under cw excitation (i.e., $dn/dt = 0$), the PL intensity $I(T)$ of each QD is written as a function of temperature $T$

$$I(T) = \frac{I(0)}{1 + a\exp(-E/k_BT)} , \tag{4.15}$$

where $a$ is a constant. The temperature dependence of dots A to C are well fitted as shown by the lines in Fig. 4.21. By examining many QDs, it was found that each QD has a different activation energy, ranging from 5 meV to 40 meV.

It should be mentioned that the experimentally obtained value of 40 meV is very close to the LO phonon energy recently reported by Kozin et al. [63], supporting this model. The small value (5 meV) of the activation energy may be due to the surface effect. Namely, if a SAD has a rough surface, the confined exciton is easily destroyed through interaction with the acoustic phonons when the temperature is increased; eventually an increase of the nonradiative relaxation of the confined exciton results. This causes the decrease of the effective activation energy. The relaxation rate of the excitons is governed by the competing processes of the LO and acoustic phonon scattering. Therefore, the effective thermal activation energy differs widely from dot to dot.

On the other hand, the unusual temperature dependence of dot D is qualitatively explained as follows. Some SADs have states below the lowest luminescing states, probably due to defects at the interfaces between the SAD and the $Ga_{0.5}In_{0.5}P$ matrix. As schematically shown in Fig. 4.22b, excitons created in the $Ga_{0.5}In_{0.5}P$ matrix (state $|e\rangle$) relax into the InP SADs, and then some of the confined excitons radiatively annihilate from state $|2\rangle$. The other excitons relax through nonradiative processes from state $|1\rangle$, which is related to the defects at the interfaces. The PL intensity of dot D at low temperature is weaker than those of dots A to C by an order of magnitude, which suggests an efficient exciton relaxation through the nonradiative process. As the temperature is raised, the excitons in state $|1\rangle$ are thermally populated to state $|2\rangle$, causing the PL intensity to increase. At the same time, the excitons in state $|2\rangle$ are thermally quenched through state $|3\rangle$. This is why the PL intensity of dot D is recovered and has a peak at about 80 K. The intensity change in this case is given by solving a set of rate equations

$$\frac{\mathrm{d}n_1}{\mathrm{d}t} = I_1 - n_1\gamma_1 \exp\left(-\frac{E_{12}}{k_\mathrm{B}T}\right) - \frac{n_1}{\tau_1} , \qquad (4.16)$$

$$\frac{\mathrm{d}n_2}{\mathrm{d}t} = I_2 + n_1\gamma_1 \exp\left(-\frac{E_{12}}{k_\mathrm{B}T}\right) - n_2\gamma_2 \exp\left(-\frac{E_{23}}{k_\mathrm{B}T}\right) - \frac{n_2}{\tau_2} , \qquad (4.17)$$

where $n_i$, $I_i$, and $E_{ij}$ are the number of the carriers having the lifetime $\tau_i$ in state $|i\rangle$, the number of the carriers which relax from $\mathrm{Ga_{0.5}In_{0.5}P}$ matrix to state $|i\rangle$, and the energy separation between the states $|i\rangle$ and $|j\rangle$, respectively. Finally we obtain

$$I(T) = \frac{I_2(0) + \rho I_1(0)}{1 + b\exp(-E_{23}/k_\mathrm{B}T)} , \qquad (4.18)$$

with

$$\rho = \frac{c\exp(-E_{12}/k_\mathrm{B}T)}{1 + a\exp(-E_{12}/k_\mathrm{B}T)} , \qquad (4.19)$$

where $a(=\gamma_1\tau_1)$, $b(=\gamma_2\tau_2)$, and $c(=\gamma_1\tau_2)$ are constants, and $I_1(0)$ $(I_2(0))$ reflects the number of excitons that relax through nonradiative (radiative) processes at 0 K. Since the real system has a complicated energy structure, both processes shown in (4.15) and (4.18) are considered to coexist.

The fit of the model to the data for dot D is shown in Fig. 4.21. In spite of the simple model, the tendency of the temperature dependence is well reproduced. This result suggests that the investigation of the surface condition of SADs is important for considering the electronic structure and the relaxation processes of the confined excitons. Note that some SADs show multiple PL peaks even under very weak excitation conditions with energy spacings of a few meV, although a normal SAD has a single PL peak, as we have seen in Figs. 4.2 and 4.15. Therefore, our assumption of the existence of multiple stable states is reasonable.

Besides the $\mathrm{InP/Ga_{0.5}In_{0.5}P}$ and $\mathrm{InAs/GaAs}$ SADs, the temperature dependence of the band gap energy shift and the thermal activation energy have also been investigated in various SADs, such as $\mathrm{In_{0.50}Ga_{0.35}Al_{0.15}As/GaAs}$

[62], CdSe/ZnSe [64], and InGaN/GaN [18]. Among them, single dot measurements of CdSe/ZnSe revealed the existence of several states having different activation energies, which has never been clarified by macroscopic observations. This is an excellent example that shows the merit of single dot spectroscopy.

## 4.6  Fluorescence Intermittency

One of the most interesting findings reported in QD systems so far has been the fluctuations of the PL peak energy and intensity with time, which would have been unobservable in macroscopic studies. The former phenomenon is called spectral diffusion or PL wandering, and has been observed in CdSe [7,31,65–69], InAlAs [70], and Ag$_2$O [71] QDs, where the PL peak energies from confined excitons and their LO side bands fluctuate during the time of measurement, as shown in Figs. 4.23 and 4.24. Notably, spectral shifts of the PL peak are so large in silver oxide nanoparticles that large-scale dynamic single-particle color changes between red, green, and yellow in random fashion are observed [71]. The latter phenomenon has been reported in many 0D systems such as CdSe [7,65,69,72–75], ZnCdSe [76], GaAs [77], and InP [6,21,78,79], where the PL intensity hops among two or more discrete levels with time. This is referred to as fluorescence intermittency or random telegraph signal (RTS). The blinking phenomenon has also been observed in other low-dimensional systems such as porous Si [80] and InGaN clusters [81]. Spectral diffusion and fluorescence intermittency are tentatively attributed to

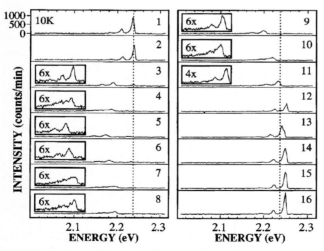

**Fig. 4.23.** An example of the spectral diffusion observed in a CdSe single QD. Sixteen consecutive 60 s spectra of the same single nanocrystal are shown in order of occurrence. The peak energy and intensity change with time (from [67])

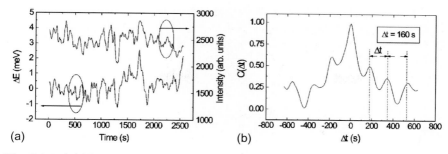

**Fig. 4.24.** (**a**) Time traces of the μ-PL peak energy and intensity of a CdSe single SAD; (**b**) Autocorrelation function of the spectral shift. Distinct satellite peaks at $\Delta t$ with a period of 160 s are observed (from [68])

photoionization, modification of the nanocrystal surface, or mobile photoactivated nonradiative recombination centers. However, the detailed mechanism of fluorescence intermittency is still under discussion. Further, as discussed in Sect. 4.6.2, quantum jumps have been reported in various systems, although the underlying physics seems to differ from system to system. For example, single molecules are one of the well-studied systems in the area of blinking and spectral diffusion phenomena [82–84], where the blinking is caused by the spectral shift [83], while the energy of the emitted photons is shifted in a single QD following a blink-off/blink-on event [69].

In this section, we discuss the origin of fluorescence intermittency in InP SADs. It is shown that the blinking is due to a local electric field generated by carriers trapped by deep localized centers in the $Ga_{0.5}In_{0.5}P$ matrix.

### 4.6.1   Micro-Photoluminescence Images of Blinking Dots

Figure 4.25a,b shows μ-PL images consecutively observed over the same area of the sample. Each bright spot corresponds to the PL from a single InP SAD. As marked by the dotted circles, it was observed that some InP SADs exhibit the blinking phenomenon, i.e., the PL intensity switches randomly between a high efficiency state (hereafter referred to as *on*) and a very weak state (*off*), while the PL from other SADs is stable.

The μ-PL images of a blinking InP SAD measured successively 30 times are shown in Fig. 4.26. We can clearly observe random jumps in intensity corresponding to the switching between the *on* and *off* states.

The number of QDs showing RTS behavior differs from sample to sample. In case of the InP SADs, the maximum number was one per about $10^3$ SADs, and the minimum was one per $10^6$ SADs (only a few blinking QDs in a small piece of the sample). Overall, the number of the blinking QDs was very small. On the other hand, in case of CdSe QDs, many QDs show the blinking phenomenon [75]. The fluorescence from individual nanocrystals is

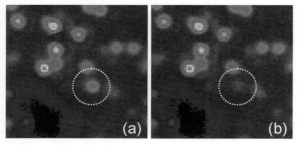

**Fig. 4.25.** Micro-images of a blinking InP SAD in (**a**) *on* and (**b**) *off* states measured at 4 K with a 200 ms integration time. The observed area is 15 μm×15 μm. The *black region* at the bottom left side of each image is a flaw on the sample surface, and thus the PL from the $Ga_{0.5}In_{0.5}P$ matrix is not observed in this region. Even when a blinking QD is in the *off* state, the μ-PL is not completely quenched, but weak μ-PL remains (from [79])

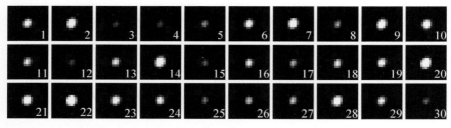

**Fig. 4.26.** Consecutive 1 s images of a blinking InP SAD. The size of each area shown is 3.1 μm×3.6 μm. The *on* and *off* switching behavior of the PL intensity is clearly seen over time

occasionally seen to flicker *on* and *off* on the time scale of one hundred milliseconds [7]. Further, Nirmal et al. have reported a dramatic increase in both average *on* and *off* times when the surface of CdSe QDs are overcoated with a shell of ZnS [65]. These results suggest that the appearance of the blinking QDs depends on the sample growth method or on the surrounding matrix.

### 4.6.2 Random Telegraph Signals in Various Systems

Besides QDs and single molecules as mentioned above, the jumping behavior of the physical quantities between two or more discrete states has so far been reported in several materials. The study of RTSs is not only of interest from the viewpoint of fundamental physics, but also of great importance for applications, because it will be a key to improve the performance of electronic devices. In some cases, the blinking is unwanted and necessarily decreases the efficiency of the devices. Consequently, silicon has been the most widely studied material with RTSs. For example, the capture and emission kinetics of the RTS in silicon metal-oxide semiconductor field-effect transistors (MOSFET's) have been investigated in detail for the last two decades, where the carrier transitions between the oxide trap and the channel are observed as fluctuations of the drain current or voltage [85,86]. Other Si-based devices, such as bipolar junction transistors and p–n junctions, are also known to show RTS. On the other hand, Lust and Kakalios have reported sharp jumps of the resistance $\Delta R/R \sim 1\%$ with switching events extending $\sim 10$ ms in an a-Si:H film [87]. They have pointed out that such large change in the resistance is not due to single electron capture and emission, but arises from inhomogeneous current filaments subject to localized diffusion processes.

The telegraph noises have been observed in device structures fabricated using III–V materials as well, such as in InGaAs/AlGaAs infrared photodetectors [88], GaAs/AlGaAs single-barrier tunnel structures [89], and split-gate transistors [90]. Further, Coppinger et al. have studied the telegraph noise to probe the magnetic moment of ErAs clusters in a GaAs matrix at very low temperature ($<1$ K) [91].

In addition to these systems, low-frequency noise has also been reported in many materials other than semiconductors as various signals, including resistance fluctuations in metallic nanoconstrictions [92], magnetization switching in individual ferromagnetic nanoparticles influenced by spin frustration of non-compensated spins [93], local dielectric fluctuations near the glass transition in polyvinyl acetate [94], and manganite colossal magnetoresistive materials owing to the coexistence of a ferromagnetic metallic phase and a phase with relatively depressed magnetic and electrical properties [95,96]. Moreover, in experiments of organic systems, the random-telegraph-like signals are observed in reactions when certain chemical reagents are introduced [97]. Therefore, we can regard RTSs as phenomena existing widely in nature.

### 4.6.3   Random or Correlated?

A time trace of the integrated μ-PL intensity from a single blinking QD is shown in Fig. 4.27a. As is clearly seen, the blinking occurs between two levels, namely, the *on* and *off* states. The intensity differences $\Delta I$ between the *on* and *off* states vary from dot to dot, as shown in Fig. 4.28a,b. In some QDs, switching among more than three levels is observed (see Fig. 4.28c). Furthermore, some show very complicated switching behaviors, where the switching frequency changes with time during the measurement (Fig. 4.28d). In order to clarify the physics of the blinking phenomena, the measurements were performed using blinking QDs showing the jumps between only two levels, and the data was statistically analyzed.

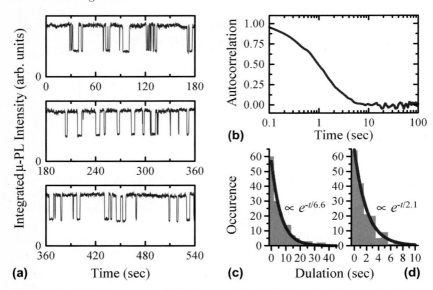

**Fig. 4.27.** (**a**) Time trace of the integrated μ-PL intensity from a single blinking SAD. Distinct change of the PL intensity between two levels is observed. The high PL efficiency state is referred to as *on*, and the low PL efficiency state is *off*. (**b**) Autocorrelation function converted from the integrated μ-PL intensity shown in (**a**). Numbers of the appearances of the (**c**) *on* and (**d**) *off* periods are plotted as a function of the duration of the events

Figure 4.27b shows an autocorrelation $C(\tau)$ converted from the time trace shown in Fig. 4.27a, which is defined by

$$C(\tau) = \frac{\sum\limits_{t=0}^{N} \left(I(t) - \bar{I}\right)\left(I(t+\tau) - \bar{I}\right)}{\sum\limits_{t=0}^{N} \left(I(t) - \bar{I}\right)^2} \, , \tag{4.20}$$

where $I(t)$ and $\bar{I}$ are the integrated μ-PL intensity at time $t$ and the average intensity, respectively. For meaningful statistics, the signal was acquired for a sufficiently long time (typically $\sim$1 h). It is clearly seen that the signal is strongly correlated only at $t \simeq 0$. The autocorrelation monotonically decreases with increasing correlation time $\tau$.

The average durations of the *on* and *off* periods are obtained by statistical analysis as follows. As an example, the numbers of occurrences $N$ of the *on* and *off* periods under 10 mW/cm² excitation are shown by histograms as a function of the duration time $t$ of the periods in Fig. 4.27c,d. As shown by the solid curves, they can be well fitted using the expression

$$N = N_0 \exp\left(-t/\tau\right) , \tag{4.21}$$

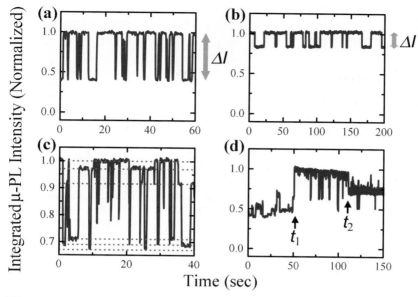

**Fig. 4.28.** Four time traces of integrated intensities of blinking InP SADs. In some QDs, large intensity change $\Delta I$ between *on* and *off* states is observed (trace **a**), while other QDs have small $\Delta I$ (trace **b**); (**c**) Some QDs jump more than three states in intensity; (**d**) Further, some QDs show very complicates intensity changes. Switching frequency also changes with time, e.g., at times $t_1$ and $t_2$

where $N_0$ is a constant. Hereafter, $\tau_{on}$ ($\tau_{off}$) is referred to as the average *on*-(*off*-) time. Based on these results, it is clear that the blinking is a random phenomenon.

It should be mentioned that the CdS SADs blink periodically [68], as shown in Fig. 4.24b. This is different from InP SADs. In order to examine if this is a common characteristic of low-dimensional systems, it will be interesting to repeat the experiments using various QDs. Further, the autocorrelation in the nanosecond region or the photon anti-bunching has been observed in several QD systems recently [75,98,99]. Observation of the autocorrelation over the nanosecond region to the second region will also be interesting.

### 4.6.4  Excitation Power Dependence

Figure 4.29a shows the integrated μ-PL observed under various excitation powers. The switching rate becomes slower as the excitation power is decreased. Figure 4.29b shows the average *on*- and *off*-times as a function of the excitation power density $P$. Here, the switching rate is defined by $1/\tau$. As shown by the solid lines, the *on*- and *off*-times are found to be proportional to $P_{on}^{-1.5}$ and $P_{off}^{-1}$, respectively. The excitation power dependence of the *on*- and *off*-times were studied using several blinking QDs, as shown in

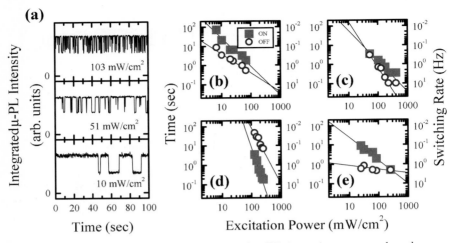

**Fig. 4.29.** (a) Time traces of the integrated μ-PL intensity measured under various excitation powers. With the increase of the excitation power, the switching frequency becomes faster; (b–e) Four examples of the excitation power dependence of the $\tau_{\text{on}}$ (*closed squares*) and $\tau_{\text{off}}$ (*open circles*)

Fig. 4.29c–e. It was found that each blinking QD shows different excitation power dependence, i.e., $P_{\text{on}}^{-1.2}$ and $P_{\text{off}}^{-2.4}$ for the QD shown in Fig. 4.29c, $P_{\text{on}}^{-4.7}$ and $P_{\text{off}}^{-2.9}$ in Fig. 4.29d, and $P_{\text{on}}^{-1.2}$ and $P_{\text{off}}^{-0.2}$ in Fig. 4.29e. The excitation power dependence of all blinking QDs can be fit by an exponential function through the origin ($\tau \propto P^{-\eta}$). This fact indicates that the switching processes are photoinduced phenomena at low temperature. This also suggests that the switching occur through higher-order interactions between the confined carriers and phonons or photons. It should be again emphasized that these results were obtained by blinking QDs showing the jumps between two states. The results also indicate that the number of elementary excitations contributing to the switching differs from dot to dot. This probably reflects some feature in the electronic structure of the blinking QD.

Figure 4.30a shows a contour plot of the μ-PL spectra of a single blinking QD successively recorded 30 times with an individual integration time of 200 msec. It was found that not only the PL intensity but also the μ-PL spectrum changes with time between the two types of spectra, as shown by thick and thin curves in Fig. 4.30b. Each of them corresponds to the μ-PL spectrum when the blinking QD is in the *on* and *off* states, respectively. The PL peak denoted by A shows the most drastic change in intensity. On the other hand, the PL intensities of peaks B and C are almost unchanged, but the peak energies shift slightly to the higher-energy side by a few meV when the state switches from *on* to *off*. This result indicates that the large intensity change observed in μ-PL images is mainly caused by the intensity change of the PL band having the highest PL peak energy. This also explains the weak PL in

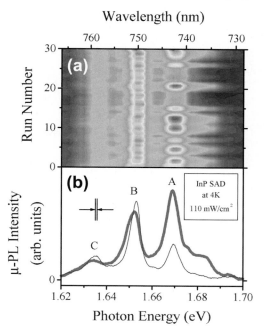

Fig. 4.30. (a) Contour map made from μ-PL spectra at 4 K of a blinking InP SAD measured successively 30 times; (b) Micro-PL spectra in the *on* (*thick curve*) and *off* (*thin curve*) states. Peak A, with the highest PL peak energy, shows a large intensity change between the *on* and *off* states. Peaks B and C shift to the higher-energy side when the system switches from the *on* state to the *off* state, while the intensity changes of these peaks are very small (from [79])

the *off* state. Namely, some PL peaks change little in intensity when the state switches from *on* to *off*. The PL intensity in the *off* state, though very weak, is not completely quenched, as shown in Fig. 4.25b.

The μ-PL spectra of a blinking QD measured under various excitation powers are shown in Fig. 4.31a. The thick and thin curves show the μ-PL spectra when the QD is in the *on* and *off* states, respectively. Under strong excitation of 200 mW/cm$^2$, peak A observed at the higher-energy side shows the largest intensity change, while the intensity changes of peaks B and C are small. When the excitation is reduced by 3 times (67 mW/cm$^2$), since the number of the confined excitons decreases and the confined exciton level is not filled up to level A, peak A is not observed. Under this excitation, peak B shows the largest change in intensity, while under strong excitation of 200 mW/cm$^2$ the PL intensity of peak B was almost constant. Therefore, the PL band that contributes to the blinking differs with the excitation power.

The magnified μ-PL spectra of the dotted region in Fig. 4.31a are shown in Fig. 4.31b. Under very weak excitation (670 μW/cm$^2$), we found that three μ-PL lines are observed, where the number of the confined exciton is about 0.01 on the average. This should be compared with the μ-PL spectrum from a normal InP SAD that shows a stable PL. Namely, under low carrier density, a normal SAD has a single PL peak reflecting a δ-function-like density of states of 0D systems, as we previously saw in Fig. 4.15. Even if the excitation

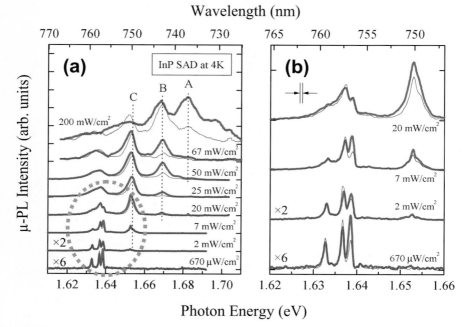

**Fig. 4.31.** Excitation power dependence of the μ-PL spectra from a single blinking QD at 4 K. The excitation power density was changed from 200 mW/cm² (*top curve* in **a**) to 670 μW/cm² (*bottom curves* of **a,b**). (**a**) The PL peaks observed at the higher-energy side in each spectrum show large intensity changes; (**b**) The magnified μ-PL spectra of the *dotted circle area* shown in (**a**). In contrast to a normal SAD, a few PL peaks remain even under very weak excitation conditions

power was significantly reduced, these three lines were still observed. Such multi-PL lines under very weak excitation were observed in all blinking InP SADs. This result indicates that blinking QDs have many stable states.

### 4.6.5   Origin of Fluorescence Intermittency in InP Self-Assembled Dots

In many cases, a blinking QD was observed near a minor flaw on the sample surface; for instance, a small flaw is observed at the bottom left side in Fig. 4.25, which provides a key clue for exploring this problem. It is reasonable to assume that there are many defect states near macroscopic flaws. Further, it was found that new blinking QDs appeared after the surface of the sample had been intentionally scratched by a needle. Therefore, fluorescence intermittency appears to be related to deep defect levels.

A possible mechanism for the switching between the *on* and *off* states can be modeled as follows. As schematically drawn in Fig. 4.32, in the vicinity of some QDs there are localized deep levels due to defects. When a carrier

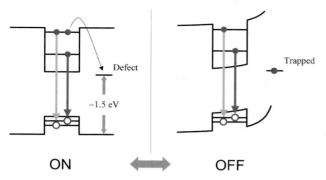

**Fig. 4.32.** Model for the blinking phenomena in InP SADs (see text)

is trapped by such a localized state, a local electric field generated by the trapped carrier is applied to the QD. In this case, the overlap of the wave function of the confined electrons and holes is decreased. If the local field is much stronger, the confined carriers escape from the QD to the matrix. Therefore, the µ-PL intensity eventually becomes weak, and this is observed as the switching from the *on* state to the *off* state. Especially since the wave functions of the excitons in higher-energy states are more spread out than those in lower-energy states, a PL peak observed at the higher-energy side shows a large intensity change. This corresponds to the quantum-confined Stark effects that have been reported in 2D systems [100,101].

The trapped carrier relaxes after a time by recombining with another carrier of opposite sign, mediated by phonons or through a photo-reabsorption process. In this case, the local field by the trapped carrier disappears. Consequently, the system returns to the initial condition. This shows the switching from the *off* state to the *on* state.[2]

### 4.6.6   Experimental Verification of the Model

In order to confirm the validity of the proposed model, it is essential to perform several additional experiments that shed light on localized trap centers and local electric fields. First, measurements of the temperature dependence of the integrated µ-PL intensity were carried out, as shown in Fig. 4.33. If the switching is due to the trapping process of carriers, the trapping probability should be explained using a thermal activation model. As is clearly seen,

---

[2] If one assumes that the trapping occurs through the Auger process as theoretically studied by Efros and Rosen [102] and the Auger coefficient is the same order as that reported by Raymond et al. [103], the duration of the *on* period can be of the same order as this observation. As for the *off* period, the relaxation time of a single trapped electron observed in several systems is of the same order as the *off* period in this case [104]. Thus, this model is reasonable to explain the duration of the *on* and *off* periods.

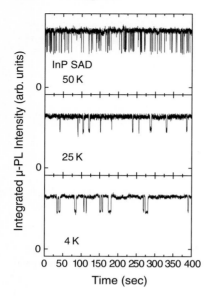

**Fig. 4.33.** Time traces of the integrated μ-PL intensity of a blinking QD measured at various temperatures. The switching rate becomes faster as the temperature is raised

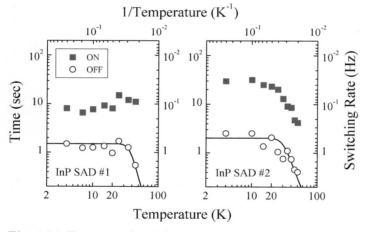

**Fig. 4.34.** Two examples of the temperature dependence of the average *on-* (*closed squares*) and *off-* times (*open circles*). The *solid curves* are fitted using a thermal activation model as defined by (4.22). The temperature dependence of the *off* duration is well reproduced (see [79])

the switching becomes faster with increasing temperature. For example, the switching rate becomes faster by about an order of magnitude when the temperature is raised from 4 K to 50 K. This result indicates that the switching can occur by the combination of thermal and photoinduced processes.

The average *on-* and *off-*times measured at various temperatures are summarized in Fig. 4.34, where the results obtained from two blinking QDs are shown. The temperature dependence of the blinking rates was examined for several blinking QDs, and found that, in all blinking QDs we studied, the average *off-*time $\tau_{\mathrm{off}}(T)$ given in (4.21) can be well reproduced as a function of temperature $T$ by a thermal activation function

$$\tau_{\mathrm{off}}(T) = \frac{\tau_{\mathrm{off}}(0)}{1 + a\exp(-E/k_{\mathrm{B}}T)} \, , \tag{4.22}$$

as shown by the solid lines in Fig. 4.34, where $a$ is a constant, and $E$ and $k_{\mathrm{B}}$ denote the activation energy and the Boltzmann constant, respectively. This result strongly supports the existence of a localized trap center that is responsible for the switching phenomena. It should be mentioned that the same *on–off* switching rate was obtained when the experiments were repeated on different days using the same blinking QD, where the sample temperature was raised to room temperature between the experiments. This behavior differs from that in copper nanoconstrictions, where, once the sample is taken above about 150 K and then recooled, the observed RTS is completely different from that observed before the sample was heated [92].

The depth of the localized center was investigated using a two-color laser excitation technique. Figure 4.35a shows the integrated μ-PL intensity under weak excitation by an Ar-ion laser. As shown in Fig. 4.35b, it was found that a drastic enhancement of the *on–off* switching rate is observed when the QDs

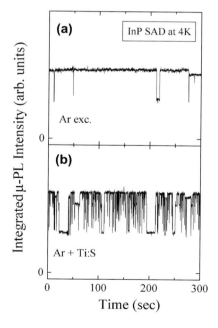

Fig. 4.35. (a) The integrated μ-PL intensity versus time trace of a blinking QD under band-to-band (an Ar-ion laser) excitation; (b) The blinking rate is drastically enhanced when a near-IR (Ti:sapphire) laser beam irradiates the sample simultaneously with a band-to-band excitation laser beam (see [79])

are resonantly excited by a near-IR (Ti:sapphire) laser of 1.75 eV in addition to the band-to-band excitation (Ar-ion) laser.

The switching rates, which are defined as reciprocals of the *on-* and *off-* times, are plotted in Fig. 4.36b as a function of the excitation energy of the near-IR laser. For comparison, the μ-PL spectrum of the blinking QD is also shown in Fig. 4.36a. The broken and dotted lines show the switching rates of the *on* and *off* states, respectively, measured without the near-IR laser irradiation. When the near-IR laser energy is above 1.5 eV, the switching

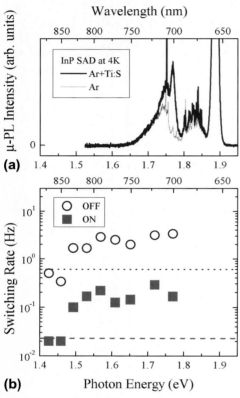

**(a)**

**(b)**

**Fig. 4.36.** (a) Micro-PL spectra of a blinking QD with (*thick curve*) and without (*thin curve*) near-IR laser irradiation. The *sharp line* observed at 1.75 eV comes from the scattered light of the near-IR laser. The PL intensity around 1.76 eV, which is higher than the near-IR laser energy, is enhanced with the near-IR laser irradiation. (b) Plot of the switching rates of the *on* (*closed squares*) and *off* (*open circles*) states. The *broken* and *dotted lines* show the switching rates of the *on-* and *off*-times, respectively, measured without the near-IR laser irradiation. When the near-IR laser beam with an energy higher than 1.5 eV is applied, the switching rates of the *on* and *off* states are drastically enhanced (see [79])

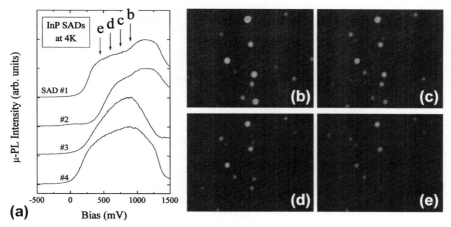

**Fig. 4.37.** (a) Four examples of the integrated μ-PL intensity in normal InP SADs in the presence of an external electric field. (b–e) Micro-PL images at various strengths of electric fields. The applied field of each image is denoted by the *arrows* in (a). The size of each image is 26 μm×33 μm

rates of the *on* and *off* states become 5 and 3 times faster than those excited only by the band-to-band laser, respectively.[3] It should be mentioned that the enhancement can be observed even when the near-IR laser energy is below the lowest PL level of the InP SAD, because the near-IR laser can directly excite a carrier to the defect level in the $Ga_{0.5}In_{0.5}P$ matrix that is deeper than the confined exciton level in the InP SADs. The switching rates were almost independent of the excitation energy of the near-IR laser above 1.5 eV. When the near-IR laser energy is lower than 1.5 eV, however, the enhancement was not observed, and the switching rate was almost the same as that measured only by the Ar laser irradiation shown in Fig. 4.35a. Therefore, we conclude that the localized center responsible for blinking has an excitation threshold energy of 1.5 eV.[4]

The second key point of the model, namely the local electric field, was then examined, where the local field is considered to cause a strong change in μ-PL intensity. Figure 4.37a shows four examples of the integrated μ-PL intensity from individual QDs in the presence of an electric field. In order

---

[3] We note that the increase of the temperature by the near-IR laser irradiation is negligible in this case (less than a few K) because the redshifts of the μ-PL peak energies of the GaAs substrate and the $Ga_{0.5}In_{0.5}P$ matrix were not observed. Thus, the observed enhancement of the switching rates are *not* due to an increase of temperature.

[4] The depth of the deep level that we observe is about 400 meV, which is close to the energy due to a Ga-vacancy defect and P-vacancy-related complexes [105]. Further, the anti-Stokes PL due to the deep states with energy 1.5 eV has recently been reported [106]. The value is consistent with that observed in the current study.

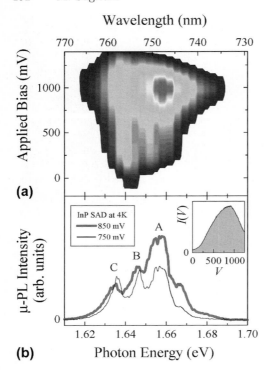

**Fig. 4.38.** (a) Contour plot of the μ-PL spectra in an external electric field. The PL intensity of the peak observed at the higher-energy side is more sensitive to the field than those observed at the lower-energy side. (b) μ-PL spectra measured by applying a bias of 850 mV (the flat band condition, *thick curve*) and 750 mV (*thin curve*), which qualitatively reproduce the μ-PL spectra of the *on* and *off* states, respectively (compare with Fig. 4.30(b)). Inset: Integrated μ-PL intensity versus the applied bias (from [79])

to perform this experiment, a semitransparent contact was fabricated by evaporating 20 nm of Au onto the sample surface. The μ-PL images in the presence of various electric field strengths are also shown in Fig. 4.37b–e. The PL intensity strongly depends on the external field, because the field changes the overlap between the electrons and the holes. The change in PL intensity is as large as that observed in a blinking QD.

Figure 4.38a shows the contour map of the μ-PL spectra of a normal single QD showing stable PL measured by applying a positive bias at the top of the sample. The PL intensity observed at the higher-energy side changes more sensitively than those observed at the lower-energy side, reflecting the weak localization of the exciton wave function in the higher-energy state. As a result, the external electric field causes a fairly large change of the integrated μ-PL intensity, as shown in the inset of Fig. 4.38b. The thick and thin curves in Fig. 4.38b were measured under the bias of 850 mV= 0 kV/cm, which corresponds to the flat band condition, and under a bias of about 10% less (750 mV=−3 kV/cm), respectively.[5]

---

[5] A single InP SAD shows a δ-function-like μ-PL line under weak excitation, as we observed in Sect. 4.4. Under the excitation shown in Fig. 4.38, however, since the number of the confined exciton is ~10, multiple μ-PL lines are observed and the μ-PL bandwidth is broad due to the exciton–exciton interaction.

A small change in the electric field induces a drastic change of the μ-PL spectra. One of the most important points in this study is that the changes in intensity and peak energy of the μ-PL spectra in the electric field qualitatively well reproduce the change of μ-PL spectra between the *on* and *off* states of a blinking QD shown in Fig. 4.30b. Namely, in Fig. 4.38b, peak A observed at the higher-energy side shows the largest change in intensity, and peaks B and C shift to the higher-energy side by a few meV, keeping their PL intensity almost unchanged when the bias is changed from 850 meV to 750 meV. This fact is direct evidence supporting the model, providing that the fluorescence intermittency is due to a local electric field induced by a trapped carrier near a single QD. It should be noted that the importance of the local electric field effect in blinking phenomena has also been pointed out in CdSe QDs [7,67].

Furthermore, the distance between a blinking QD and a trap center can be estimated from this result; if one assumes a static electric field in the sample layer and that one localized center can accept only one carrier, the distance is estimated to be 20 nm. This result indicates that the trap center located very close to (but spatially separated by a potential barrier from) the QD plays an important role in blinking phenomena.

As observed in Fig. 4.28, the depth of the *off* state differs from dot to dot, and in some QDs the PL intensity changes among more than three levels. These phenomena can also be qualitatively explained in the framework of our model, as schematically drawn in Fig. 4.39. The strength of the local electric field depends on the distance between a blinking QD and a localized

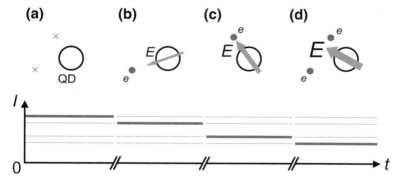

**Fig. 4.39.** Model for a blinking QD showing the jump among more than three states. There are several localized centers near the QD. (**a**) There is no carrier trapped near the QD. The crosses indicate localized centers. (**b**) When a carrier is trapped at one of the localized centers near the QD, an electric field is applied to the QD. The integrated μ-PL intensity $I$ becomes weak. (**c**) When a carrier is trapped at another localized center that is closer than the center shown in (**b**), $I$ becomes much weaker because the electric field applied to the QD is stronger. (**d**) Sometimes both trapped centers are occupied by carriers. This causes a significant decrease in μ-PL intensity due to the strong electric field generated by two carriers. Since the trapping processes of each localized center are independent, any quantum jumping from one state to another state is possible

center. For example, when the distance is short (long), a strong (weak) field is applied to a QD, resulting in a large (small) change in the μ-PL intensity. Thus, the depth of the *off* state $\Delta I$ differs from dot to dot. In the vicinity of some blinking QDs, it is inevitable that a few localized centers are formed, and the distances between the QDs and each localized center have different values. The electric field applied to the blinking QD changes with time in a way determined by the number of carriers occupying the localized centers. Namely, the electric field changes among discrete values. This is why the jumping of μ-PL intensity among more than two levels is observed in some QDs. Since the trapping processes of each localized center are independent, any quantum jumping from one state to another state is possible. This is consistent with the experimental results shown in Fig. 4.28c.

It should be mentioned that the quadratic Stark coefficient of the confined excitons in the ground state is of the order of $\sim 10^{-31}$ Fcm$^2$, as explained in Sect. 4.7.1. Since the local electric field generated by the trapped carrier is a few kV/cm, the energy shift or spectral diffusion of the PL peaks under weak excitation cannot be observed between the *on* and *off* states within our resolution (see the μ-PL spectra of a blinking SAD measured under very weak excitation, shown in Fig. 4.31b).

These new findings reinforce the fact that the optical properties of 0D systems are strongly influenced by the interface between the QD and the matrix.

## 4.7   Some Other Interesting Phenomena

The recent upsurge in μ-PL measurements has provided much valuable knowledge. In particular, single-dot spectroscopy is indispensable when a system has such a complicated energy structure that its energy separations are much smaller than the inhomogeneous broadening. In this section, various interesting phenomena revealed by observing μ-PL spectra are briefly discussed.

### 4.7.1   External Electric Field Effects

Studying external field effects of 0D systems is very important for determining the exciton wave functions [8,14,67,107–109]. As shown in Sect. 4.4, since the energy separation between the μ-PL peaks from the exciton and biexciton is only a few meV, the observation of individual QDs is necessary to obtain the exciton and biexciton parameters independently. Figure 4.40a shows a series of μ-PL spectra of an InP SAD measured by systematic changes in an external electric field, where the offset of each spectrum to the vertical direction indicates the applied electric field. The μ-PL spectrum measured under the open circuit situation is reproduced when the bias of $F_{FB}=+1.06$ V is applied. With the decrease of the electric field, distinct blueshifts of the μ-PL lines from the excitons and biexcitons are observed. The PL intensities

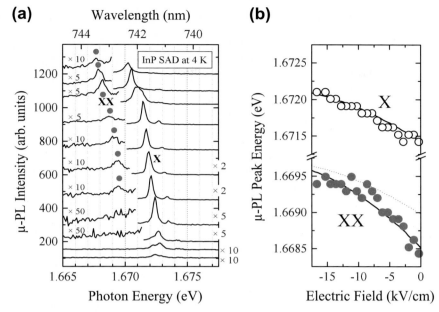

**Fig. 4.40.** (a) Applied electric field dependence of the μ-PL spectra from a single InP SAD. The main peak X of each spectrum comes from the confined exciton in the ground state. The peak denoted by XX is due to radiative decay of the biexcitons. The offset of each μ-PL spectrum shows the applied electric field. (b) Plots of the μ-PL peak energies of the exciton (X, *open circles*) and biexciton (XX, *closed circles*). The *lines* are the fits to the experimental results by (4.23)

of these lines decrease with the field and can be explained as follows: in the electric field, the valence and conduction bands become tilted and the overlap of the wave functions between electrons and holes is reduced, resulting in a decrease in PL intensities.

The PL peak energies of the exciton and biexciton lines are plotted by open and closed circles, respectively, as a function of the internal electric field $F$ in Fig. 4.40b. As shown by the solid curves, the transition energy $E(F)$ of each μ-PL line is well fitted using the expression,

$$E(F) = E_0^{(F)} + pF + \beta F^2 \qquad (4.23)$$

where $E_0^{(F)}$ is the energy at $F=0$. Single dot spectroscopy allowed one to evaluate the quadratic Stark coefficients $\beta$ of both excitons and biexcitons ($\beta \sim 10^{-31}$ Fcm$^2$).

Further, as shown in Fig. 4.40b, the linear coefficient $p$ is *not* zero. This suggests the existence of a non-zero permanent dipole moment in the InP SAD. From these values of the dipole moments, the separation between the electron and hole is estimated to be about 6 Å.

## 4.7.2    Magnetic Micro-Photoluminescence Spectra

The next topic is the effect of magnetic field on μ-PL spectra. Figure 4.41a shows two examples of the μ-PL peaks measured under various magnetic fields. To observe the μ-PL spectra from the individual QDs set in a pair of super-conducting magnets, the number of QDs to be excited was reduced using a mesa sample as introduced in Sect. 4.2. This measurement was performed in the Faraday configuration ($z\|k\|B$, where the growth direction was

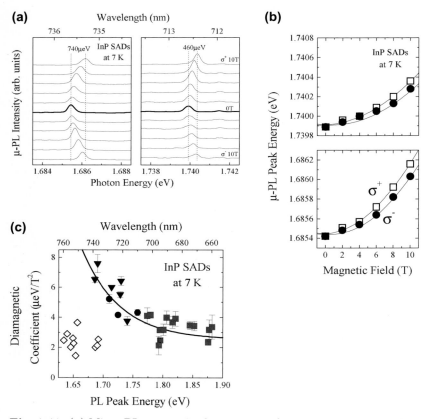

**Fig. 4.41.** (**a**) Micro-PL spectra in the presence of an external magnetic field. Two QDs with different PL peak energies (i.e., different sizes) are shown. The *upper five curves* show the PL components having the $\sigma^+$ polarization, while the *lower five curves* show those having the $\sigma^-$ polarization; (**b**) Plots of the PL peak energies of the spectra shown in (**a**). They are well fitted by (4.24). The PL peaks observed at the higher-energy side (i.e., a small QD) have a smaller diamagnetic coefficient than those observed at the lower-energy side (i.e., a large QD); (**c**) Size dependence of the diamagnetic coefficient. The diamagnetic coefficient increases on decreasing the μ-PL peak energy, i.e., with the increase of the dot size. The *solid curve* shows a simple model fitting

taken to be $z$). The experiment showed that the PL peaks blueshifted with increasing magnetic field.

The PL peak energies in Fig. 4.41b are plotted as a function of the applied magnetic field. They are well fitted by

$$E(B) = E_0^{(B)} \pm \frac{1}{2}\mu_{\mathrm{B}} g^* B + \alpha B^2 \tag{4.24}$$

where $E_0^{(B)}$, $\mu_{\mathrm{B}}$, and $g^*$ are the PL peak energy at 0 T, the Bohr magneton, and the effective $g$-value of the confined exciton, respectively. As clearly seen, the energy shifts of the PL peaks observed at the lower-energy side are greater than those shown at the higher-energy side when the field is increased from 0 T to 10 T.

A summary of the experimentally obtained diamagnetic coefficients $\alpha$ is plotted as a function of their PL peak energies at 0 T in Fig. 4.41c, where the closed and open symbols are observed in the Faraday and Voigt ($\boldsymbol{z}\|\boldsymbol{k}$, $\boldsymbol{z}\perp\boldsymbol{B}$) configurations, respectively. In the Faraday configuration, a systematic decrease of the diamagnetic coefficient was found with the increase of the μ-PL peak energy. In other words, the magnetic response becomes less sensitive with decreasing dot size. The diamagnetic coefficient asymptotically approaches a constant value of $\sim$3 μeV/T$^2$, which is almost the same as the value measured in the Voigt configuration. From this experiment, the effective exciton radius showing the μ-PL peak at 1.7 eV is estimated to be $\sqrt{\langle\rho^2\rangle}\sim$5 nm.

The carriers are confined along the lateral ($x$–$y$) direction by the field when measured in the Faraday configuration, while they are confined along the growth ($z$) direction in the Voigt configuration. Reflecting the dot shape, the quantum confinement along the $z$-direction is stronger than that along the $x$–$y$-direction, and thus the exciton wave function has already been highly shrunk along the $z$-direction, even at 0 T. This explains why the diamagnetic coefficient is smaller when observed in the Voigt configuration. A small energy shift in the Voigt configuration has also been reported in GaAs/AlAs quasi-0D excitons [8] and macroscopically in InAlAs/AlGaAs SADs [114]. From a simple analysis, it was found that the gradual change of the diamagnetic coefficient measured in the Faraday configuration reflects the strength of the confinement as drawn by the solid curve in Fig. 4.41c.

The magnetic μ-PL spectra also contain important information that can be used to determine the exciton and biexciton wave functions [14,56,58,110–113]. Bayer et al. have obtained the $g$-values of the neutral and charged excitons in In$_{0.60}$Ga$_{0.40}$As SADs [113]. In addition, a fine splitting due to dynamic polarization of lattice nuclei has been observed in a weak magnetic field, which is called the Overhauser effect [11].

### 4.7.3   Fine Splitting by Anisotropic Strain

Figure 4.42 shows the fine splitting of the exciton ground state due to the asymmetric anisotropy observed in InP SADs in a $Ga_{0.5}In_{0.5}P$ matrix grown by a GS-MBE system [45]. In the case of the sample grown by a MBE system, the composition modulation planes of gallium and indium with a period of a few tens of nanometers are formed in the $Ga_{0.5}In_{0.5}P$ matrix along the $[1\bar{1}0]$ direction. Spectra a and d in Fig. 4.42 are observed without a polarizer, while spectra b and e are observed for the $[1\bar{1}0]$ polarization, and c and f for the $[110]$ polarization. The most important feature seen in the μ-PL spectrum is the doublet nature of each peak. Each constituent of the doublet is fully polarized, i.e., one is along the $[110]$ direction and the other is along the $[1\bar{1}0]$ direction.

A doublet with an energy separation of $\Delta E \simeq 400$ μeV at 1.851 eV is clearly resolved in the polarized spectrum (Fig. 4.42b,c). The observed splitting energy differs from dot to dot. For example, even if the PL is observed at 1.870 eV as a single peak in Fig. 4.42d, it is composed of two constituents and they are selected by using a polarizer (see Fig. 4.42e,f). Such fine splitting resolved by polarization was also observed at the lower-energy edge of the PL

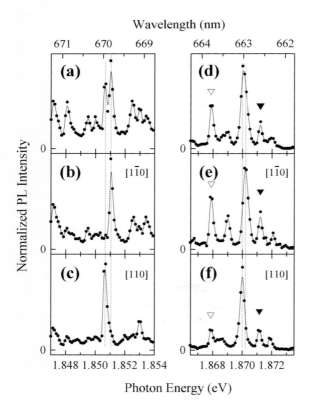

**Fig. 4.42.** Normalized μ-PL spectra at 8 K obtained without (spectra **a,d**) and with a polarizer along the $[1\bar{1}0]$ (**b,e**) and $[110]$ (**c,f**) directions. The *solid lines* are to guide the eyes. The separations between the *broken lines* are 400 μeV (*left*) and 190 μeV (*right*) (from [45])

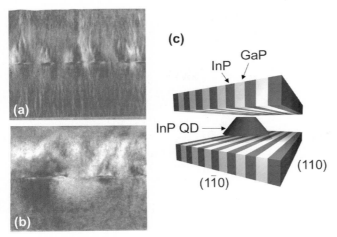

**Fig. 4.43.** Cross-sectional bright-field TEM images of the sample observed along the (**a**) [110] and (**b**) [$1\bar{1}0$] directions. The image in the [$1\bar{1}0$] direction shows periodic contrast, while that in [110] shows a uniform plane. A higher-than-average In content exists in the *dark domains* and a higher-than-average Ga content exists in the *white domains* [44]. (**c**) Schematic drawing of the sample structure. InP SADs are embedded in an $Ga_{0.5}In_{0.5}P$ natural superlattice formed due to lateral composition modulation, i.e., alternative Ga-rich and In-rich domains are naturally formed along the [110] direction with a period of a few tens of nanometers

band ($\sim$1.64 eV) [45]. On the other hand, the polarization-dependent energy shifts of some PL peaks represented by open and closed triangles in Fig. 4.42 were not observed within the resolution limit of the detection system. As shown in Fig. 4.43, cross-sectional bright-field TEM measurements revealed that the optical anisotropy is caused by the anisotropic strain due to the formation of the lateral composition modulation planes in the $Ga_{0.5}In_{0.5}P$ matrix [44].

Theoretical analysis shows that the fine splitting arises from the electron–hole exchange interaction in the presence of structural asymmetry [45]. In this case, the difference in oscillator strength between the constituents of a doublet $\Delta f$ is considered to be proportional to $\Delta E$. This is qualitatively in good agreement with the observed spectrum. Figure 4.44 shows the relation between the energy separation of the doublets $\Delta E$ and the PL intensity difference $\Delta I$ normalized by the intensity of the lower-energy side component. The difference $\Delta I$ is considered to reflect $\Delta f$ because of the high quantum efficiency of the QD system. In almost all cases, the PL intensity of the higher-energy side component is stronger than that of the lower one. When $\Delta E$ is small, the PL intensity of each component is almost the same. The difference $\Delta I$ is dispersed in the large $\Delta E$ region, since the actual magnitude of the splitting depends on the electron–hole envelope functions and hence on

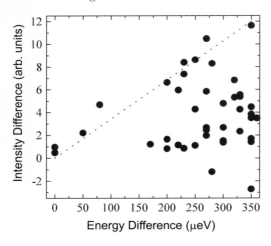

**Fig. 4.44.** Relation between the energy separation of the doublets $\Delta E$ and the PL intensity difference $\Delta I$ normalized by the intensity of the lower-energy side constituent. The theoretically expected relation is plotted by the *dotted line* (from [45])

the size and strain distribution of each QD. However, the tendency roughly agrees with the theoretical consideration, i.e., doublets with larger $\Delta E$ tend to have larger $\Delta I$ as well.

It is important to note that the energy splitting between the [110] and [1$\bar{1}$0] polarized exciton emissions is caused by the combined effect of the exchange interaction and the asymmetric strain; the latter causes the symmetry reduction. This effect is similar to the stress-exchange interplay observed in bulk semiconductors [115]. Again, the actual magnitude of the splitting depends on the electron–hole envelope functions owing to the size and strain distribution of each QD. The observed dispersion in the energy splitting, therefore, implies that the structure of the QDs is affected by the long-range inhomogeneities in the composition-modulated $Ga_{0.5}In_{0.5}P$ matrix.

The fine splittings of the μ-PL lines have been reported in several systems [10,13,15,116]. Figure 4.45 shows the linear polarization dependence observed in a quasi-0D structure fabricated using a CdSe/ZnSe quantum well. In this case, the splitting is attributed to the formation of QDs with a noncircular shape. Both the sign and the magnitude of the splitting vary from dot to dot because the asymmetry of each individual QD is different, and thus there is no preferential dot orientation. On the other hand, in the case of the InP SADs in the $Ga_{0.5}In_{0.5}P$ matrix, since the optical anisotropy is mainly due to the structural anisotropy in the matrix, the polarizations of almost all SADs are aligned.

It is worth pointing out that a fine splitting of the μ-PL lines from biexcitons has been observed in CdSe single quantum dots [116]. The PL from biexcitons is due to the optical transition from the biexciton state to a single exciton state. Since the biexciton state has a $\Gamma_1$ symmetry (spin-singlet), a shape anisotropy cannot induce a fine splitting of its energy structure. On the other hand, a large shape anisotropy lifts the degeneracy of the bright

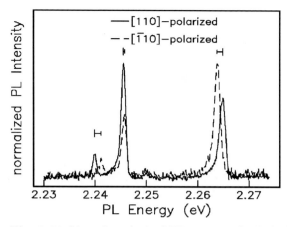

**Fig. 4.45.** Linearly polarized PL spectra of a CdSe/ZnSe quantum well in a 60 nm mesa for the [110] (*solid line*) and [$\bar{1}$10] (*dashed line*) directions. The ground state of the confined exciton splits into a doublet with an energy spacing up to 1.5 meV. This is attributed to a reduction of the $D_{2d}$ symmetry of an ideal quantum well due to the formation of QDs with noncircular shapes (from [15])

and dark exciton states. Consequently, the doublet structures appear in the µ-PL lines from both the excitons and biexcitons.

We emphasize that although optical anisotropy is observed in both InP SADs grown by MBE and MOVPE systems, the crystal structures and symmetry axes of the $Ga_{0.5}In_{0.5}P$ matrix differ between the samples grown by these systems, as discussed in Sect. 4.3. The main point is that QDs embedded in an anisotropic matrix are subjected to an anisotropic strain, and the strain becomes stronger with the increase of the degree of symmetry reduction in the matrix, resulting in a strong optical anisotropy in the QDs.

This technique can be applied in various kinds of QDs by using a wide variety of ternary alloys to control their band gaps and degrees of polarization.

### 4.7.4  Time Domain and Nonlinear Measurements

Carrier dynamics is also of great interest and importance for the optical characterization of QDs. However, because of the low signal level from a single exciton and the background noise, performing successful measurements is challenging. Therefore, in contrast to the large number of reports on time-resolved PL spectra by macroscopic methods, very few studies regarding µ-PL decay times have been successfully reported [25,51,117–119]. Detailed dynamical properties of the multiexciton and charged exciton states, which were surveyed under cw excitation in Sect. 4.4, will be one of the most important subjects of future study.

Another important topic is to clarify the coherence dynamics of excitons. The first measurements of the coherent nonlinear optical response have been reported by Bonadeo et al. [120]. They measured the energy relaxation and dephasing rates, and moreover, suggested the possibility of energy transfer between QDs.

Toda et al. have reported the temporal coherence of the carrier wave function in single InGaAs SADs [121]. This result gives important information regarding the relaxation paths of the confined carriers through the interaction with phonons. It is well known that LO phonons play an important role in the relaxation of excitons. Among various kinds of bulk semiconductors, the exciton–LO-phonon coupling in II–V and I–V compound semiconductors is strong because they have higher ionicities than III–V semiconductors. Therefore, it will be interesting to repeat this experiment using QD systems having higher ionicities.

## 4.8  Summary

Single QD spectroscopy has revealed a wealth of novel information hidden behind the inhomogeneously broadened photoluminescence band. In this chapter, we have observed the rich physics of the 0D system investigated by μ-PL spectroscopy and imaging. The optical anisotropy of the PL spectra and the μ-PL images due to the Cu–Pt$_\mathrm{B}$-type long-range ordering of the $Ga_{0.5}In_{0.5}P$ matrix were presented. Many carrier effects on a QD were studied in detail under various excitation power densities. The existence of stable multiexciton states due to 3D confinement was shown. Detailed thermal effects of the band gap energy shift were investigated. The μ-PL intensities of some QDs recovered when the temperature was raised, which was explained by a simple model. From the observations of μ-PL images and the temperature dependence of the switching rate, it was shown that the blinking phenomenon in InP SADs is due to the trapping and subsequent delocalization processes of photoexcited carriers. The intensity change and peak energy shift observed in the μ-PL spectra of a blinking QD can be well reproduced artificially by applying an external electric field, which indicates that the fluorescence intermittency is intimately related to a local electric field. These results reinforce the fact that the energy structure and the optical properties of QDs are strongly affected by the symmetry of the matrix surrounding the QDs and the interfaces between the QDs and the matrix. We expect that new potential properties hidden in QD systems will be elicited and manipulated in the future by controlling the matrix.

Further, the external field effects provide important information to determine the exciton wave functions. Study of QDs is a fast-moving field. We have obtained a lot of information in the last ten years. However, satisfactory consistency of the wavefunctions of QD states, especially multiexciton states, has not been met between the experiments and the theories. In order to reveal concealed information and for comprehensive understanding, combining external field experiments, such as magnetic field and electric field experiments, will be effective to further research in this area. This experiment will also provide novel information about the optically forbidden states at zero external fields.

The μ-PL data has so far been mainly obtained by integration of the emitted photons for a moderate time. In the near future, the dynamical properties, including carrier transportation between neighboring QDs, will attract more attention. In this case, the temporal change of a few photon numbers must be detected. Therefore, new techniques and sophisticated equipment will be required to significantly increase the sensitivity. This technology will also allow an absorption spectrum from a single QD to be obtained. To carry out these experiments, μ-PL spectroscopy has potential to uncover new phenomena and will be used as a base of new techniques.

# References

1. *Proc. 24th Int. Conf. on the Phys. of Semicond.*, ed. by D. Gershoni (World Scientific, Singapore 1999)
2. A.D. Yoffe: Adv. Phys. **42**, 173 (1993); A.D. Yoffe: Adv. Phys. **50**, 1 (2001)
3. S. Schmitt-Rink, D.A.B. Miller, D.S. Chemla: Phys. Rev. B **35**, 3113 (1987)
4. U. Woggon: *Optical Properties of Semiconductor Quantum Dots* (Springer-Verlag, Berlin, Heidelberg, New York 1997)
5. H. Kamada, H. Ando, J. Temmyo, T. Tamamura: Phys. Rev. B **58**, 16243 (1998)
6. M. Sugisaki, H.-W. Ren, S.V. Nair, J.-S. Lee, S. Sugou, T. Okuno, Y. Masumoto: J. Lumin. **87–89**, 40 (2000)
7. S.A. Empedocles, R. Neuhauser, K. Shimizu, M.G. Bawendi: Adv. Mater. **11**, 1243 (1999)
8. A. Zrenner, L.V. Butov, M. Hagn, G. Abstreiter, G. Böhm, G. Weimann: Phys. Rev. Lett. **72**, 3382 (1994); A. Zrenner, A. Schaller, M. Markmann, M. Hagn, M. Arzberger, D. Henry, G. Abstreiter, G. Böhm, G. Weimann: Appl. Surf. Sci. **123–124**, 356 (1998)
9. K. Brunner, G. Abstreiter, G. Böhm, G. Tränkle, G. Weimann: Phys. Rev. Lett. **73**, 1138 (1994)
10. D. Gammon, E.S. Snow, D.S. Katzer: Appl. Phys. Lett. **67**, 2391 (1995); D. Gammon, E.S. Snow, B.V. Shanabrook, D.S. Katzer, D. Park: Science **273**, 87 (1996); D. Gammon, E.S. Snow, B.V. Shanabrook, D.S. Katzer, D. Park: Phys. Rev. Lett. **76**, 3005 (1996)
11. S.W. Brown, T.A. Kennedy, D. Gammon, E.S. Snow: Phys. Rev. B **54**, R17339 (1996)
12. M. Lowisch, M. Rabe, B. Stegemann, F. Henneberger, M. Grundmann, V. Türck, D. Bimberg: Phys. Rev. B **54**, R11074 (1996)
13. V. Nikitin, P.A. Crowell, J.A. Gupta, D.D. Awschalom, F. Flack, N. Samarth: Appl. Phys. Lett. **71**, 1213 (1997)
14. W. Heller, U. Bockelmann: Phys. Rev. B **55**, R4871 (1997); W. Heller, U. Bockelmann, G. Abstreiter: Phys. Rev. B **57**, 6270 (1998)
15. T. Kümmell, R. Weigand, G. Bacher, A. Forchel, K. Leonardi, D. Hommel, H. Selke: Appl. Phys. Lett. **73**, 3105 (1998)

16. H.M. Cheong, A. Mascarenhas, J.F. Geisz. J.M. Olson, M.W. Keller, J.R. Wendt: Phys. Rev. B **57**, R9400 (1998)

17. U. Kops, P.G. Blome, M. Wenderoth, R.G. Ulbrich, C. Geng, F. Scholz: Phys. Rev. B **61**, 1992 (2000)

18. O. Moriwaki, T. Someya, K. Tachibana, S. Ishida, Y. Arakawa: Appl. Phys. Lett. **76**, 2361 (2000)

19. U. Dörr, R. Lutz, J. Schuler, H. Kalt, W. Send, D. Gerthsen: J. Lumin. **87–89**, 718 (2000); U. Dörr, R. Lutz, E. Tsitsishvili, H. Kalt: Phys. Rev. B **62**, 15745 (2000)

20. M. Sugisaki, H.-W. Ren, K. Nishi, S. Sugou, Y. Masumoto: Phys. Rev. B **61**, 16040 (2000)

21. As a review article, A. Gustafsson, M.-E. Pistol, L. Montelius, L. Samuelson: J. Appl. Phys. **84**, 1715 (1998)

22. H.F. Hess, E. Betzig, T.D. Harris, H.N. Pfeiffer, K.W. West: Science **264**, 1740 (1994)

23. A. Chavez-Pirson, J. Temmyo, H. Kamada, H. Gotoh, H. Ando: Appl. Phys. Lett. **72**, 3494 (1998)

24. K. Matsuda, T. Matsumoto, H. Saito, K. Nishi, T. Saiki: Physica E **7**, 377 (2000)

25. D. Hessman, P. Castrillo, M.-E. Pistol, C. Pryor, L. Samuelson: Appl. Phys. Lett. **69**, 749 (1996)

26. S. Fafard, R. Leon, D. Leonard, J.L. Merz, P.M. Petroff: Phys. Rev. B **50**, 8086 (1994)

27. J.-Y. Marzin, J.-M. Gérard, A. Izraël, D. Barrier, G. Bastard: Phys. Rev. Lett. **73**, 716 (1994); J.-M. Gérard, J.-Y. Marzin, G. Zimmermann, A. Ponchet, O. Cabrol, D. Barrier, B. Jusserand, B. Sermage: Solid-State Electron. **40**, 807 (1996)

28. M. Sugisaki, H.-W. Ren, S. Sugou, K. Nishi, Y. Masumoto: Solid-State Electron. **42**, 1998 (1325)

29. G. Schedelbeck, W. Wegscheider, M. Bichler, G. Abstreiter: Science **278**, 1792 (1997)

30. M. Grundmann, J. Christen, N.N. Ledentsov, J. Böhrer, D. Bimberg, S.S. Ruvimov, P. Werner, U. Richter, U. Gösele, J. Heydenreich, V.M. Ustinov, A.Y. Egorov, A.E. Zhukov, P.S. Kop'ev, Z.I. Alferov: Phys. Rev. Lett. **74**, 4043 (1995)

31. V. Türk, S. Rodt, O. Stier, R. Heitz, R. Engelhardt, U.W. Pohl, D. Bimberg, S. Steingrüber: Phys. Rev. B **61**, 9944 (2000); V. Türk, S. Rodt, O. Stier, R. Heitz, U.W. Pohl, R. Engelhardt, D. Bimberg: J. Lumin. **87–89**, 337 (2000)

32. H.-W. Ren, M. Sugisaki, J.-S. Lee, S. Sugou, Y. Masumoto: In: *Excitonic Processes in Condensed Matter*, Vol. 98-25, ed. by R.T. Williams, W.M. Yen (The Electro-chemical Society Proceedings Series, Pennington, NJ 1998) pp. 292-297

33. H.-W. Ren, M. Sugisaki, J.-S. Lee, S. Sugou, Y. Masumoto: In: *Proc. 24th Int. Conf. on the Phys. of Semicond.*, ed. by D. Gershoni (World Scientific, Singapore 1999), PDF No. 1191; H.-W. Ren, M. Sugisaki, J.-S. Lee, S. Sugou, Y. Masumoto: Jpn. J. Appl. Phys. **38**, 507 (1999)

34. M. Sugisaki, H.-W. Ren, K. Nishi, Y. Masumoto: Solid State Commun. **117**, 679 (2001)

35. H. Saito, K. Nishi, S. Sugou, Y. Sugimoto: Appl. Phys. Lett. **71**, 590 (1997)

36. Y. Nabetani, T. Ishikawa, S. Noda, A. Sasaki: J. Appl. Phys. **76**, 347 (1994)
37. M. Henini, S. Sanguinetti, S.C. Fortina, E. Grilli, M. Guzzi, G. Panzarini, L.C. Andreani, M.D. Upward, P. Moriarty, P.H. Beton, L. Eaves: Phys. Rev. B **57**, R6815 (1998)
38. S. Noda, T. Abe, M. Tamura: Phys. Rev. B **58**, 7181 (1998)
39. W. Yang, H. Lee, T.J. Johnson, P.C. Sercel, A.G. Norman: Phys. Rev. B **61**, 2784 (2000)
40. T. Suzuki, A. Gomyo, S. Iijima, K. Kobayashi, S. Kawata, I. Hino, T. Yuasa: Jpn. J. Appl. Phys. **27**, 2098 (1988)
41. M. Kondow, H. Kakibayashi, S. Minagawa: J. Cryst. Growth **88**, 291 (1988)
42. J.E. Fouquet, V.M. Robbins, S.J. Rosner: Appl. Phys. Lett. **57**, 1566 (1990)
43. P. Ernst, C. Geng, G. Hahn, F. Scholz, H. Schweizer, F. Phillipp, A. Mascarenhas: J. Appl. Phys. **79**, 2633 (1996); F. Scholz, C. Geng, M. Burkard, H.-P. Gauggel, H. Schweizer, R. Wirth, A. Moritz, A. Hangleiter: Physica E **2**, 8 (1998)
44. H.-W. Ren, M. Sugisaki, S. Sugou, K. Nishi, A. Gomyo, Y. Masumoto: Jpn. J. Appl. Phys. **38**, 2438 (1999)
45. M. Sugisaki, H.-W. Ren, S.V. Nair, K. Nishi, S. Sugou, T. Okuno, Y. Masumoto: Phys. Rev. B **59**, R5300 (1999); M. Sugisaki, H.-W. Ren, S.V. Nair, K. Nishi, S. Sugou, T. Okuno, Y. Masumoto: In: *Excitonic Processes in Condensed Matter, Vol. 98-25*, ed. by R.T. Williams, W.M. Yen (The Electrochemical Society Proceedings Series, Pennington, NJ 1998) pp. 298-303
46. A. Gomyo, T. Suzuki, S. Iijima: Phys. Rev. Lett. **60**, 2645 (1988)
47. See, for example, P. Ils, C. Gréus, A. Forchel, V.D. Kulakovskii, N.A. Gippius, S.G. Tikhodeev: Phys. Rev. B **51**, 4272 (1995); H. Akiyama, T. Someya, H. Sakaki: Phys. Rev. B **53**, R4229 (1996); F. Vouilloz, D.Y. Oberli, M.-A. Dupertuis, A. Gustafsson, F. Reinhardt, E. Kapon: Phys. Rev. B **57**, 12378 (1998)
48. M.C. DeLong, P.C. Taylor, J.M. Olson: Appl. Phys. Lett. **57**, 620 (1990)
49. M.C. DeLong, D.J. Mowbray, R.A. Hogg, M.S. Skolnick, J.E. Williams, K. Meehan, S.R. Kurtz, J.M. Olson, R.P. Schneider, M.C. Wu, M. Hopkinson: Appl. Phys. Lett. **66**, 3185 (1995)
50. M. Sugisaki, H.-W. Ren, S.V. Nair, K. Nishi, Y. Masumoto: Solid State Commun. **117**, 435 (2001)
51. P. Castrillo, D. Hessman, M.-E. Pistol, S. Anand, N. Carlsson, W. Seifert, L. Samuelson: Appl. Phys. Lett. **67**, 1905 (1995); V. Zwiller, M.-E. Pistol, D. Hessman, R. Cederström, W. Seifert, L. Samuelson: Phys. Rev. B **59**, 5021 (1999)
52. M. Sugisaki, H.-W. Ren, E. Tokunaga, K. Nishi, S. Sugou, T. Okuno, Y. Masumoto: In: *Proc. 24th Int. Conf. on the Phys. of Semicond.*, ed. by D. Gershoni (World Scientific, Singapore 1999), PDF No. 1180
53. T. Okuno, H.-W. Ren, M. Sugisaki, K. Nishi, S. Sugou, Y. Masumoto: Solid-State Electron. **42**, 1319 (1998); T. Okuno, H.-W. Ren, M. Sugisaki, K. Nishi, S. Sugou, Y. Masumoto: Phys. Rev. B **57**, 1386 (1998); T. Okuno, H.-W. Ren, M. Sugisaki, K. Nishi, S. Sugou, Y. Masumoto: Jpn. J. Appl. Phys. **38**, 1094 (1999)
54. S.V. Nair, Y. Masumoto: J. Lumin. **87–89**, 438 (2000); S.V. Nair, Y. Masumoto: Phys. Status Solidi A **178**, 303 (2000); S.V. Nair, Y. Masumoto: Phys. Status Solidi B **224**, 739 (2001)

55. S. Charbonneau, L.B. Allard, A.P. Roth, T. Sudersena Rao: Phys. Rev. B **47**, 13918 (1993)
56. A. Kuther, M. Bayer, A. Forchel, A. Gorbunov, V.B. Timofeev, F. Schäfer, J.P. Reithmajer: Phys. Rev. B **58**, R7508 (1998)
57. Y. Toda, O. Moriwaki, M. Nishioka, Y. Arakawa: Phys. Rev. Lett. **82**, 4114 (1999)
58. F. Findeis, A. Zrenner, G. Böhm, G. Abstreiter: Phys. Rev. B **61**, R10579 (2000)
59. A. Hartmann, Y. Docommun, M. Dächthold, E. Kapon: Physica E **7**, 461 (2000); A. Hartmann, Y. Docommun, E. Kapon, U. Hohenester, E. Molinari: Phys. Rev. Lett. **84**, 5648 (2000)
60. Y.P. Varshni: Physica (Utrecht) **34**, 149 (1967)
61. W.J. Turner, W.E. Reese, G.D. Pettit: Phys. Rev. A **136**, 1467 (1964)
62. H. Htoon, H. Yu, D. Kulik, J.W. Keto, O. Baklenov, A.L. Holmes, B.G. Street-man, C.K. Shih: Phys. Rev. B **60**, 11026 (1999)
63. I.E. Kozin, I.V. Ignatiev, S.V. Nair, H.-W. Ren, S. Sugou, Y. Masumoto: J. Lumin. **87-89**, 441 (2000); I.V. Ignatiev, I.E. Kozin, S.V. Nair, H.-W. Ren, S. Sugou, Y. Masumoto: Phys. Rev. B **61**, 15633 (2000)
64. J.C. Kim, H. Rho, L.M. Smith, H.E. Jackson, S. Lee, M. Dobrowolska, J.K. Furdyna: Appl. Phys. Lett. **75**, 214 (1999); S. Lee, J.C. Kim, H. Rho, C.S. Kim, L.M. Smith, H.E. Jackson, J.K. Furdyna, M. Dobrowolska: Phys. Rev. B **61**, R2405 (2000)
65. M. Nirmal, B.O. Dabbousi, M.G. Bawendi, J.J. Macklin, J.K. Trautman, T.D. Harris, L.E. Brus: Nature (London) **383**, 802 (1996)
66. S.A. Blanton, M.A. Hines, P. Guyot-Sionnest: Appl. Phys. Lett. **69**, 3905 (1996)
67. S.A. Empedocles, D.J. Norris, M.G. Bawendi: Phys. Rev. Lett. **77**, 3873 (1996); S.A. Empedocles, M.G. Bawendi: Science **278**, 2114 (1997)
68. J. Seufert, R. Weigand, G. Bacher, T. Kümmell, A. Forchel, K. Leonardi, D. Hommel: Appl. Phys. Lett. **76**, 1872 (2000)
69. R.G. Neuhauser, K.T. Shimizu, W.K. Woo, S.A. Empedocles, M.G. Bawendi: Phys. Rev. Lett. **85**, 3301 (2000)
70. H.D. Robinson, B.B. Goldberg: Phys. Rev. B **61**, R5086 (2000)
71. L.A. Peyser, A.E. Vinson, A.P. Bartko, R.M. Dickson: Science **291**, 103 (2001)
72. U. Banin, M. Bruches, A.P. Alivisatos, T. Ha, S. Weiss, D.S. Chemla: J. Chem. Phys. **110**, 1195 (1999)
73. F. Koberling, A. Mews, T. Basché: Phys. Rev. B **60**, 1921 (1999)
74. M. Kuno, D.P. Fromm, H.F. Hamann, A. Gallagher, D.J. Nesbitt: J. Chem. Phys. **112**, 3117 (2000)
75. P. Michler, A. Imanoğlu, M.D. Mason, P.J. Carson, G.F. Strouse, S.K. Bu-ratto: Nature (London) **406**, 968 (2000)
76. B.P. Zhang, Y.Q. Li, T. Yasuda, W.X. Wang, Y. Segawa, K. Edamatsu, T. Itoh: Appl. Phys. Lett. **73**, 1266 (1998)
77. D. Bertram, M.C. Hanna, A.J. Nozik: Appl. Phys. Lett. **74**, 2666 (1999)
78. M.-E. Pistol, P. Castrillo, D. Hessman, J.A. Prieto, L. Samuelson: Phys. Rev. B **59**, 10725 (1999); M.-E. Pistol: Phys. Rev. B **63**, 113306 (2001)
79. M. Sugisaki, H.-W. Ren, K. Nishi, Y. Masumoto: Phys. Rev. Lett. **87**, 4883 (2001); M. Sugisaki, H.-W. Ren, I.S. Osad'ko, K. Nishi, Y. Masumoto: Phys. Status Solidi B **224**, 67 (2001)

80. M.D. Mason, G.M. Credo, M.D. Weston, S.K. Buratto: Phys. Rev. Lett. **80**, 5405 (1998)
81. T. Aoki, Y. Nishikawa, M. Kuwata-Gonokami: Appl. Phys. Lett. **78**, 1065 (2001)
82. R.M. Dickson, A.B. Cubitt, R.Y. Tsien, W.E. Moerner: Nature (London) **388**, 355 (1997)
83. T. Baschè: J. Lumin. **76–77**, 263 (1998)
84. *Single-Molecule Optical Detection, Imaging and Spectroscopy*, ed. by T. Baschè, W.E. Moerner, M. Orrit, U.P. Wild (Weinheim, New York 1997)
85. As a review article of RTSs in MOSFETs, M.J. Kirton, M.J. Uren: Adv. Phys. **38**, 367 (1989)
86. See, for example, E. Simoen, C. Claeys: Appl. Phys. Lett. **66**, 598 (1995); N.B. Lukyanchikova, M.V. Petrichuk, N.P. Garbar, E. Simoen, C. Claeys: Appl. Phys. Lett. **73**, 2444 (1998); J.H. Scofield, N. Borland, D.M. Fleetwood: Appl. Phys. Lett. **76**, 3248 (2000); H.M. Bu, Y. Shi, X.L. Yuan, J. Wu, S.J. Gu, Y.D. Zheng, H. Majima, H. Ishikuro, T. Hiramoto: Appl. Phys. Lett. **76**, 3259 (2000)
87. L.M. Lust, J. Kakalios: Phys. Rev. Lett. **75**, 2192 (1995)
88. L. Kore, G. Bosman: J. Appl. Phys. **86**, 6586 (1999)
89. P.M. Campbell, E.S. Snow, W.J. Moore, O.J. Glembocki, S.W. Kirchoefer: Phys. Rev. Lett. **67**, 1330 (1991)
90. D.H. Cobden, A. Savchenko, M. Pepper, N.K. Patel, D.A. Ritchie, J.E.F. Frost, G.A.C. Jones: Phys. Rev. Lett. **69**, 502 (1992)
91. F. Coppinger, J. Genoe, D.K. Maude, U. Gennser, J.C. Portal, K.E. Singer, P. Rutter, T. Taskin, A.R. Peaker, A.C. Wright: Phys. Rev. Lett. **75**, 3513 (1995)
92. K.S. Ralls, R.A. Buhrman: Phys. Rev. Lett. **60**, 2434 (1988)
93. W. Wernsdorfer, E. Bonet Orozco, K. Hasselbach, A. Benoit, B. Barbara, N. Demoncy, A. Loiseau, H. Pascard, D. Mailly: Phys. Rev. Lett. **78**, 1791 (1997)
94. E. Russell, N.E. Israeloff: Nature (London) **408**, 695 (2000)
95. R.D. Merithew, M.B. Weissman, F.M. Hess, P. Spradling, E.R. Nowak, J. O'Donnell, J.N. Eckstein, Y. Tokura, Y. Tomioka: Phys. Rev. Lett. **84**, 3442 (2000)
96. B. Raquet, A. Anane, S. Wirth, P. Xiong, S. von Molnár: Phys. Rev. Lett. **84**, 4485 (2000)
97. See, for example, D.K. Marciano, M. Russel, S.M. Simon: Science **284**, 1516 (1999); M.A. Valverde, P. Rojas, J. Amigo, D. Cosmelli, P. Orio, M.I. Bahamonde, G.E. Mann, C. Vergara, R. Latorre: Science **285**, 1929 (1999); D.K. Marciano, M. Russel, S.M. Simon: Science **284**, 1516 (1999); L.S. Premkumar, G.P. Ahern: Nature (London) **408**, 985 (2000)
98. P. Michler, A. Kiraz, C. Becher, W.V. Schoenfeld, P.M. Petroff, L. Zhang, E. Hu, A. Imamoğlu: Science **290**, 2282 (2000); C. Becher, A. Kiraz, P. Michler, A. Imamoğlu, W.V. Schoenfeld, P.M. Petroff, L. Zhang, E. Hu: Phys. Rev. B **63**, 121312(R) (2001)
99. C. Santori, M. Pelton, G. Solomon, Y. Dale, Y. Yamamoto: Phys. Rev. Lett. **86**, 1502 (2001)
100. D.A.B. Miller, D.S. Chemla, T.C. Damen, A.C. Gossard, W. Wiegmann, T.H. Wood, C.A. Burrus: Phys. Rev. B **32**, 1043 (1985)

101. H.-J. Polland, L. Schultheis, J. Kuhl, E.O. Göbel, C.W. Tu: Phys. Rev. Lett. **55**, 2610 (1985)
102. A.L. Efros, M. Rosen: Phys. Rev. Lett. **78**, 1110 (1997)
103. S. Raymond, K. Hinzer, S. Fafard, J.L. Merz: Phys. Rev. B **61**, R16331 (2000)
104. See, for example, X. Jiang, M.A. Dubson, J.C. Garland: Phys. Rev. B **42**, 5427 (1990); H.H. Mueller, M. Schulz: J. Appl. Phys. **83**, 1734 (1998); J.H. Scofield, N. Borland, D.M. Fleetwood: Appl. Phys. Lett. **76**, 3248 (2000)
105. J.R. Dekker, A. Tukiainen, R. Jaakkola, K. Väkeväinen, J. Lammasniemi, M. Pessa: Appl. Phys. Lett. **73**, 3559 (1998); Z.C. Huang, C.R. Wie, J.A. Varriano, M.W. Koch, G.W. Wicks: J. Appl. Phys. **77**, 1587 (1995)
106. I.V. Ignatiev, I.E. Kozin, H.-W. Ren, S. Sugou, Y. Masumoto: Phys. Rev. B **60**, R14001 (1999)
107. J. Lindahl, M.-E. Pistol, L. Montelius, L. Samuelson: Appl. Phys. Lett. **68**, 60 (1996)
108. S. Raymond, J.P. Reynolds, J.L. Merz, S. Fafard, Y. Feng, S. Charbonneau: Phys. Rev. B **58**, R13415 (1998)
109. H. Gotoh, H. Kamada, H. Ando, J. Temmyo: Appl. Phys. Lett. **76**, 867 (2000)
110. U. Bockelmann, W. Heller, G. Abstreiter: Phys. Rev. B **55**, 4469 (1997)
111. Y. Toda, S. Shinomori, K. Suzuki, Y. Arakawa: Appl. Phys. Lett. **73**, 517 (1998)
112. M. Sugisaki, H.-W. Ren, K. Nishi, S. Sugou, T. Okuno, Y. Masumoto: Physica B **256–258**, 169 (1998)
113. M. Bayer, A. Kuther, F. Schäfer, J.P. Reithmaier, A. Forchel: Phys. Rev. B **60**, R8481 (1999)
114. P.D. Wang, J.L. Merz, S. Fafard, R. Leon, D. Leonard, G. Medeiros-Ribeiro, M. Oestreich, P.M. Petroff, K. Uchida, N. Miura, H. Akiyama, H. Sakaki: Phys. Rev. B **53**, 16458 (1996)
115. See, for example, *Excitons*, ed. by K. Cho (Springer-Verlag, Berlin, Heidelberg, New-York 1979)
116. V.D. Kulakovskii, G. Bacher, R. Weigand, T. Kümmell, A. Forchel, E. Borovit-skaya, K. Leonardi, D. Hommel: Phys. Rev. Lett. **82**, 1780 (1999)
117. U. Bockelmann, W. Heller, A. Filoramo, P. Roussignol: Phys. Rev. B **55**, 4456 (1997)
118. J. Bellessa, V. Voliotis, R. Grousson, X.L. Wang, M. Ogura, H. Matsuhata: Phys. Rev. B **58**, 9933 (1998)
119. G. Bacher, R. Weigand, J. Seufert, V.D. Kulakovskii, N.A. Gippius, A. Forchel, K. Leonardi, D. Hommel: Phys. Rev. Lett. **83**, 4417 (1999)
120. N.H. Bonadeo, G. Chen, D. Gammon, D.S. Kratzer, D. Park, D.G. Steel: Phys. Rev. Lett. **81**, 2759 (1998)
121. Y. Toda, T. Sugimoto, M. Nishioka, Y. Arakawa: Appl. Phys. Lett. **76**, 3887 (2000)

# 5 Persistent Spectral Hole Burning in Semiconductor Quantum Dots

Yasuaki Masumoto

## 5.1 Introduction

Nanometer-size semiconductor crystals or semiconductor nanocrystals are known as zero-dimensional quantum dots [1–6]. Their optical properties have been characterized by the quantum confinement effect, and the lowest excited states show blue shifts depending on their size. Quantum dots act independently, when they are independently embedded in host materials for the isolation of each dot. Otherwise, the quantum confinement effect is weakened or modified. Quantum dots are sharply different from other low-dimensional quantum structures such as quantum wells and quantum wires because quantum dots are made of as few as $10^3$–$10^6$ atoms. A considerable fraction of atoms face the surface or the interface of quantum dots in the surrounding materials. Therefore, it is quite natural to consider that the electronic states of quantum dots should not be treated by themselves, but should be treated together with the real surfaces or interfaces and the surrounding materials. This consideration is correct, in general. However, it is not clearly noted that the electronic states of quantum dots are seriously affected by the surrounding host.

A semiconductor quantum dot is so small that its electronic excitation energy strongly depends on the number of excitons in the quantum dots as well as on the spatial geometry of electrons and holes in or out of the quantum dots. This conjecture was verified by the observation of the two exciton-pair and the three exciton-pair states in quantum dots in recent years [3,4,7,8] and the observation of the persistent spectral hole-burning phenomena in many kinds of quantum dots [9]. If we consider the case where an electron (hole) is in the quantum dot and a hole (electron) is outside the quantum dot, in other words, the case where the quantum dot is ionized, the photon energy that creates an electron–hole pair in the quantum dot differs from the photon energy of the quantum dots that do not hold any electrons and holes. Photoionization causes the persistent spectral hole-burning phenomenon. This was proved by the observation of the negatively and positively charged exciton states by luminescence hole-burning spectroscopy [10].

The quantum dots showing the persistent spectral hole-burning phenomena are composed of $10^3$–$10^4$ atoms and are much larger than the ions, defects, and molecules that show persistent spectral hole burning. If the guest size is much smaller than the quantum dot system, such as the molecular or ionic guests embedded in glass, crystal, or polymer hosts, we can naturally understand that the guests would be greatly influenced by the surrounding hosts, and that the energy levels of the guest–host system would be inhomogeneously broadened. A persistent spectral hole-burning phenomenon has been observed in a number of these materials [11]. Various types of photophysical and photochemical processes cause persistent spectral hole-burning phenomena. The quantum dots showing persistent spectral hole-burning phenomena are much larger than the ions, defects, and molecules that show persistent spectral hole burning. The Coulomb interaction between an electron (hole) in the quantum dot and a hole (electron) outside the quantum dot has a larger range than the size of the quantum dot. This explains why a quantum dot changes its energy as a result of photoionization and why such large quantum dots show persistent spectral hole burning. A considerable fraction of atoms face the surface or the interface of quantum dots in the surrounding materials, resulting in the finite probability of photoionization.

So far, long-lived photoluminescence [12], persistent spectral hole burning, photoluminescence blinking [13,14], and spectral diffusion [15] of quantum dots are observed. The latter three phenomena are not observed in bulk semiconductors or even quantum wells or quantum wires, and therefore are considered to be unique for quantum dots. All the phenomena are considered to arise from the presence of traps near quantum dots, and the energy of quantum dots is affected by the carriers captured by traps. These phenomena are considered to be due to the photoionization of quantum dots and the succeeding carrier capture at the traps near the quantum dots [9,16]. Trapped carriers are excited to the conduction or valence band by the second illumination, and anti-Stokes luminescence radiates as a result of electron and hole recombination. This process causes photostimulated luminescence, which is the anti-Stokes luminescence of materials pre-excited by light with higher energy than the luminescence.

When the spectrally narrow light excites an ensemble of semiconductor quantum dots having inhomogeneously broadened absorption bands, a spectral hole is formed at the excitation photon energy in the absorption band. Before the discovery of the persistent spectral hole-burning phenomena of quantum dots, inhomogeneous broadening was believed to come from the size and shape distribution; spectral hole burning observed in quantum dots was believed to be transient. Nevertheless, persistent spectral hole-burning phenomena were discovered in many kinds of semiconductor quantum dots embedded in crystals, glass, or polymers [9,16–22]. The observation of persistent spectral hole-burning phenomena in many kinds of quantum dots proves that persistent spectral hole-burning phenomena take place widely in quantum dots. The observation of the phenomenon requires additional inhomogeneous broadening for the quantum dots coming from the various ground state configurations of the dot-matrix system.

Persistent spectral hole-burning phenomena enable us to precisely investigate the size-dependent energy levels in "laser-marked" quantum dots by observing their site selectively burned energies in the inhomogeneously broadened absorption spectra. As a result of persistent spectral hole burning, the electronic and excitonic quantum states, including the excited states, are burned. This allows us to investigate the size-dependent quantized energies of quantum dots systematically [22–24]. This approach is effective in studying the size dependence of not only the quantized electronic levels but also of the phonons in quantum dots. It is known that the optical phonon energy in the electronic excited state differs from that in the ground state in many molecular systems and localized states. Similarly, 10% softening of the longitudinal optical phonon energy in the photoexcited dot was found by means of persistent spectral hole-burning spectroscopy [25,26]. This observation is striking, because the quantum dots we studied consisted of several thousands of atoms and are much larger than the molecular or localized systems. The acoustic phonon is confined in quantum dots, and its energy is inversely proportional to the size of the dots. The confined acoustic phonon was observed in persistent spectral hole-burning spectroscopy [27,28].

Persistent spectral hole-burning phenomena make the accumulated photon echo measurements possible under an ideally weak excitation. In recent years, a very sharp optical spectrum whose width is comparable to or less than 0.1 meV has been reported for several kinds of quantum dots by means of single quantum dot spectroscopy and site-selective laser spectroscopy [29–35], but the linewidth is too sharp to be studied with sufficient spectral resolution. This experimental difficulty has masked the intrinsic physical properties of quantum dots. An alternative approach in the time domain has potential. A very sharp homogeneous width indicates a long dephasing time, which is easily accessible by time-domain measurements, such as photon echo. The Fourier transform of the time trace observed by a photon echo gives a homogeneous optical spectrum. Application of persistent spectral hole burning to this direction is discussed in Chap. 8.

In this chapter, we review persistent spectral hole burning and related phenomena in semiconductor quantum dots and their applications.

## 5.2   Precursor and Discovery of the Persistent Spectral Hole-Burning Phenomenon

The nonlinear optical properties of semiconductor quantum dots have been extensively investigated experimentally [3–5]. The research was sparked by the theoretical prediction of the enhancement of optical nonlinearity [36,37]. In experimental publications before 1994, we can find long-lived nonlinear effects that are related to persistent spectral hole burning [5]. However, the remarkable precursor to persistent spectral hole burning was not considered noticeable. In CdSe quantum dots, pump-and-probe experiments showed that the bleaching structure lasts for more than 40 ns [38]. Time-resolved luminescence of CdSe quantum dots has been investigated under picosecond size-selective excitation of the lowest excited state [39]. The emission showed two decay components with time constants of 110 ps and on the order of microseconds. The microsecond decay component is ascribed to the localization of the excited carriers. In CdTe quantum dots, time-resolved pump-and-probe spectroscopy showed microsecond bleaching, suggesting carrier trapping at the dot surface or in the glass host [40]. This observation is not the long-lived hole burning but the long-lived spectral broadening of the absorption spectrum. A similar long-lived spectral change had been reported by other groups [41,42].

The discovery of the persistent spectral hole-burning phenomenon in semiconductor quantum dots originated from two precursory strange phenomena observed in nanosecond pump-and-probe experiments. One is the long delayed recovery of the spectral hole observed in CuCl quantum dots. The other is the spectral hole trace made by the pump pulses irradiated previously in the absorption spectrum of CuCl quantum dots. After the strange pump-and-probe data suggesting fairly long decay times of the spectral holes was noted, a rigorous experimental test was performed in the millisecond and second time regimes in CdSe and CuCl quantum dots. The differential absorption spectra of these samples showed clear hole structures in the millisecond and second time regimes, indicating the definitive proof of the persistent spectral hole-burning phenomenon [17,18]. After that, persistent spectral hole-burning phenomena were clearly observed in several kinds of quantum dots, as is shown in the following section. In addition, the hole-burning efficiency was found to be as high as 9.7% for CuCl quantum dots embedded in glass at 2 K [43]. However, efficient hole filling had masked the observation of persistent spectral hole burning phenomena for a long time. This is why the discovery of persistent spectral hole-burning phenomena was delayed.

## 5.3    Persistent Spectral Hole Burning, Hole Filling, and Their Mechanism

Following the discovery of persistent spectral hole burning [17,18], systematic investigation of these phenomena in quantum dots was made [16,19–22]. Samples examined were nanometer-size CdSe quantum dots in $GeO_2$:$Na_2O$ glass, CdSSe quantum dots in glass, CdS quantum dots in polyvinyl alcohol polymers, CuCl quantum dots in glass, and NaCl crystals, CuBr quantum dots in glass, and NaBr crystals and CuI quantum dots in glass, where the host glass was aluminoborosilicate glass, except where $GeO_2$ glass is noted. The sizes of the quantum dots were changed by heat treatment. Their average radii $R$ were determined by small-angle X-ray scattering experiments. The radius distribution of quantum dots in typical samples was measured by transmission electron microscopy, and the average radius derived from the distribution agrees with that estimated by the small-angle X-ray scattering experiments. The hole-burning measurement was done in a simplest mode. A narrow-band nanosecond dye laser was used as a pump source. A halogen lamp was used as a probe source, because its good stability gives an improved signal-to-noise ratio and because its low power reduces hole filling. A coaxial pump-and-probe configuration was adopted for the reliable overlap of the pump beam and the probe beam. First, the absorption spectrum was obtained. Next, the sample was excited by dye laser pulses. After that, the laser exposure was stopped, and the absorption spectrum was taken. The absorption spectral change $-\Delta\alpha d$ is defined as the difference in the absorbance of the exposed sample from that of the virgin sample.

Absorption spectra of quantum dots show the blueshift from the bulk energy depending on their size. Figure 5.1a shows the absorption spectrum of CdSe quantum dots embedded in $GeO_2$ glass at 2 K. Three structures corresponding to A-exciton, B-exciton, and C-exciton bands are indicated. The

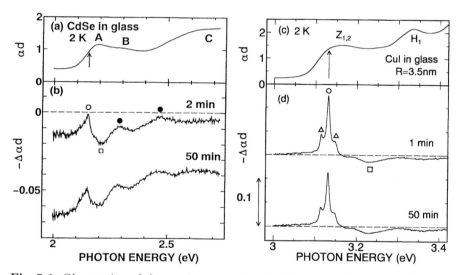

**Fig. 5.1.** Observation of the persistent spectral hole burning phenomenon in CdSe and CuI quantum dots embedded in glass at 2 K. Both the samples belong to the strong-confinement regime. The *upper figures*, (**a**) and (**c**), show the absorption spectrum of the virgin samples, CdSe ($R=2.4$ nm) and CuI ($R=3.5$ nm) in glass, respectively. The *lower figures*, (**b**) and (**d**), show the spectral change observed at 1 min or 2 min and 50 min after the burning laser exposure was stopped. The CdSe sample was excited by 1800 shots of dye laser pulses with photon energy of 2.148 eV and energy density of 0.5 mJ/cm$^2$. The CuI sample was excited by 9000 shots of dye laser pulses with photon energy of 3.130 eV and energy density of 80 μJ/cm$^2$. The spectral hole and associated structure were preserved for more than several hours at 2 K (from Masumoto et al. [16,20], Masumoto [9])

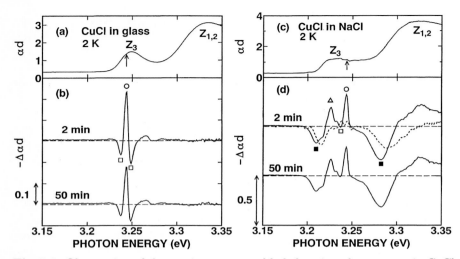

**Fig. 5.2.** Observation of the persistent spectral hole burning phenomenon in CuCl quantum dots embedded in glass and NaCl. Both samples belong to the weak-confinement regime. The *upper figures*, (**a**) and (**c**), show the absorption spectrum of the virgin samples, CuCl ($R$=2.5 nm) in glass and CuCl ($R$=3.5 nm) in NaCl, respectively. The *lower figures*, (**b**) and (**d**), show the spectral change observed at 2 min and 50 min after the burning laser exposure was stopped. The samples were excited by 1800 shots of dye laser pulses with photon energy of 3.245 eV and energy density of 33 μJ/cm$^2$. The spectral hole and the associated structure were preserved for more than several hours at 2 K. A *dashed line* in (**d**) shows the derivative of absorption spectrum (**c**) (from Masumoto et al. [16], Masumoto [9])

spectrum has a pronounced A-exciton peak at 2.19 eV, which is higher than the band gap energy of bulk CdSe by 0.35 eV. The blueshift is the indication of electron and hole quantum confinement. This strong-confinement model holds not only for CdSe quantum dots, but also for CdSSe quantum dots, CdS quantum dots, and CuI quantum dots. Figure 5.1c shows the absorption spectrum of CuI quantum dots embedded in glass. The absorption spectrum shows the $Z_{1,2}$ exciton structure whose energy increases with decreasing size. The blueshift was explained by the strong-confinement model [20]. Besides the $Z_{1,2}$ exciton structure, the $H_1$ exciton structure of hexagonal CuI quantum dots was observed. Figures 5.2a and c show absorption spectra of CuCl quantum dots embedded in glass and a NaCl crystal. The broad peaks indicated by $Z_3$ are due to the absorption of $Z_3$ excitons. These peaks show blueshifts compared with that of bulk CuCl due to exciton quantum confinement [1,44]. The weak-confinement model holds for CuCl quantum dots. Figure 5.3a shows the absorption spectrum of CuBr quantum dots embedded in glass. The spectrum has a pronounced peak at 3.073 eV, which is higher than the $Z_{1,2}$ exciton energy of bulk CuBr by 109 meV. The blueshift observed in case of CuBr quantum dots in glass cannot be explained by both

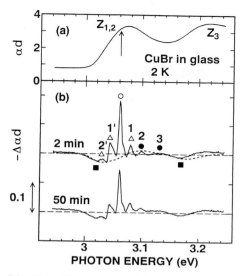

**Fig. 5.3.** Observation of the persistent spectral hole burning phenomenon in CuBr quantum dots embedded in glass at 2 K. The *upper figure*, (**a**), shows the absorption spectrum of the virgin sample, CuBr ($R$=3.6 nm) in glass. The *lower figure*, (**b**), shows the spectral change observed at 2 min and 50 min after the burning laser exposure is stopped. The sample was excited by 9000 shots of dye laser pulses with photon energy of 3.061 eV, and energy density of 5 μJ/cm$^2$. The spectral hole and associated structure were preserved for more than several hours at 2 K. A *dashed line* in (**b**) shows the derivative of the absorption spectrum (**a**) (from Masumoto et al. [19,16])

strong- and weak-confinement models [45], but probably can be explained by the size-dependent Coulomb energy between trapped carriers and excitons.

In both strong- and weak-confinement regimes, persistent spectral hole-burning phenomena were observed. The persistent spectral hole-burning phenomena in the strong-confinement regime were typically observed in CdSe quantum dots in GeO$_2$ glass [9,18] and CuI quantum dots in glass [20]. The observation of persistent spectral hole-burning phenomena in the weak-confinement regime were best demonstrated in CuCl quantum dots in glass and a NaCl crystal [16]. Persistent spectral hole-burning phenomena were also observed in CdTe quantum dots in glass [22], CdS quantum dots in a polymer [21], and CuBr quantum dots in glass [16,19]. After the samples were excited by dye laser pulses, the absorption spectra of the samples were changed, as shown in Figs. 5.1b,d, 5.2b,d, and 5.3b. This spectral change was preserved for long time at low temperatures. At 2 K, the spectral holes were preserved for more than 8 h after the excitation. These spectra show a main spectral hole (○), satellite holes (●), and an induced absorption structure (□) which are superposed on the photodarkened spectra (Fig. 5.1b); a main spectral hole ($\circ$), phonon sideband holes ($\triangle$), and an induced absorption structure ($\square$) (Fig. 5.1d); a main spectral hole ($\circ$), phonon sideband holes ($\triangle$), satellite holes ($\bullet$), or an antihole ($\square$), which are superposed on the broad wavy structure ($\blacksquare$) approximated by the first derivative of the absorption spectrum (Figs. 5.2d and 5.3b); and a sharp spectral hole ($\circ$) and antiholes ($\square$) (Fig. 5.2b). The first derivative of the absorption spectrum shows the small redshift of the absorption spectrum. In the time regime ranging from minutes to hours, the hole depth decays spontaneously in proportion to the logarithm of time, indicating the broad distribution of spontaneous decay rates [16,19].

Persistent spectral hole-burning phenomena require that the guests (quantum dots) should be greatly affected by the surrounding host materials. The requirement tells us that persistent spectral hole-burning phenomena should depend strongly on the size of the quantum dots, and should disappear with increasing size. This conjecture was found to hold from the size-dependent persistent spectral hole-burning experiments of CdSe quantum dots in $Ge_2O$ glass, CuCl quantum dots in glass, and CuBr quantum dots in glass. The absorption spectral changes observed in CdSe quantum dots and CuBr quantum dots consist of hole structures and wavy structures having correlation with the absorption spectra, as is shown in Figs. 5.4 and 5.5. The wavy structure becomes dominant, and the hole structure decreases with increasing size of

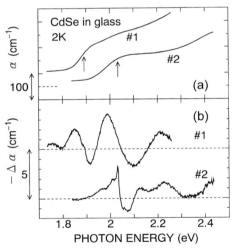

**Fig. 5.4.** (a) Absorption spectra of virgin samples, CdSe quantum dots embedded in $GeO_2$ glass. Average radii of quantum dots in samples #1 and #2 are 11.5 nm and 6.7 nm, respectively. *Upward arrows* show the excitation photon energies. (b) The absorption spectral change of the samples after spectrally narrow laser exposure. The samples were excited by 4500 shots of dye laser pulses with the pump photon energy arrowed and excitation energy density of 40 µJ/cm$^2$. *Dashed lines* show the zero base (from Masumoto [9])

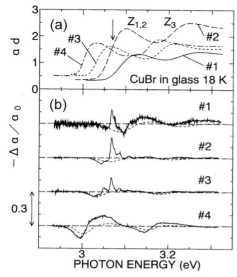

**Fig. 5.5. (a)** Absorption spectra of virgin samples, CuBr quantum dots embedded in glass. Average radii of quantum dots in samples #1, #2, #3, and #4 are 2.6 nm, 2.9 nm, 3.8 nm, and 6.5 nm, respectively. A *downward arrow* shows the excitation photon energy. **(b)** The absorption spectral change of the samples after spectrally narrow laser exposure. The absorption spectral change is normalized by the absorbance of the samples at the excitation photon energy. The samples were excited by 9000 shots of dye laser pulses with the pump photon energy arrowed (3.169 eV) and excitation energy density of 50 μJ/cm$^2$. *Dashed lines* show the derivatives of the absorption spectra (from Masumoto [9])

the quantum dots. The persistent spectral hole burning was hardly observed in CuCl quantum dots in glass whose radii were larger than 10 nm. A similar size-dependent efficiency of persistent spectral hole burning was also observed in CuBr quantum dots embedded in borosilicate glass [46]. Persistent spectral hole burning was observed in CuBr quantum dots whose average radius is less than 4 nm.

The redshift of the absorption spectrum was observed in CuCl quantum dots in NaCl and CuBr quantum dots in glass simultaneously with the burned hole, as is seen in Figs. 5.2d, 5.3b and 5.5. The hole and the red shift simultaneously grow logarithmically as a function of the laser exposure time or the laser fluence, as is shown in Fig. 5.6. This dependence indicates the broad distribution of the burning rate [47] and the presence of a common origin to produce the hole burning and the redshift [16,19]. The broad distribution of the burning rate suggests the persistent spectral hole burning takes place via tunneling between the two-level system through a potential barrier with a broadly distributed barrier height and thickness [47]. Thermally annealing hole filling and light-induced hole filling are observed, as are shown in

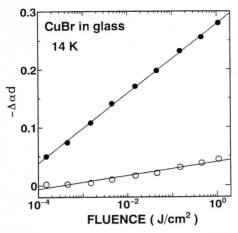

**Fig. 5.6.** Growth of the hole depth of CuBr quantum dots as a function of the burning laser fluence at 14 K. *Solid circles* show the hole depth as a function of the laser fluence. *Open circles* show the maximum amplitude of the wavy structure approximated by the derivative of the absorption spectrum. The sign of the amplitude was changed for convenience. Both data are taken by changing the exposure time with an excitation energy density of $5~\mu J/cm^2$. Both hole depth and amplitude of the wavy structure increase in proportion to the logarithm of the laser fluence (from Masumoto et al. [19,16], Masumoto [9])

Figs. 5.7 and 5.8. The observation of hole filling denies the possibility of permanent laser damage of quantum dots and suggests a slight photophysical or photochemical change of the guest–host system.

Experimental annealing-temperature dependence of hole filling is well explained by the thermally activated barrier-crossing mechanism [16,19,48]. The rate of this mechanism is represented by $\nu=\nu_0\exp(-V/kT)$, where $\nu_0$ is the frequency factor whose order is given by $kT/h(10^{11}–10^{13}~s^{-1})$ and $V$ the potential barrier. During the holding time $t$ at the annealing temperature $T$, the hole is filled if the condition $\nu t>1$ holds. Therefore the hole is filled, if $V<kT\ln(\nu_0 t)$. From the rather uniform distribution of the tunneling parameter, we assume that the distribution of the potential barrier $P(V)$ is represented by $P(V)\propto 1/\sqrt{V}$ with a maximum barrier height $V_{max}$ [48]. The normalized hole depth observed after the temperature cycle between the burning temperature $T_b$ and the annealing temperature $T$ can be expressed by a functional form of $(1 - \sqrt{kT\ln(\nu_0 t)/V_{max}})/(1 - \sqrt{kT_b\ln(\nu_0 t)/V_{max}})$ for $kT\ln(\nu_0 t)<V_{max}$. The experimental results of CuI quantum dots in glass are well fitted by this functional form. The maximum barrier height is 0.5 eV for CuI quantum dots in glass [20]. The experimental results of CuCl quantum dots in glass, CuCl quantum dots in a NaCl crystal, and CuBr quantum dots in glass are well fitted by the expression, if we consider two distributions of potential barriers. A fitting example for the CuBr sample is shown in Fig. 5.7.

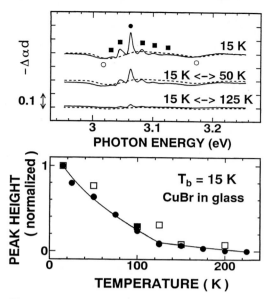

**Fig. 5.7.** Temperature cycle experiment for the hole-burning structures. The hole was burned at $T_b$=15 K by 9000 shots of dye laser pulses with energy density of 5 µJ/cm$^2$. Sample temperature was elevated to annealing temperature $T$=50 or 125 K and was maintained at $T$ for 5 min. After that, the sample temperature was cooled to $T_b$ and the absorption spectral difference was measured. The temperature rise to $T$ indicated at the right side partially erases the spectral holes and associated structures. The spectral hole and associated structures remain up to the annealing temperature of 220 K. The *lower figure* shows the normalized hole depth (*solid circles*) and the normalized amplitude of the wavy structure (*open squares*) approximated by the derivative of the absorption spectrum as a function of the annealing temperature. The *solid line* is the fit of the hole depth by the expression $0.7(1 - 0.089T)/(1 - 0.089T_b)$ for $T$<125 K, plus $0.3(1 - 0.067T)/(1 - 0.067T_b)$ for $T$<225 K, where $T_b$ is the burning temperature and $T$ is the annealing temperature (from Masumoto et al. [19,16], Masumoto [9])

The obtained maximum barrier heights for two distributions are 0.36 eV and 0.59 eV for CuCl quantum dots in glass, 0.15 eV and 0.41 eV for CuCl quantum dots in a NaCl crystal, and 0.36 eV and 0.65 eV for CuBr quantum dots in glass.

The hole structures are also erasable by light exposure. Figure 5.8 shows the light-induced hole-filling data for the CuBr sample. Hole-filling data show that the erased spectrum, on the whole, returns to the spectrum of the virgin sample when the erase light is below the absorption edge of the quantum dots. On the other hand, when the absorption band of the quantum dots is excited by broadband light, it causes hole filling more efficiently than light

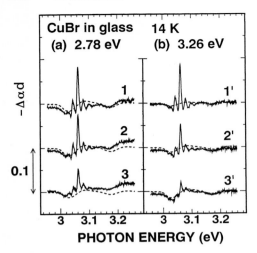

**Fig. 5.8.** Light-induced hole-filling experiments for case where (**a**) the erase broadband light is below the absorption edge of the CuBr quantum dots; and (**b**) the erase broadband light is above the absorption edge. Exposure time for spectra 1(1′), 2(2′) and 3(3′) is 0 s, 120 s and 3000 s, respectively. *Dashed lines* show the derivatives of the absorption spectra. They can well approximate the wavy persistent spectral change spectra (from Masumoto et al. [19,16], Masumoto [9])

below the absorption edge, but the persistent wavy spectral change coming from the redshift of the absorption spectrum is enhanced more.

To explain the accumulated experimental data of persistent spectral hole-burning phenomena of quantum dots, the photoionization of quantum dots and carrier trapping into the hosts were considered as the mechanism to induce persistent spectral hole burning [16,19]. To derive this conclusion, similar behavior of the burned hole and the redshift in both burning and filling processes was a key point. The proposed mechanism giving rise to persistent spectral hole-burning phenomena is shown schematically in Fig. 5.9. A photoexcited exciton in a quantum dot is localized, and an electron or a hole is trapped at the surface of the quantum dot. A hole or an electron escapes from the quantum dot, tunnels through the potential barrier in the host and is trapped in the host. The potential barrier height between quantum dots and traps in the host is considered to be broadly distributed. In ionized quantum dots, Coulomb interaction in the quantum dots causes the blue- or redshift of the exciton energy, which results in hole burning at the laser excitation energy and induced absorption at the high- or low-energy side. This explains the observed main hole and the associated induced absorption. Spatially separated electrons and holes apply the local electric field to quantum dots and cause the quantum-confined Stark effect. The effect gives the redshift of the exciton structure. Not only photoexcited quantum dots but also non-photoexcited quantum dots feel the electric field because

**(a)**     **Quantum Dot**

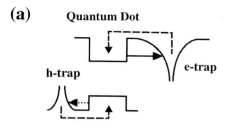

**(b)**     **Photoionized Quantum Dots**

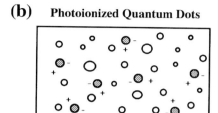

**Fig. 5.9. (a)** Schematic model of the photoionization of quantum dots embedded in glass and crystals, which is the persistent spectral hole-burning mechanism. Photogenerated electrons or holes tunnel through barriers and are captured by traps in the host. The tunneling process is shown by *solid* and *dotted arrow lines*. Hole filling due to thermal annealing and light-induced hole filling processes are shown by *dashed arrow lines*. **(b)** Schematic view of the photoionized quantum dots and carriers in traps. Various configurations between photoionized quantum dots and spatial charge distribution in the host give another inhomogeneous broadening that satisfies the requirement of persistent spectral hole burning (from Masumoto et al. [16], Masumoto [9])

of the long-range character of the Coulomb field. The ensemble of quantum dots shows the quantum-confined Stark shift. Therefore, the redshift of the inhomogeneously broadened absorption spectrum is observed.

Spectral holes and the broad wavy structure behave similarly in hole burning and thermally annealing hole filling. These observations are clearly understood by the photoionization model. The only difference is observed in light-induced hole filling. When the erase light energy is below the absorption edge of the quantum dots, trapped holes or electrons are excited and recombine with electrons or holes localized at the interface of the quantum dots. This process erases both the burned hole structures and the wavy structures coming from the redshift of the absorption spectrum because the photoionized quantum dots return to deionized quantum dots. On the other hand, broadband erase light above the absorption edge of quantum dots may deionize the burned quantum dots, but ionizes many unburned quantum dots. The former process fills the burned hole and reduces the redshift. However, the latter process fills the burned hole, gives the photoproduct that fills the hole burned previously, and increases the redshift because the trapped carrier density increases.

Photoionization of quantum dots was reported for CdS quantum dots embedded in silicate glass [49]. Photoionized quantum dots and trapped holes are stable enough to give persistent spectral hole burning at low temperatures. There are many spatial arrangements for electrons and holes in both quantum dots and glass. A variety of the spatial arrangements induce additional inhomogeneous broadening of the absorption spectra of quantum dots. The presence of inhomogeneous broadening from the carrier distribution satisfies the requirement of persistent spectral hole burning. Persistent spectral hole burning was observed in CdS quantum dots embedded in polyvinyl alcohol [21]. The persistence time depends on the synthesis conditions of the quantum dots. In the samples with over 10% $Cd^{2+}$ ions, the observed hole structure lasted for several hours. In the samples whose $Cd^{2+}/S^{2-}$ ratio is equal to or less than 1, the hole persistence time was less than 1 min. This observation supports the explanation that persistent spectral hole burning takes place by the photoionization of quantum dots. Some of the photogenerated electrons escape from the quantum dots and are captured by the trap centers formed by the excess $Cd^{2+}$ ions around the quantum dots in the polymer host.

It was reported that photoionization of quantum dots induced permanent photochemical spectral hole burning for CdS quantum dots embedded in a polymer film [5,50,51]. At the first stage, irradiation by an ultraviolet narrowband laser caused selective photoionization of quantum dots and made local electric fields around the dots. This broadened the optical spectrum of the dots, inducing a quantum-confined Stark shift toward lower energy. At the second stage, holes localized at the surface were considered to cause irreversible oxidation of the CdS surface. As a result, permanent spectral holes with full width of 100–120 meV were observed at room temperature.

There are several reports about long-lived spectral changes [40–42] and the Stark effect [52] in semiconductor quantum dots. The observed spectral changes are due to the broadening of the inhomogeneously broadened absorption spectra. They were explained by the trapped-carrier-induced quantum-confined Stark effect on the basis of their similarity to the Stark effect data [40,42] or the simulated results [41]. In CuCl quantum dots in NaCl and CuBr quantum dots in glass, a small redshift of the exciton absorption spectrum was observed. In CuCl quantum dots in glass, on the other hand, a small blueshift of the exciton absorption spectrum was observed. There remains an unresolved question about the spectral change of quantum dots induced by the Stark effect. Further study is necessary to resolve the question.

## 5.4   Luminescence Hole Burning and Charged Exciton Complexes

The presence of ionization of quantum dots and succeeding carrier trapping in the host glass is supported by the following unusual phenomena of luminescence. These are luminescence elongation following the increase in the light exposure, which is observed in CuI quantum dots in glass [20,53], and luminescence hole burning observed in CuCl quantum dots in NaCl crystals [10,53], which are observed simultaneously with the persistent spectral hole burning.

Luminescence elongation was observed in CuI quantum dots in glass. An absorption spectrum of the sample shows the $Z_{1,2}$ exciton structure of zincblende CuI and the $H_1$ and $H_2$ exciton structures of hexagonal layered CuI, as is shown in Fig. 5.10a. The $Z_{1,2}$ exciton structure shows a blueshift of 175 meV from its bulk energy of 3.06 eV. The observed blueshift of 175 meV

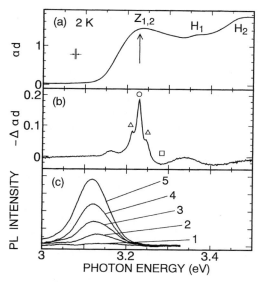

**Fig. 5.10.** (a) Absorption spectrum of a virgin sample, CuI quantum dots ($R$=3.0 nm) embedded in glass at 2 K. (b) The absorption spectral change of the sample at 10 min after the narrowband laser exposure was stopped. The sample was exposed by 3.6 mW frequency-doubled output of Ti:sapphire laser pulses at photon energy of 3.229 eV with accumulated energy density of 11 J/cm². (c) Luminescence spectra of CuI quantum dots embedded in glass at 2 K. Luminescence signals were accumulated while the sample was excited by 3000 shots of 355 nm (3.49 eV) pulses with energy density of 2 μJ/cm². Spectra 1, 2, 3, 4, and 5 correspond to luminescence of the sample exposed by accumulated energy densities of 0, 0.75, 2.25, 8.25, and 53 J/cm² of the 355 nm (3.49 eV) pulses, respectively (from Masumoto et al. [53], Masumoto [9])

corresponds to a radius of 3.0 nm in the simplest strong-confinement model. Figure 5.10b shows the absorption spectral change $-\Delta\alpha d$. The absorption spectral change consists of a spectral hole (o), phonon sideband holes ($\triangle$), and an induced absorption structure ($\square$). The absorption spectral change is conserved for long times at 2 K [20].

With the increase of the integrated intensity of laser exposure, the luminescence intensity increases, as is shown in Fig. 5.10c. Luminescence spectra of CuI quantum dots show a Stokes shift from the absorption peak of the $Z_{1,2}$ exciton. The Stokes shift suggests the luminescence comes from the localized or bound exciton. It was also found that the luminescence intensity is quenched after the annealing cycle [20]. These characteristic phenomena of luminescence are explained, if the carrier trapping at the capture centers in the host glass is saturated with an increase in the laser fluence, and if trapped carriers are released as the temperature rises. Localized or bound excitons formed from free excitons in quantum dots are radiatively annihilated or decay nonradiatively through ionization, tunneling, and capture by traps in the host glass. These processes are shown schematically in Fig. 5.11. This leads to the persistent spectral hole-burning phenomena. The number of traps that are easily accessible by tunneling from quantum dots is finite, and traps seized by carriers become ineffective. For this reason carrier trapping at the capture centers is saturated. A temperature rise causes the thermal activation of the trapped carriers at the capture centers. Activated carriers are radiatively or nonradiatively annihilated. As a result, traps in the host be-

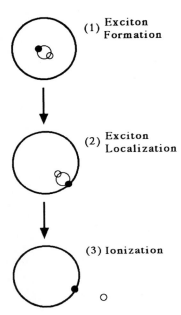

(1) **Exciton Formation**

(2) **Exciton Localization**

(3) **Ionization**

Fig. 5.11. Schematic drawing of the proposed time sequence of exciton dynamics to account for luminescence hole burning, absorption hole burning, and luminescence elongation (from Masumoto et al. [53], Masumoto [9])

come effective again. In this way, the luminescence elongation with increasing light exposure is explained by the photoionization model.

As a result of photoionization of a quantum dot, an electron or a hole remains in the quantum dot. Then additional photoexcitation of an exciton may form a charged exciton in a quantum dot. A negatively charged exciton is a complex composed of two electrons and a hole, analogous to $H^-$. A positively charged exciton is a complex composed of an electron and two holes, analogous to $H_2^+$. Both negatively and positively charged excitons were observed in CuCl quantum dots in NaCl crystals [10]. Figure 5.12a shows the absorption spectrum of CuCl quantum dots embedded in a NaCl crystal. The $Z_3$ exciton absorption peak shows blueshift of 20 meV from the bulk value. It corresponds to the blueshift of 3.0-nm-radius CuCl quantum dots

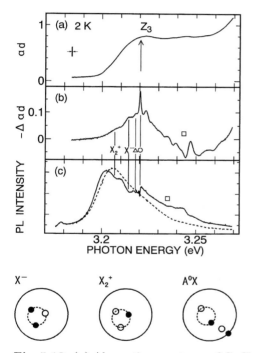

**Fig. 5.12.** (a) Absorption spectrum of CuCl quantum dots ($R$=3.0 nm) embedded in a NaCl crystal at 2 K. The *arrow pointing up* shows the energy of the dye laser pulses. (b) Absorption spectral change of the sample exposed to the narrowband dye laser at 3.220 eV. (c) Luminescence spectra before (*dashed line*) and after (*full line*) the sample was exposed at 3.220 eV. The luminescence spectrum was taken under excitation at 355 nm with excitation density of 0.5 $\mu J/cm^2$, after the sample was exposed at 3.220 eV with excitation density of 10 $\mu J/cm^2$. In the *lowermost part*, a negatively charged exciton $X^-$, a positively charged exciton $X_2^+$, and an exciton bound to a neutral acceptor $A^0X$ are schematically drawn (from Masumoto et al. [53], Masumoto [54])

in the exciton confinement model [44]. Luminescence spectra of the sample are shown in Fig. 5.12c. The dashed line shows the spectrum before the sample was excited by the narrowband 3.220 eV laser pulses. The peak position of the luminescence is located at 3.206 eV, which is lower than the absorption peak of the $Z_3$ exciton, 3.223 eV. The luminescence spectrum shows a Stokes shift. After the sample was excited by the narrowband 3.220 eV laser pulses, the luminescence spectrum changes, as is shown by the solid line in Fig. 5.12c. The luminescence spectrum shows a sharp hole (○), much larger satellite holes ($X_2^+$, $X^-$), which are observed at the low-energy side of the burning laser energy, and a phonon sideband hole (△). Simultaneously, the luminescence intensity increases at the high-energy side of the burning laser energy (□). After the luminescence hole-burning experiment, the absorption spectrum was taken and the absorption spectrum change was obtained, as shown in Fig. 5.12b. It consists of a main hole (○), satellite holes ($X_2^+$, $X^-$), phonon sideband holes (△), and an induced absorption structure (□). The luminescence spectral change and the absorption spectral change are correlated with each other. In fact, a main hole, a phonon sideband hole, satellite holes, and broad increased absorption or luminescence structures are observed in both spectra at the same energy.

Quantum confinement of the translational motion of the charged exciton explains the luminescence hole-burning structures. The translational masses of negatively and positively charged excitons, $M^-$ and $M_2^+$, are given by $M^-=2m_e^* + m_h^*$ and $M_2^+=m_e^* + 2m_h^*$, where $m_e^*=0.5m_0$ and $m_h^*=1.8m_0$ are the effective masses of an electron and a hole in a CuCl crystal, respectively. The energy shifts due to the quantum confinement effect are represented by $\Delta E^-=\hbar^2\pi^2/2M^- R^{*2}$ for the negatively charged exciton $X^-$, and $\Delta E^+=\hbar^2\pi^2/2M_2^+ R^{*2}$ for the positively charged exciton $X_2^+$, where $R^*$ is the effective radius of the quantum dot defined by the quantum dot radius minus half the exciton Bohr radius [44]. This model nicely explained the experimental energy shift of the structures with a change in the burning photon energy. Therefore, the structures $X_2^+$ and $X^-$ in Fig. 5.12 originate from the recombination of the positively charged exciton and the negatively charged exciton, respectively. The burning photon energy dependence of the $I_1$ bound-exciton structure resembles that of $X_2^+$. The $I_1$ line denoted also by $A^0X$ originates from the recombination of the exciton bound to a neutral acceptor. Therefore, $A^0X$ can be regarded as an excitonic molecule whose constituent electron is pinned and can be considered as the complex $X_2^+$ going around the pinned electron. In this sense, the $I_1$ bound exciton resembles the positively charged exciton $X_2^+$. Similar to the positively charged exciton $X_2^+$, the translational motion of the $I_1$ bound exciton is confined in the quantum dots. Therefore, the burning energy dependence of the $I_1$ bound-exciton structure resembles that of $X_2^+$.

In this way, the luminescence spectral change was understood by the photoionization of the quantum dots and the creation of charged exciton states. Lower-energy satellite holes due to charged excitons are much larger than the resonant hole in the luminescence spectrum, but the reverse is the case in the absorption spectrum. This fact shows that most of photoexcited excitons in quantum dots are localized rather than radiatively annihilated as they are. Therefore, the main path of exciton relaxation is localization before radiative annihilation. The ionization of quantum dots takes place via the exciton localization process, and the ionization causes the persistent spectral hole-burning phenomena. Femtosecond pump-and-probe measurement gives us the fastest ionization time of excitons. It was found to be on the order of picoseconds or tens of picoseconds for CuCl quantum dots in glass and CuBr quantum dots in glass [55].

The energy of photoionized quantum dots is changed from the original energy, and their new energies depend on the spatial arrangement of trapped electrons and holes. Quantum confinement of carriers and the resulting strong Coulomb interaction between confined carriers and trapped carriers are essential for the energy change. The long-range Coulomb interaction explains the reason why such large quantum dots are affected by the surrounding host. Photoionized quantum dots and trapped carriers are stable enough to give persistent spectral hole burning at low temperatures because the thermal activation energy for trapped carriers is lower than the potential barrier for hole filling. There are many spatial arrangements for electrons and holes in both the quantum dots and the host. A variety of spatial arrangements induce additional inhomogeneous broadening of the absorption spectra besides the inhomogeneous broadening produced by the size distribution.

Bound exciton complexes are observed in the low-temperature luminescence spectra of bulk semiconductors [56]. They are excitons bound to ionized donors ($D^+X$), ionized acceptors ($A^-X$), neutral donors ($D^0X$), and neutral acceptors ($A^0X$). Let us consider bound-exciton complexes in quantum dots. Because electron(s) and hole(s) are confined in quantum dots, both the kinetic energy and the Coulomb energy are enhanced. Resonant energies of bound-exciton complexes depend on the size of quantum dots. This is the quantum size effect of bound-exciton complexes [57]. Further, they depend on the position of the trapped carriers. Quantum dots are not perfect spheres, therefore resonant energies of bound-exciton complexes depend not only on the size but also on the position of the trapped carriers in the quantum dots. As a result, resonant energies of bound-exciton complexes are inhomogeneously broadened by not only the size distribution but also by the variety of bound positions in the quantum dots. This is the reason why the luminescence holes of the bound excitons are much broader than the resonantly burned luminescence holes.

An exciton bound to an ionized donor ($D^+X$) or an ionized acceptor ($A^-X$) resembles $H_2^+$ or $H^-$, respectively. When the mass ratio of electron to hole, $m_e^*/m_h^*$, is less than 0.4, the exciton is bound to an ionized donor, but is not bound to an ionized acceptor [56]. In the case of CuCl, $m_e^*=0.50$

$m_0$, $m_h^*=2.0m_0$, and $m_e^*/m_h^*=0.25$ [58]. In the case of CuBr, from the values of $m_e^*=0.28m_0$ and $m_h^*=1.4m_0$, $m_e^*/m_h^*=0.2$; and in the case of CuI, $m_e^*=0.33m_0$, $m_h^*=1.40m_0$ and $m_e^*/m_h^*=0.24$ [58]. Therefore an exciton is not bound to an electron trapped at the surface of CuCl, CuBr, or CuI quantum dots. In other words, creating an exciton in a quantum dot that has an ionized donor needs lower photon energy than creating an exciton in a neutral quantum dot, while creating an exciton in a quantum dot that has an ionized acceptor needs higher photon energy than creating an exciton in a neutral quantum dot. An ionized donor causes the redshift of the exciton absorption spectrum of quantum dots, while an ionized acceptor causes its blueshift. In other words, ionized quantum dots raise their resonant energy, if electrons remain at the surface of quantum dots. This explains the appearance of the increase of the luminescence and absorption at the higher-energy side of the burning laser energy observed in Fig 5.12. The increase of the absorption at the higher-energy side of the burning laser energy is also observed for CuI and CuBr quantum dots in Figs. 5.1d and 5.3b.

Luminescence hole-burning phenomena were also observed in other quantum systems. Figure 5.13 shows the persistent luminescence hole-burning spectrum of porous Si [59]. In porous Si, a photoluminescence suppression band is formed by the resonant excitation of the narrowband laser and is extended to the lower-energy side with the appearance of step structures separated by the transverse optical phonon energy. The luminescence suppression is considered to take place as a result of Auger processes in photoionized Si quantum dots. It persists for long times at low temperatures. Two trans-

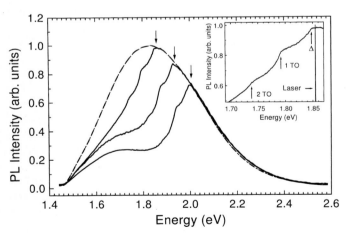

**Fig. 5.13.** Luminescence persistent hole-burning spectra of porous Si at different burning energies indicated by *arrows*. For comparison, the photoluminescence spectrum of a virgin sample is shown by a *dashed line*. In the *inset*, photoluminescence spectrum in the vicinity of the burning energy is displayed in an enlarged scale (from Kovalev et al. [59])

verse optical phonon structures observed in the luminescence suppression band show that the luminescence of porous Si comes from radiative indirect transitions betweeen quantum-confined states in the Si quantum dots.

Persistent change of the photoluminescence spectra was also observed in InP quantum wires embedded in chrysotile asbestos in the presence of strong resonant narrowband excitation [60]. Three onsets related to the longitudinal acoustic phonons of InP were observed in the changed spectrum. The spectral changes were preserved for long times at 2 K. The observed spectral changes, the enhancement and suppression of photoluminescence of InP quantum wires, were considered to be caused by trapping of the carriers at the surfaces of the quantum wires and by tunneling into traps in the matrix.

Luminescence hole burning was observed at 10 K in GaAs quantum dots formed by $Ga^+$ and $As^+$ implantation into $SiO_2$ followed by thermal annealing [61]. A resonant hole and its optical phonon sideband were clearly observed.

## 5.5  Photostimulated Luminescence, Luminescence Blinking, and Spectral Diffusion

In Sects. 5.3 and 5.4, the photoionization of quantum dots and the succeeding carrier capture at the traps near the quantum dots are ascribed to the origin of persistent spectral hole-burning phenomena. Photostimulated luminescence, luminescence blinking, and spectal diffusion are observed in quantum dots. These phenomena, together with persistent hole burning, come from the same origin, photoionization of quantum dots, and are strongly related to each other. In this section, photostimulated luminescence, luminescence blinking, and spectral diffusion are discussed in connection with persistent spectral hole burning.

Photostimulated luminescence is the anti-Stokes luminescence of materials pre-excited by light with higher energy than the luminescence. Many phosphors, such as alkali halides doped with rare earth ions and semiconductors doped with metal ions, show photostimulated luminescence [62]. The most famous and valuable photostimulable phosphor is $BaFX:Eu^{2+}$ (X=Cl, Br, I), which is widely used as a photostimulable phosphor for radiographic imaging [63]. In this system, F centers in BaFX work as deep electron traps, and $Eu^{2+}/Eu^{3+}$ ions work as hole traps. Similarly to this example, photostimulable phosphors need traps for carriers, and trapped carriers are stored by the pre-excitation. Trapped carriers are excited to the conduction or valence band by the second illuminatio, and anti-Stokes luminescence radiates as a result of electron and hole recombination.

The photoionization of quantum dots and the succeeding carrier capture at the traps near the quantum dots cause persistent spectral hole burning. If the carriers at the traps are preserved for a long time as a result of the first light exposure, and if the second illumination energy is higher than the trap

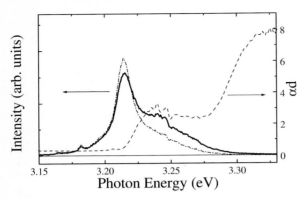

**Fig. 5.14.** Absorption spectrum (*dashed line*), photoluminescence spectrum (*dash-dotted line*), and photostimulated luminescence spectrum (*solid line*) of CuCl quantum dots embedded in a NaCl crystal at 2 K (from Masumoto et al. [64])

depth, photostimulated luminescence is expected for quantum dots formed in glass, crystals, and polymers. This conjecture may inspire a new class of photostimulated luminescence materials. In fact, photostimulated luminescence was observed in CuCl quantum dots embedded in NaCl crystals [64]. The observation of photostimulated luminescence gives the definitive conclusion that quantum dots are photoionized and that carriers are captured by traps in the host. The study of photostimulated luminescence is valuable to understanding carrier trapping and recombination processes in the quantum dot system.

The absorption spectrum, the luminescence spectrum observed under continuous-wave HeCd laser excitation, and the photostimulated luminescence of CuCl quantum dots in a NaCl crystal at 2 K are shown in Fig. 5.14. The absorption peak in the wavy $Z_3$ exciton band shows a blueshift of 37 meV from the size confinement of excitons in quantum cubes whose side length is 4.3 nm, corresponding to 16 times $a/2=0.27$ nm, where $a$ is a lattice constant of bulk CuCl [23]. The Stokes-shifted luminescence indicates that localized excitons are dominant. Photostimulated luminescence was detected under the excitation of a red semiconductor laser (670 nm) for 1 s for the sample pre-excited by the HeCd laser (325 nm). The photostimulated luminescence spectrum is dominated by localized exciton luminescence, but confined exciton luminescence reflecting the wavy structure in the absorption spectrum is enhanced more than the photoluminescence spectrum. With increasing temperature, localized exciton luminescence decreases first and confined exciton luminescence decreases later in the photostimulated luminescence spectrum. The photostimulated luminescence is observable up to 120 K. This quenching temperature is almost equal to the temperature where thermally activated hole filling is completed [65], suggesting the same thermal activation of

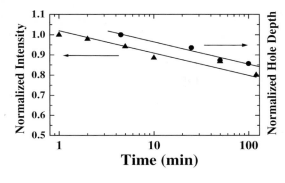

**Fig. 5.15.** Decay of the photostimulated luminescence intensity after pre-excitation (*solid triangles*) and decay of the persistent hole depth after hole burning (*solid circles*) [16] (from Masumoto et al. [64])

trapped carriers is responsible for photostimulated luminescence quenching and hole filling.

The long-term decay experiments for photostimulated luminescence were done by means of a HeCd laser and a semiconductor laser with mechanical shutters. The results are shown in Fig. 5.15. With a further increase in time delay, the photostimulated luminescence was observed for a few hours. The photostimulated luminescence intensity decreases in proportion to logarithm of time, similarly to the spontaneous decay of the persistent spectral hole burning observed in CuCl quantum dots embedded in NaCl crystals [16]. Photoionization and subsequent carrier trapping at nearby host quantum dots are considered as the persistent spectral hole-burning mechanism. The similar decay suggests that both photostimulated luminescence and persistent spectral hole burning in CuCl quantum dots originate from the same carrier trapping at the same trap sites. The logarithmic decay comes from the broad distribution of decay rates.

The photostimulated luminescence intensity does not increase in proportion to the exposure of the ultraviolet pre-excitation. The photostimulated luminescence intensity grows, but its growth slows with the increase of the pre-excitation fluence. To explain the broadly distributed growth rate of persistent spectral hole burning, a tunneling model with broadly distributed barrier height and thickness was used [16]. The same model was used to explain the dependence of pre-excitation fluence on photostimulated luminescence. The agreement between the experimental growth of photostimulated luminescence and the fitted result is good. The good agreement and similarity show that carriers formed in quantum dots tunnel through the barrier and are captured at the trap sites in NaCl crystals.

Photostimulated luminescence excitation spectra give the trap levels. To explain the experimental excitation spectrum, it was assumed that the traps form a Gaussian band where the peak energy depth of the Gaussian band was measured from the band of NaCl, the standard deviation of the band, and the energy depth, respectively. It is assumed that photostimulation excites carriers to the band of NaCl and that excited carriers recombine with carriers remaining in the quantum dots. The photostimulated luminescence intensity is proportional to the photoexcitation of the trapped carriers to the band in NaCl. Therefore, its energy dependence is proportional to the integral of the Gauss function from zero to the photostimulation photon energy $E$. The agreement between the experimental excitation spectrum and the calculation is good when the peak energy and the width of the Gaussian band are 2.45 eV and 0.53 eV, respectively.

In the samples, CuCl quantum dots are considered to be surrounded by $Cu^+$ monomers and dimers because the quantum dots are formed by aggregation of $Cu^+$ ions. In fact, we observed photoluminescence of $Cu^+$ dimers at 2.1 eV in similar samples to our samples at low temperature [65] and photostimulated luminescence of $Cu^+$ monomers at 3.5 eV for X-ray irradiated samples at room temperature [66]. $Cu^+$ monomers and dimers can work as hole traps and become $Cu^{2+}$.

Photoluminescence blinking was reported for chemically synthesized CdSe quantum dots [13,14]. Luminescence intensity frequently turns on and off. About half of all the atoms of dot surfaces are passivated by organic capping groups. Further, a process of the chemical synthesis can overcoat CdSe quantum dots by ZnS, whose bandgap is higher than the bandgap of CdSe

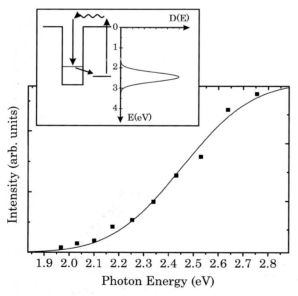

**Fig. 5.16.** Excitation spectrum of photostimulated luminescence of CuCl quantum dots embedded in a NaCl crystal at 2 K. The *solid line* was fitted by assuming a Gaussian distribution with parameters of trap depth and its standard deviation. In the *upper part*, schematic diagrams showing the photostimulated luminescence processes and the obtained trap energy distribution are shown (from Masumoto et al. [64])

by 2 eV. With the increase of overcoated layer thickness of quantum dots, blinking frequency is reduced. Blinking frequency is also dependent on the excitation intensity. The average on time is inversely proportional to the excitation intensity, while the off time is independent of the excitation intensity. Auger ionization is believed to cause ionization of quantum dots and subsequent carrier trapping at the surrounding materials. In ionized quantum dots, Coulomb repulsive force between carriers increases the Auger nonradiative recombination. The simulation based on the model can reproduce the blinking features [67].

Photoluminescence blinking was also reported for InP quantum dots [68,69,14,70]. They are self-assembled quantum dots grown in the Stranski–Krastanow mode. A small fraction of the dots (about one dot in a thousand dots) shows photoluminescence blinking. The luminescence switches between two states, on and off, in [68,69] but switches among more than three states in [14,70]. The off state is not the perfectly dark state. The luminescence intensity in the off state is reduced from that in the on state. The blinking frequency increases with increasing temperature as well as excitation intensity [68,69]. The blinking dots are observed in the vicinity of formed scratches [14]. The blinking frequency is drastically enhanced when near infrared light with photoenergy higher than 1.5 eV is irradiated simultaneously with the band-to-band excitation. The luminescence spectra at the on and off states are the same as those without and with an external electric field. These observations show that luminescence blinking originates from the local electric field due to carrier trapping at a deep localized center with an energy of 1.5 eV [70]. Detailed discussion of luminescence blinking is found in Chap. 4. The presence of deep traps at an energy of 1.5 eV causes anti-Stokes luminescence of InP quantum dots [71].

Spectral diffusion was observed in chemically synthesized CdSe quantum dots by means of single quantum dot spectroscopy [72,15]. The exciton energy of CdSe quantum dots shifted toward lower energy quadratically with the increase of the external electric field plus some local electric field. This means the quantum-confined Stark effect holds fully for CdSe quantum dots. The local electric field changed abruptly, and the change caused the spectral diffusion of CdSe quantum dots. The estimated local electric field increased from 90 to 150 kV/cm with the decrease of the radius of the dot from 3.8 nm to 2.2 nm.

Spectral diffusion was also observed for excitonic transitions in CdSe quantum dots epitaxially grown by metal organic chemical-vapor deposition in a $ZnS_{0.06}Se_{0.94}$ matrix [73]. A metal mask with small windows was used to restrict the number of quantum dots. Low-temperature cathodoluminescence spectra show a set of sharp lines diffusing synchronously. The amplitude of the spectral diffusion depended on the excitation intensity and hence on the density of excited electrons. The spectral diffusion was assigned to a quantum-confined Stark effect induced by randomly fluctuating electric fields induced by charges localized at defects in the vicinity of quantum dots.

Small ensembles of $In_{0.55}Al_{0.45}As$ self-assembled quantum dots grown by molecular bean epitaxy showed spectral diffusion [74]. It was observed by means of near-field scanning optical microscopy at 4.2 K. With the increase of the excitation intensity, the spectral diffusion of individual dot emission lines took place. The redshift was at most 1.1 meV and a discrete spectral jump smaller than 0.5 meV was explained by a Hartree simulation, taking into account energy losses of the electron and hole due to their displacement from the potential of the charge, and energy gain due to the polarization of the exciton.

## 5.6    Application of Persistent Spectral Hole Burning to Site-Selective Spectroscopy

As an application of the persistent spectral hole-burning phenomenon in semiconductor quantum dots, the energy-domain optical storage is obvious [18,75,43]. CuCl quantum dots in a NaCl crystal were selected for the demonstration of multiple optical data storage. Figure 5.17 shows the demonstra-

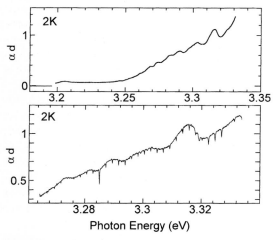

**Fig. 5.17.** An example of optical multiple recording by persistent hole burning. The sample is CuCl quantum dots embedded in a NaCl crystal. Twenty-nine holes were burned between 3.268 eV and 3.33 eV. The *upper spectrum* shows the absorption spectrum before burning, and the *lower figure* shows the spectrum after burning. A deep hole was made at 3.283 eV as a result of burning by accumulated intensity 3 times, while no hole was made at 3.328 eV because of the absence of burning. These holes were preserved for more than several hours at low temperatures (from Masumoto [76])

tion of multiple optical data storage. Twenty-nine positions in the inhomogeneously broadened $Z_3$ exciton band were burned in turn. After that, the absorption change was observed. It is clearly observed that the absorption change consists of 29 holes. This experiment shows the application capabilities of semiconductor quantum dots for the optical memory.

Formation efficiencies of persistent holes in CuCl quantum dots, CuBr quantum dots and CuI quantum dots embedded in glass and crystals have been measured [43]. The highest quantum efficiency of 0.097 was obtained at 2 K for CuCl quantum dots in glass. The persistent absorption change of quantum dots caused by a photon is larger than that of the most efficient ionic or molecular persistent hole burning materials by three orders of magnitude. The easiest way to understand the surprisingly high efficiency of the quantum efficiency is that the absorbance of a quantum dot is larger than that of a molecule or an ion by three orders of magnitude. This is because a 2-nm-radius quantum dot consists of $10^3$ molecules.

Many complex conditions, such as the size, shape, strain, surroundings of quantum dots and spatial configuration of trapped carriers give inhomogeneous broadening of the optical spectra of quantum dots, and mask the intrinsic properties of quantum dots. Persistent spectral hole-burning phenomena can be applied not only to multiple optical data storage, but also to site-selective spectroscopy of quantum dots. Size-dependent energy levels in "laser-marked" quantum dots are seen as clear persistent holes in the inhomogeneously broadened absorption spectra. Quantum-confined electronic or excitonic excited states and phonons confined in quantum dots are observed as persistent sideband holes. They are observed at anti-Stokes as well as Stokes sides of the resonantly burned holes and sometimes show a complicated spectrum. They are classified, and successful examples of persistent spectral hole-burning spectroscopy of quantum dots are shown in this section. The sideband hole structures are schematically illustrated in the upper portion of Fig. 5.18. On the Stokes side, sideband holes coming from acoustic phonons (1), optical phonons (2), and lowest electronic or excitonic states (3) are observed. Phonon sidebands arise when quantum dots whose resonant energy is equal to the laser energy minus the phonon energy absorb laser light simultaneously with the emission of phonons. When the excited electronic or excitonic states of quantum dots absorb laser light, fast relaxation to the lowest state gives their sideband holes. On the anti-Stokes side, acoustic phonon sideband holes (1), optical phonon sideband holes, and the sideband holes coming from the excited electronic or excitonic states (4) are observed. The latter arise because photoionization of quantum dots more or less changes all the excited electronic or excitonic states. The acoustic phonon and optical phonon sidebands arise because the burned quantum dots change their energy and because their absorption band arising from the light absorption with the emission of phonons is reduced.

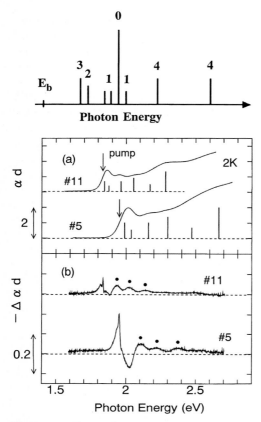

**Fig. 5.18.** *Upper figure*: Schematic persistent hole-burning spectrum of semiconductor quantum dots. The numbers 0–4 denote the resonantly burned main hole (0), acoustic phonon sideband holes (1), an optical phonon sideband hole (2), sideband holes coming from the ground (3), and excited (4) electronic or excitonic states of quantum dots. A ground electronic or excitonic state in a bulk crystal is denoted by $E_b$. *Lower figure*: (a) Optical absorption spectra and (b) absorption change spectra of CdTe quantum dots embedded in $GeO_2$ glass at 2 K. The average radii of dots in samples #11 and #5 are 3.7 nm and 2.9 nm, respectively. *Vertical arrows* show the burning photon energies of 1.8336 eV and 1.9526 eV, respectively. The absorption change spectra were measured after the samples were exposed to 1800 shots of dye laser pulses with excitation energy of 640 $\mu J/cm^2$. *Positions of vertical bars* in (a) represent the energy positions of transitions deduced from the experiments (from Masumoto et al. [22], Masumoto [54])

Quantum confinement of holes gives an additional angular momentum $L$ for the envelope function, so that the $L$–$J$ coupling is considered to result in complicated energy levels for the hole band. Persistent spectral hole burning is effective in studying the size-dependent quantum confinement in this case. Figure 5.18a shows the inhomogeneously broadened absorption spectra of CdTe quantum dots embedded in GeO$_2$ glass. The absorption spectra show a notable blueshift with the decrease in dot size. After the samples are exposed to narrowband dye laser pulses whose photon energy corresponds to the lowest absorption band, they show persistent higher-energy sideband holes, which move with the change of the burning photon energy (Fig. 5.18b) [22]. Plotting these transition energies as a function of the excitation photon energy, we can observe that straight lines fit the experimental

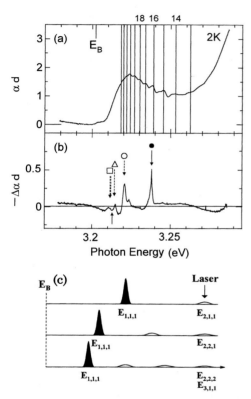

**Fig. 5.19.** (**a**) Optical absorption spectrum of CuCl quantum dots embedded in a NaCl crystal. (**b**) The absorption spectrum changes. A *solid circle* shows the spectral position where the sample was excited by a laser. An *open circle*, a *triangle*, and a *square* show the satellite holes observed at $E=E_b+(E_l-E_b)/i$, where $E_l$ is the laser energy, $E_b$ is the bulk exciton energy, and $i$ is 2, 3, 3.67, or 4 for an *open circle*, a *triangle*, and a *square*, respectively. An *upward arrow* shows the position at $E=E_b+(E_l-E_b)/3.35$. A *short vertical line* indicates the energy position of $E_b$. *Long vertical lines* show the calculated energies of the Z$_3$ exciton confined in quantum cubes. (**c**) A laser excites the excited states (2, 1, 1), (2, 2, 1), (3, 1, 1), or (2, 2, 2), and hole burning takes place at the (1, 1, 1) state (from Sakakura et al. [24], Masumoto [54])

data. Good straight-line fitting meets the simplest electron–hole quantum-confinement model and shows that intermixing of valence bands does not take place in these CdTe quantum dots. CdTe has the largest spin-orbit splitting $\Delta$=0.927 eV, and the smallest band gap energy $E_g$=1.606 eV among CdS, CdSe, and CdTe. As a result, the split-off band is expected to mix with the topmost valence band only weakly, and the quantum-confined hole levels become simpler than those in CdSe quantum dots, which were site-selectively studied by means of hole burning spectroscopy [42,77].

CuCl quantum dots in NaCl crystals sometimes show a multistructured absorption band of $Z_3$-excitons [23,24], as shown in Fig. 5.19. Each structure is explained by the size-quantized energy of a $Z_3$ exciton confined in a CuCl quantum cube whose side changes stepwise in a unit of $a/2$, where $a$ is the lattice constant of the CuCl crystal [78]. Persistent spectral hole-burning spectra of CuCl quantum dots in NaCl crystal show a resonantly burned hole and lower energy satellite holes. The latter are observed at $E$, where $E$ satisfies the equation $E_l-E_b=i(E-E_b)$. Here $E_l$ is the laser energy, $E_b$ is the bulk energy of exciton, and $i$ is 2, 3, 3.67, or 4. The experimental results can be explained if the 2nd, 3rd, 4th, and 5th excited states, i.e., $(l,m,n)$=(2, 1, 1), (2, 2, 1), (3, 1, 1), and (2, 2, 2) states of a quantum cube, are excited and the hole burning is observed in the (1, 1, 1) state. Note that $i$ should be 2.05, 3.35, and 4, respectively, if quantum dots are assumed to be spheres. This observation confirms that the exciton energy levels are described by $E=E_b+[\hbar^2\pi^2/2M(L-a_B)^2](l^2+m^2+n^2)$, where $M$ is the translational mass of the exciton, $a_B$ is its Bohr radius, and $L$ is the side length of a cube. Thus, this hole-burning spectrum clearly demonstrates the excited states of an exciton in quantum cubes.

Persistent spectral hole-burning phenomena are also applicable to the study of the phonon energies in quantum dots [25–28,79]. Phonon energies are detected as the energy separation between the resonant hole and its phonon sidebands. Figure 5.20 shows persistent spectral hole-burning spectra of CuCl quantum dots that show the resonant hole, its LO phonon sideband, and the hole made by the 2S exciton formation followed by the relaxation to the 1S exciton. The LO phonon sideband is formed as a result of light absorption along with the emission of the excited state LO phonon. The energy separation between the resonant hole and the LO phonon sideband, 23.5 meV, is smaller than that of the bulk value, 26 meV, by 10%. Excitons can interact with LO phonons strongly via Fröhlich interaction. If the energy separation between the 2S confined exciton and the 1S confined exciton approaches the LO phonon energy with decreasing quantum dot radius, the Fröhlich interaction gives two normal modes, and the LO phonon energy is modified from the bulk value. The experimental results clearly show the renormalization of excited LO phonons in quantum dots [25].

The size dependence of excitonic states and LO phonons in CuCl cubic quantum dots embedded in NaCl crystals was also studied by means of site-

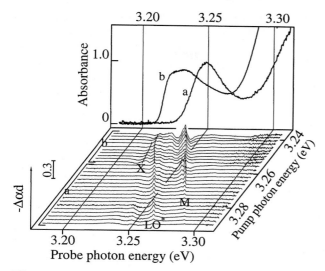

**Fig. 5.20.** The absorption spectra and the persistent hole-burning spectra of CuCl quantum dots in glass. Data for two samples with average radii 2.5 nm (**a**) and 4 nm (**b**) are shown. The lowest energy peak for each size corresponds to the exciton ground state. Persistent holes corresponding to the excitation of the acoustic, TO, and LO phonon sidebands, and the 2S exciton are also seen. The holes corresponding to the LO phonon (LO$^*$) and the quantum confined 2S–1S exciton split energy (X) are mixed (from Zimin et al. [25], Masumoto [54])

selective persistent spectral hole-burning spectroscopy. At the Stokes side of the resonantly burned hole, a LO phonon satellite hole and holes in the (1, 1, 1) state formed by the relaxation from the resonantly excited higher quantum states were observed. The Stokes shifts of the observed holes $E_1$, $E_2$, $E_3$, LO, and P as a function of the exciton confinement energy are shown in Fig. 5.21. Here holes $E_1$, $E_2$, and $E_3$ were made by the hole burning at the (1, 1, 1) state under the excitation of (2, 1, 1), (3, 1, 1), and (3, 3, 1) states, respectively. A LO phonon hole showed an almost size-independent Stokes shift of 23.1 meV, which is slightly smaller than the phonon energy of 23.5 meV in the excited state of CuCl quantum dots in glass [25]. It was shown that the interaction of the LO phonon with the exciton results in the formation of coupled exciton–phonon modes when the energy of the LO phonon approaches the energy spacings between the exciton ground state (1, 1, 1) and the exciton excited states, (3, 1, 1) and (3, 3, 1). The energy anticrossings of the LO phonon modes with two higher optically allowed excited states of the exciton in CuCl quantum dots were clearly observed at confinement energies of about 10 and 6 meV, respectively.

Acoustic phonons confined in CuCl quantum dots were investigated precisely as a function of the radius of quantum dots [27,28,79]. Acoustic phonons

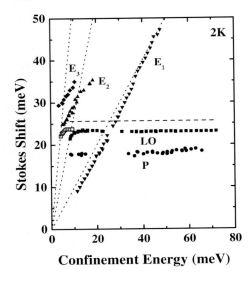

**Fig. 5.21.** Stokes shifts of the satellite holes $E_1$, $E_2$, LO, and P in CuCl quantum dots in NaCl crystals as a function of the exciton confinement energy. The *dashed line* represents the energy of LO phonons in a bulk CuCl crystal. The *dotted lines* represent the size dependence of the energy spacings between the ground and excited states in the CuCl quantum dots on the quantum cube model (from Zhao et al. [26])

confined in the quantum dots were observed as Stokes-shifted sidebands of the resonantly burned hole. The lowest energy torsional and spheroidal modes were observed for CuCl quantum dots embedded in glass and KCl crystals. Their energies were inversely proportional to the radius of dots and were well explained by the theoretical calculation with the free boundary condition. For CuCl quantum dots embedded in NaCl crystals, the observed Stokes shifts were almost half of those for CuCl quantum dots embedded in glass and KCl crystals and were assigned as acoustic phonons confined in the quantum cubes with the free boundary condition.

## 5.7   Summary

This chapter explains the phenomena, mechanism, and application of persistent spectral hole burning and its related phenomena in semiconductor quantum dots embedded in glass, crystals, or polymers. The spectral change of CdSe, CdTe, CdS, CuCl, CuBr, and CuI quantum dots induced by narrowband laser excitation at 2 K consists of the narrow burned hole and the associated spectral changes which are conserved for a long time after the laser irradiation. Hole-burning and hole-filling phenomena were studied. Hole depth increases in proportion to the logarithm of the fluence of the burning pulses. Hole filling due to thermal annealing and light-induced hole filling were observed. These observations suggest that photoionization of the quantum dots is the photochemical mechanism to induce the persistent spectral hole-burning phenomenon. Unusual luminescence behaviors were observed in strong connection with persistent spectral hole burning in quantum dots. One is luminescence elongation with increasing light exposure observed in

CuI quantum dots embedded in glass. Another is luminescence hole burning observed in CuCl quantum dots embedded in a NaCl crystal. These behaviors are explained by exciton localization and the ionization of quantum dots, followed by the carrier tunneling into traps in hosts. The energy of photoionized quantum dots is changed from the original energy, and their new energies depend on the spatial arrangement of carriers trapped in hosts as well as in quantum dots. Quantum confinement of carriers and the resulting strong Coulomb interaction between confined carriers and trapped carriers are essential for the energy change. Luminescence hole burning was also observed in porous Si, InP quantum wires, and GaAs quantum dots. The photoionization of quantum dots and the succeeding carrier trapping at the defects near the quantum dots cause not only persistent spectral hole burning, but also photostimulated luminescence, photoluminescence blinking, and spectral diffusion. Photostimulated luminescence was observed in CuCl quantum dots embedded in NaCl crystals at low temperature up to 120 K. The traps form a Gaussian band located below the NaCl band by 2.45 eV with width of 0.53 eV and are considered to be $Cu^+$ in NaCl. Photoluminescence blinking was observed in CdSe and InP quantum dots, and spectral diffusion was observed in CdSe and InAlAs quantum dots. Applications of persistent spectral hole burning to precise high-resolution spectroscopy were discussed. Size-dependent excited exciton quantum states and electron–hole quantum states in quantum dots were clarified. Size-dependent confined acoustic phonons and renormalization of longitudinal optical phonons confined in quantum dots were revealed by studying sideband holes.

# References

1. A.I. Ekimov, A.L. Efros, A.A. Onushchenko: Solid State Commun. **56**, 92 (1985)
2. A.D. Yoffe: Adv. Phys. **42**, 173 (1993)
3. L. Bányai, S.W. Koch: *Semiconductor Quantum Dots* (World Scientific, Singapore 1993)
4. U. Woggon: *Optical Properties of Semiconductor Quantum Dots* (Springer, Berlin 1997)
5. S.V. Gaponenko: *Optical Properties of Semiconductor Nanocrystals* (Cambridge Univ. Press, Cambridge 1998)
6. D. Bimberg, M. Grundmann, N.N. Ledentsov: *Quantum Dot Heterostructures* (John Wiley & Sons, New York 1999)
7. M. Ikezawa, Y. Masumoto, T. Takagahara, S.V. Nair: Phys. Rev. Lett. **79**, 3522 (1997)
8. M. Ikezawa, Y. Masumoto: Jpn. J. Appl. Phys. **36**, 4191 (1997)
9. For review, see Y. Masumoto: J. Lumin. **70**, 386 (1996)
10. T. Kawazoe, Y. Masumoto: Phys. Rev. Lett. **77**, 4942 (1996)
11. *Persistent Spectral Hole-Burning: Science and Applications*, ed. by W.E. Moerner (Springer-Verlag, Berlin 1988)

12. Y. Masumoto, L.G. Zimin, K. Naoe, S. Okamoto, T. Kawazoe, T. Yamamoto: J. Lumin. **64**, 213 (1995), and references therein

13. M. Nirmal, B.O. Dabbousi, M.G. Bawendi, J.J. Macklin, J.K. Trautman, T.D. Harris, L.E. Brus: Nature **383**, 802 (1996)

14. M. Sugisaki, H.-R. Ren, S.V. Nair, J.-S. Lee, S. Sugou, T. Okuno, Y. Masumoto: J. Lumin. **87–89**, 40 (2000)

15. S.A. Empedocles, M.G. Bawendi: Science **278**, 2114 (1997)

16. Y. Masumoto, S. Okamoto, T. Yamamoto, T. Kawazoe: Phys. Status Solidi B **188**, 209 (1995)

17. Y. Masumoto, L.G. Zimin, K. Naoe, S. Okamoto, T. Arai: Mater. Sci. Eng. B **27**, L5 (1994)

18. K. Naoe, L.G. Zimin, Y. Masumoto: Phys. Rev. B **50** 18200 (1994)

19. Y. Masumoto, T. Kawazoe, T. Yamamoto: Phys. Rev. B **52**, 4688 (1995)

20. Y. Masumoto, K. Kawabata, T. Kawazoe: Phys. Rev. B **52** 7834 (1995)

21. J. Qi, Y. Masumoto: Solid State Commun. **99**, 467 (1996)

22. Y. Masumoto, K. Sonobe: Phys. Rev. B **56**, 9734 (1997)

23. N. Sakakura, Y. Masumoto: Phys. Rev. B **56**, 4051 (1997)

24. N. Sakakura, Y. Masumoto: Jpn. J. Appl. Phys. **36**, 4212 (1997)

25. L.G. Zimin, S.V. Nair, Y. Masumoto: Phys. Rev. Lett. **80**, 3105 (1998)

26. J. Zhao, S.V. Nair, Y. Masumoto: Phys. Rev. B **63**, 33307 (2001)

27. S. Okamoto, Y. Masumoto: J. Lumin. **64**, 253 (1995)

28. J. Zhao, Y. Masumoto: Phys. Rev. B **60**, 4481 (1999)

29. K. Brunner, G. Abstreiter, G. Böhm, G. Tränkle, G. Weimann: Appl. Phys. Lett. **64**, 3320 (1994)

30. J.-Y. Marzin, J.-M. Gérard, A. Izraël, D. Barrier, G. Bastard: Phys. Rev. Lett. **73**, 716 (1994)

31. M. Grundmann, J. Christen, N.N. Ledentsov, J. Böhrer, D. Bimberg, S.S. Ruvimov, P. Werner, U. Richter, U. Gösele, J. Heydenreich, V.M. Ustinov, A.Y. Egorov, A.E. Zhukov, P.S. Kop'ev, Z.I. Alferov: Phys. Rev. Lett. **74**, 4043 (1995)

32. S.A. Empedocles, D.J. Norris, M.G. Bawendi: Phys. Rev. Lett. **77**, 3873 (1996)

33. T. Wamura, Y. Masumoto, T. Kawamura: Appl. Phys. Lett. **59**, 1758 (1991)

34. T. Itoh, M. Furumiya: J. Lumin. **48–49**, 704 (1991)

35. Y. Masumoto, T. Kawazoe, N. Matsuura: J. Lumin. **76–77**, 189 (1998)

36. E. Hanamura: Phys. Rev. B **37**, (1988) 1273

37. S. Schmitt-Rink, D.A.B. Miller, D.S. Chemla: Phys. Rev. B **35**, 8113 (1987)

38. M.G. Bawendi, W.L. Wilson, L. Rothberg, P.J. Carroll, T.M. Jedju, M.L. Steigerwald, L.E. Brus: Phys. Rev. Lett. **65**, 1623 (1990)

39. M.G. Bawendi, PJ. Carroll, W.L. Wilson, L.E. Brus: J. Chem. Phys. **96**, 946 (1992)

40. V. Esch, B. Fluegel, G. Khitova, H.M. Gibbs, Xu Jiajin, K. Kang, S.W. Koch, L.C. Liu, S.H. Risbud, N. Peyghambarian: Phys. Rev. B **42**, 7450 (1990)

41. K. Kang, A.D. Kepner, Y.Z. Hu, S.W. Koch, N. Peyghambarian, C.-Y. Li, T. Takada, Y. Kao, J.D. Mackenzie: Appl. Phys. Lett. **64**, 1487 (1994)

42. D.J. Norris, A. Sacra, C.B. Murray, M.G. Bawendi: Phys. Rev. Lett. **72**, 2612 (1994)

43. T. Kawazoe, Y. Masumoto: Jpn. J. Appl. Phys. **37**, L394 (1998)

44. T. Itoh, Y. Iwabuchi, M. Kataoka: Phys. Status Solidi B **145**, 567 (1988)

45. A.I. Ekimov, A.L. Efros, M.G. Ivanov, A.A. Onushchenko, S.K. Shumilov, Solid State Commun. **69**, 565 (1989)

46. J. Valenta, J. Moniatte, P. Gilliot, B. Hönerlage, J.B. Grun, R. Levy, A.I. Ekimov: Phys. Rev. B **57**, 1774 (1998)
47. R. Jankowiak, R. Richert, H. Bässler: J. Phys. Chem. **89**, 4569 (1985)
48. W. Köhler, J. Meiler, J. Friedrich: Phys. Rev. B **35**, 4031 (1987)
49. V.Y. Grabovskis, Y.Y. Dzenis, A.I. Ekimov, I.A. Kudryavtsev, M.N. Tolstoi, U.T. Rogulis: Sov. Phys. Solid State **31**, 149 (1989); see also A.I. Ekimov: Phys. Scripta T **39**, 217 (1991)
50. M.V. Artemyev, S.V. Gaponenko, I.N. Germanenko, A.M. Kapitonov: Chem. Phys. Lett. **243**, 450 (1995)
51. S.V. Gaponenko, I.N. Germanenko, A.M. Kapitonov, M.V. Artemyev: J. Appl. Phys. **79**, 7139 (1996)
52. For example, F. Hache, D. Ricard, C. Flytzanis: Appl. Phys. Lett. **55**, 1504 (1989)
53. Y. Masumoto, T. Kawazoe: J. Lumin. **66–67**, 142 (1996)
54. Y. Masumoto: Jpn. J. Appl. Phys. **38**, 570 (1999)
55. T. Okuno, H. Miyajima, A. Satake, Y. Masumoto: Phys. Rev. B **54**, 16952 (1996)
56. D.C. Reynolds, T.C. Collins: *Excitons, Their Properties and Uses* (Academic, New York 1981)
57. A.I. Ekimov, I.A. Kudryavtsev, M.G. Ivanov, A.L. Efros: J. Lumin. **46**, 83 (1990)
58. *Numerical Data and Functional Relationships for Science and Technology, Physics of II–VI and I–VII Compounds, Semimagnetic Semiconductors, Landolt-Börnstein, Vol. 17/b*, ed. by O. Madelung (Springer-Verlag, Berlin, Heidelberg 1982)
59. D. Kovalev, H. Heckler, B. Averboukh, M. Ben-Chorin, M. Schwartzkopff, F. Koch: Phys. Rev. B **57**, 3741 (1998)
60. E.A. Zhukov, Y. Masumoto, E.A. Muljarov, S.G. Romanov: Solid State Commun. **112**, 575 (1998)
61. Y. Kanemitsu, H. Tanaka, Y. Fukunishi, T. Kushida, K.S. Min, H.A. Atwater: Phys. Rev. B **62**, 5100 (2000)
62. K. Takahashi: In: *Phosphor Handbook*, ed. by S. Shionoya, W.M. Yen (CRC Press, Boca Raton 1998) p. 553
63. Y. Amemiya, J. Miyahara: Nature **336**, 89 (1988)
64. Y. Masumoto, S. Ogasawara, Jpn. J. Appl. Phys. **38**, L623 (1999)
65. S. Okamoto, Y. Masumoto: Phys. Rev. B **56**, 15729 (1997)
66. H. Nanto, T. Usuda, K. Murayama, H. Sokooshi, S. Nakamura, K. Inabe, N. Takeuchi: Appl. Phys. Lett. **59**, 1838 (1991)
67. A.L. Efros, M. Rosen: Phys. Rev. Lett. **78**, 1110 (1997)
68. P. Castrillo, D. Hessman, M.-E. Pistol, J.A. Prieto, C. Pryor, L. Samuelson: Jpn. J. Appl. Phys. **36**, 4188 (1997)
69. M.-E. Pistol, P. Castrillo, D. Hessman, J.A. Prieto, L. Samuelson: Phys. Rev. B **59**, 10725 (1999)
70. M. Sugisaki, H.-W. Ren, K. Nishi, Y. Masumoto: Phys. Rev. Lett. **86**, 4883 (2001)
71. I.V. Ignatiev, I.E. Kozin, H.-W. Ren, S. Sugou, Y. Masumoto: Phys. Rev. B **60**, 14001 (1999)
72. S.A. Blanton, M.A. Hines, P. Guyot-Sionnest: Appl. Phys. Lett. **69**, 3905 (1996)
73. V. Türck, S. Rodt, O. Stier, R. Heitz, R. Engelhardt, U.W. Pohl, D. Bimberg, R. Steingrüber: Phys. Rev. B **61**, 9944 (2000)

74. H.D. Robinson, B.B. Goldberg: Phys. Rev. B **61**, R5086 (2000)
75. S. Okamoto, Y. Masumoto: Jpn. J. Appl. Phys. **34**, Suppl. 34-1, p. 128 (1995)
76. Y. Masumoto: Butsuri **54**, 431 (1999) (in Japanese)
77. D.J. Norris, M.G. Bawendi: Phys. Rev. B **53**, 16338 (1996)
78. T. Itoh, S. Yano, N. Katagiri, Y. Iwabuchi, C. Gourdon, A.L. Ekimov: J. Lumin. **60–61**, 396 (1994)
79. J. Zhao, M. Ikezawa, A.V. Fedorov, Y. Masumoto: J. Lumin. **87–89**, 525 (2000)

# 6  Dynamics of Carrier Relaxation in Self-Assembled Quantum Dots

Ivan V. Ignatiev and Igor E. Kozin

## 6.1  Introduction

Relaxation of hot carriers in quantum dots (QDs) has been widely discussed since the beginning of research of self-assembled QDs. Understanding of carrier relaxation mechanisms is important from both the practical and the fundamental points of view. The carrier relaxation in QDs controls, to a great extent, the efficiency of luminescence from ground radiative states. This is highly important, for example, for laser applications of QD structures where the relaxation must be as fast as possible [1–3]. On the other hand, QD lasers in the mid-IR range can be devised on inter-sub-band transitions if a relatively slow carrier relaxation takes place [4]. To reach these limiting points, a careful study of various processes of carrier relaxation should be done.

From the fundamental point of view, carrier relaxation in QDs should differ drastically from that in structures of higher dimension (quantum wires and quantum wells) or in bulk materials. The discrete energy structure of the QDs imposes a number of restrictions on the relaxation processes. This is why each particular relaxation process should be thoroughly revised in the case of QDs independently of the results obtained for other types of structures. Realization of a specific relaxation process depends on properties of the QD system and experimental conditions, including energy structure of the QDs, electron–hole interaction (excitonic effect), phonon energy spectrum, electron–phonon interaction, number of carriers in a QD, temperature, electric current through the heterostructure, and so on.

The QD energy level structure is usually represented by a set of discrete energy levels for the electron and hole separately (Fig. 6.1). This approach is well justified when the processes occurring with only one carrier (electron or hole) are treated – its capture by a QD from the barrier and its energy relaxation inside the QD. Note, however, that this representation seems to be of little use when relaxation of an electron–hole pair (e–h pair) is analyzed.

The point is that the Coulomb interaction between the electron and the hole allows them to exchange energy fairly fast. The interaction energy in the range of several meV, typical for QDs, corresponds to sub-picosecond times for the electron–hole energy exchange process. In the scale of the carrier energy relaxation times, which can reach hundreds of picoseconds, this process is virtually instantaneous. For this reason, one can expect that an

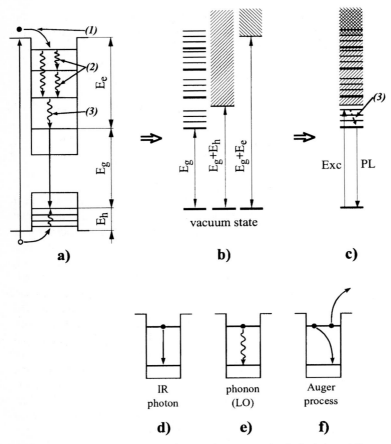

**Fig. 6.1.** The energy diagram for an electron and a hole (**a**) and for an e–h pair (**c**) in a QD. Diagram (**b**) shows how the e–h pair spectrum is being formed. Different steps of relaxation of single carriers are shown in (**a**): step (1) is the carrier capture by a QD from the barrier or the wetting layer, step (2) is the relaxation through intermediate states, step (3) is the final step of the relaxation to the ground state. Diagram (**c**) shows steps (2) and (3) of relaxation of the e–h pair created in the QD by quasiresonant excitation. Diagrams (**d**–**f**) illustrate different possible mechanisms of the carrier relaxation: (**d**) shows the relaxation with emission of an infrared photon; (**e**) with emission of a phonon; and (**f**) shows relaxation of one carrier with excitation of another (Auger-like process)

electron and a hole in a QD relax as a single quantum mechanical system – a correlated e–h pair or exciton. The energy levels of the e–h pair are actually represented by a combination of the electron and hole levels as shown in Fig. 6.1. The spacing between adjacent levels of the e–h pair is usually controlled by the hole energy structure and can be sufficiently small for large

QDs. Apart from the discrete spectrum, the energy structure of the e–h pairs shows two continuous-spectrum bands corresponding to above-barrier motion of the electron or hole. The lower edge of these bands is raised above the lowest state of the e–h pair by the electron and hole binding energy in their lowest states, denoted in Fig. 6.1 by $E_e$ and $E_h$, respectively. The quantities $E_e$ and $E_h$ can be sufficiently small for small quantum dots due to the size-quantization effect. The total spectrum of the e–h pair is a superposition of the discrete and continuous spectra as shown in Fig. 6.1c. As seen, this spectrum can be sufficiently dense both for large and for small QDs.

The carrier relaxation mechanisms strongly depend on the method of the excitation and the number of carriers in a QD. For a single carrier, the relaxation can be divided in a few steps as shown in Fig. 6.1a.

The first step is the capture of the carriers created by light absorption in the barrier layer or by electric current flowing through the heterostructure. In this case, the carrier has to move from a state in the continuous spectrum to a discrete energy state. This transition occurs usually above some potential barrier and therefore takes a finite time of a few tens of picoseconds in any relaxation mechanism (see, for example, [5,6] and references therein).

The second step is the cascade relaxation over discrete energy levels in a QD. Different relaxation channels through different intermediate states, including the mixed electron–phonon states, are possible to reach the lowest energy state. This complicated process has not yet been studied sufficiently well.

As a separate (third) step, the final relaxation of the carrier to its ground state, can be taken. This separation is expedient because, on the one hand, this is usually the slowest step of the relaxation that mostly determines the relaxation rate and, on the other hand, modern experimental techniques are well developed to study this step of relaxation in detail [7–10].

Steps (2) and (3) can also be considered for the case of the e–h pair relaxation as shown in Fig. 6.1c. When the barrier band-to-band excitation is used, the e–h pair is formed if a carrier (e.g., electron) is captured into the QD, which already contains a carrier of the other sign (hole). Because of the statistical nature of this process, it is usually difficult to determine experimentally which of the two relaxation processes – relaxation of a single carrier or of the carrier bound to an e–h pair – takes place. For this reason, the purest conditions for studying the e–h pair relaxation are organized when the e–h pairs are generated inside the QD by a quasiresonant optical excitation, i.e., by light with photon energy smaller than the bandgap of the barrier and wetting layers. It is exactly this case that is considered in this chapter in great detail.

Generally, it is sufficient to consider a few mechanisms of the carrier relaxation. They are shown schematically in Fig. 6.1d–f for a single carrier. The first one is the radiative relaxation of the hot carrier (Fig. 6.1d) when the carrier emits a photon in the mid-IR range. This process and the discrete energy spectrum of QDs are those intrinsic properties that allow one to consider the QDs as artificial atoms. The mid-IR radiative relaxation of carriers has been observed experimentally in several studies [4,11,12]. The efficiency of this process is, however, extremely low because it is proportional to cube of the optical frequency $\omega$. As a result, the probability of mid-IR radiative relaxation proves to be a few orders of magnitude smaller than that of radiative recombination of the hot e–h pairs. So, in most cases it can be neglected.

The second process is phonon-assisted relaxation (Fig. 6.1e), where the hot carrier relaxes to its ground state with spontaneous emission of one or several phonons. This process should be considered as the intrinsic relaxation mechanism because it can occur in the absence of any other quasiparticles (carriers, phonons) at low temperature and under weak excitation. It determines the lower limit of the relaxation rate.

In the polar semiconductors, electrons and holes can efficiently interact with the longitudinal optical (LO) phonons. The LO phonons are known to produce an electric field to which the electrons and holes, as charged particles, are sensitive. The energy spectrum of the LO phonons is typically rather narrow and can be roughly considered as consisting of one line. For this reason, fast relaxation can occur only in QDs with energy spacings equal to the energy of one or several LO phonons. The carriers in other QDs must relax with the emission of acoustic phonons. This process is inherently slow because of the weakness of the deformation potential interaction. This effect is well known for bulk materials, where the relaxation with emission of acoustic phonons is two orders of magnitude slower than that with emission of LO phonons [13,14].

Theoretical considerations predict that the acoustic-phonon-assisted relaxation in QDs should be slowed even more. According to calculations, the acoustic-phonon-assisted relaxation time in QDs should be in the order of several nanoseconds, which is slower than the typical radiative decay times lying in the subnanosecond range [15]. Only the longitudinal acoustic (LA) phonons with wavelengths of the order of the QD diameter and energy equal to an interlevel spacing may cause one-phonon relaxation. Although two-phonon relaxation is not limited by such strict selection rules, this process is also expected to be slow because of the weakness of the electron–phonon interaction [16]. Poor results coming from the slow phonon-assisted relaxation in QDs are commonly referred to as the "phonon bottleneck" [17].

The last process is the relaxation caused by scattering of the carrier by another carrier located in the QD or in the surrounding material. In this process, one of the carriers loses its excess energy and another one acquires it (see Fig. 6.1f). This process is usually referred to as the Auger-like process [18,19]. Relaxation due to the Auger process is known to be fast because the carriers as charged particles interact with each other more strongly than with phonons. However, this mechanism of relaxation can evidently be efficient only when several carriers are present in or near the QD.

Similar mechanisms can be considered for the e–h pair relaxation. The hot e–h pair can move to the lowest radiative state by emitting a photon or a phonon (one or several) or by carrier scattering. One should expect that the probability of the latter two processes is higher for e–h pairs than for a single carrier because the energy conservation law can be easier satisfied in these processes due to a denser energy level system.

Experimental study of carrier relaxation can be performed by various methods. The most direct one is to study the photoluminescence (PL) kinetics after a short laser excitation. The PL kinetics includes its rise and decay, determined by radiative recombination of the e–h pairs. The front edge (rise) of the PL pulse reflects the time evolution of the radiative state population due to the carrier relaxation from excited states. Therefore, the study of the PL front edge allows one to study the carrier relaxation process. However, studies of this kind require high-sensitivity experimental setups with high spectral and time resolution. A synchroscan-streak camera is frequently used as the fast photodetector in such setups. An example of an experimental setup of this type is described in Sect. 6.2. The time resolution in such setups is longer than a few picoseconds.

Better time resolution limited only by the pulsewidth can be achieved by means of the PL upconversion technique, where PL of the QDs is directed into a nonlinear crystal and upconverted by mixing it with a delayed laser pulse. This technique has a lower sensitivity and therefore needs a much higher pump power density on the sample, which usually modifies contributions of different relaxation mechanisms in favor of the Auger-like process. Another technique is based on study of differential transmission [20] (also referred to as transient absorption [21]) of a sample when a pump beam changes absorbance of the QDs and a delayed probe beam is used for detection of time evolution of the absorbance. This technique also requires a relatively strong pumping of the sample (as any nonlinear technique) and frequently does not allow one to study the intrinsic relaxation processes.

An original experimental method based on electric-field-controlled nonradiative losses is described in [8–10]. The electric field allows one to increase the PL quenching rate in a wide range, making it successively comparable with the relaxation rate via acoustic phonon emission, LO phonon emission, or carrier–carrier scattering. Competition of the relaxation and nonradiative-loss channels gives rise to pronounced modifications of the PL spectrum measured under steady state conditions. As a result, this spectrum reflects spectral dependence of the carrier relaxation rate in the QD ensemble. In some sense, the nonradiative process serves here as an effective optical gate with a variable exposure for the QD PL. The described method allows one to study the spectral dependence of the carrier relaxation rates with high spectral and time resolution at low power density of the optical excitation.

The carrier relaxation in self-assembled QDs has been studied by direct measurements of the PL kinetics in many works [5–7,20,22–29]. Almost all the measurements revealed a relatively short rise of the PL from the lowest energy states of the QDs (usually shorter than 100 ps), which is considered as evidence of fast capture of the carriers from a barrier or wetting layer with their subsequent relaxation in the QDs [30]. This general result of the experiments contradicts the theoretically predicted phonon bottleneck effect for carrier relaxation in QDs. This is why many efforts were made to study experimentally the physical mechanism of the relaxation.

The PL kinetics of the InGaAs QDs was studied under the band-to-band excitation of the barrier layer [5]. The carrier relaxation times at low temperature and low pump power density were shown not to exceed several tens of picoseconds. An increase of the pump power density above some threshold value $P_0 \approx 4$ W/cm$^2$ caused a gradual reduction of the PL pulse rise time. This was attributed to the Auger-like processes, which accelerated both the carrier capture and the carrier relaxation. An important result was obtained in studies of temperature dependence of the carrier relaxation under weak excitation. Considerable acceleration of the relaxation with an increase of the temperature to 70 K is well explained by phonon-assisted relaxation with the thermoactivated process of stimulated phonon emission. The effective phonon energy obtained from analysis of the temperature dependence is about 3 meV. This allowed the authors to explain the temperature-dependent relaxation by a multiphonon process that involves specific LA phonons with the wavelength of about the QD base diameter. It should be stressed that, though the obtained experimental results were explained by invoking the Auger-like and multiphonon processes, no direct evidence of the proposed relaxation mechanisms is available.

More direct study of the relaxation mechanism in InP QDs was presented by Vollmer et al. [7]. The authors studied the time-resolved PL spectra under excitation slightly above the PL band of the QDs (hereafter referred to as quasiresonant excitation). The PL spectra revealed, within a few tens of picoseconds after the pump pulse, the features (resonances) shifted from the excitation energy by the energies of one or two LO phonons. This is a direct observation of the fast carrier relaxation via the LO phonon emission in QDs with the interlevel spacing equal to the LO phonon energy. A limited spectral and time resolution did not allow the authors to determine experimentally the relaxation mechanism in other QDs. They proposed this mechanism to be the acoustic-phonon-assisted relaxation without discussing the type of the phonons and nature of the carrier–phonon interaction.

Much indirect evidence of the phonon-assisted carrier relaxation was found in studies of the PL spectra under quasiresonant excitation and of the PL excitation (PLE) spectra in selected samples with QDs [24,31–36]. In the spectra of these samples, the LO phonon resonances are clearly seen in steady state conditions. Heitz et al. [32] explained this observation by the existence of nonradiative losses of excited carriers in the studied samples. If the carrier escapes the QD quickly enough, this process can compete with the relaxation process. In this case, only fast relaxation with emission of the LO phonons "saves" the QD emission. A detailed study performed by Heitz et al. [24] allowed the authors to determine a few types of the LO phonons participating in the relaxation process in the InAs QDs. No evidence for participation of phonons of other types in the relaxation process was found.

To conclude the discussion on these experiments, it should be stressed that only the LO-phonon-assisted relaxation was directly identified as an intrinsic relaxation mechanism at low power density of the excitation. Some indirect evidence of the relaxation with emission of the low-energy LA phonons is found in studies of temperature dependence of the PL rise time [5,27]. The role of phonons with energies widely ranging from 3 meV (LA phonons with a wavelength of about QD base diameter) to the LO phonons' energy (30 meV in the InAs QDs and 45 meV in the InP QDs) was not studied.

The described situation in the experimental study stimulates many theoretical speculations about the intrinsic mechanisms of carrier relaxation [16,37–42]. Inoshita and Sakaki [16] discussed the two-phonon processes involving the LO $\pm$ LA combination that widens to several meV the energy window for fast relaxation around the LO phonon energy. Another theoretical idea is based on a presumably strong electron–LO-phonon interaction resulting in Rabi splitting of the mixed electron–phonon states [39,41]. Then, the anharmonic decay of the LO phonons should give rise to a large broadening of the states and to fast carrier relaxation in a wide energy window of about 20 meV in the case of the InAs QDs [41]. As shown in [10], this prediction is not supported by the experiment.

Apart from the intrinsic mechanisms of relaxation, a defect-related mechanism has also been considered [37,38,40]. This mechanism assumes some special energy structure and special location of the defect that allow an efficient tunneling of the carrier to the defect, its subsequent fast relaxation on the defect, and its final tunneling back to the QD. It is noteworthy that defects should really exist around the QDs due to the lattice mismatch between the QD and the host materials. Some indirect evidence for the presence of the defects was found in studies of nonradiative losses [7,24,31–33] and anti-Stokes PL of QDs [43]. However, there was no experimental evidence for such a relaxation process.

One more relaxation process proposed by Toda et al. [42] should be mentioned. In this paper, a continuous background in the PLE spectrum of a single InGaAs QD was observed. It was assumed that the QDs had 2D-like continuum states as their intrinsic property. The authors have proposed that the carriers can relax easily within continuum states by resonant emission of localized phonons. The authors did not discuss the nature of the continuous spectrum of the QD energy states. It is clear from the above discussion, however, that most likely they have observed a continuum of the e–h pair energy states shown in Fig. 6.1b. The lower bound of the continuum is determined by the lowest energy level of the electron in the QD and by the energy of highly-excited states of holes, corresponding to their delocalization in the wetting layer. The potential well depth for holes is about 50 meV, as can be derived from the data presented in [42]. This is the energy where the continuous background of the PLE spectrum starts. It should be stressed that optical transitions from the highly excited hole states to the ground electronic state can be partially allowed due to low symmetry of the QDs.

The Auger-like processes of carrier relaxation have also been extensively studied. These are the processes of practical importance, in particular, for laser applications, when many carriers are created in QDs and their vicinity. A typical study of the Auger process involves measurements of the pump power dependence of the relaxation time [5,6,27–29]. In all these papers, a considerable shortening of the relaxation time up to a subpicosecond time scale at strong enough pumping has been found.

A few specific Auger-like processes, including the electron–electron, hole–hole, and electron–hole scattering were considered theoretically [18,19,44–46]. In particular, the electron–hole scattering was extensively treated in [19,45]. However, as was already noted, the energy transfer from an electron to a hole and back is not accompanied, by itself, by a change of the e–h pair energy. Therefore, this process cannot be considered as energy relaxation of the e–h pair without involving any additional process like emission of phonons or scattering of the e–h pair by some excess carrier.

One more Auger-like process proposed by Bockelmann and Eleger [18] is the interaction of carriers with the electron–hole plasma around the QD created by strong pumping. This process is considered to be important for carrier capture from the barrier layer into the QD [5,6]. It should be noted again that the experiments performed so far did not give direct proof of these particular Auger-like processes.

In this chapter, carrier relaxation in InP and InAs or InGaAs self-assembled QDs is considered. Heterostructures with this type of QDs are extensively studied due to their highly promising properties for device applications [2,3]. Experimental data obtained by two experimental methods, namely, by time-resolved PL and by the controlled nonradiative losses mentioned above, are discussed in detail. These methods can be successfully applied to studies of the last step of the relaxation (see Fig. 6.1a–c). They allow one to determine the experimental conditions when a particular relaxation process becomes efficient. It is demonstrated that LO-phonon-assisted relaxation is characterized by a very short relaxation time, lying in the range of a few picoseconds or even in a subpicosecond range. There is also clear experimental evidence that acoustic-phonon-assisted relaxation is not suppressed as strongly as was predicted theoretically. Interaction with the high-energy acoustic phonons is fairly strong and gives rise to a relatively fast relaxation, which is only an order of magnitude slower than the LO-phonon-assisted relaxation. A possible physical mechanism underlying the strength of the interaction of the e–h pair with the high-energy acoustic phonons is discussed. All the experimental data show convincingly that there is no phonon bottleneck effect in QDs.

Several specific Auger-like processes of carrier relaxation and the experimental conditions under which they can occur are also discussed. In particular, carrier relaxation in the presence of excess carriers in QDs and also in the presence of electric current flowing through a layer of QDs are considered. Clear experimental evidence for most of the processes under consideration is demonstrated.

Some details of the experimental methods and theoretical approaches to analysis of experimental data are also discussed. In particular, a model for the PL quenching in the presence of an electric field is considered. An essential point of this model is tunneling of carriers from the QDs into the barrier layer. In the framework of the model, a quantitative analysis of the experimental data allows one to evaluated the depth of the potential well for holes in InP QDs.

## 6.2  Experimental Details

Before proceeding to the discussion of experimental data confirming one or another relaxation mechanism, it makes sense to consider in more detail some structural properties of the samples with QDs and the experimental methods of their studies. In this chapter, the data obtained for two types of structures containing InP and InGaAs are discussed [8–10,47,48]. Comparison of the results obtained for two structures allows one to make a clear generality and to clarify characteristics of relaxation processes in these structures. A specific feature of the InP QDs is that the potential well for a hole is rather shallow [10,49,50]. For this reason, the energy spectrum of the e–h pair proves to be fairly dense: the spacing between the discrete levels lies in the range of a few meV, with the continuous spectrum edge lying only $\approx 15$ meV above the radiative level. The InGaAs quantum dots are characterized by a more resolved spectrum of discrete states of the e–h pair, with the continuous spectrum starting 50 meV or more above the radiative level.

The samples with InP QDs were grown by gas source molecular beam epitaxy on an $n^+$ GaAs substrate [51]. A 300 nm GaAs buffer layer was grown at 600°C. In the middle of the buffer layer, a superlattice with three layers of AlAs 2-nm-thick separated by GaAs layers 10-nm-thick was grown to suppress dislocations. A thin (2 nm) AlAs layer on the buffer layer was grown to prevent the compositional interdiffusion between the GaAs and $In_{0.5}Ga_{0.5}P$ layers. One layer of the InP QDs with a nominal thickness of 4 monolayers was grown between the 100 nm $In_{0.5}Ga_{0.5}P$ barrier layers. The growth rate was 0.5 monolayer/s for $In_{0.5}Ga_{0.5}P$ and 0.25 monolayer/s for InP. The interruption times before and after the InP growth were 2 and 20 s, respectively.

The sample with the $In_{0.35}Ga_{0.65}As$ QDs was grown by metal organic vapor phase epitaxy on the (711)B $n^+$ substrate [52]. The QD layer with a nominal thickness of 4.5 monolayers was sandwiched between the 250 nm and 100 nm GaAs layers containing 2 nm AlAs layers as stop layers for photogenerated carriers. Further details of the growth procedure can be found in the preceding chapter.

Various experimental methods are currently available to characterize the sample structure. Determination of the surface density and approximate dimensions of the quantum dots is usually made using atomic force microscopy on a reference sample, grown in the same conditions without the top barrier layer. For the InP QD structures under discussion, the surface density found in this way is about $10^{10}$ $cm^{-2}$. The average base diameter of the QDs is 35–40 nm and the height is $\approx 5$ nm. For one of the samples, these values were found using cross-sectional transmission electron microscopy with spatial resolution higher than that of atomic force microscopy. This technique also allows one to determine dimensions of the quantum dots directly in the structures under study with the upper barrier layer. This is important since burying the QD layer in the barrier layer can be accompanied by certain changes in size, shape, and composition of the QDs due to interdiffusion of the atoms.

For the structures with the $In_{0.35}Ga_{0.65}As$ QDs under discussion, the surface density is approximately half that for the InP QDs. The average base diameter is $\approx 70$ nm and the height is $\approx 10$ nm, as determined by atomic force microscopy of the reference sample.

The above data about the surface density correspond to the mean distance between the quantum dots of the order of 100–150 nm, which exceeds by several times the base diameter of the QDs. For this reason, the relaxation processes can be assumed to occur in isolated QDs, i.e., the excitation transfer between the QDs is negligible. It is noteworthy also that the quantum dots are fairly large in both types of the structures. For this reason, the energy levels of the photogenerated e–h pair are arranged in a sufficiently dense fashion, as was mentioned in Sect. 6.1. This allows one to study the relaxation processes occurring with assistance of different types of phonons with different energy. In small quantum dots, the low-energy phonons are excluded from the relaxation process because of a large energy gap between the radiative and first excited states of the e–h pair.

One layer of the self-assembled QDs is an extremely thin film of the material with the effective thickness in the range of units of nm. This layer does not noticeably affect optical properties of the sample, such as its refractive index and absorption coefficient. This is why the most reliable optical methods of studying the QDs are based on techniques where the luminescence intensity serves as the output signal. Two methods of this kind successfully employed in a number of papers to study relaxation in QDs will be described here.

The first method is based on studying the PL spectra under quasiresonant optical excitation in presence of an external electric field. A quasiresonant excitation generates an e–h pair in one of the excited states close to the radiative one, so that relaxation of the pair to the radiative state can occur in one step (step (3) in Fig 6.1c). An electric field causes quenching of the PL due to tunneling of the carriers from the QDs to the barrier layers, as has been shown in [10] and will be discussed in detail in Sect. 6.3. The tunneling of carriers from the excited states competes with their relaxation, which results in spectrally selective PL quenching. As a result, the PL spectrum of an inhomogeneously broadend ensemble of QDs reflects, under certain conditions, the relaxation rate spectrum, i.e., the dependence of the relaxation rate on the value of the energy separation between the excited and radiative levels of the e–h pair (see Sects. 6.3–6.4).

For study of the PL spectra in an external electric field, the samples are provided with a semitransparent gold or indium–tin-oxide Schottky contact on the top surface and with an ohmic contact on the back surface. In the study of phonon-assisted relaxation, the PL is excited quasiresonantly by a cw laser with small enough pump power density, typically less than 100 W/cm$^2$, to avoid creation of more than one e–h pair per QD. The excitation can be considered as small, while the PL intensity varies linearly with the pump power without any change of the PL band shape.

The spectral data discussed below were obtained on a setup with a tunable Ti:sapphire laser as an excitation source. The spectral width of the laser line was smaller than 0.01 meV without any sidebands. The PL was dispersed by a double monochromator U1000 (focal length 1 m, numerical aperture 1:8, linear dispersion 0.36 nm/mm) with extremely low background of scattered light ($10^{-14}$ at a range of 3 meV from the laser line). The spectral resolution of the whole setup was approximately 0.15 meV. A photon counting system with a cooled GaAs or InGaAs photomultiplier tube was used for detection of the PL signal.

The second method of studying the relaxation consists in measuring the kinetics of the QD PL under quasiresonant excitation by a short laser pulse. The kinetics of the PL rise (the leading edge of the PL pulse) is determined only by the relaxation rate of the excited e–h pair to the radiative level, provided that other processes of depletion of the excited level (e.g., tunneling of the photocreated carriers from the QD leading to nonradiative losses, or radiative recombination of the carriers before their energy relaxation) are sufficiently slow.

Studying the PL kinetics at different spectral points also makes it possible to determine spectral dependence of the relaxation rate. This method, unlike the previous one, allows one to obtain absolute values of the relaxation rates, provided that the time resolution is sufficiently high. However, because of the pulsed mode of excitation, in this method it is more difficult to achieve high spectral resolution while retaining the low level of excitation. The trade-off between time resolution and sensitivity of PL detection also plays a role.

The PL kinetics discussed in Sect. 6.3 was studied using a picosecond Ti:sapphire laser with a pulsewidth ranging from 1 to 5 ps and a repetition rate of 82 MHz. In most experiments, a relatively low pump power density of about 50 W/cm$^2$ was used to prevent creation of more than one electron–hole pair in a QD. This made it possible to avoid relaxation due to carrier–carrier scattering. A 0.25 m double subtractive-dispersion monochromator (spectral resolution 0.5 nm) and a streak camera were used for accumulation of the signal in the selected spectral points. The time resolution of the setup was about 6 ps. Most measurements were done at sample temperatures from 2 to 5 K to exclude any processes involving absorption or stimulated emission of phonons.

## 6.3   Photoluminescence Spectra
## in External Electric Field

Photoluminescence spectra measured at nonresonant excitation usually show an intense band of QD emission predominates over the emission bands of the barrier and wetting layers. At sufficiently low pump power density, the PL band of the QDs show a smooth bell-like shape with the full width at half maximum (FWHM) of a few tens of meV. This band is mainly formed due to radiative recombination of the e–h pairs from the lowest excited state. The emission intensity from intermediate states is small since the relaxation rate considerably exceeds that of radiative recombination, as was already noted in Sect. 6.1. The shape of the QD PL band is mainly determined by statistical spread of the QD sizes and shapes, leading to a spread in the energy of the lowest excited state of the e–h pair and, thus to inhomogeneous broadening of the optical transition.

Under quasiresonant excitation, when the e–h pairs are generated just in the QDs, the intensity of the QD PL band is smaller by 2 to 3 orders of magnitude than under nonresonant excitation with the same pump power density, which results from the small absorption in a single QD layer. The shape of the QD PL band remains smooth and bell-like for high-quality structures, but exhibits various spectral features (resonances) in structures with large nonradiative losses.

As an example, Fig. 6.2 shows PL spectra of a sample with InP Qs under both nonresonant and resonant excitation. As is seen, the most intense band in the PL spectra is that of the QDs. The PL profile does not show sharp features for any photon energy of excitation. The PLE spectrum is also smooth.

The QD PL band changes drastically when an external electric field is applied. The behavior of the band is strongly different for the positive and negative bias applied to the sample surface.

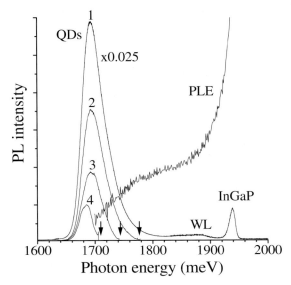

**Fig. 6.2.** The PL and PLE spectra of a sample with InP QDs. PL spectrum 1 was recorded under interband excitation of the InGaP barrier. Its intensity is multiplied by 0.025. The labels InGaP, WL, and QDs denote the PL bands of the InGaP barriers, wetting layer, and InP QDs, respectively. PL spectra 2, 3, and 4 were recorded under quasiresonant excitation of the InP QDs with photon energies marked by *arrows*. The PLE spectrum was recorded with PL detection at maximum of the QD band, $E_{\mathrm{PL}}=1700$ meV (from Ignatiev et al. [10])

At small positive bias, the PL intensity decreases with increasing bias. However, for the bias $U_{\mathrm{bias}}>U_{\mathrm{Schottky}}$, the electric current flowing through the sample sharply increases. Here, $U_{\mathrm{Schottky}}$ is the potential of the Schottky barrier formed by a semitransparent electrode on the sample surface: $U_{\mathrm{Schottky}}=0.5$ V for the ITO electrode and $U_{\mathrm{Schottky}}=0.7$ V for the gold one. The electric current leads to a fast carrier relaxation that is discussed in Sect. 6.7. For structures with InP QDs, an interesting phenomenon of anti-Stokes PL caused by this current is observed [43], which is beyond the scope of this chapter.

When a negative bias is applied, the total PL intensity decreases. At the same time, the decrease of the PL in different spectral points is different and, as a result, pronounced resonances arise in the PL spectra as shown in Fig. 6.3. The most prominent resonance is shifted from the excitation line by approximately $E_{\mathrm{LO}}=45$ meV, which is close to the LO phonon energy of bulk InP crystals (43.5 meV [53]).

A very similar behavior of the resonances is observed in the PLE spectra of InP QDs, as shown in Fig. 6.4. The intense 1LO and 2LO resonances shifted by $E_{\mathrm{LO}}$ and $2E_{\mathrm{LO}}$, respectively, from the spectral point of the PL

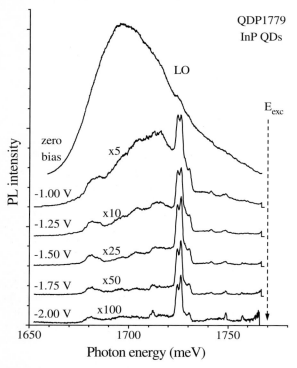

**Fig. 6.3.** PL spectra of InP QDs at different applied bias. The spectral position of the excitation is marked by an *arrow*. The most intense resonance shifted from the laser line by the LO phonon energy is marked LO. The applied bias is shown for each curve. For clarity, the spectra are scaled and shifted vertically (from Ignatiev et al. [9])

detection become clearly distinguished at a sufficiently strong negative bias.

The energy positions of the resonances strictly follow the photon energy of excitation as shown in Fig. 6.5. Under excitation at higher photon energy, the 2LO resonance can be also observed. As is seen, the relative intensities of the 1LO and 2LO resonances strongly vary with the excitation photon energy. Intensities of the resonances become weak when they go out of the PL band measured at zero bias.

As was already mentioned in Sect. 6.1, some authors observed LO phonon resonances in the PL or PLE spectra of samples with a low PL efficiency of the QDs. The spectra observed by these authors are qualitatively similar to those shown in Figs. 6.3–6.5 for negative bias. Observation of the LO phonon resonances is direct evidence of participation by the LO phonons in the relaxation process. The qualitative model describing the behavior of the resonances is discussed in Sect. 6.4.

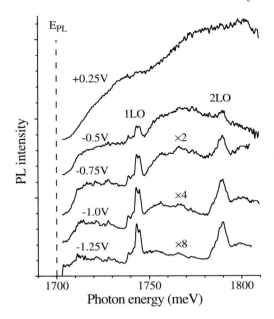

**Fig. 6.4.** PLE spectra of InP QDs at different applied bias. Spectral position of PL detection $E_{PL}$ is marked by a *dashed line*. The most intense resonances 1LO and 2LO are shifted from $E_{PL}$ by the energy of one or two LO phonons, respectively. The applied bias is shown for each curve. For clarity, the curves are scaled and shifted vertically (from Ignatiev et al. [9])

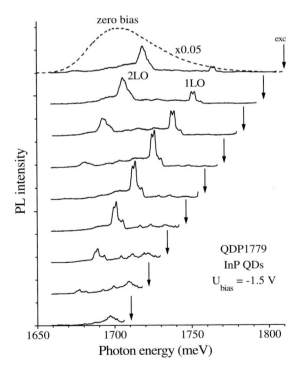

**Fig. 6.5.** PL spectra of a sample with the InP QDs at different photon energies of excitation. Photon energies of the excitation are indicated by *arrows*. $U_{bias} = -1.5$ V. For clarity, the spectra are shifted vertically. The *top curve, shown by a dashed line*, is the spectrum of the sample at zero bias. It is scaled by a factor of 0.05. The resonances shifted from the laser line by the energy of one or two LO phonons are marked by 1LO and 2LO, respectively (from Ignatiev et al. [10])

Several more weak resonances are observed in the spectral region between the excitation line and the 1LO resonance (hereafter referred to as the acoustic region of the spectrum) and also between the 1LO and 2LO resonances (see Fig. 6.5). Their energy shifts from the excitation or 1LO resonance line coincide with energies of the high-energy edges of transverse acoustic (TA) and longitudinal acoustic (LA) phonon bands in InP material. These resonances show up as broad peaks with Stokes shifts of 13 meV and 22.5 meV, respectively. Appearance of such resonances indicates effective participation of high-energy acoustic phonons in the relaxation process.

PL spectra of a sample with InGaAs QDs are changed in electric field in a similar way, as shown in Fig. 6.6. Without bias, the PL spectrum consists of two smooth peaks (see inset), probably related to two sets of QDs with different sizes present in this sample. A number of new maxima (resonances) appear in the spectrum at negative bias. The most pronounced resonances can be assigned to the GaAs-like and InAs-like LO phonons of the InGaAs

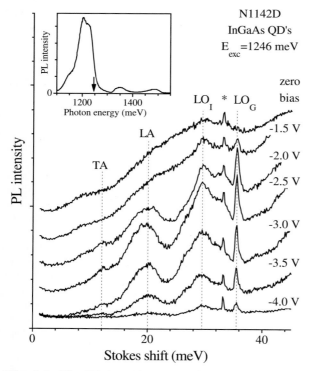

**Fig. 6.6.** The PL intensity versus Stokes shift for the sample with InGaAs QDs at different biases indicated for each curve and $E_{exc}$=1246 meV. For clarity, the spectra are shifted vertically. The energies of the resonances marked by TA, LA, $LO_I$, and $LO_G$ are 12 meV, 20 meV, 30 meV, and 35.5 meV, respectively. *Inset*: PL spectrum of the sample without bias (from Ignatiev et al. [10])

QDs because their energy shifts from the excitation line are close to energies of the LO phonons in GaAs and InAs crystals, respectively. In Fig. 6.6, they are marked $LO_G$ and $LO_I$. The broad intense resonance with a Stokes shift of about 20 meV can be assigned to LA phonons of the InGaAs solid solution. A narrow peak with a Stokes shift of 33.2 meV marked by * in Fig. 6.6 is probably caused by Raman scattering from the GaAs barrier layers because the intensity of this peak does not depend on bias.

The LO phonon resonances for all the samples have a rather complicated structure consisting of a few narrow peaks. Some of the resonances observed in the spectra of the InP and InGaAs QDs are shown in Fig. 6.7 in more detail. A fit of the resonances by the set of Lorentzians allows one to estimate the full width at half maximum (FWHM), which proves to be about 1.2 meV for the InP QDs and 0.5 meV for the InGaAs QDs. This observation contradicts the large broadening of the LO resonance due to electron–phonon interaction predicted theoretically [41].

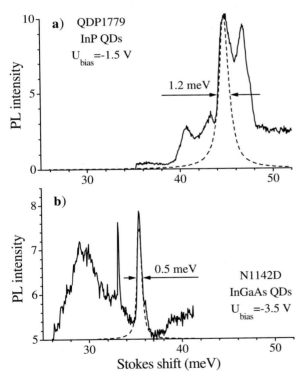

**Fig. 6.7.** Fragments of PL spectra with LO resonances for samples with **(a)** InP and **(b)** InGaAs QDs at biases −1.5 V and −3.5 V, respectively. A Lorentzian fit of the narrowest peaks in the LO resonances is shown by *dashed lines*. The values of FWHM of the Lorentzians are indicated against each peak (Fig. 6.7(a) is redrawn from Ignatiev et al. [10])

## 6.4  Physical Mechanisms

As was mentioned in Sect. 6.1, the LO phonon resonances observed in samples with a low PL efficiency were ascribed to fast carrier relaxation with emission of LO phonons [32]. However, due to the random value of the nonradiative losses in those samples, this mechanism was not proved. The controllable variation of nonradiative losses allows one to analyze qualitatively and semi-quantitatively the mechanism, as discussed in [9,10].

Generally, a few processes leading to formation of the resonances in the PL spectra are possible – resonance Raman scattering, phonon-assisted electron–hole recombination, phonon-assisted absorption, and phonon-assisted relaxation in the presence of nonradiative losses.

The resonance Raman scattering for which the incident laser light is in resonance with the absorbing transition (incoming resonance), shows a rather low efficiency. The results of studies of a heterostructure with InP QDs by Sirenko et al. [54] serve as an example. To obtain a measurable signal of Raman scattering, the authors had to excite the heterostructure along the layer of QDs. This fact alone suggests that the data presented in Figs. 6.2–6.6 cannot be explained by resonance scattering. In addition, a number of other arguments can be adduced. In particular, the dependence of the phonon resonance intensities on the applied bias (see Figs. 6.3, 6.4, 6.6) and strong variations of the ratio of the 1LO and 2LO resonance intensities with spectral position of excitation within the PL band (see Fig. 6.5) also allow one to rule out resonance Raman scattering as the process responsible for the phonon resonances. This point is further supported by a slow decay time of the QD emission at the phonon resonances and by a strong temperature dependence of the resonance intensities discussed below.

Resonance Raman scattering with double resonance (incoming and outgoing), when both the incident and scattered light are in resonance with optical transitions in the QD, cannot be distinguished from PL.

Phonon-assisted electron–hole recombination can be also ruled out because the electron–phonon interaction is weak in these structures and is able to produce only very small phonon sidebands in the PL spectra [55,56]. Besides, the ratio of intensities of the one-phonon and two-phonon sidebands must be almost independent of the excitation photon energy, which is in contrast with the observed behavior of the 1LO and 2LO resonances.

The phonon resonances in the spectra can also be caused by phonon-assisted absorption when an absorbed photon creates an electron and a hole in their ground states and a phonon. The probability of this process is equal to the probability of phonon-assisted PL and is also small. Phonon-assisted absorption in $In_{0.4}Ga_{0.6}As$ QDs was observed by Findeis et al. [57]. Note a particularly interesting case when the transition with emission of a phonon is close in energy to zero-phonon transition into an excited state of the e–h pair. The probability of the phonon-assisted transition, in this case, essentially increases due to "borrowing" of probability from the zero-phonon transition.

This problem was discussed in great detail in [47,58]. At exact coincidence of energies of the transitions, the zero-phonon transition predominates, and the e–h pair is generated in the excited state with its subsequent relaxation.

It is evident from the aforesaid that the main process giving rise to the phonon resonances is the selective phonon-assisted relaxation of hot carriers in the presence of nonradiative losses. Appearance of the phonon resonances due to this process is illustrated by Fig. 6.8a and is explained as follows [32]. The quasiresonant excitation creates electrons and holes in the excited states. The QDs of the ensemble have slightly different sizes and shapes, and therefore the interlevel energy spacing, $\Delta E$, has some distribution. The spacing $\Delta E$ can match the LO phonon energy $E_{LO}$ only in some subset of the QDs. Carrier relaxation in these QDs is fast due to high efficiency of this process with emission of an LO phonon. The relaxation of the rest QDs occurs via

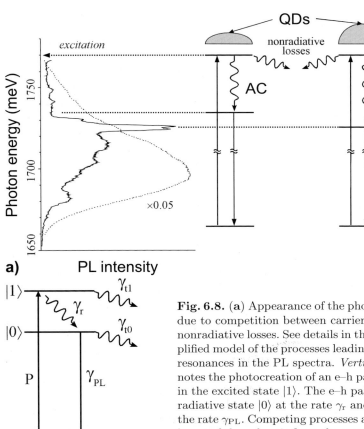

**Fig. 6.8.** (a) Appearance of the phonon resonances due to competition between carrier relaxation and nonradiative losses. See details in the text. (**b**) Simplified model of the processes leading to the phonon resonances in the PL spectra. *Vertical arrow* P denotes the photocreation of an e–h pair at the rate P in the excited state $|1\rangle$. The e–h pair relaxes to the radiative state $|0\rangle$ at the rate $\gamma_r$ and recombines at the rate $\gamma_{PL}$. Competing processes are nonradiative losses of the e–h pair from the states $|1\rangle$ and $|0\rangle$ at the rates $\gamma_{t1}$ and $\gamma_{t0}$, respectively, due to carrier escape from the QD

emission of acoustic phonons, i.e., via a much slower process. If the electron or the hole can efficiently leave the QD (this process is usually referred to as nonradiative losses) before relaxation via acoustic phonon emission, the PL does not appear at any spectral point except the point where it is "saved" by the fast LO-phonon-assisted relaxation. In this case, a narrow peak shifted from the excitation line by the energy of the LO phonon $E_{LO}$ should be observed in the time-integrated PL spectrum. However, if there are no significant nonradiative losses, electrons and holes in *any* QD eventually relax to the lowest levels and recombine. In this case, the PL spectrum must reproduce the energy distribution of the lowest optical transition that usually has a bell-like smooth shape. Behavior of the spectra presented in Figs. 6.3–6.6 agrees with this scenario.

To illustrate this, consider a simple three-level energy diagram for the e–h pair shown in Fig. 6.8b. Quasiresonant excitation generates an e–h pair in excited energy levels labeled by $|1\rangle$. The rate of the e–h pair generation is denoted by $P$. It is assumed that the pump rate is slow enough so that not more than one e–h pair can be found in each QD at a time. The main process of relaxation of the hot carriers to their ground states in this case is phonon-assisted relaxation. The case of strong pumping is discussed in Sect. 6.7. The hot e–h pair relaxes at the rate $\gamma_r$ to the radiative state $|0\rangle$ and then recombines at the rate $\gamma_{PL}$. Both these processes compete with the processes of nonradiative losses due to carrier escape from the QD. The rate of the nonradiative losses from states $|1\rangle$ and $|0\rangle$ is denoted by $\gamma_{1t}$ and $\gamma_{0t}$, respectively.

Dynamics of the electron–hole pair populations $n_1$ and $n_0$ at the levels $|1\rangle$ and $|0\rangle$, respectively, is described, in the framework of this model, by the equations

$$\frac{dn_1}{dt} = P - (\gamma_r + \gamma_{1t})n_1 , \qquad \frac{dn_0}{dt} = \gamma_r n_1 - (\gamma_{PL} + \gamma_{0t})n_0 . \tag{6.1}$$

For a cw excitation ($P$=const), the steady-state solution of (6.1) yields the expression for the bias dependence of the radiative state population $n_0$. As a result, the bias dependence of the PL intensity, $I_{PL} = \gamma_{PL} n_0$, is given by

$$I_{PL}(\Delta E) = NP \frac{1}{[1 + \gamma_{1t}/\gamma_r(\Delta E)]} \frac{1}{[1 + \gamma_{0t}/\gamma_{PL}]} . \tag{6.2}$$

where $N$ is the number of QDs capable of emitting light at a given spectral point under a given excitation; $\Delta E$ is the energy separation between the $|1\rangle$ and $|0\rangle$ states.

If the nonradiative losses are absent, i.e., $\gamma_{1t}=0$ and $\gamma_{0t}=0$, then $I_{PL}=NP$. This means that the PL intensity does not depend on the relaxation rate. In the other extreme case, when the nonradiative losses are large, i.e., $\gamma_{1t} \gg \gamma_r$ and $\gamma_{0t} \gg \gamma_{PL}$,

$$I_{PL}(\Delta E) = NP \frac{\gamma_r(\Delta E)}{\gamma_{1t}} \frac{\gamma_{PL}}{\gamma_{0t}} . \tag{6.3}$$

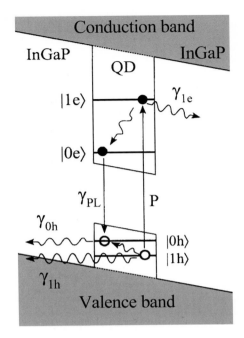

**Fig. 6.9.** The model of the PL quenching in the electric field. Only a couple of energy states for an electron ($|1e\rangle$ and $|0e\rangle$) and a hole ($|1h\rangle$ and $|0h\rangle$) are considered. The *vertical arrows* marked P and $\gamma_{PL}$ denote the pumping and the radiative recombination, respectively. Relaxation of the hot carriers to their ground states is shown by *wavy arrows*, which are not labelled. The *wavy arrows* marked by $\gamma_{1h}$, $\gamma_{0h}$, and $\gamma_{1e}$ indicate tunneling of the hole from both the states and tunneling of the electron from the excited state, respectively (from Ignatiev et al. [10])

Equation (6.3) describes the main regulations in behavior of the PL spectra in the strong electric field. The last ratio in (6.3) $\gamma_{PL}/\gamma_{0t}$ describes the PL quenching due to nonradiative loss of the excitation from the ground state $|0\rangle$. This quenching does not depend on the relaxation rate and is, therefore, the same for the whole PL band. The first fraction $\gamma_r/\gamma_{1t}$ describes selective suppression of the PL. For the slow acoustic phonon relaxation, this ratio is smaller and, therefore, the PL quenching is stronger than that for the fast LO phonon relaxation. Thus, the PL spectrum is determined by spectral dependence of the relaxation rate $\gamma_r(\Delta E)$. In other words, the PL spectrum of QDs with large radiative losses reflects the relaxation rate spectrum. The process leading to nonradiative losses, in a certain sense, forms an effective time gate for the PL signal, which makes it possible to study fast relaxation processes in steady state conditions.

### 6.4.1 Model of Selective Photoluminescence Quenching

For a quantitative description of the PL suppression in an electric field, a more detailed model is needed. This model is schematically illustrated in Fig. 6.9. It is assumed that the quasiresonant excitation generates an electron and a hole on excited energy levels labeled by $|1e\rangle$ and $|1h\rangle$, respectively. Relaxation of the electron and hole to their ground states, for simplicity, is shown in Fig. 6.9 by separate arrows. However, in accordance with Sect. 6.1, the relaxation of the e–h pair is supposed to be characterized by the rate $\gamma_r$. The treatment of the electron and hole separately is needed to quantitatively describe the nonradiative losses.

The relaxation process competes with the nonradiative losses. It is assumed that the external electric field activates the tunneling of the photocreated carriers from the QD into the barrier layer. The depth of the potential well for holes is usually smaller than that for electrons. As for the InP QDs, it is assumed that the holes are weakly localized in the dots or even in the surrounding material [49,50]. Real potential profiles for carriers may be complicated due to strain or composition fluctuations [59]. Nevertheless, it can be assumed for simplicity that the potential barrier for holes acquires a triangular shape in the electric field, as shown in Fig. 6.9.

In the semiclassical approach, the tunneling rate, $\gamma_i(U)$, from the state $|i\rangle$ through the triangular barrier is given by the expression [60]

$$\gamma_i = \gamma_i^0 e^{-U_i/U} \,, \tag{6.4}$$

with

$$U_i = (4/3e\hbar)\sqrt{2m^*}E_i^{3/2}d \,, \tag{6.5}$$

where $m^*$ is the effective mass of a carrier, $E_i$ is the depth of the potential well for the carrier in the state $|i\rangle$, and $d$ is the effective thickness of the insulating layer to which the bias is applied ($d \approx 0.5$ μ). The dependence of $E_i$ on $U$ is neglected for simplicity.

The well depth for electrons in the InP QDs is about 200 meV, which prevents the electron from tunneling into the conduction band of the barrier layer at moderate biases. However, the electrons may tunnel to deep levels present in the vicinity of the QDs even in high-quality heterostructures [43,61,62]. An analysis shows that the electron tunneling from the excited state $|1e\rangle$ affects the PL of the InP QDs at $U_{\text{bias}} < -1$ V. To simplify the discussion, the rate of this process $\gamma_{1e}$ is also described by (6.4). Also neglected are variations of the carrier relaxation and optical transition probabilities in the electric field, possible built-in electric field [61], and intrinsic dipole moment of the QDs [59]. All these effects influence the PL of the QDs much less than the carrier tunneling.

The PL behavior in an electric field can be described using (6.2) with explicit expressions for the tunneling rates $\gamma_{1t} = \gamma_{1h} + \gamma_{1e}$ and $\gamma_{0t} = \gamma_{0h}$ from the states $|1\rangle$ and $|0\rangle$, respectively. The electron and hole tunneling rates are calculated using (6.4). In Fig. 6.10, the bias dependence of the PL intensities of the sample with the InP QDs is plotted for the LO resonance and for a neighboring spectral point marked by 2AC in the inset. The PL in the 2AC spectral point is contributed by those QDs in which the carriers cannot relax with the emission of a single phonon because of the energy gap between the acoustic and the LO phonons. The intensities $I_{\text{LO}}(U)$ and $I_{2\text{AC}}(U)$ at the LO and 2AC spectral points, respectively, are fitted by (6.2). The ratio of the intensities normalized to unity at zero bias may be used as a measure

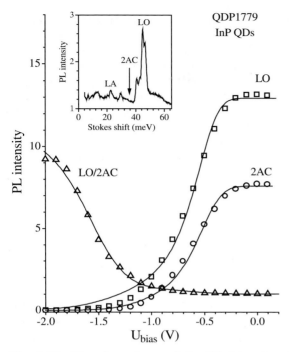

**Fig. 6.10.** Bias dependence of the PL intensity at the LO (*squares*) and 2AC (*circles*) spectral points for the sample with InP QDs. $E_{exc}=1771$ meV. *Solid curves* are fitted by (6.2). A bias dependence of the ratio of the LO and 2AC intensities is shown by *triangles*. A *solid line over the triangles* is fitted by (6.6). The fitting parameters for all the curves are: $U_{1h}=2$ V, $U_{0h}=4$ V, $U_{1e}=16$ V, $\gamma_{1e}^0/\gamma_{1h}^0=3.8\times10^4$, $\gamma_{0h}^0/\gamma_{PL}=150$, $\gamma_{LO}/\gamma_{2AC}=12$, and $\gamma_{1h}^0/\gamma_{LO}=0.2$. *Inset*: The PL spectrum at $U_{bias}=-1.5$ V. The 2AC spectral point is indicated by an *arrow* (from Ignatiev et al. [10])

of the contrast between the features in the PL spectrum at strong bias. The bias dependence of this ratio is also plotted in Fig. 6.10 and is fitted by the equation

$$\frac{I_{LO}(U)}{I_{2AC}(U)} = \frac{(1+\gamma_{1t}/\gamma_{2AC})}{(1+\gamma_{1t}/\gamma_{LO})} , \tag{6.6}$$

which follows directly from (6.2). Here, $\gamma_{LO}$ and $\gamma_{2AC}$ are the relaxation rates with emission of an LO phonon and two acoustic phonons, respectively.

As is seen from the figure, the calculated curves reproduce the experimental values reasonably well. Thus the model, in spite of its simplicity, describes adequately, as a whole, the behavior of the PL intensity of the InP QDs versus the applied voltage. Therefore, it can be stated that the tunneling of the carriers, in particular of the holes, from the QDs into the InGaP barrier is the main process that leads to quenching of the PL in the electric field. The tunneling of electrons becomes of importance for sufficiently high negative bias.

From the fit, the values $U_{1h}=2$ V, $U_{0h}=4$ V, and $U_{1e}=16$ V can be determined. These values allow one to estimate the depth of the potential wells for carriers by means of (6.5). The calculated values are $E_{0h}=13$ meV, $E_{1h}=9$ meV, and $E_{1e}=65$ meV. These values are obtained using the heavy-hole effective mass $m^*_{hh}=0.65m_0$ and the electron effective mass $m^*_e=0.08m_0$, in the InP crystal [63]. Here $m_0$ is the electron mass. The results thus obtained show that the potential well for holes in the InP QDs is small and positive, i.e., the holes are localized in the dots. This conclusion is further supported by a relatively high probability of the optical transitions in the InP QDs, as discussed in Sect. 6.5.

Accuracy of the values of $E_{0h}$ and $E_{1h}$ is limited by many simplifications used in the model. The accuracy is estimated to be about 50% by analysis of all sets of experimental data as well as by comparing the results using various modifications of the described model. It should be stressed that the actual accuracy can be verified only by independent measurements of these values with a different technique.

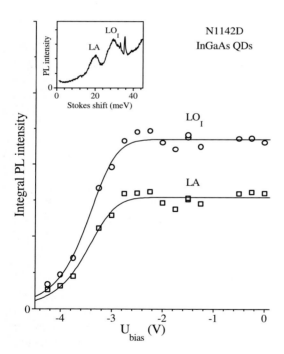

**Fig. 6.11.** The bias dependence of the PL intensity at the $LO_I$ (*open circles*) and LA (*open squares*) spectral points for the sample with InGaAs QDs. The data are taken from the spectra presented in Fig. 6.6. *Solid curves* are fitted by (6.7). The fitting parameters are: $U_0=40$ V and $\gamma^0_0/\gamma_{PL}=9\times10^4$. *Inset*: PL spectrum at $U_{bias}=-3$ V (from Ignatiev et al. [10])

In a similar way, the PL data for the InGaAs QDs presented in Fig. 6.6 can be analyzed. The bias dependence of the $LO_I$ and LA resonances is shown in Fig. 6.11. Both dependencies are fitted well by the equation

$$I_{PL}(U) = \frac{P}{(1 + \gamma_{0t}/\gamma_{PL})} , \tag{6.7}$$

which accounts only for tunneling from the radiative state $|0\rangle$. This means that the tunneling from the excited state $|1\rangle$ at biases used in the experiment is much slower than the carrier relaxation with emission of the $LO_I$ and LA phonons. Assuming that the hole-tunneling process is responsible for the PL quenching, the energy of the lowest hole state $E_{0h}=70$ meV can be obtained from the fitting procedure using the heavy hole effective mass $m_{hh}^*=0.43m_0$. Actually, the type of carriers, electron or hole, that tunnel from the QD cannot be uniquely determined from the experimental data presented above. In the case of electron tunneling, a similar estimation using the electron mass $m_e^*=0.05m_0$ gives rise to the energy of the lowest electron state $E_{0e}=140$ meV. Both the obtained values agree with typical values used for InGaAs QDs [3].

## 6.5   Kinetics

The kinetic PL measurements allow one to study the time evolution of the radiative state population $n_0$. The radiative level is depleted due to recombination of the e–h pairs (provided that there are no nonradiative losses of excitation), with characteristic times in the range of several hundreds of picoseconds. Population of the radiative level occurs due to relaxation of the e–h pairs from the excited states via different processes characterized by different rates. The slowest relaxation of the e–h pairs with emission of acoustic phonons takes no more than 100 ps. The times of relaxation with emission of LO phonons take a few or fractions of picoseconds. For this reason, the population of the radiative level can be changed very fast, and a high time resolution is needed to study this kinetics. As was mentioned in Sect. 6.1, it is rather difficult to achieve subpicosecond time resolution holding the same low level of excitation. The experimental data discussed here were obtained with a time resolution of 6 ps using the technique described in Sect. 6.2. This time resolution is sufficient to distinguish between the fast and slow relaxation processes in the e–h pairs and to study in real time the acoustic-phonon-assisted relaxation.

The PL kinetics of the InP QDs for two spectral points within the PL band at quasiresonant excitation is shown in Fig. 6.12. As seen, the kinetics depends on the bias applied to the sample. At zero bias, the kinetics consists of a short rising part and a rather long decay. The PL pulse leading edge is different for different spectral points, as shown in the low inset in Fig. 6.12. It is shorter than the time resolution of the setup at the LO resonance and much longer at the other spectral point. At negative bias, the decay becomes faster and the amplitude of the signal decreases.

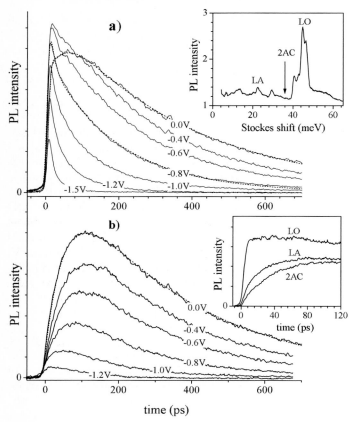

**Fig. 6.12.** The PL kinetics of the InP QDs at the (**a**) LO and (**b**) 2AC spectral points (marked in the *upper inset*) for different biases indicated for each curve. Examples of fitting of the PL kinetics by (6.8)–(6.10), respectively, are shown by *dotted lines*. Values of the fitting parameters $\gamma_{PL}+\gamma_{0t}=\tau_{PL}(U)^{-1}$ and $\gamma_r+\gamma_{1t}=\tau_r(U)^{-1}$ for LO and 2AC kinetics, respectively, can be extracted from Fig. 6.13. The fitting parameter for 2AC kinetics at $U=0$ V, $\gamma_{PL}+\gamma_{0t}=(358\text{ ps})^{-1}$. Values of the other parameters for LO kinetics are: $\gamma_{ac}+\gamma_{1t}=(43.3\text{ ps})^{-1}$ and $\alpha=0.67$ for $U=0$ V, $\gamma_{ac}+\gamma_{0t}=(9.9\text{ ps})^{-1}$ and $\beta=0.60$ for $U=-0.8$ V. Values of the parameter $\gamma_{LO}+\gamma_{1t}$ are chosen to be 1 ps$^{-1}$ for all biases. The fitting curves are convoluted by Gaussian with a FWHM of 6 ps. $E_{exc}=1771$ meV. *Upper inset*: PL spectrum at $U_{bias}=-1.5$ V. *Lower inset*: initial part of the PL kinetics at zero bias for a few spectral points indicated in the *upper inset* (redrawn from Ignatiev et al. [9])

Time evolution of the PL intensity can be described by an equation that follows from solution of (6.1). Under the assumption that the excitation pulse is shorter than any other process in the system, i.e., $P=P_0\delta(t)$, the time-dependent PL intensity $I_{PL}(t)=\gamma_{PL}n_0$ is described by

$$I_{\mathrm{PL}}(t) = I_0 \left( \mathrm{e}^{-(\gamma_{\mathrm{PL}}+\gamma_{0t})t} - \mathrm{e}^{-(\gamma_{\mathrm{r}}+\gamma_{1t})t} \right) , \tag{6.8}$$

where

$$I_0 = \gamma_{\mathrm{r}} P_0 / \left[ (\gamma_{\mathrm{r}} - \gamma_{\mathrm{PL}}) + (\gamma_{1t} - \gamma_{0t}) \right] . \tag{6.9}$$

The two terms of (6.8) describe, respectively, the decay and rise of the PL pulse.

The function $I_{\mathrm{PL}}(t)$ given by (6.8) fits the kinetics for acoustic resonances well. An example of the fit is shown in Fig. 6.12b. The behavior of the PL kinetics at the LO resonance is, however, more complicated. At zero bias, one can clearly see fast and slow components of the PL rise (Fig. 6.12a). The kinetics is well fitted by the equation

$$I_{\mathrm{LO}}(t) = I_0 \left( \mathrm{e}^{-(\gamma_{\mathrm{PL}}+\gamma_{0t})t} - \left[ \alpha \mathrm{e}^{-\gamma_{\mathrm{LO}}t} + (1-\alpha)\mathrm{e}^{-\gamma_{\mathrm{ac}}t} \right] \mathrm{e}^{-\gamma_{1t}t} \right) . \tag{6.10}$$

The slow PL rise component is probably related to the multistep carrier relaxation with emission of acoustic phonons in a fraction $(1-\alpha)$ of the QDs.

At negative bias, the carrier tunneling compensates for slow relaxation, and another process, responsible for the relatively fast PL decay component, becomes observable (see Fig. 6.12a). This process is probably related to cascade relaxation of the carriers in large QDs where the lowest optical transition is redshifted relative to the spectral point of observation. In this case the PL kinetics is described by the equation

$$I_{\mathrm{LO}}(t) = I_0 \left( \left[ \beta \mathrm{e}^{-\gamma_{\mathrm{ac}}t} + (1-\beta)\mathrm{e}^{-\gamma_{\mathrm{PL}}t} \right] \mathrm{e}^{-\gamma_{0t}t} - \mathrm{e}^{-(\gamma_{\mathrm{LO}}+\gamma_{1t})t} \right) , \tag{6.11}$$

where $\beta$ is the fraction of large QDs.

A good agreement of the fitting with the PL kinetics allows one to determine the bias dependence of the PL decay time $\tau_{\mathrm{PL}}(U)$ and the PL rise time $\tau_{\mathrm{r}}(U)$. Figure 6.13 shows the bias dependencies of $\tau_{\mathrm{PL}}(U)$ for the LO resonance and of $\tau_{\mathrm{r}}(U)$ for the 2AC spectral point. According to (6.8)–(6.11), these dependencies are given by

$$\tau_{\mathrm{PL}}(U) = \frac{\tau_{\mathrm{PL}}}{1+\tau_{\mathrm{PL}}\gamma_{0t}}, \qquad \tau_{\mathrm{r}}(U) = \frac{\tau_{\mathrm{r}}}{1+\tau_{\mathrm{r}}\gamma_{1t}} . \tag{6.12}$$

The data in Fig. 6.13 are fitted by these functions using (6.4), and (6.5) for the tunneling rate, $U_0$=4 V determined above, and $U_1$ as a fitting parameter. The carrier tunneling from the e–h pair excited state is described, in this case, by an effective rate $\gamma_{t1}$, in contrast to the more complicated model considered in the previous section where the electron and hole tunneling are treated separately. Both the error of fitting the kinetics and the limiting time resolution do not allow one to determine the parameters of a more sophisticated model with higher accuracy. The value found using the fitting procedure is $U_1$=5.5.

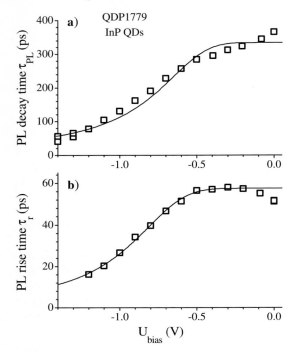

**Fig. 6.13.** Bias dependence of (**a**) the PL decay time $\tau_{\mathrm{PL}}$ and (**b**) of the PL rise time $\tau_{\mathrm{r}}$. *Open squares* represent the data obtained by fitting of the PL kinetics. Fitting by (6.12) is shown by *solid lines*. The fitting parameters are: (**a**) $\tau_{\mathrm{PL}}$=330 ps, $\gamma_{0t}^{0}$=0.2 ps$^{-1}$, and $U_0$=4 V; (**b**) $\tau_{\mathrm{r}}$=58 ps, $\gamma_{1t}^{0}$=2.5 ps$^{-1}$, and $U_1$=5.5 V (from Ignatiev et al. [10])

As seen from Fig. 6.13, the fitting curves well reproduce general behavior of the rise and decay times. The PL decay time at small bias is about 300 ps. The experiments show that this time as well as the PL intensity of the QDs are almost independent of the sample temperature up to several tens of Kelvins. This is an indication that the PL decay is determined by radiative recombination of the electron–hole pairs, i.e., the nonradiative losses of excitation at small biases are negligible. A smooth profile of the QD PL band observed at any excitation energy (see Fig. 6.2) provides further evidence for this conclusion. The obtained value of $\tau_{\mathrm{PL}}$ agrees well with the model calculations of radiative lifetime for the disk-shaped QDs [64]. Thus, one can speak about a relatively high probability of optical transitions in the InP QDs.

The PL rise time at the 2AC spectral point is about 60 ps at small bias. Similar kinetic measurements at the acoustic resonances give faster PL rise times by a factor of two. The PL rise time at the LO resonance is shorter than the time resolution of 6 ps. The lower limit for this time can be deduced from the FWHM of separate peaks of the LO resonances. It is about 1.2 meV

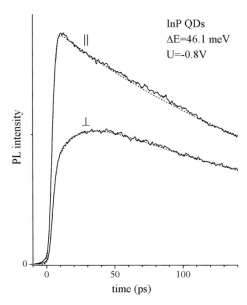

**Fig. 6.14.** PL kinetics of InP QDs at LO resonance in two linear polarizations of PL that are parallel ($\parallel$) and perpendicular ($\perp$) to polarization of excitation light. The *dotted lines* are fitted by (6.8)–(6.11)

for the InP QDs (see Fig. 6.7), which corresponds to a relaxation time of 0.6 ps. The corresponding value for the InGaAs QDs is about 1.5 ps.

It should be noted that PL kinetics depends not only on the energy relaxation of carriers, but also on their *spin dynamics*. This point is illustrated by Fig. 6.14 where the PL kinetics in two linear polarizations, parallel and orthogonal to polarization of excitation, is shown. As seen, the PL rise in crossed polarization is *slower* than in parallel polarization. This interesting question about spin dynamics is beyond the scope of this chapter. To minimize the effects of spin dynamics, all the PL data discussed in this chapter were obtained in linear polarization parallel to that of excitation. The PL delay in crossed polarization related to spin dynamics of the excited carriers is about 30 ps in the case of the InP QDs [65].

## 6.6  Acoustic Phonon Resonances

As mentioned in Sect. 6.1, the acoustic-phonon-assisted relaxation was predicted theoretically to be inefficient [15–17]. The PL spectra of the biased samples as well as the PL kinetics data presented in Sect. 6.5 demonstrate that the carrier relaxation with emission of high-energy acoustic phonons, in contradiction with theoretical predictions, is efficient. To make this contradiction evident, the PL spectra presented in Fig. 6.5 are replotted in Fig. 6.15 versus the Stokes shift. It is seen that the acoustic phonon resonances are clearly observed not only between the laser line and the 1LO resonance but also between the 1LO and 2LO resonances. One can see the resonances shifted

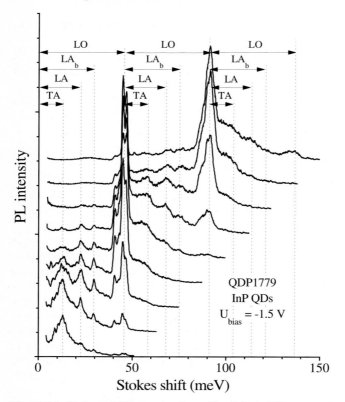

**Fig. 6.15.** Enlarged PL spectra from Fig. 6.5 at different excitation photon energies versus the Stokes shift. $U_{bias}=-1.5$ V. The spectra are shifted vertically for clarity. Similar phonon peaks of different spectra are marked by *vertical lines*. The *arrows labelled* TA, etc. indicate the energy shift of the resonances from the laser line or from the 1LO or 2LO resonances (from Ignatiev et al. [10])

from the laser line or from the 1LO resonance by the energy of the high-frequency TA or LA phonons of the InP crystal, or LA phonons of the barrier InGaP layer. For the InGaAs QDs, similar resonances are distinctly seen in Fig. 6.6.

The spectrum of the phonon resonances reflects the phonon density spectrum of the bulk crystal [9]. Acoustic phonons forming the phonon-density peaks in the bulk crystal are characterized by large wave vectors, i.e., their wave functions show a high spatial oscillation frequency. Contrary to phonons, the envelope wave functions for electrons and holes in the QDs are considered to be smooth, i.e., to have a low spatial oscillation frequency. For this reason, the interaction of the e–h pair with such phonons is considered to be negligible [15–17]. However, the spectra presented in Figs. 6.6 and 6.15 give clear evidence that high-energy acoustic phonons can efficiently interact with the e–h pairs.

Further evidence that the fast relaxation occurs via acoustic phonon emission is given by the kinetic data. As discussed in Sect. 6.5, analysis of the PL kinetics allows one to determine absolute values of the relaxation rates. For the QDs under discussion, spectral dependencies of the relaxation rates were obtained by measuring the PL kinetics in different spectral points starting from the close vicinity of the laser line. Special measures were taken to strongly suppress the scattered light. The amplitude of the signal from the scattered laser light was checked by measuring the signal with the PL completely eliminated by a strong negative bias. In the case of the InP QDs, the amplitude was as small as 5% of the PL amplitude at the Stokes shift $\Delta E = 2.5$ meV and rapidly decreased with increasing Stokes shift. In the case of the InGaAs QDs, the PL intensity and the sensitivity of the setup were much smaller; hence the smallest Stokes shift at which the PL kinetics could be reliably detected was about 8 meV. To avoid the Auger-like processes due to charging of the QDs (see Sect. 6.7) as well as phonon-assisted absorption [47], the PL kinetics was measured for samples with a negative bias ($U_{bias} = -0.8$ V for the sample with the InP QDs and $U_{bias} = -2.0$ V for the sample with the InGaAs QDs). A deconvolution procedure that takes into account the real profile of the laser pulse allowed one to enhance the time resolution down to 2 ps and 8 ps for the InP and InGaAs QDs, respectively.

The PL kinetics was analyzed using (6.8), and also (6.11) to model the two-exponential PL decay when necessary. By this analysis, the average PL rise rate was determined. It should be emphasized that this rate is the total rate of *depletion* of the excited state. This rate is controlled not only by relaxation but also by the carrier tunneling rate and by the rate of radiative relaxation of the e–h pair from the excited state. One should expect, however, that the two latter rates vary smoothly with the energy and cannot produce spectral singularities. For the biases used in the experiments, the tunneling rate is $\gamma_{t1} \approx 1/300$ ps$^{-1}$ for the InP QDs and is negligibly small for the InGaAs QDs. The radiative recombination rates are of the order of 1 ns$^{-1}$. These values are small compared with the relaxation rate and are further neglected.

The determined PL rise rates are plotted versus the Stokes shift in Fig. 6.16 for the InP QDs and in Fig. 6.17 for the InGaAs QDs. Examples of the PL kinetics for selected spectral points are also shown in these figures.

The presented spectral dependencies of the relaxation rates clearly show resonant features at Stokes shifts corresponding to energies of the high-frequency TA and LA phonons. The relaxation time with emission of a single TA or LA phonon is about 50 ps for the InP and InGaAs QDs. The carrier relaxation with emission of two acoustic phonons (spectral point 2AC in Fig. 6.16) takes less than 100 ps. Spectral dependence of the relaxation rates resembles the PL spectrum of the sample at a sufficiently strong bias.

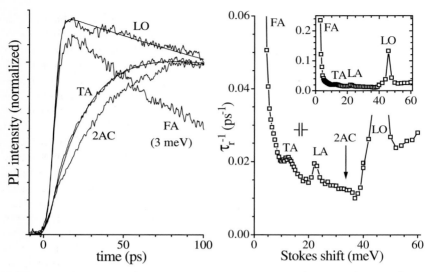

**Fig. 6.16.** The PL kinetics for selected spectral points (*left panel*) and the PL rise rate $\tau_r^{-1}$ versus Stokes shift (*right panel*) for a sample with InP QDs. The spectral features discussed in the text are marked (redrawn from Davydov et al. [47])

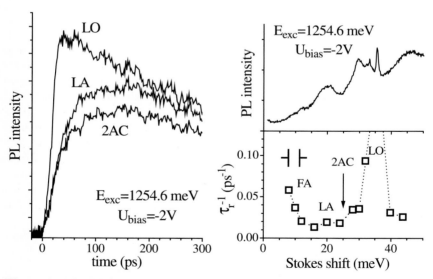

**Fig. 6.17.** The PL kinetics for selected spectral points (*left panel*) and the PL rise rate $\tau_r^{-1}$ versus Stokes shift (*lower curve in right panel*) for a sample with InGaAs QDs. The *upper curve* in the right panel is the PL spectrum versus Stokes shift at $U_{bias} = -2$ V

One more interesting feature, marked by FA in Figs. 6.16 and 6.17, is present in the relaxation rate spectra. One can assume that the fast relaxation observed at a small Stokes shift is due to interaction with low-energy LA phonons. So, this observation supports theoretical predictions about efficient carrier relaxation via emission of LA phonons with wavelengths close to the QD diameter [15]. The presented data show that the phonon-assisted relaxation for any interlevel spacing is considerably faster than the radiative recombination.

One more evidence for phonon-assisted relaxation is provided by the temperature dependence of the PL spectra of the biased sample. A few spectra of the InP QDs are shown in Fig. 6.18. It is clearly seen that the phonon resonances become smaller relative to the structureless pedestal of the spectrum at $T=100$ K. The temperature increase at first destroys the low-energy phonon resonance (TA), then the other acoustic resonances, and finally the LO resonance.

This behavior can be easily understood. Heating of the sample creates phonons. These phonons can cause carrier relaxation due to stimulated emission of phonons into the same phonon modes. The number of phonons with

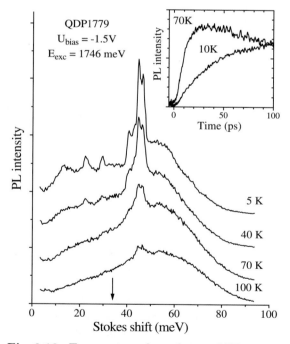

**Fig. 6.18.** Temperature dependence of PL spectra of a biased sample with InP QDs. The excitation photon energy is $E_{exc}=1746$ meV and the electric bias is $U_{bias}=-1.5$ V. The temperature is indicated for each spectrum. (from Ignatiev et al. [10])

small energy is always larger than that with higher energy. Therefore, the stimulated emission is more efficient for the low-energy phonons. Temperature dependence of the phonon-assisted relaxation rate is described by a simple equation

$$\tau_r(T) = \tau_{r0}(n_B + 1) , \tag{6.13}$$

where $n_B = [\exp(E_{ph}/kT) - 1]^{-1}$ is the Bose distribution function for the phonons of energy $E_{ph}$. A considerable change of the PL spectra occurs when the temperature increases up to several tens of Kelvins (see Fig. 6.18). This indicates that the effective energy of the phonons responsible for the thermo-stimulated acceleration of the relaxation lies in the range of several meV. These are probably the long-wavelength LA phonons mentioned earlier in this section. A similar result was obtained by Heitz et al. [32].

An important result for device applications is that, at a temperature of about 100 K, the relaxation rate observed experimentally becomes comparable with that via emission of the LO phonons independent of the spectral point. Thus, one can conclude that no retarded relaxation of the hot carriers in QDs should be expected at elevated temperatures.

At present, no quantitative description of the interaction between the e–h pairs and high-frequency acoustic phonons is available. However, a few points important for future modeling can be noted. First of all, there are no grounds to suggest that the acoustic phonons can be essentially localized in the QDs under consideration, because the phonon bands of the materials of the QDs and the barrier layers overlap fairly well, with their elastic properties being nearly identical. For this reason, the acoustic phonons can be considered, to a good approximation, as freely propagating through the QDs and characterized by a certain wave vector.

The high-energy acoustic phonons have a large momentum near the Brillouin zone boundary. Interaction of carriers with phonons of large momentum is efficient if the carrier eigenstate is characterized by the same momentum. The eigenstates of carriers in QDs are formed by mixing of the bulk states due to spatial confinement. In other words, the wave function of a quantum-confined state in a QD is a linear combination of wave functions of states in the Brillouin zone. To explain the observed resonances, it should be assumed that this linear combination contains a considerable contribution from states with large momentum, such as, for example, the states of the $X$ or $L$ valleys.

This assumption is supported by the fact that the energy shift of the lowest electronic state in the InP and InGaAs QDs due to confinement is about a few hundreds of meV. This value is comparable with the energy separations between the minima of the $\Gamma$ and $L$ valleys in these materials [63]. It should be also taken into account that the effective mass of an electron in the valleys near the Brillouin zone boundary is large; hence the energy shift of the states due to confinement is much smaller than that at the $\Gamma$ point. As a result, the energy separation between the states of the $\Gamma$ and $L$ valleys decreases with decreasing size of the QD. That may be the reason why the mixing of these states in the QDs is not small. A significant decrease of the energy separation between the $\Gamma$ and $L$ minima in InP QDs was observed by Menoni et al. [66]. This decrease can also be caused by strain due to lattice mismatch of the QDs and the surrounding material. Observation of the $\Gamma$–$X$ crossover under high pressure in InP QDs [67] and in InAs QDs [68] supports this assumption.

To model this situation, it is necessary to go beyond the widely used effective mass approximation, in the framework of which the QD eigenstates are constructed from the Bloch functions in the neighborhood of the $\Gamma$-point. In fact, the empirical pseudopotential calculations have already shown the importance of the $\Gamma$–$X$–$L$ valley mixing in QDs (see, e.g., [69]). Similar calculations for the structures discussed here would be of interest.

Some experimental evidence for this assumption seems to be provided by the spectra shown in Fig. 6.3 for the case of strong bias. A part of the spectrum for $U_{\text{bias}}$=2 V is enlarged in Fig. 6.19. It is clearly seen that the TA, LA, and LO+TA resonances look like narrow peaks with a FWHM of about 2–3 meV. Formation of such narrow peaks is possible only if a small portion of the acoustic phonons causes a sufficiently strong relaxation. According to the

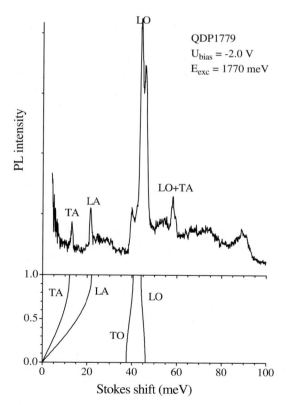

**Fig. 6.19.** Enlarged PL spectrum from Fig. 6.3 at $U_{\text{bias}}$=$-2$ V versus Stokes shift. The *marked narrow peaks* are presumably caused by interaction with high-momentum phonons. *Lower part* of the figure shows schematically the dispersion curves for the phonons of the InP crystal (redrawn from Ignatiev et al. [10])

above assumption, selection of these acoustic phonons may result from the wavenumber selection rule. It should be also taken into account that phonons with wavenumbers near the Brillouin zone boundary show small dispersion.

The interaction with high-energy acoustic phonons may be enhanced for two more reasons. First, a small group velocity of the phonons lengthens the interaction time. Second, the piezoelectric effect in these crystals gives rise to long-range electric fields associated with acoustic phonons. Modeling of these interactions at large wave vectors $q$ of the phonons remains an open problem.

## 6.7   Auger-Like Processes

The phonon-assisted relaxation can be observed only under specific experimental conditions. Carrier–carrier scattering, usually referred to as the Auger process, can considerably accelerate the relaxation [18,19,45]. Relaxation due to the Auger process must be fast because the carriers as charged particles interact with each other much more strongly than with phonons. A few particular Auger processes discussed in this section are illustrated schematically in Fig. 6.20. This figure shows a conventional energy diagram for electrons and holes separately. Such representation is illustrative and convenient for qualitative treatment of processes occurring with the participation of several carriers.

For simplicity, optical excitation is assumed to create an electron in the excited state and a hole in the ground state. This assumption can be partially justified by the low symmetry of the QDs because the standard selection rules for optical transitions are broken in this case. In the presence of an electric field, the indirect transitions ($|0h\rangle{\rightarrow}|1e\rangle$ or $|1h\rangle{\rightarrow}|0e\rangle$) become more allowed [47]. Exactly the same analysis can be performed for the case when

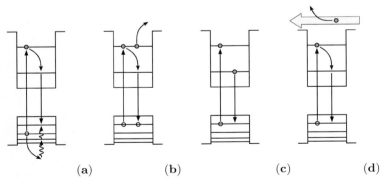

$$\text{(a)}\qquad\qquad\text{(b)}\qquad\qquad\text{(c)}\qquad\qquad\text{(d)}$$

**Fig. 6.20.** Various Auger-like processes providing fast carrier relaxation: (**a**) electron–hole scattering; (**b**) carrier–carrier scattering in the presence of two e–h pairs; (**c**) relaxation in a charged QD; (**d**) electric–current-induced relaxation (from Masumoto et al. [48])

the electron is created in the ground state, while the hole is in the excited state. In reality, due to fast energy exchange between the electron and hole (see Sect. 6.1), it is difficult to distinguish between these two situations.

The simplest process of the Auger-like relaxation is the electron–hole scattering shown schematically in Fig. 6.20a. In this process, excess energy of a hot electron is transferred to a hole. As a result, the electron goes down to its ground state and the hole is excited up to a higher-energy level. Then, the hole relaxes with emission of phonons. This process was extensively discussed in [19,45,46]. It should be noted, however, that the process of energy exchange between an electron and a hole, by itself, does not lead to energy relaxation in the e–h pair. The relaxation is brought about only by phonon emission, and therefore this process is indistinguishable from the process considered in previous sections.

As was mentioned in Sect. 6.1, the energy exchange between an electron and a hole is assumed to be very fast, with characteristic times of fractions of picoseconds, provided that the distance between the hole energy levels $\Delta E_{\mathrm{h}}$ coincides with that between the electron levels $\Delta E_{\mathrm{e}}$. Otherwise the energy transfer between the two carriers should be accompanied by emission of a phonon with energy $\hbar\omega=|\Delta E_{\mathrm{e}}-\Delta E_{\mathrm{h}}|$. This is evidently just a relaxation of the e–h pair with relaxation times dependent on the type of the emitted phonon, as discussed previously. Variations in the electron and hole energies in such a process can be much larger than the variation of the e–h pair energy. However, nothing definite can be said about the energy state of each separate carrier from optical experiments.

The next process (Fig. 6.20b) occurs when the excitation creates more than one e–h pair. This process is studied experimentally using optical excitation by either a cw or a pulsed laser source. In the first case, a short-wavelength excitation in the range of the barrier layer absorption is used. Then, many free electrons and holes are generated in the barrier that are randomly captured by QDs. Under such excitation, the number of electrons in a QD can evidently differ from the number of holes. For this reason, "purer" experimental conditions can be created using quasiresonant excitation of a pulsed laser. Despite the small magnitude of the resonant absorption coefficient ($\sim 10^{-4}$), a greater peak power density allows one to produce several e–h pairs in a QD.

Figure 6.21 shows PL spectra of a biased sample with InP QDs measured under quasiresonant excitation. Pulses of 5 ps with a repetition period of about 12 ns were used to increase the peak power density of the excitation. As seen, the acoustic phonon resonances disappear, and the LO resonance becomes barely observable under strong pumping. This fact means that the relaxation rate for all interlevel spacings becomes comparable with the LO phonon-assisted relaxation rate. The PL kinetics reveals shortening of the PL rise time with increasing pump power (see inset in Fig. 6.21), which also indicates acceleration of the carrier relaxation.

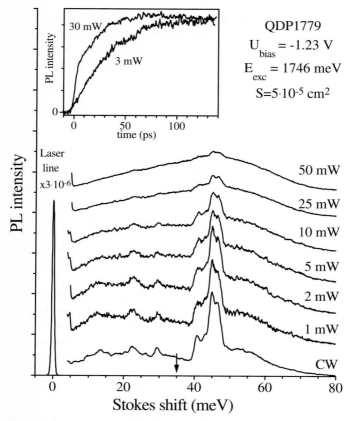

**Fig. 6.21.** PL spectra of a biased sample with InP QDs under cw excitation (*bottom curve*) and pulsed excitation with power indicated for each curve. The laser spot area on the sample is $S=5\times10^{-5}$ cm$^2$, pump photon energy is $E_{exc}=1746$ meV, and electric bias is $U_{bias}=-1.23$ V. The *arrow* indicates the spectral point where the PL kinetics was measured. *Inset*: initial part of the PL kinetics at different excitation power indicated for each curve (from Ignatiev et al. [10])

The observed behavior of the PL spectra and kinetics can be explained by invoking the Auger-like process schematically shown in Fig. 6.20b, where one carrier loses its energy and another acquires it [18,19,26,45,70]. This process, as in the previous case, is not phonon-assisted and is resonantly enhanced when separation between the energy levels for the first carrier coincides with that for the second one. In the case of the InP QDs, this condition can be easily satisfied for electron–hole and hole–hole scattering because of the large density of high-energy levels for holes. Also, the hole can be ejected out of the QD into the continuum of the barrier layer, so that the energy conservation law is easily satisfied. In addition, one can expect that the multi-particle scattering, accompanied by a change of the energy of three or more carriers, should be also efficient in QDs due to localization of carriers in a small volume [45]. Multi-particle scattering allows one to satisfy the energy conservation law more easily due to a great number of combinations of different transitions of the carriers.

Because of these processes, one electron and one hole may be found in their ground states very shortly after their creation. For this reason, the leading edge of the PL pulse should be extremely short. Because of the inhomoheneity of the QDs, no spectral features (resonances) should be expected due to relaxation via these processes. It is also clear that the efficiency of these processes should depend on the number of carriers in the QD, because each carrier can be scattered by any other one.

The described features of the relaxation process agree qualitatively with the experiment. Therefore, one can conclude that strong pumping really activates an alternative relaxation mechanism that is much faster than the acoustic-phonon-assisted relaxation. The relaxation rate achieved under experimental conditions of Fig. 6.21 is comparable with the rate of the LO-phonon-assisted relaxation, i.e., is of the order of $1$ ps$^{-1}$.

A quantitative analysis of the data presented in Fig. 6.21 is difficult because the absorption coefficient of the InP QDs at the laser line wavelength is unknown. The number of photogenerated e–h pairs in an InP QD $N_c$ can be roughly estimated using the Fermi golden rule for optical transitions between the discrete energy states of the QD, assuming that the spectral width of the laser pulse is much larger than that of the optical transition, but smaller than the spectral distance between the nearest optical transitions. The dipole moment of the optical transition can be estimated from the radiative lifetime of the e–h pair excited state. This lifetime, to a good approximation, coincides with the radiative lifetime of the lowest state of the e–h pair and can be easily determined experimentally. The estimate gives $N_c=8$ for the top curve in Fig. 6.21. Due to saturation of the absorption, the actual number of the carriers must be smaller.

The dependence of the PL intensity on pump power density for the data shown in Fig. 6.21 is more complicated than quadratic. This is probably related to some additional processes at strong pumping, such as saturation of absorption, which should lead to a sublinear dependence, or saturation of deep trap levels, which should lead to a superlinear dependence.

The third process shown in Fig. 6.20c is related to charging the QDs. It should be noted that a QD is a potential well for electrons and holes. For this reason, free carriers that are present in the heterostructure can be captured by QDs. The possibility of such a process can be vividly illustrated by the experiment whose results are presented in Fig. 6.22.

The sample with the InGaAs QDs was excited by two lasers as shown on the diagrams of Fig. 6.22. The quasiresonant excitation of the pulsed laser was used to generate e–h pairs directly in the QDs. Excitation of the cw

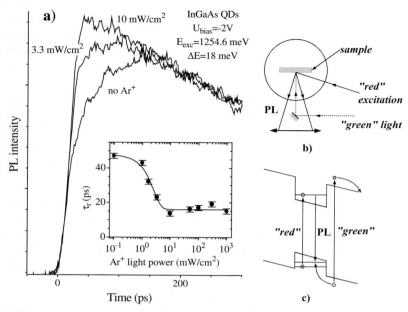

**Fig. 6.22.** (a) PL kinetics of InGaAs QDs in the presence of "green" cw excitation with power density $P_g$ indicated for each curve. The dc component of the PL (pedestal) excited by the "green" pump is subtracted. $\Delta E=18$ meV, $U_{bias}=-2$ V. In the *inset* is shown a dependence of the PL rise time $\tau_{PL}$ on $P_g$. (**b**) Configuration of excitation of the sample by two lasers. (**c**) A simplified diagram of the processes in the heterostructure in the presence of pulsed quasiresonant (red) and cw nonresonant (green) optical excitation and external electric field

Ar$^+$ laser was used to generate free electrons and holes in the barrier InGaP layer. Kinetics of the PL excited by the pulsed laser was detected at different densities of the cw "green" pump produced by the Ar$^+$ laser. In addition, the sample was subjected to an external electric field to remove excess charges from the QDs.

Figure 6.22a shows the PL kinetics at the Stokes shift $\Delta E=18$ meV, corresponding to the relaxation of the e–h pair with emission of a LA phonon. In the absence of the "green" pump, the PL rise kinetics is relatively slow, with the effective rise time 50 ps, characteristic of the acoustic-phonon-assisted relaxation. However, a weak "green" pump with the power density in the range of few mW/cm$^2$ results in dramatic shortening of the PL rise time, as shown in the inset of Fig. 6.22, so that this time becomes shorter than the time resolution of the setup.

The results of this experiment can be explained as follows. The carriers generated by the "green" pump are captured, in a random fashion, by QDs. The e–h pairs formed in the QDs recombine, producing continuous emission of the sample (pedestal). Since this process is random, it may happen that one or a few excess carriers of the same sign are present in a QD at a time. In other words, the QD may appear to be charged. In the presence of excess carriers, an e–h pair generated by the quasiresonant pulse can relax very fast by carrier–carrier scattering without emission of phonons. This process should evidently result in shortening of the PL rise time. So one can assume that the experimental data shown in Fig. 6.22 is evidence of charging of QDs.

The excess charges may appear in a QD without any optical excitation. There is always some amount of impurities in the heterostructure, giving rise to shallow donor or acceptor levels. Under elevated temperatures of the sample, e.g., at 300 K, the carriers are ejected from these levels into the conduction band, become free, and are captured by QDs. The charged QDs have been discussed in [57,71,72]. The type and number of the excess carriers in the QD depend on the type and amount of impurities in the heterostructure and also on whether the electric field (built-in or external) is present or not.

The number of excess carriers can be changed using the external electric field. This effect is illustrated in Fig. 6.23, which shows PL kinetics of the InP QDs measured at various biases at the wavelength of the LA phonon resonance. The PL rise kinetics is seen to vary from very fast (step-like) at positive voltages, to rather slow at $U_{\mathrm{bias}}=-0.8$ V with times typical for the acoustic-phonon-assisted relaxation. The fast PL rise is related to the

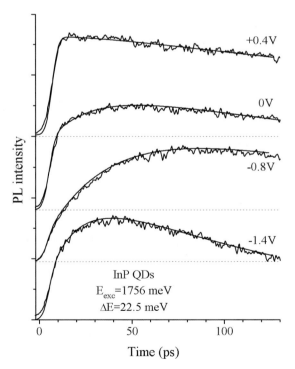

**Fig. 6.23.** PL kinetics of InP QDs at LA resonance under various applied biases indicated for each curve (*noisy curves*). *Smooth curves* are fitted by (6.14) (from Masumoto et al. [48])

presence of excess carriers in the QDs. These are presumably holes [43]. The holes are generated in the barrier InGaP layer in the vicinity of the QDs by optical excitation, which gives rise to electron transitions from the valence band to deep levels. The photocreated holes are then captured by the QDs.

Under negative voltage, the holes tunnel out of the QDs [10]. This results in discharging of the QDs, and therefore in slowing down the relaxation of an e–h pair. Therefore $U_{\text{bias}}=-0.8$ V at which the PL rise time for the InP QDs is close to maximum gives rise to neutral QDs. Further increases of the negative bias lead to further shortening of the PL rise kinetics, which is caused by the effect discussed below.

The PL kinetics shown in Fig. 6.23 can be approximated by the equation

$$I(t) = I_0 \left[ e^{-\gamma_{\text{PL}}t} - \left( \alpha e^{-\gamma_{\text{A}}t} + (1-\alpha)e^{-\gamma_{\text{ph}}t} \right) \right] , \qquad (6.14)$$

which includes a fast component of the PL rise with the Auger relaxation rate $\gamma_{\text{A}}$ in some fraction $\alpha$ of the QDs, and a slow rise component characterized by the phonon-assisted relaxation rate $\gamma_{\text{ph}}$ in the other QDs. The fitting curves were convoluted with a Gaussian with the halfwidth 5 ps, modeling the instrumental function of the setup. The rate $\gamma_{\text{A}}$ was chosen equal to 1 ps$^{-1}$, i.e., the Auger process was considered to be instantaneous in the time scale under consideration.

Dependence of the fraction $\alpha$ on $U_{\text{bias}}$ is shown in Fig. 6.24. The fast PL rise component is seen to increase with increasing positive voltage. Therefore,

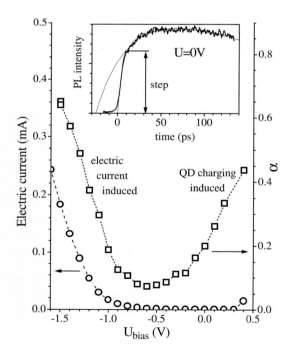

Fig. 6.24. Bias dependence of the step in the PL rise at LA resonance with Stokes shift $\Delta E=22.5$ meV (*squares*) and of the electric current (*circles*) for a sample with InP QDs. *Inset*: PL kinetics at zero bias (*noisy curve*). A *solid line* is the fit of the slow PL rise by (6.12) with $\alpha=0$. The step-like PL rise is clearly seen (from Masumoto et al. [48])

as the voltage increases, the fraction of the charged dots increases, as well as the number of excess carriers in each QD. As shown in [65], in the QDs under study, on average one excess hole is contained in each dot at $\approx -0.2$ V.

A similar effect is observed in the InGaAs QDs as shown in Fig. 6.25. The time resolution of the setup, in this case, was not enough to distinguish between the slow and the fast components. Therefore the effective PL rise rate was determined by fitting the PL kinetics with (6.14) with $\alpha = 0$. One can see that the PL rise rate sharply increases at positive voltage, indicating charging of the QDs. The sign of the excess carriers for these QDs is not known.

The last Auger process that is considered in this chapter is the electric-current-induced relaxation of e–h pairs shown schematically in Fig. 6.20d. Flowing carriers can interact with the carriers inside the QDs and cause their relaxation. This process is illustrated by the data in Figs. 6.24 and 6.25 which were obtained under sufficiently strong negative bias when no other Auger process was involved. As is seen, the increase of the negative voltage results in a drastic increase of electric current through the sample, which is accompanied by shortening of the effective PL rise time. PL kinetics of the InP QDs in this case is also well fitted by (6.14) with the step-like component of the PL rise. Quantitative description of this process is an unresolved problem that needs simulation of the carrier motion through the layer of QDs. One

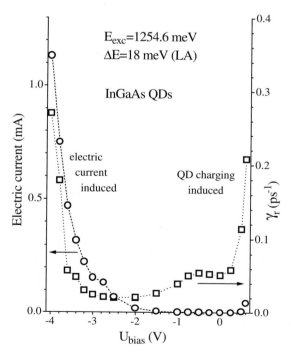

**Fig. 6.25.** Bias dependence of the PL rise rate at LA resonance with Stokes shift $\Delta E = 18$ meV (*squares*) and of the electric current (*circles*) for sample with InGaAs QDs

can only assert that the cross section of the interaction between the moving carriers and the carriers in the QDs is not small. This is supported also by the data of [43].

## 6.8    Conclusion

In this chapter, possible mechanisms of relaxation of the e–h holes in InP and InGaAs self-assembled QDs have been considered in detail. The discussion was restricted to the last stage of relaxation from an excited level of the e–h pair to its lowest state. For this purpose, there has been thorough analysis of the experimental data obtained upon generation of e–h pairs directly inside QDs using quasiresonant optical excitation.

The experimental data discussed in this section were obtained mainly by two methods. The first is based on activation of the process of nonradiative losses of excitation in the QD by controlling the process by means of an external electric field. In this process, the tunneling of the carriers from the excited states competes with their relaxation to the lowest states, which provides a sort of time gate for the PL. As a result, the spectral dependence of the relaxation rate can be measured under steady state conditions. The second method is based on direct measurement of the PL rise kinetics under pulsed excitation. In the absence of nonradiative losses, the PL rise time is determined by the rate of the e–h pair relaxation from the excited state.

The above experimental data show fairly convincingly that the e–h pair relaxation in the InP and InGaAs QDs is fast. The relaxation times are shorter than 100 ps for any energy distance $\Delta E$ between the excited and lowest states of the e–h pair. This is a few orders of magnitude smaller than the value predicted theoretically.

There are many different relaxation mechanisms of the e–h pairs. Predomination of one or another mechanism depends on the sample structure and experimental conditions. The slowest relaxation is provided by the process assisted by spontaneous phonon emission. This process can be observed only under special conditions. The main conditions are: (i) low level of excitation, so each QD has no more than one e–h pair at a time, (ii) sufficiently low temperature of the sample, so the thermal energy $kT$ is smaller than the energy of phonons involved into the relaxation, and (iii) the absence of both excess carriers in the QDs and electric current through the sample.

As seen from the data, the phonon-assisted relaxation of e–h pairs is efficient at any $\Delta E$. The relaxation times with emission of LO phonons lie in the range of units to fractions of picoseconds. The energy interval where this relaxation occurs is about 5 meV around $\Delta E=45$ meV for the InP QDs and about 10 meV around $\Delta E=35$ meV for the InGaAs QDs. In the range of small $\Delta E$ of about units of meV, the e–h pairs relax with emission of long-wavelength LA phonons, i.e., of phonons with the wavelength of the order of the base diameter of the QDs. This relaxation time is also of the order of units to fractions of picoseconds.

The presented experimental data provide convincing evidence that the e–h pair relaxation with emission of high-frequency acoustic phonons is also efficient. These results strongly contradict the present-day theoretical views. The energy interval where this relaxation occurs extends to the edge of the acoustic phonon spectrum, approximately equal to 25 meV for both the InP and InGaAs QDs. These relaxation times are of the order of tens of ps. The relaxation times with emission of two acoustic phonons are approximately twice as slow as those with emission of a single acoustic phonon.

The problem of theoretical description of the relaxation of the e–h pairs with emission of high-frequency acoustic phonons remains at present unsettled. In this chapter, we have discussed only a few reasons why this relaxation proves to be efficient. One of the most important reasons is likely to be the $\Gamma$–$X$–$L$ valley mixing, which is essential for the electron and hole states in QDs. This mixing can be caused by strong localization of carriers in the small volume of the QD (spatial confinement) and by internal stresses due to the mismatch between the crystal lattices of the QD and its environment. The $\Gamma$–$X$–$L$ mixing enhances the efficiency of interaction between the localized e–h pairs and high-frequency acoustic phonons with large wave vectors. The second important reason is the overall increase of electron–phonon coupling in QDs.

When studying phonon-assisted relaxation, an e–h pair should be considered as a single correlated system. This is indicated by the narrow phonon resonances observed experimentally. This is because of a high rate of energy exchange between the electrons and holes, which, as charged particles, interact with each other fairly strongly. This rate is much higher than that of the phonon-assisted relaxation, which is determined by the value of the carrier–phonon interaction. The system of the e–h pair energy levels is a combination of energy levels of electrons and holes and, for this reason, is denser than the energy level system of each of the carriers. In particular, the discrete energy spectrum of the e–h pair starts to overlap its continuous spectrum at a distance of $\Delta E \approx 15$ meV and $\Delta E \approx 50$ meV from the lowest level for the InP and InGaAs QDs, respectively. The dense energy spectrum of the e–h pair allows many more channels for phonon-assisted relaxation than in the case of relaxation of separate carriers with a sparser spectrum.

The above data taken as a whole allow one to conclude that *the bottleneck effect does not play any role* in the QDs under consideration. The phonon-assisted relaxation rate is at least a few times faster than the radiative recombination rate of the e–h pair from the excited state. It should be emphasized, however, that this conclusion cannot be considered as universal. For "large" QDs with lateral dimensions of the order of the light wavelength, the radiative recombination time can approach the time typical for the quantum wells and be equal to tens of picoseconds. At the same time, the phonon-assisted relaxation of the e–h pair in "large" QDs can be retarded due to a decrease in the electron–phonon interaction.

The increase in the sample temperature gives rise to a sharp acceleration of the phonon-assisted relaxation. This acceleration is caused by the stimulated emission of phonons due to the presence of thermal phonons in the same phonon modes. At temperatures above 100 K, the relaxation rate of e–h pairs with emission of acoustic phonons becomes of the same order as the LO-phonon-assisted relaxation rate. Thus, in QD-based devices operating at room temperature, there is no reason to expect any problems related to the finite relaxation time of the e–h pairs, because this time always lies in the range of units to fractions of picoseconds.

In this chapter we have also considered the Auger-like processes, which occur when several carriers interact with each other. In these processes, due to carrier–carrier scattering one of the carriers loses its energy, while the other acquires it. As a result, the carriers may relax very quickly to the lowest states without emission of phonons. Relaxation via Auger-like processes is fast because the carriers as charged particles interact with each other much more strongly than with phonons.

The simplest process of carrier–carrier scattering, most frequently treated in the literature, is the process of energy exchange between an electron and a hole in the e–h pair. This process, however, does not lead to relaxation of the e–h pair provided that no phonon is emitted. True irreversible relaxation via the Auger-like process occurs in the case when the e–h pair interacts with excess carriers in the QD or in its vicinity, or when a few e–h pairs are generated. In all these cases, the relaxation times are shortened to units or fractions of picoseconds. It should be emphasized that the presence of excess charges in the QD or barrier layer, an electric current flowing through the heterostructure, and strong excitation of the QDs – all these are usual conditions for operation of QD-based devices. For this reason, the carrier relaxation in these devices should be essentially affected by Auger-like processes.

# References

1. Y. Arakawa, H. Sakaki: Appl. Phys. Lett. **40**, 939 (1982)
2. D. Bimberg, M. Grundmann, N.N. Ledenstov: *Quantum Dot Heterostructures* (Wiley, New York 1999)
3. M. Sugawara: *Self-Assembled InGaAs/GaAs Quantum Dots: Semiconductors and Semimetals* **60** (Academic Press, London 1999)
4. A. Weber, K. Goede, M. Grundmann, V.M. Ustinov, A.E. Zhukov, N.N. Ledentsov, P.S. Kop'ev, Z.I. Alferov: Phys. Status Solidi B **224**, 833 (2001)
5. B. Ohnesorge, M. Albrecht, J. Oshinowo, A. Forchel, Y. Arakawa: Phys. Rev. B **54**, 11532 (1996)
6. S. Raymond, K. Hinzer, S. Fafard, J.L. Merz: Phys. Rev. B **61**, R16331 (2000)
7. M. Vollmer, E.J. Mayer, W.W. Rühle, A. Kurtenbach, K. Eberl: Phys. Rev. B **54**, 17292 (1996)

8.  I.E. Kozin, I.V. Ignatiev, H.-W. Ren, S. Sugou, Y. Masumoto: J. Lumin. **87–89**, 441 (2000)
9.  I.V. Ignatiev, I.E. Kozin, S.V. Nair, H.-W. Ren, S. Sugou, Y. Masumoto: Phys. Rev. B **61**, 15633 (2000)
10. I.V. Ignatiev, I.E. Kozin, V.G. Davydov, S.V. Nair, J.-S. Lee, H.-W. Ren, S. Sugou, Y. Masumoto: Phys. Rev. B **63**, 075316 (2001)
11. S. Sauvage, P. Boucaud, T. Brunhes, A. Lemaître, J.M. Gérard: Phys. Rev. B **60**, 15589 (1999)
12. D. Wassermann, S.A. Lyon: Phys. Status Solidi B **224**, 585 (2001)
13. S.A. Permogorov: Phys. Status Solidi B **68**, 9 (1975)
14. V.F. Gandmacher, Y.B. Levinson: *Carrier Scattering in Metals and Semiconductors* (North-Holland, Amsterdam 1987)
15. U. Bockelmann, G. Bastard: Phys. Rev. B **42**, 8947 (1990)
16. T. Inoshita, H. Sakaki: Phys. Rev. B **46**, 7260 (1992)
17. H. Benisty, C.M. Sotomayor-Torrès, C. Weisbuch: Phys. Rev. B **44**, 10945 (1991)
18. U. Bockelmann, T. Egeler: Phys. Rev. B **46**, 15574 (1992)
19. A.L. Efros, V.A. Kharchenko, M. Rosen: Solid State Commun. **93**, 281 (1995)
20. T.S. Sosnowski, T.B. Norris, H. Jiang, J. Singh, K. Kamath, P. Bhattacharya: Phys. Rev. B **57**, R9423 (1998)
21. V.I. Klimov, A.A. Mikhailovsky, D.W. McBranch, C.A. Leathedrale, M.G. Bawendi: Phys. Rev. B **61**, R13349 (2000)
22. F. Adler, M. Geiger, A. Bauknecht, F. Scholz, H. Schweizer, M.H. Pilkuhn, B. Ohnesorge, A. Forchel: J. Appl. Phys. **80**, 4019 (1996)
23. S. Grosse, J.H. H. Sandmann, G. von Plessen, J. Feldmann, H. Lipsanen, M. Sopanen, J. Tulkki, J. Ahopelto: Phys. Rev. B **55**, 4473 (1997)
24. R. Heitz, M. Veit, N.N. Ledentsov, A. Hoffmann, D. Bimberg, V.M. Ustinov, P.S. Kop'ev, Z.I. Alferov: Phys. Rev. B **56**, 10435 (1997)
25. T. Okuno, H.-W. Ren, M. Sugisaki, K. Nishi, S. Sugou, Y. Masumoto: Phys. Rev. B **57**, 1386 (1998)
26. M. Braskén, M. Lindberg, M. Sopanen, H. Lipsanen, J. Tulkki: Phys. Rev. B **58**, R15993 (1998)
27. S. Marsinkevičius, R. Leon: Phys. Rev. B **59**, 4630 (1999)
28. D. Morris, N. Perret, S. Fafard: Appl. Phys. Lett **75**, 3593 (1999)
29. C. Lobo, N. Perret, D. Morris, J. Zou, D.J.H. Cockayne, M.B. Johnston, M. Gal, R. Leon: Phys. Rev. B **62**, 2737 (2000)
30. There are a few papers (see [22] and also F. Adler et al.: J. Appl. Phys **83**, 1631 (1998), H. Born et al.: Phys. Status Solidi B **224**, 487 (2001), where a relatively long PL rise time in a few hundred picoseconds were reported. Each particular case of this slow PL rise should be carefully investigated before concluding about a slow carrier relaxation
31. S. Fafard, R. Leon, D. Leonard, J.L. Merz, P.M. Petroff: Phys. Rev. B **52**, 5752 (1995)
32. R. Heitz, M. Grundmann, N.N. Ledentsov, L. Eckey, M. Veit, D. Bimberg, V.M. Ustinov, A.Y. Egorov, A.E. Zhukov, P.S. Kop'ev, Z.I. Alferov: Appl. Phys. Lett. **68**, 361 (1996)
33. K.H. Schmidt, G. Medeiros-Ribeiro, M. Oestreich, P.M. Petroff, G.H. Döhler: Phys. Rev. B **54**, 11346 (1996)

34. M. Grundmann, R. Heitz, N. Ledentsov, O. Stier, D. Bimberg, V.M. Ustinov, P.S. Kop'ev, Z.I. Alferov, S.S. Ruvimov, P. Werner, U. Gosele, J. Heydenreich: Superlatt. Microstruct. **19**, 81 (1996)
35. C. Guasch, C.M. Sotomayor-Torrès, N.N. Ledentsov, D. Bimberg, V.M. Ustinov, P.S. Kop'ev: Superlatt. Microstruct. **21**, 509 (1997)
36. A.V. Baranov, V. Davydov, H.-W. Ren, S. Sugou, Y. Masumoto: J. Lumin. **87–89**, 503 (2000)
37. P.C. Sercel: Phys. Rev. B **51**, 14532 (1995)
38. D.F. Schroeter, D.J. Griffiths, P.C. Sercel: Phys. Rev. B **54**, 1486 (1996)
39. T. Inoshita, H. Sakaki: Phys. Rev. B **56**, R4355 (1997)
40. X.-Q. Li, Y. Arakawa: Phys. Rev. B **56**, 10423 (1997)
41. X.-Q. Li, H. Nakayama, Y. Arakawa: Phys. Rev. B **59**, 5069 (1999)
42. Y. Toda, O. Moriwaki, M. Nishioka, Y. Arakawa: Phys. Rev. Lett. **82**, 4114 (1999)
43. I. Ignatiev, I. Kozin, H.-W. Ren, S. Sugou, Y. Masumoto: Phys. Rev. B **60**, R14001 (1999)
44. A.V. Uskov, F. Adler, H. Schweizer, M.H. Pilkuhn: J. Appl. Phys. **81**, 7895 (1997)
45. S. Nair, Y. Masumoto: J. Lumin. **87–89**, 408 (2000); S. Nair, Y. Masumoto: Phys. Status Solidi B **178**, 303 (2000)
46. R. Ferreira, G. Bastard: Appl. Phys. Lett. . **74**, 2818 (1999); R. Ferreira, G. Bastard: Phys. Status Solidi A **178**, 327 (2000)
47. V. Davydov, I.V. Ignatiev, I.E. Kozin, S.V. Nair, J.-S. Lee, H.-W. Ren, S. Sugou, Y. Masumoto: Phys. Status Solidi B **224**, 493 (2001)
48. Y. Masumoto, I.V. Ignatiev, I.E. Kozin, V.G. Davydov, S.V. Nair, J.-S. Lee, H.-W. Ren, S. Sugou: Jpn. J. Appl. Phys. **40**, 1947 (2001)
49. C. Pryor, M-E. Pistol, L. Samuelson: Phys. Rev. B **56**, 10404 (1996)
50. M. Hayne, R. Provoost, M.K. Zundel, Y.M. Manz, K. Eberl, V.V. Moshchalkov: Phys. Rev. B **62**, 10324 (2000)
51. H.-W. Ren, M. Sugisaki, S. Sugou, K. Nishi, A. Gomyo, Y. Masumoto: Jpn. J. Appl. Phys. Part 1 **38**, 2438 (1999)
52. J.-S. Lee, K. Nishi, Y. Masumoto: J. Cryst. Growth **221**, 586 (2000)
53. E. Bedel, G. Landa, R. Charles, J.P. Redoulés, J.B. Renussi: J. Phys. C **19**, 1471 (1986)
54. A.A. Sirenko, M.K. Zundel, T. Ruf, K. Eberl, M. Cardona: Phys. Rev. B **58**, 12633 (1998)
55. R. Heitz, I. Mukhametzhanov, O. Stier, A. Madhukar, D. Bimberg: Phys. Rev. Lett. **83**, 4654 (1999)
56. M. Bissiri, G.B.H. von Högersthal, A.S. Bhatti, M. Capizzi, A. Frova, P. Frigeri, S. Franchi: Phys. Rev. B **62**, 4642 (2000)
57. F. Findeis, A. Zrenner, G. Bohm, G. Abstreiter: Phys. Rev. B **61**, R10579 (2000)
58. L. Zimin, S. Nair, Y. Masumoto: Phys. Rev. Lett. **80**, 3105 (1998)
59. P.W. Fry, I.E. Itskevich, D.J. Mowbray, M.S. Skolnick, J.J. Finley, J.A. Barker, E.P.O'Reilly, L.R.Wilson, L.A. Larkin, P.A. Maksym, M. Hopkinson, M. Al-Khafaji, J.P.R. David, A.G. Gullis, G. Hill, J.C. Clark: Phys. Rev. Lett. **84**, 733 (2000)
60. L.D. Landau, E.M. Lifshitz: *Quantum Mechanics, 3rd edn.* (Pergamon, Oxford 1977), Sect. 50

61. V. Davydov, I. Ignatiev, H.-W. Ren, S. Sugou, Y. Masumoto: Appl. Phys. Lett. **74**, 3002 (1999)
62. P.C. Sercel, A.L. Efros, M. Rosen: Phys. Rev. Lett. **83**, 2394 (1999)
63. *Numerical Data and Functional Relationships in Science and Technology, Landolt-Börnstein, Vol. 22a* (Springer-Verlag, Heidelberg, Berlin 1987), pp. 120, 141, 351
64. M. Sugawara: Phys. Rev. B **51**, 10743 (1995)
65. V. Davydov, A.V. Fedorov, I.V. Ignatiev, I.E. Kozin, H.-W. Ren, M. Sugisaki, S. Sugou, Y. Masumoto: Phys. Status Solidi B **224**, 425 (2001)
66. C.S. Menoni, L. Miao, D. Patel, O.I. Mic'ic', A.J. Nozik: Phys. Rev. Lett. **84**, 4168 (2000)
67. C. Ulrich, S. Ves, A.R. Goñi, A. Kurtenbach, K. Syassen, K. Eberl: Phys. Rev. B **52**, 12212 (1995)
68. I.E. Itskevich, M.S. Skolnick, D.J. Mowbray, I.A. Trojan, S.G. lyapin, L.R. Wilson, M.J. Steer, M. Hopkinson, L. Eaves, P.C. Main: Phys. Rev. B **60**, R2185 (1999)
69. H. Fu, A. Zunger: Phys. Rev. B **57**, R15064 (1998); L.W. Wang, J. Kim, A. Zunger: Phys. Rev. B **59**, 5678 (1999); L.W. Wang, A. Zunger: Phys. Rev. B **59**, 15806 (1999)
70. R. Ferreira, G. Bastard: Appl. Phys. Lett. **74**, 2818 (1999)
71. A. Hartmann, Y. Ducommun, E. Kapon, U. Hohenester, E. Molinary: Phys. Rev. Lett. **84**, 5648 (2000)
72. J.J. Finley, A. Lemaître, K.L. Schumacher, A.D. Ashmore, D.J. Mowbray, I. Itskevich, M.S. Skolnick, M. Hopkinson, T.F. Krauss: Phys. Status Solidi B, **224**, 373 (2001)

# 7 Resonant Two-Photon Spectroscopy of Quantum Dots

Alexander Baranov

## 7.1 Introduction

In the last decade, a variety of linear and non-linear optical spectroscopic methods has been used for studying the energy structure of electronic and vibrational elementary excitations and their interactions in semiconductor nanocrystals embedded in dielectric matrices, or quantum dots (QDs). Among the methods, the techniques of resonant two-photon spectroscopy, such as two-photon absorption (TPA), two-photon excited resonant luminescence (TPL), resonant hyper-Rayleigh scattering, or second-harmonic scattering (RSHS), and hyper-Raman scattering (RHRS) play a significant role. The corresponding selection rules (including those related to polarization) for electronic and phonon transitions differ from the selection rules in the case of one-photon excitation. Therefore, the above non-linear techniques provide information on the electronic and phonon states unobservable in resonant one-photon processes, including resonant Raman scattering (RRS) [1–4]. For this reason the methods of two-photon spectroscopy are complementary to the traditional one-photon spectroscopy. Because of the low efficiency of two-photon processes, the related experiments involve the complete resources of modern spectroscopic techniques. Nevertheless, the unique spectroscopic possibilities of RHRS and RSHS have been demonstrated in the last decade by the experimental studies of organic molecules [3], bulk solids [5], and semiconductor epitaxial layers [6]. A theoretical model for RHRS by LO phonons has been recently developed for polar semiconductors [4].

The RSHS and RHRS phenomena are three-photon processes wherein the interaction between a system and two incident photons of energy $E_L$ results in scattering of one photon with energy $2E_L$ (RSHS) or $2E_L - \hbar\Omega$ (RHRS), where $\hbar\Omega$ is the phonon energy. A strong resonant enhancement of the processes is observed when twice the energy of the incident photons $2E_L$ or the energy of scattered photons approach the energy of a certain electronic transition, $E_0$. This offers a possibility to analyze the electronic structure by using the RSHS and RHRS two-photon excitation spectra. Under the resonance condition, $2E_L = E_0$, the two-step TPL process, which consists of a real population of the resonant state via TPA followed by phononless (with energy $2E_L$) or phonon-assisted luminescence (with energy $2E_L - \hbar\Omega$), becomes very efficient. The resonant state can decay radiatively to the same states as in the RSHS and RHRS processes, forming a similar spectrum of secondary emission with the same resonant peculiarities. The relation between the RHRS and TPL integral intensities $(P)$ is analogous to that between the RRS and resonant luminescence intensities: their relative contribution to the total secondary emission under two-photon excitation is expressed as $P_{\mathrm{RHRS}}/P_{\mathrm{TPL}} = T_2^*/T_1$, where $T_2^*$ and $T_1$ are the pure phase relaxation time and the resonant state lifetime, respectively [3]. In some cases, the processes can be distinguished by spectroscopic measurements, since the widths of the lines in the RHRS and TPL spectra are different. In the former case the linewidth is defined by the widths of the final states and in the latter case by the widths of the lumines-cence transitions. Discrimination is also possible in time-resolved experiments since the RHRS or RSHS proceed only during the laser pulse excitation, while the TPL decays with the time $T_1$. Finally, the steady state spectra of RHRS and TPL supply information on the energy spectrum of electronic and phonon excitations and their interaction in the systems, whereas the time evolution of the spectra provides data on the dynamics of the excitations.

In the case of semiconductor QDs, three-dimensional confinement results in quantization of the translation motion of quasiparticles (electrons, holes, excitons, and phonons) inside the nanocrystal, thus forming unique size-dependent discrete energy states. Besides, the symmetry of the states and hence the selection rules for optical and phonon transitions between them depend on the shape of the confinement potential. In particular, the alter-native selection rules for one- and two-photon interband transitions were predicted for spherical semiconductor QDs in the pioneering work by Efros and Efros [7]. Therefore, two-photon spectroscopy can provide a complete set of data on the energy structure of QDs to check the degree of adequacy of the theoretical models. From the experimental standpoint, two-photon spec-troscopy was shown to be a promising size-selective spectroscopic method for QDs systems with a number of advantages. The signals of two-photon excited secondary emission from the QDs can be easily obtained in most cases by us-ing a conventional spectroscopic technique. The intensity and shape of the RSHS (phononless TPL) line as well as the RHRS (TPL) spectra in the range of small Stokes shifts can be measured in detail because the incident light can be completely cut off by an appropriate filter. Furthermore, the second harmonic generation is forbidden in commonly used isotropic host matrices. Since the sample is usually transparent to incident photons, the correction of the signals for reabsorption, which is very important in the detailed analysis of two-photon excitation spectra, becomes easier as compared to one-photon excitation spectroscopy. Moreover, in many cases the two-photon excitation spectra of QD systems can be obtained at an excitation power density below the threshold for persistent hole burning [8], photobleaching, and photodark-ening phenomena typical for one-photon experiments on II–VI and I–VII semiconductor nanocrystals embedded in dielectric matrices.

In this chapter we will focus our attention mainly upon the latest experimental results obtained by resonant two-photon spectroscopy for the electron and phonon energy structure and the electron–phonon interaction in II–VI and I–VII semiconductor QDs embedded in dielectric matrices. For recent theoretical models of TPA and RHRS in spherical QDs, we refer the readers to [9] and [10], respectively.

## 7.2   Electronic Structure of CdS(Se) Quantum Dots

Pioneering work on resonant two-photon spectroscopy of QD systems was carried out in the late 1980s [3,11]. The $CdS_xSe_{1-x}$-based nanocrystals embedded in a silicate glass matrix were investigated at 300 K. The sizes of the nanocrystals corresponded to the strong-confinement regime. The spectra were excited at the two-photon resonance of incident light (the 1064 nm pulse radiation of a Q-switched $Nd^{3+}$:YAG laser) with optical transitions corresponding to the generation of electron–hole pairs in low-energy confined states of the QDs. Except for the intense line at twice the energy of the incident photons ($2E_L$), the spectra showed lines with the Stokes shifts relevant to the CdS- and CdSe-like LO phonons, 200 cm$^{-1}$ and 290 cm$^{-1}$, respectively, and their overtones and sum tones (Fig. 7.1).

To ensure that the two-photon excited secondary emission of QDs shows a resonant behaviour, the spectra of series of samples (Fig. 7.1a) different in the QD mean radius $R_0$ and hence in the energies of the lowest optical transition in the QDs were measured. It was clearly shown that the line intensities strongly increase when $2E_L$ comes close to the energy of the lowest energy optical transition of the QD system (Fig. 7.1c). A comparison (Fig. 7.1b) with the resonant Raman spectra of the same samples under the second harmonic laser excitation (532 nm) allowed the Stokes-shifted lines in the two-photon excited spectra to be assigned to RHRS by optical phonons in the QDs. The presence of the $B_1$ optical phonon line with a Stokes shift of 325 cm$^{-1}$, allowed only in the HRS process, supports the RHRS origin of the spectra. Particular attention was placed on the behaviour of the $2E_L$ line. Its resonant enhancement in the range of the QD's optical transitions, practically isotropic index, and large depolarization ratio suggest that the line is due to RSHS by randomly oriented non-interacting nanocrystals. In other words, it is a spatially incoherent analogue of the process of second harmonic generation (SHG). An important point is that the spectral resonances in the RSHS two-photon excitation spectra coincide with the resonances in the corresponding susceptibilities of QDs. This fact allows the resonances to be used for analysis of the electronic structure of a QD. It is worthwhile to note that phase-matching conditions make it difficult to use the SHG resonances for electronic structure studies in the case of bulk crystals [13]. Another advantage is that the SHG process is forbidden in commonly used isotropic

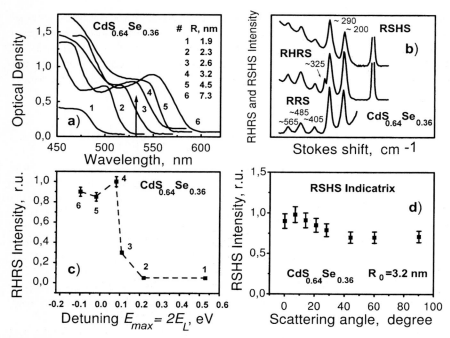

**Fig. 7.1.** (**a**) One-photon absorption spectra of samples with different mean radius $R_0$ of $CdS_{0.64}Se_{0.36}$ nanocrystals. An *arrow* shows the wavelength corresponding to twice the incident photon energy. (**b**) *Upper two traces* are the two-photon excited secondary emission spectra of samples 3 and 4. The RSHS line and the lines of RHRS by the CdS- and CdSe-like LO phonons of QDs are clearly seen. *Lowest trace* is the RRS spectrum of sample 3 presented for comparison. The Stokes shifts of the lines are shown in units of wavenumbers. (**c**) Resonant enhancement of the RHRS integral intensity. $E_{max}$ and $E_L$ are the energy of the lowest energy maximum in the one-photon absorption spectra of the samples shown in (**a**) and the incident photon energy, respectively. (**d**) RSHS integral intensity as a function of the angle between the direction of the incident beam and that of the RSHS detection [12]

host matrices, so that the related signal does not mask the RSHS signal from QDs at their small densities.

## 7.2.1   Two-Photon Absorption Techniques

The first detailed study of the electronic structure of QDs by a comparative analysis of their one- and two-photon absorption (OPA and TPA) spectra was carried out on CdS QDs embedded in a glass matrix [14]. It was noticed that the comparison between the OPA and TPA spectra of spherical QDs could clarify the adequacy of theoretical models for the electronic structure of QDs. In the simplest model to describe spherical QDs in a strong-confinement regime (the effective mass approximation, a spherical infinite confinement potential, a single parabolic conduction band, and a single parabolic valence band, while ignoring the Coulomb interaction), the confinement induces a splitting of each continuous conduction (valence) band to discrete levels [7]. Each discrete level corresponds to definite numbers of angular momentum $l$ and submagnetic moment $m$. The wave function and energy of the electronic state are expressed as

$$\Psi_{nlm}^{e(h)} = \mathcal{R}_{nl}(r)Y_{lm}(\varphi,\theta), \quad \mathcal{R}_{nl}(r) = \sqrt{\frac{2}{R_0^3}}\,\frac{j_l(\phi_{nl}r/R_0)}{j_{l+1}(\phi_{nl})}\ ,$$

$$E_{nl}^{e(h)} = \pm\left(\frac{E_g}{2} + \frac{\hbar^2\phi_{nl}^2}{2R_0^2 m_{e(h)}}\right)\ , \tag{7.1}$$

where $Y_{lm}(\varphi,\theta)$, $j_l(x)$ are the spherical harmonics and spherical Bessel functions, $\phi_{nl}$ is the $n$-th root of the equation $j_l(\phi_{nl})=0$, $m_e$ and $m_h$ are the electron and hole effective masses, $E_g$ is the energy gap of the bulk crystal.

The predicted low-energy states for the confined electrons and holes are shown in Fig. 7.2a. The labels s and p correspond to the states with the angular momenta $l=0$ and 1, while the indices e and h denote the electrons and holes. In accordance with the model, one-photon transitions between the states are allowed for $\Delta l=0$, whereas two-photon transitions for $\Delta l=\pm1$, as shown by vertical arrows in Fig. 7.2a. Although the OPA and TPA bands of real QD systems are inhomogeneously broadened to a great extent as shown in Fig. 7.2b, it was expected that the differences in energies between the low-energy OPA and TPA bands would be observed. It follows from (7.1) that the

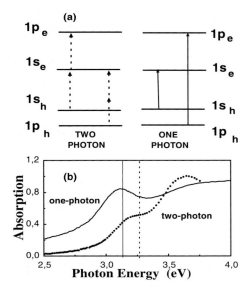

Fig. 7.2. (a) One- and two-photon allowed transitions between the discrete states in the spherical potential well; (b) Theoretical calculation of one-and two-photon absorption based on a parabolic band model [14]

expected differences are related to the ratio of the electron to hole effective masses by

$$\frac{E_{(1p_h - 1s_e)} - E_{(1s_h - 1s_e)}}{E_{(1s_h - 1p_e)} - E_{(1s_h - 1s_e)}} = \frac{E_{(1p_h - 1p_e)} - E_{(1s_h - 1p_e)}}{E_{(1p_h - 1p_e)} - E_{(1p_h - 1s_e)}} = \frac{m_e}{m_h} , \qquad (7.2)$$

where $E$ is the energy of the subscripted transitions, and the first TPA peak for CdS QDs was expected to be placed between the first and second OPA peaks.

In the samples studied, the mean radii of CdS nanocrystals range from 0.9 to 8.0 nm, overlapping the ranges of both the strong- and weak-confinement regimes (the exciton Bohr radius $R_{ex}$ is 3.2 nm for bulk CdS). A pulsed radiation from a Ti:sapphire laser or a DCM dye laser was used for the excitation. The TPA spectra of QD systems at 10 K have been obtained through the two-photon excited luminescence whose intensity was detected as a function of twice the incident photon energy in the range of electronic transitions in the QDs (2.5–4.0 eV). The origin of the detected TPL band with a peak at about 800 nm was not discussed. The band comes, most likely, from nanocrystal surface defects.

The experiments, however, revealed very smooth TPA spectra without any noticeable differences between the OPA and TPA peculiarities. This allowed for the conclusion that the simple model is insufficient to explain the optical transitions in CdS QDs. Then an improved theoretical model [15,16], which involves Coulomb interaction and mixing of the heavy- and light-hole valence bands induced by the spherical confining potential, was applied to analyze the experimental data. The model predicts several new one- and two-photon transitions in CdS QDs. Moreover, the two lowest-energy hole states may be roughly degenerated due to the mixing, resulting in a near-degeneracy of the one- and two-photon transition energies. The mixing strongly depends on the coupling constant $\mu = (4\gamma_2 + 6\gamma_3)/5\gamma_1$, where $\gamma_1$, $\gamma_2$, and $\gamma_3$ are the Luttinger parameters. The OPA and TPA transitions are degenerate (non-degenerate), if $\mu$ is higher (lower) than a certain critical value dependent on the nanocrystal size. For CdS QDs under study, the $\mu$ value has been evaluated at 0.75. With this value of $\mu$, the numerically calculated OPA and TPA spectra were found to peak at close energies and to fit well the experimental spectra. As a result, it should be concluded that the valence-band mixing and Coulomb interaction control the electronic energy structure in CdS QDs.

### 7.2.2    The Line-Narrowing Technique

Resonant two-photon spectroscopy was also used for studying the electronic structure of highly monodisperse colloidal CdSe QD systems with the mean nanocrystal radii correspondending to the strong-confinement regime [17,18]. The fluorescence line-narrowing technique was adopted for two-photon excitation. The two-photon excited luminescence of the QDs was detected in several narrow ranges (4 nm bandwidth) at the red edge of the lowest-energy TPL band. The two-photon excitation spectra, the analogue of the TPA spectra, were measured by using a tunable pulse laser source (1.65–1.00 eV) with a spectral bandwidth of 10 meV. The combined excitation and detection resolution was about 20 meV. The two-photon excitation spectra of the QDs have been analyzed in comparison with the predictions of a theory that involves a spherical potential well, a valence band degeneracy, and a nonparabolicity of the conduction band [15,16,19]. According to the theory, the total angular momentum $F=L\pm1/2$, its projection, $M$ and the lowest angular momentum of the constitutive envelope spherical harmonics $L$ are good quantum numbers for the discrete electronic states. Parity is also a good quantum number, and hence one-photon and two-photon transitions satisfy the alternative selection rules: $\Delta L=0, \pm2$ and $\Delta F=0, \pm1$, and $\Delta L=\pm1, \pm3$ and $\Delta F=0, \pm1, \pm2$, respectively. Therefore, a more complete picture of the electronic energy states in the QDs was expected to be obtained by probing the two-photon transitions and correlating them with the one-photon spectra. A comparison between the two-photon and one-photon excitation spectra of the QDs showed a marked difference in the range of high-energy transitions. The two-photon spectra were fitted to a sum of four to seven Gaussians whose energies were calculated for the case of an infinite hole barrier and a finite electron barrier. The other parameters of the model were found from the best fitting of the one-photon spectra. The simulated two-photon spectra were in qualitative agreement with the observation. On this base, the observed spectral features were satisfactorily attributed to the following definite two-photon transitions: $1P_{3/2}-1S_e$, $1P_{5/2}-1S_e$, and $2P_{5/2}-1S_e$ or $1S_{3/2}-1P_e$ (listed in order of increasing energy). However, a theoretically predicted shift of 20 meV between the lowest-energy one- and two-photon transitions ($1S_{3/2}-1S_e$ and $1P_{3/2}-1S_e$) was not found in the experiments on CdSe QDs, similarly to the case of CdS QDs described above. Although the experimental observation of the shift looks problematic because of poor spectral resolution, it was concluded that the model describes the band-edge electron–hole states inaccurately, since some mixing of the states due to crystal polarity or deviations from spherical symmetry was not taken into account in the model.

## 7.2.3  Analysis of RHRS and RSHS Excitation Spectra

The RHRS and RSHS spectroscopic techniques were applied for studing the electronic structure of spherical CdS QDs at 300 K [20,21]. The mean radius of the nanocrystals embedded in a silicate glass matrix corresponded to the strong-confinement conditions ($R_0=1.8$ nm, $R_{ex}=3.2$ nm). The spectra were excited by a pulsed radiation of a wavelength-tunable Ti:sapphire laser with twice the incident photon energies $2E_L$ falling into the range of low-energy electron–hole transitions. The secondary emission from the sample was collected in the $90^0$ direction and dispersed with the use of a single-grating monochromator. The spectral resolution of the measurements was

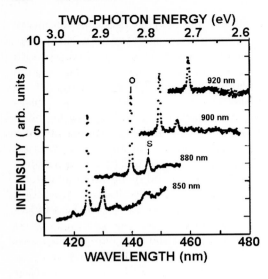

**Fig. 7.3.** Examples of resonant hyper-Raman signals (S) due to LO phonon in CdS dots, observed at 300 K for several excitation wavelengths. The *line* (O) at the lower energy side in each spectrum is a second-harmonic scattering signal [21]

about 4 meV (25 cm$^{-1}$). Figure 7.3 shows a representative set of the spectra obtained with several incident photon energies.

The line with the energy $2E_L$ (marked O in each spectrum) is the RSHS signal. The line with a Stokes shift of 305 cm$^{-1}$ equal to the LO phonon energy in bulk CdS (marked S) is the RHRS signal. The observation of solely LO phonon lines in the RHRS spectra clearly indicates that the Fröhlich-type electron–phonon interaction prevails in CdS QDs. A comparison between the two-photon excitation spectra of the RSHS and RHRS lines and the OPA spectrum of the sample (Fig. 7.4) clearly demonstrates the resonant character of the two-photon excitation.

At the same time, it was found that the energies of the resonances in the RSHS and RHRS excitation spectra deviate significantly from those in the OPA spectrum. Namely, the lower-energy resonances show a blueshift

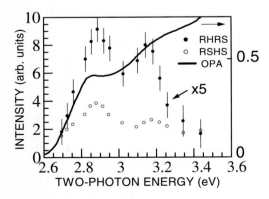

**Fig. 7.4.** Two-photon excitation spectra of resonant hyper-Raman (RHRS) and second-harmonic (RSHS) signals in CdS dots. One-photon absorption spectrum is also shown for comparison [21]

($\sim$33 meV), whereas the higher-energy resonances show a redshift ($\sim$40 meV) with respect to the related peaks in the OPA spectrum. The possible channels of the scattering processes and the selection rules for optical transitions were analyzed in the framework of the effective mass approximation for a simple two-band model of QDs with an infinite spherical confining potential [7]. As a result, the peaks in the RSHS and RHRS excitation spectra were assigned to the resonant two-photon generation of electron–hole pairs in the confined states with a total angular momentum $l$ of 1. These states were not observed in the OPA spectra of spherical QDs because the selection rules allow the one-photon excitation of electron–hole pairs only to the confined states with $l$=0. It was concluded that the experimental findings support the confined electronic structure of spherical QDs predicted by Efros and Efros [7] for the strong confinement regime. From the measured values of the shifts (see (7.1)), the effective masses of the electrons and holes in the QDs under study were estimated: $m_e/m_0$=0.3 and $m_h/m_0$=1.4, where $m_0$ is the free electron mass. The results are not too different from the values for the bulk material. However, the estimation may be rather crude, because it was done in the framework of a strongly simplified model.

A semiquantitative analysis of the possible contributions of different channels in the RHRS and RSHS efficiency was carried out. The analysis showed that the off-diagonal matrix elements of the Fröhlich electron–phonon interaction made a major contribution to RHRS by LO phonons in CdS QDs, resulting in the comparable intensities of the RHRS and RSHS signals. The conclusion is in accordance with the predictions of the RHRS theory recently developed for spherical CdS QDs [10].

## 7.3   Energy Structure of Low-Energy Confined Excitons in CuCl Quantum Dots

Resonant two-photon spectroscopy was used to study the confinement-modified exciton energy structure and exciton–phonon interaction in QDs in the regime of weak confinement. The CuCl and CuBr nanocrystals embedded in glassy or NaCl matrices provide examples of such QD systems because the exciton Bohr radii in the bulk CuCl (0.7 nm) and CuBr (1.3 nm) crystals are smaller than the nanocrystal radii. Since the parameters of elementary excitations (excitons, phonons, exciton–phonon interaction, etc.) in these bulk semiconductors are well established [22], the effects of three-dimensional confinement can be studied in detail.

Both the CuCl and CuBr materials are cubic semiconductors of $T_d$ symmetry with the band structures shown in Fig. 7.5. In CuCl the lowest-energy exciton (traditionally denoted as $Z_3$) involves a hole from a spin-orbit split-off valence band. In CuBr the lowest-energy exciton state corresponds to a two-fold degenerated exciton in the center of the Brillouin zone, which involves light and heavy holes (the so-called $Z_{1,2}$ exciton). In the framework of a simple

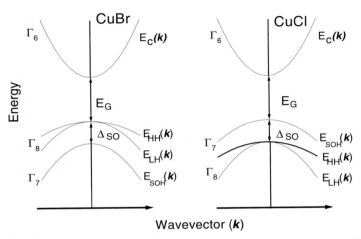

**Fig. 7.5.** Band structure of CuCl and CuBr semiconductors. The symmetries of the conduction band (c), the heavy- and light-hole bands (HH and LH), and the spin-orbit split-off band (SOH) are shown

two-band model for cubic semiconductors, with the effective mass approximation and a spherical potential barrier, the three-dimensional confinement results in a decomposition of continuous exciton bands to discrete confined exciton states [7]. The states are characterized by the quantum numbers $nlm$, where $n=1,2,3...$ is the principal quantum number, $l=0,1,2,...$ (S,P,D...) is the angular momentum, and $m$ is the projection of the angular momentum. In the model, the confined exciton states are degenerated with respect to $m$, and denoted as $n$S, $n$P, $n$D, etc. In spherical semiconductor QDs in the weak-confinement regime, the energies of the confined exciton states depend on the nanocrystal size as [7]:

$$E_{n,l} = E_{\mathrm{B}} + \frac{\hbar^2 \alpha_{n,l}^2}{2MR^2} , \qquad (7.3)$$

where $E_{\mathrm{B}}$ is the exciton energy in the bulk crystal, $M$ is the translational mass of exciton, $R$ is the radius of nanocrystals, and $\alpha_{n,l}$ is the $n$th root of the spherical Bessel functions of the $l$th order, $J_l(\alpha_{n,l})=0$ ($\alpha_{1,0}=\pi$, $\alpha_{1,1}=4.49$, $\alpha_{1,2}=5.76$, and $\alpha_{2,0}=2\pi$ for the low-energy confined exciton states in sequence). The model describes quantitatively the confinement-induced blueshift of the 1S $Z_3$ exciton absorption peak in CuCl QDs [23].

At the same time, it is known that the two-band model is oversimplified when dealing with the description of the low-energy exciton structure of cubic semiconductors, in particular, CuCl [22,24]. The exchange interaction splits the degenerated exciton states: its short-range part results in a singlet-triplet splitting while its long-range (non-analytic) part induces a longitudinal–transverse (L–T) splitting of the triplet exciton state. For non-centrosymmetric materials, e.g., those of the $T_{\mathrm{d}}$ symmetry, spin-orbit interaction produces a splitting of the conduction and valence bands, i.e., removes the Kramer's degeneracy of the electron states with a nonzero wave vector [25,26]. This results in a splitting of the transverse exciton into the two states, $T_1$ and $T_2$. It is expected that three-dimensional confinement modifies these interactions, producing changes in the parameters of exciton states, exciton–phonon interaction, and in the dynamics of the excitations. Studies of the low-energy electronic structure of semiconductor QDs have led to the discovery of new fine details of the confinement effect. A fine exciton structure due to the short-range exchange interaction ("dark" and "bright" electron–hole pair states) was observed in the low-temperature photoluminescence spectra of CdSe QDs in the strong-confinement regime [27–29]. The results are consistent with the theoretically predicted enhancement of the interaction in QDs [30]. The exciton–phonon interaction initiates the exciton–phonon-coupled states, e.g., 1S exciton–polaron [31] and 1S exciton–acoustic-phonon [8,32] states, with energies different from the energy of a free confined exciton by several meV. Of great interest is the long-range exchange interaction in CuCl QDs, which yields the L–T splitting of Wannier excitons in the bulk crystal.

The selection rules allow both one- and two-photon generation of 1S $Z_{1,2}$ and 1S $Z_3$ excitons in spherical QDs. At the same time, the generation of 1P excitons is allowed only by two-photon transitions [9]. Therefore, all these excitons are expected to manifest themselves in the TPL spectra of QDs, in contrast to the one-photon excited photoluminescence spectra. Resonant two-photon spectroscopy was used to study the low-energy confined $Z_3$ exciton states in spherical CuCl QDs with the mean radii of 2.4 and 2.9 nm [33]. The exciton energy structure was established from the analysis of the energies of the TPL lines as functions of the incident photon energy and from the two-photon excitation spectra of the lines. The analysis was based on the size-selective behaviour of the excitation. Actually, in an inhomogeneously broadened QD ensemble, the incident photons of specified energy can generate excitons only in QDs specified in size, and thus are responsible for the relevant TPL spectra. The requirement of a small homogeneous width of the resonant transitions as compared with the inhomogeneous broadening for the QD ensemble is fulfilled for the CuCl QD systems [34]. A diagram of certain transitions between the confined exciton states is shown in Fig. 7.6 to illustrate the formation of size-selective TPL spectra. Once the confined exciton energies satisfy (7.3), the following relation is fulfilled:

$$E_{n,l}^{ex} - E_{n',l'}^{em} = \left(E_{n,l}^{ex} - E_{ex}\right)\left[1 - (\alpha_{n',l'}/\alpha_{n,l})^2\right] , \qquad (7.4)$$

where $E_{n,l}^{ex} \equiv 2E_L$ is the energy of the state excited by light and $E_{n',l'}^{em}$ the energy of the state that the luminescence comes from. Then, the dependence of the Stokes shift of the luminescence line $\Delta = (2E_L - E_{n',l'}^{em})$ on twice the incident phonon energy $2E_L$ or on the confinement shift of the 1S exciton, $2E_L - E_B \equiv E_{1S} - E_B$ allows one to find the $\alpha_{n',l'}/\alpha_{n,l}$ ratio and thus to establish the exciton states involved in the process. In this way the lines with $2E_L$-,

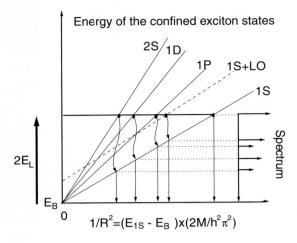

**Fig. 7.6.** Scheme of some transitions forming the TPL spectrum of the system of spherical QDs with broad size distribution. At definite incident photon energy $E_L$, the TPL lines belong to the quantum dots of different sizes. *Solid lines* and a *dashed line* are related to the energies of the confined exciton states and a coupled 1S-exciton–LO-phonon state, respectively. $E_B$ is the exciton energy in bulk CuCl

and hence the $R$-dependent Stokes shifts are correlated with those confined exciton states, whose energies depend on $R$ according to (7.3). Once some energy states are not accounted for in the framework of the simple model involved, additional data must be invoked for their identification. On the other hand, it is evident that the maximal intensities of TPL lines correspond to the excitation of QDs at the maximum of their size distribution. With a direct light excitation of different exciton states, this condition is satisfied at different $2E_L$. This allows use of the two-photon excitation spectra of TPL lines for identification of the energy states in QDs.

Figure 7.7 shows a set of TPL spectra obtained with different incident photon energies for a specimen with the nanocrystal mean radius $R_0=2.9$ nm [33]. At least five lines that differ in spectral behaviour have been found in the spectra. A narrow line marked in Fig. 7.7 as RL, with the energy equal to twice the incident photon energy, shows a resonant enhancement when $2E_L$ is close to the 1S confined state energy. This line corresponds to either the two-photon excited resonant luminescence or RSHS.

The two-photon excitation spectra of the lines are shown in Fig. 7.8. Figure 7.9 shows the Stokes shift of the L1, L2, and L3 lines as a function of the confinement energy of the 1S exciton state, $\Delta_{1S}=E_{1S}-E_B$. A weak band with a Stokes shift of 25.5 meV equal to the CuCl bulk LO phonon energy [36] and independent of the incident photon energy was assigned to RHRS by the LO phonons. The RHRS signal emerges at the same incident photon energies as the RL line. An increase in the incident photon energy results in a consecutive appearance of three strong luminescence bands marked in Fig. 7.7 as L1, L2, and L3, respectively. The intensities of these bands are strongly dependent on $2E_L$. It was found that the Stokes shift of the L1 and L2 bands $\Delta L1(\Delta L2)=2E_L-E_{L1(L2)}$ increased with $2E_L$, where $E_{L1(L2)}$ is the peak energy of the L1(L2) band. The peak energy of the broad band L3,

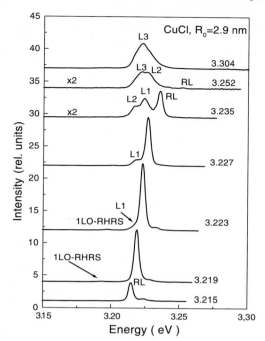

**Fig. 7.7.** Two-photon excited luminescence (TPL) spectra of CuCl dots with $R_0 = 2.9$ nm. Resonance luminescence and three luminescence bands are marked by RL, L1, L2, and L3, respectively. A weak band of RHRS by LO phonons is also shown. The intensity of two spectra are multiplied by two ($\times 2$). Twice the incident photon energies are indicated [33]

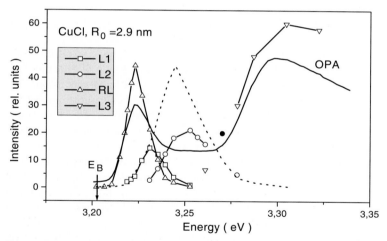

**Fig. 7.8.** Excitation profiles of the TPL bands and OPA spectrum (*solid line*) for CuCl dots with $R_0 = 2.9$ nm. Calculated position of the L2 excitation spectrum is shown by the *dotted curve*. A *solid circle* corresponds to the sum of the intensities of the L2 and L3 bands when the band intensities cannot be reliably measured due to coincidence of their peak positions. An *arrow* shows the exciton energy $E_B$ of bulk CuCl [33]

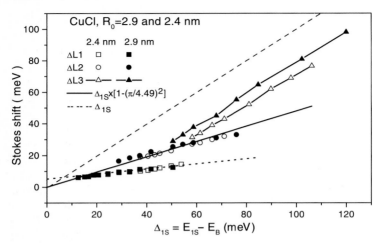

**Fig. 7.9.** The Stokes shift of the luminescence bands, L1, L2, and L3, as a function of the confinement energy, $\Delta_{1S}$. The *solid line* is the calculated dependence for L2($\Delta_{1S}$). The *short-dashed line* is a straight line fitted to the result of $\Delta$L1($\Delta_{1S}$). The *dashed line* shows that the L3 peak position does not depend on the incident photon energy [33]

which appears at higher $2E_L$ and dominates the spectra, does not depend on the excitation wavelength. Analysis of the Stokes shift dependencies on $2E_L$, with invoking the diagram of Fig. 7.6, shows that the lines are due to annihilation of the lowest-energy 1S exciton created directly by incident light (the RL line) or by relaxation from the higher-energy confined exciton states (the L1–L3 lines). Since the L3 peak energy is independent of the incident photon energy, it is ascribed to thermalized emission due to recombination of the exciton in QDs at the band edge. For the incident photons of high energies, the shape of the L3 line reflects the size distribution of QDs because the nanocrystals of all sizes can be excited. It was concluded that one of the higher-energy states was the 1P exciton, which was generated by the light and, after relaxation to the 1S state with its subsequent annihilation, gave rise to the L2 line. Indeed, the $2E_L$-dependence of the Stokes shift of the L2 line (Fig. 7.9) satisfies well the relation $\Delta L2 = \Delta_{1S} \times [1-(\pi/4.49)^2]$ that follows from (7.3) for the 1P state ($\alpha_{1,1} = 4.49$). Moreover, the two-photon excitation spectrum of the line is shifted to a higher energy with respect to that of the RL line, which is consistent with the calculated excitation spectrum of the L2 line (Fig. 7.8). It is significant that the excitation spectrum has a peculiarity. Namely, a resonant enhancement of the L2 band intensity was found for those QDs for which the energy separation between the 1P and 1S states $E_{1P}-E_{1S}$ is close to the LO phonon energy. The increase in the L2 line intensity was interpreted as a result of a new 1P-to-1S relaxation channel that involves the LO phonon. According to Itoh et al. [31], the confined LO phonons with angular momentum $l=1$, or the $p$-type LO phonons, are most likely involved in this process via the deformation potential interaction.

Analysis of the L1 band parameters allowed a proposal that the band can be associated with the lowest-energy confined state of a longitudinal (L) exciton ($1S_L$). Its energy is higher than the lowest-energy exciton state in QDs, which is supposed to be the $1S_T$ state of a transverse (T) exciton. Actually, the L1 band was observed only with a two-photon excitation. This is natural for the L exciton since, similarly to the bulk material [35], the two-photon generation of the L exciton is allowed in QDs while the one-photon generation is forbidden. Then the L1 band is due to the two-photon generation of $1S_L$ excitons, followed by subsequent relaxation of L excitons to the 1S state of T excitons and radiative annihilation of the T excitons (Fig. 7.6). The shift of the excitation spectrum of the L1 line with respect to that of the RL line (Fig. 7.8) shows that the L1 line results from photogeneration of excitons in a state higher in energy than the $1S_T$ state. Any other high-energy confined states of the T exciton (1P, 1D, 2S, etc.) cannot be responsible for the L1 band. Then, the observed Stokes shift $\Delta L1$ represents the variation in the L–T exciton splitting energy $\Delta_{LT}$ as a function of $\Delta_{1S}$. The result reveals that it increases with decreasing QD size. The size dependence of $\Delta L1$ was fitted by a straight line $\Delta L1 = 4.95 + 0.175 \times \Delta_{1S}$, which yields the L–T exciton splitting energy of 4.95 meV at $\Delta_{1S}=0$. This value is in exellent agreement with the CuCl bulk value of 5.0–5.7 meV [36]. From the observed relation, the effective translational mass of the L exciton $M_L=(0.77\pm0.7)\times M_T$, where $M_T$ is the effective translational mass of the T exciton, has been evaluated, supposing the same size dependence for both the L and T 1S excitons in QDs. The value thus obtained differs significantly from the bulk value $1.35\times M_T$ measured by hyper-Raman scattering experiments [37]. This was explained by a nonparabolicity of the L exciton band far from the $\Gamma$ point in the Brillouin zone.

The recent theoretical work on the effect of L–T splitting on weakly confined excitons in QDs showed that the assignment of these confined exciton states in CuCl QDs to the pure T and L excitons was not correct [38]. The L–T splitting was considered in [38] in terms of the Coulomb interaction of the induced charge densities accompanying the confined exciton. This approach is quite common and can be applied to the both bulk and confined systems, and both electronic and phonon excitations. It has been shown for spherical QDs that, except for a small number of pure T and L modes, most confined states are of a mixed L–T character for finite radii, but they tend to the bulk T or L branches in the limit of large QD radii. In particular, the lowest-energy exciton state of QDs is a "mixed L–T" mode rather than a pure T mode, and the lowest-energy exciton state from those states that tend to the bulk L exciton branch is of a "mixed L–T" character as well. The latter exciton state is believed to manifest itself in the TPL spectra of CuCl QDs as the L1 line.

The observation of the L–T exciton splitting in CuCl QDs indicates that the dipole–dipole interaction in nanocrystals containing only 8–10 layers of elementary units does not noticeably differ from that in the bulk. The L–T splitting of optical phonons found in similarly sized CuBr nanocrystals [39] supports the conclusion. It follows that such macroscopic parameters of solids as dielectric constants can be used for description of small nanocrystals.

## 7.4    Exciton–Phonon Interaction in CuBr and CuCl Quantum Dots

The exciton–phonon interaction is one of the basic problems in the physics of QDs and is the subject of comprehensive theoretical and experimental studies [9,21,31,40–44]. It is known that there are four types of optical phonons in finite ionic crystals: two types of pure transversal divergence-free modes ($T_1O$ and $T_2O$), pure longitudinal curl-free modes (LO), and curl- and divergence-free surface modes (SO) [45]. In spherical nanocrystals, each of the modes is characterized by a set of quantum numbers with the same meanings as those for 3D-confined electronic excitations: the principal quantum number $n$, the angular momentum $l$, and its projection $m$. However, for TO phonons the minimal value of $l$ is 1, for SO phonons there are only the $l$ and $m$ quantum numbers, with the minimum $l=1$. Due to phonon dispersion, the energies of TO and LO phonons with different $n$ and $l$ should be different. These facts cause a strong modification of the optical processes involving phonons in QDs as compared to bulk crystals. A detailed theoretical consideration of optical transitions involving optical phonons and exciton–phonon coupling in spherical QDs of cubic semiconductors for the weak-confinement regime can be found in [9]. In this work the matrix elements of the transitions between arbitrary confined exciton states, the exciton–phonon interaction of both the Fröhlich and deformation types, and the selection rules were derived. An increase in the strength of polar-exciton–LO-phonon interaction (Fröhlich type) was predicted with decreasing CuBr and CuCl QD size.

Resonant two-photon spectroscopy was shown to be a promising method for studying the interaction between a confined exciton and optical phonons in CuBr and CuCl QDs [21,46–48]. The experiments demonstrated that the Fröhlich-type interaction is rather strong in small CuBr and CuCl QDs and increases as the nanocrystals decrease in size. This gives rise to strongly coupled exciton–LO-phonon states and to a pronounced size-dependent softening of the LO phonon in the presence of an exciton.

### 7.4.1    CuBr Quantum Dots: Coupled Exciton–LO-Phonon States

The resonant two-photon excited spectra of secondary emission from CuBr QDs of different radii in a silicate glass matrix were obtained at 2 K [21,46]. The spectra were excited using pulsed radiation from a wavelength-tunable

Ti:sapphire laser pumped by a Q-switched $Nd^{3+}$:YAG laser. The pulse width was 40 ns, the repetition rate 3 kHz, and the peak power 1 kW. The incident phonon energy was varied in such way that $2E_L$ swept the inhomogeneously broadened low-energy transitions of the confined $Z_{1,2}$ and $Z_3$ excitons, as shown in Fig. 7.10a. The secondary emission around $2E_L$ was collected in a quasiforward direction, dispersed by a single-grating monochromator, and detected by a cooled optical multichannel detector. A combined spectral resolution of 4 meV (25 $cm^{-1}$) was adopted.

It was found that up to fifth-order LO phonon lines with comparable intensities consecutively appeared in the spectra when $2E_L$ passed through the inhomogeneously broadened 1S $Z_{1,2}$ exciton band from its low-energy side to higher energies. The spectra of Fig. 7.10 (right) illustrate the features observed for the sample with $R_0$=3.2 nm.

The energy separation between the phonon lines was determined to be 20.2 meV, close to the energy of the LO phonon 21.5 meV in bulk CuBr. This suggests that the lines are due to the LO phonons of CuBr QDs.

In order to clarify the origin of the lines, their two-photon excitation spectra were measured. The excitation spectrum of the phononless line with

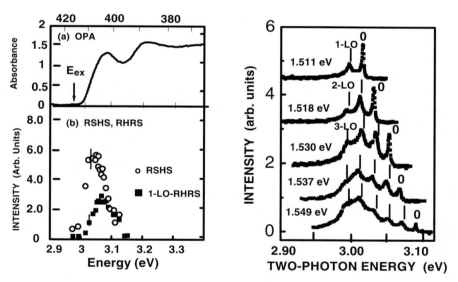

**Fig. 7.10.** *Left*: **(a)** The OPA spectrum of CuBr QDs with a mean radius of 3.2 nm observed at 2 K: the energy position of the $Z_{1,2}$, S exciton for bulk CuBr is marked by an *arrow*; **(b)** Two-photon excitation spectra of the RSHS and one-LO RHRS signals. *Right*: Examples of the RHRS spectra at 2 K for the same CuBr QD sample for several incident photon energies. The line at $2E_L$ marked by O is the RSHS signal, and the signals separated by multiples of the LO phonon energy from $2E_L$ are RHRS or hot luminescence. A broad spectrum is the band edge luminescence [46]

energy $2E_L$ (marked O in Fig. 7.10) was consistent with the lowest-energy OPA band corresponding to the 1S $Z_{1,2}$ exciton (see Fig. 7.10b). The spectrum shows the major contribution of the resonance with the lowest-energy pure confined exciton state to the signal, which has been attributed to resonant second-harmonic scattering (RSHS) or hot resonance luminescence. At the same time, the excitation spectra of LO phonon lines were found to be shifted to the higher energy side in such a way that the maximum of the spectrum of the Nth LO phonon line ($E_N$) satisfied the condition: $E_N = E_0 + N\hbar\Omega_{LO}$, where $E_0$ is the peak energy of the RSHS excitation spectrum and $\hbar\Omega_{LO}$ is the LO phonon energy, as shown in Fig. 7.11. Analysis of the experimental data allowed a conclusion that the perturbation theory, with the pure exciton states as unperturbed base functions and the exciton–phonon interaction as a perturbation, may not apply to the QD system. It can be concluded also that the strongly coupled states of the confined exciton and LO phonon are elementary excitations with good quantum numbers for the CuBr QDs, just as are the vibronic states in molecules and impurity centers with a strong electron-vibrational coupling [49]. Then successive (cascading) real transitions between these vibronic states followed by hot luminescence should be responsible for the observed spectral features, rather than the virtual transitions in the RHRS process. Otherwise, a set of all multiphonon lines must appear in the spectra at the two-photon resonance with the pure exciton state. The successive appearance and coexistence of multiphonon lines are evident in the light of size-selective resonant two-photon excitation of a subensemble of nanocrystals in a system with a broad nanocrystal size dispersion. In this case, radiation with a definite energy can be in resonance with the pure exciton state (1S) of relatively large QDs and with the vibronic states formed with different numbers of LO phonons in nanocrystals smaller

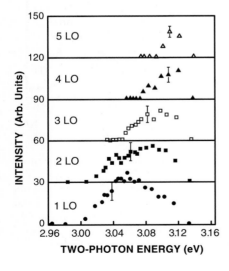

**Fig. 7.11.** Comparison of the respective two-photon excitation spectra for one- to five-LO-phonon signals in a CuBr QD sample with mean radius of 3.2 nm. The *horizontal axes* are shifted relative to each other [46]

in size. It was concluded that the observed multiphonon structure was due to hot luminescence from the different vibronic states (the pure exciton state included) populated by real cascade LO phonon transitions from resonantly excited higher-energy vibronic states. It should be pointed out that TPA to the pure exciton state and to the exciton–LO-phonon coupled states with the LO phonon of $s$- and $p$-types and with the TO phonons is allowed [9]. At the same time, the observation of LO phonons alone in the spectra shows that the Fröhlich-type interaction dominates in the exciton–phonon coupling in the CuBr QDs.

It has been emphasized that, due to the discrete energy spectrum of confined exciton states in QDs, the origin of multiphonon structures in their spectra is quite different as compared to the well-known multiphonon features observed in the secondary emission spectra of bulk polar crystals, such as CdS, wherein the continuum band state was excited [50]. In the latter case, the continuous energy spectrum of an exciton is a necessary condition for the origin of the multiphonon spectral features [51], whereas, in the present case, only one or two discrete exciton states may be involved. Indeed, for CuBr QDs, the size-dependent energy gap between the 1S and 1P confined states of the $Z_{1,2}$ exciton $\Delta E_{\mathrm{SP}}$ is around 110 meV for $R_0 = 3.2$ nm, much more than the LO phonon energy.

In the framework of the proposed model, the multiphonon structure reflects the energy spectrum of coupled exciton–LO-phonon states. It allows an explanation of the smaller energy separation between the phonon lines, or the energy of the LO phonon in the excited state of QDs, as compared to the LO phonon energy in bulk CuBr, by the effect of exciton–phonon interaction [52]. This relative shift can be used for a quantitative estimation of the interaction. However, because of inadequate spectral resolution and low intensities of the relevant signals, the size dependence of the interaction could not be obtained quantitatively. Nevertheless, a comparison between the two-photon excited spectra of samples with different $R_0$ (3.2 nm, 4.3 nm, and 5.2 nm) showed that the number of multiphonon lines and their relative intensities (in comparison with the phononless line) decreased with increasing $R_0$. These qualitative data reveal that the Fröhlich interaction increases with decreasing nanocrystal size in CuBr QDs.

### 7.4.2 CuCl Quantum Dots: Size Dependence of the Exciton–LO-Phonon Interaction

The size dependence of the exciton–LO-phonon interaction in QDs was obtained at a quantitative level by resonant size-selective two-photon spectroscopy of spherical CuCl QDs embedded in a glass matrix [47]. In the experiments, five samples with the mean nanocrystal radii 1.8, 2.0, 2.2, 2.9, and 3.6 nm were studied. A highly sensitive liquid-nitrogen-cooled CCD camera and an apparatus with a spectral resolution of 1 meV was adopted. The experimental setup and the measurement procedures were similar to those

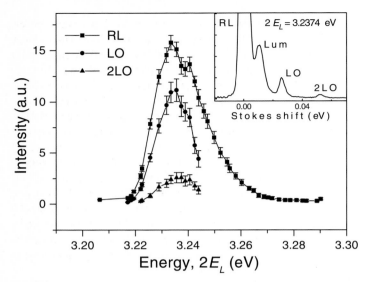

**Fig. 7.12.** Two-photon excitation profiles of the RL, LO, and 2LO bands for CuCl spherical nanocrystals with mean radius of 2.2 nm. *Inset* shows spectrum of two-photon excited light emission containing resonant luminescence (RL), luminescence (Lum), and RHRS bands (LO and 2LO); twice the incident photon energy $2E_L$ is shown [47]

described in [33]. In the secondary emission spectra of the samples excited in two-photon resonance with the lowest 1S $Z_3$ exciton, the intense lines attributed to RHRS by the LO and 2LO phonons of CuCl QDs with the energies 25.6 and 52 meV, respectively, were observed (Fig. 7.12).

It has been shown that the ratio between the 2LO and LO line integral intensities ($\rho = I_{2LO}/I_{LO}$) and its size dependence can be used to determine the parameter characterizing the exciton–LO-phonon interaction, i.e., the so-called Huang–Rhys factor $S$ and its size dependence. This is based on the assumption that RHRS by LO phonons in CuCl QDs can be described in the framework of the vibronic theory of RHRS for molecular systems, if the RHRS is excited in resonance with a strong transition, well-allowed for both one- and two-photon absorption, and described by the term similar to the so-called Albrecht A-term in RRS [3]. Note, both the one- and two-photon generation of the 1S $Z_3$ exciton are allowed in CuCl QDs. Then the values of the Franck–Condon factors, which determine the intensities of different-order RHRS lines, can be found on the basis of the offset harmonic-oscillator model of electron-vibrational coupling. In other words, $\rho$ is a function of $\Delta$, where $\Delta$ is the dimensionless displacement of the harmonic oscillator potentials in the ground and excited states. On the contrary, the $\Delta(R)$ values can be calculated from the experimental data for $\rho(R)$, where $R$ is the radius of QDs resonantly interacting with the incident photons. For the Fröhlich-type exciton–phonon interaction with one LO phonon mode, the well-known relation $S=\Delta^2$ allows the determination of $S(R)$. It was established experimentally that the system under study showed several important features. In particular, the coincidence of the two-photon excitation spectra for the LO and 2LO lines with that for the phononless line revealed a single two-photon resonance with the transition to the pure 1S $Z_3$ exciton state (Fig. 7.12). Besides, the homogeneous width of the transition was much smaller than the LO phonon energy. Moreover, the excitation at the low-energy side of the excitation spectra made it possible to eliminate a contribution to the signals from the luminescence of the 1S $Z_3$ exciton generated via LO-phonon-assisted absorption. Taking into consideration these facts, the complicated relation between the measured value $\rho(R)$ and $S(R)$ could be reduced to the simple formula: $\rho=S/2$. As a result, the dependence $S(R)$ was found that demonstrates a prominent increase in the exciton–LO-phonon interaction from 0.22 to 0.7 with a decrease in the CuCl QD sizes from 3.6 nm to 1.6 nm (Fig. 7.13). It should be mentioned that the dependence $\rho(2E_L)$ was measured in the experiments and (7.3) was used to obtain $\rho$ as a function of $R$.

Figure 7.13 shows also the calculated results for $S(R)$ in the model accounting for the size-quantization of LO phonons in CuCl QDs [9]. The size dependence of the Huang–Rhys factor for CuCl QDs is in qualitative agreement with the theoretical prediction, however, a quantitative discrepancy that is especially pronounced for nanocrystals of greater sizes is clearly seen.

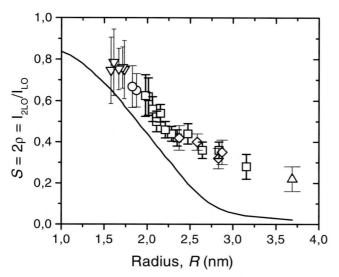

**Fig. 7.13.** The size-dependence of the Huang–Rhys factor $S$ for CuCl QDs. *Different shapes* of the experimental points correspond to different samples. The error bars are shown [33]. A *solid line* is the theoretical dependence calculated for the CuCl QDs accordingly to [9]

One possible reason is that the Huang–Rhys factor does not describe the total exciton–phonon coupling in QDs in the weak-confinement regime. The contribution of the off-diagonal Fröhlich interaction involving intraband exciton transitions is expected to be more pronounced for QDs larger in size [9]. This interesting problem is a subject for future investigation.

### 7.4.3   CuCl Quantum Dots: Softening of LO Phonons in the Presence of an Exciton

An elementary electronic excitation in small semiconductor nanocrystals containing a few hundreds of atoms is expected to result in softening of lattice vibration frequencies. Especially, the softening is pronounced in systems with strong electron–phonon coupling. As mentioned above, the experimental evidence for this feature characteristic of molecular systems and impurity centers in solids was found in CuBr QDs by resonant two-photon spectroscopy [46]. The detailed experimental [52] and theoretical [52,53] studies of the effect were performed recently on CuCl QDs, wherein the sharp structures in the persistent hole burning and photoluminescence spectra showed a ∼2 meV deviation of their Stokes shifts from the LO phonon energy (25.6 meV) in the bulk. The structures were assigned to the phonon-assisted absorption involving LO phonons in the excited state of the QDs. The reduction of the phonon energy was described theoretically in terms of phonon renormalization in the presence of an exciton [52,53]. At the same time, the softening of the LO phonon energy in CuCl QDs was found from the analysis of quantum beats in a femtosecond degenerate four-wave mixing experiment [54]. Its value was found to be twice that observed by Zimin and co-workers [52]. Therefore, a correct assignment of the observed phonon structures to the processes involving LO phonons in the excited state of CuCl QDs was a problem in question.

Recently, the results of two-photon excitation experiments on small spherical CuCl nanocrystals embedded in a glass matrix clearly showed a softening of the excited-state LO phonon induced by exciton–phonon coupling [48]. The technique allows for obtaining new proof of softening of the LO phonon in the presence of an exciton. In the work, rather small nanocrystals with mean radius 1.8 nm were studied since the exciton–LO-phonon coupling increased with decreasing QD size [47]. (Note that the lattice constant of CuCl is 0.54 nm.) The energy separation between lowest, 1S and 1P, confined $Z_3$ exciton states in the studied QDs was more than twice the LO phonon energy. To provide reliable assignment, the two-photon excitation spectra of TPL lines related to the processes involving LO phonons in the ground or excited states of QDs were measured.

In the experiments, a conventional setup for steady state resonant two-photon spectroscopy [33] was used. Temporal changes in the TPL lines were measured by the 2 ps pulse radiation of a mode-locked Ti:sapphire laser with a repetition rate of 82 MHz and an average power of ∼300 mW. The technique

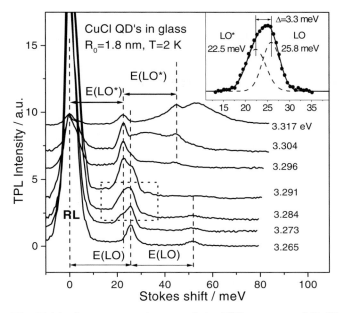

**Fig. 7.14.** A representative set of the TPL spectra of CuCl QDs plotted against their Stokes shifts for various incident photon energies, $E_L$ ($2E_L$ indicated in eV). RL denotes the resonance luminescence. E(LO*) and E(LO) denote the energies of the LO phonons in excited (LO*) and ground (LO) state of QDs, respectively. Enlarged part of the spectrum at $2E_L$=3.284 eV is shown at the *top right corner* where the line-shape fit by two Gaussian peaks demonstrates the coexistence of the LO and LO* phonon signals in the TPL spectra [48]

of time-correlated single-photon counting with a time resolution of 150 ps was used to record the signals.

Figure 7.14 shows a representative set of the TPL spectra plotted against their Stokes shifts $2E_L$–$E_{TPL}$ at different $2E_L$. The resonance luminescence (RL) line with the energy $2E_L$ results from the direct two-photon creation of 1S $Z_3$ excitons followed by their annihilation (Fig. 7.15b, scheme 1) with a decay time of ∼0.6 ns. The RL excitation spectrum shown in Fig. 7.15a by filled squares corresponds to the TPA of the 1S excitons. The TPA peak reflects the size distribution of the nanocrystals and shows a shift from the corresponding one-photon absorption peak due to a different size dependence ($R^6$ vs. $R^3$). Simultaneously, the LO phonon sidebands, with a Stokes shift $E_{LO}$ of 25.8±0.3 meV equal to the CuCl bulk LO phonon energy and its overtone were observed. The excitation spectrum of the LO line and its decay time were found to coincide with those of the RL line. It is consistent with scheme 1 in Fig. 7.15b: the LO-phonon-related lines correspond to the hyper-Raman-like phonon-assisted luminescence involving LO phonons in the ground state of the QDs. At higher $E_L$, a new line with the Stokes shift $E_{LO*}$

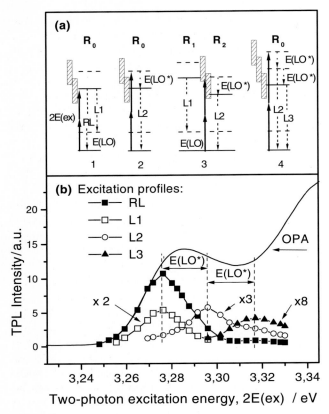

**Fig. 7.15.** (a) Excitation spectra of different lines in the two-photon excited emission of the CuCl QDs showing that the spectrum of the LO line coincides with that of the RL signal, whereas those of the LO* and 2LO* bands are shifted to the high-energy side by the E(LO*) and 2E(LO*) energies. RL denotes the resonant luminescence. L1, L2, and L3 are the luminescence processes showing Stokes shifts of E(LO), E(LO*), and 2E(LO*), respectively. Scale factors for different excitation spectra are shown. A one-photon absorption (OPA) spectrum of the sample is shown for comparison; (b) Schematic showing the excitation and relaxation processes in two-photon-excited luminescence. *Gray parallelepipeds* schematically show inhomogeneous broadening of the transitions due to the nanocrystal size distribution. $R_0$ denotes the mean radius of the nanocrystals, $R_1$ and $R_2$ refer to radii of the nanocrystals such that the resonant conditions of excitation shown in scheme 1 and 2 are satisfied simultaneously [48]

equal to 22.5±0.3 meV is produced (Fig. 7.15). The excitation spectrum of the line is shifted to higher energies from that of the RL line by ∼22.5 meV as well (Fig. 7.15a). The line dominates in the TPL spectrum at $2E_L > 3.296$ eV. Because there are no other confined exciton states in this energy range for the QDs, the LO* band comes from the process shown by scheme 2 in Fig. 7.15b.

Light generates a 1S exciton and an LO* phonon in the excited state of the QDs, and the LO* line corresponds to annihilation of the 1S exciton with a decay time of about 0.6 ns. Coexistence of the LO and LO* lines in the spectra comes from the fact that the resonant conditions shown in schemes 1 and 2 are satisfied simultaneously for QDs of different $R$ (Fig. 7.15b, scheme 3, $R_1 < R_0 < R_2$). Crude estimations show that the efficiencies of the processes via schemes 1 and 2 should be close. That is why the the LO and LO* line intensities in the maximum of the excitation spectra are approximately the same.

A further increase in $E_L$ results in the appearance of a 2LO* line whose excitation spectrum (Fig. 7.15a, filled triangles) is shifted from that of the RL line by $2E_{LO^*}$. A resonant enhancement of the signal occurs when $2E_L$ becomes equal to the sum of the energies of the exciton and two LO* phonons. The 2LO* line again corresponds to the annihilation of the 1S exciton (Fig. 7.15b, scheme 4). Note that the excitation spectra of the LO* line is asymmetrical because of an additional spectral feature close to the maximum of the excitation spectrum of the 2LO* line (Fig. 7.15a). Detailed analysis of the RL excitation spectrum also shows the same spectral feature. Most likely it arises from the fact that real (not virtual) coupled 1S-exciton–$n$LO*-phonon states are formed in small CuCl QDs, analogously to CuBr QDs described previously, allowing a cascade LO* phonon intraband relaxation. Then, the direct excitation of the 1S–2LO* state followed by the LO* phonon relaxation and annihilation of the 1S–LO* state produces a signal that contributes to the LO* line. At the same time, RL from the 1S–2LO* state contributes to the RL signal, and hence a relevant feature appears in the excitation spectra of the RL line. A similar contribution would be expected in the range of the 1S–LO* state resonance, but it was not clearly observed because of overlap with a strong RL signal from the 1S exciton state of QDs of smaller sizes.

The 13% reduction in the LO phonon energy in the excited state of CuCl QDs as compared to the ground state agrees well with the theoretical results reported in [53]. The 1.5 times greater reduction than that observed for 2.5 nm CuCl QDs [52] is due to a stronger exciton–LO-phonon coupling in smaller nanocrystals [47]. It is worth noting that the width of the LO and LO* phonon lines (FWHM$\simeq$4.8 meV) is much broader than that observed in the one-photon luminescence experiment [52]. It is probably due to some unresolved internal structure of the lines. For instance, under a two-photon excitation not only confined s-like LO* phonons but also p-like LO phonons with a slightly different energy may be involved in the processes, giving rise to the broadening. The similarity of the LO phonon softening in CuCl and CuBr QDs suggests that the softening is a common feature of ionic semiconductor nanocrystals possessing an efficient polar-exciton–phonon coupling.

## 7.5    Determination of the Orientation of CuCl Nanocrystals in a NaCl Matrix

As mentioned above, CuCl is a direct-gap semiconductor that crystallizes in the zinc blende structure (the point-group $T_d$). The electronic band structure near the band gap is characterized by a $\Gamma_6$ conduction band and two valence bands. The uppermost and lower valence bands correspondingly have $\Gamma_7$ and $\Gamma_8$ symmetry. An electron in the conduction band and a hole in the $\Gamma_7$ valence band form a $Z_3$ exciton, whereas an electron and a hole in the $\Gamma_8$ valence band form a $Z_{1,2}$ exciton. The 1S orthoexcitons have the $\Gamma_5$ symmetry. They are dipole-allowed in both one- and two-photon absorption. One-photon absorption is isotropic in cubic crystals for electric dipole transitions, whereas two-photon absorption of excitons depends on the polarization direction of the laser beam with respect to the crystal axes. This peculiarity of the TPA polarization selection rules was used for experimental examination of the assumption that CuCl nanocrystals embedded in a NaCl matrix are cubic and show a preferable orientation in the host matrix [55]. Indeed, if CuCl nanocrystals have a definite orientation with respect to the crystal axes of the NaCl host crystal, it should be possible to determine the orientation in the polarization experiment on TPA. The polarization dependence of the two-photon generation rate for $\Gamma_5$ excitons is given by the symmetric vector product of the polarization vectors:

$$\epsilon_x^2 \epsilon_z^2 + \epsilon_z^2 \epsilon_y^2 + \epsilon_y^2 \epsilon_x^2 , \tag{7.5}$$

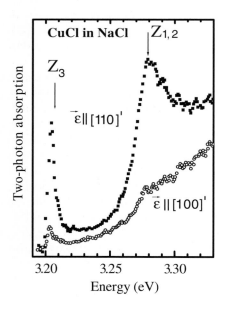

**Fig. 7.16.** Two-photon absorption spectrum of CuCl nanocrystals in a NaCl matrix at a temperature of 8 K for two different polarizations. The incoming light is parallel to the [001]′ axis of the NaCl matrix [55]

where $\epsilon_i (i=x,y,z)$ are the components of the polarization vector of the incoming laser beam. Thus the TPA signal is expected for polarization parallel to the [110] direction in CuCl nanocrystals, whereas the polarization selection rules forbid TPA for $\bar{\epsilon} \| [100]$. The polarized TPA spectra were measured through two-photon excited free-exciton luminescence. The experiment was performed with a tunable dye laser of a $\sim 5$ ns pulse duration and a 10 Hz repetition rate. The plane-polarized laser beam illuminated the sample in a certain direction relative to the NaCl crystal axes. A pronounced TPA polarization dependence was observed, as shown in Fig. 7.16. There was a strong signal for $\bar{\epsilon} \| [110]'$ and a very weak signal for $\bar{\epsilon} \| [100]'$ (primed indices refer to the NaCl host crystal). Therefore, the polarization dependence gave strong evidence that the [100] crystal axes of the CuCl nanocrystals were parallel to the [100]' crystal axes of the NaCl host crystal. Moreover, the orientation favors the assumption of cubic nanocrystals.

## 7.6  Single Nanocrystal Luminescence by Two-Photon Excitation

Two-photon excitation has an advantage for the study of a single QD by the use of a conventional technique of confocal microscopy. Since TPA is proportional to the incident light intensity squared, small volume selection can be achieved without a pinhole. This advantage was used for the study of capped CdS and CdSe QDs in organic glasses or PMMA at low temperatures [17,18]. It was found that an effective volume of about 1 $\mu m^3$ was selected with a high-numerical-aperture objective ($100\times$, NA=1.3) for focusing an 800 nm incident light beam and collecting a 400 nm photoluminescence. For a system of CdS QDs with a low QD density at 7 K, single lines at different positions within the inhomogeneous envelope were resolved, with a 15 meV average width corresponding to the spectral resolution adopted. The narrow singularities attributed to single QDs were observed at a higher QD density than expected and only after significant photobleaching. They were long-lived and exhibited slow variations in intensity. It was supposed that the photobleaching removed defective QDs from interaction with the incident light, and hence only defect-free QDs contributed to the single lines. For ZnS-capped CdSe QDs, similar single lines of about 25 meV width were observed at 10 K. The lines were slightly asymmetric: some unresolved feature could be distinguished at the low-energy side of the line. It was supposed that it was due to the LO phonon sideband with the Stokes shift of 25.4 meV. Despite poor spectral resolution, the Huang–Rhys factor for the LO phonon coupling was estimated at $S=1.2$ in the framework of the shifted harmonic oscillator model.

The effect of the reduction of the excitation volume at two-photon excitation to 0.1–1 $\mu m^3$ was used in the microphotoluminescence study of CuCl QDs in thin NaCl crystals at 5 and 30 K [56,57]. Due to the effect, the sharp

TPL lines at $2E_L$ originating from individual QDs under resonant two-photon excitation could be observed. The spectral widths of the lines were controlled by the spectral bandwidth (0.4 nm) of the picosecond Ti:sapphire laser used for the excitation.

The detected signals of two-photon excited luminescence from single CdSe QDs embedded in organic glasses [17] and CuCl QDs embedded in NaCl crystal [56,57] showed that the QDs decayed irreversibly within several minutes under irradiation of the excitation laser. The effect was assigned to the cascade absorption process: under the intense fundamental laser irradiation, the two-photon population of the lowest state was followed by effective cascade one-photon transitions to the higher excited states, with subsequent destructive chemical processes [17] or ionization [56,57] of the QDs.

## 7.7    Conclusion

Resonant two-photon spectroscopy has proved itself a promising size-selective method complementary to one-photon spectroscopy for exploring the electronic and phonon structure as well as the electron–phonon interaction in II–VI and I–VII semiconductor QDs embedded in dielectric matrices.

## References

1. S.J. Cyvin, J.E. Rauch, J.C. Decius: J. Chem. Phys. **43**, 4083 (1965)
2. V.N. Denisov, B.N. Mavrin, V.B. Podobedov: Phys. Rep. **151**, 1 (1987)
3. A.V. Baranov, Y.S. Bobovich, V.I. Petrov: Sov. Phys. Usp. **33**, 812 (1990)
4. A. García-Cristóbal, A. Cantarero, C.Trallero-Giner, M. Cardona: Phys. Rev. B **58**, 10443 (1998)
5. K. Inoue, K. Watanabe: Phys. Rev. B **39**, 1977 (1989); K. Watanabe, K. Inoue: Phys. Rev. B **46**, 2024 (1992)
6. K. Inoue, K. Yoshida, F. Minami, Y. Kato: Phys. Rev. B **45**, 8807 (1989); S. Matsushita, F. Minami, K. Yoshida, K. Inoue, H. Akinaga: J. Lumin. **66–67**, 414 (1996)
7. Al.L. Efros, A.L. Efros: Sov. Phys. Semicond. **16**, 772 (1982)
8. Y. Masumoto: J. Lumin. **70**, 386 (1996)
9. A.V. Fedorov, A.V. Baranov: Phys. Rev. B **54**, 8627 (1996)
10. E. Menéndes-Proupin, C. Trallero-Giner, A. García-Cristóbal: Phys. Rev. B **60**, 5513 (1999)
11. A.V. Baranov, Y.S. Bobovich, N.I. Grebenshchikova, V.I. Petrov, M.A. Tsenter: Opt. Spectrosc. **60**, 685 (1986)
12. A.V. Baranov: *Resonant Two-Photon Spectroscopy of Organic Molecules and Quasi-Zero-Dimensional Semiconductor Nanostructures*, D. Sci. Thesis, State Optical Institute, St.Petersburg (1998)
13. D.C. Haueisen, H. Mahr: Phys. Rev. B **8**, 734 (1973)
14. K.I. Kang, B.P. McGinnis, Sandalphon, Y.Z. Hu, S.W. Koch, N. Peyghambarian, A. Mysyrowicz, L.C. Liu, S.H. Risbud: Phys. Rev. B **45**, 3465 (1992)
15. J.B. Xia: Phys. Rev. B **40**, 8500 (1989)

16. P.C. Sercel, K.J. Vahala: Phys. Rev. B **42**, 1990 (1990)
17. S.A. Blanton, M.A. Hines, M.E. Schmidt, P. Guyot-Sionnest: J. Lumin. **70**, 253 (1996)
18. M.E. Schmidt, S.A. Blanton, M.A. Hines, P. Guyot-Sionnest: Phys. Rev. B **53**, 12629 (1996)
19. A.I. Ekimov, F. Hache, M.C. Schanne-Klein, D. Richard, C. Flytzanis, I.A. Kudryavtsev, T.V. Yazeva, A.V. Rodina, Al.L. Efros: J. Opt. Soc. Am. B **10**, 100 (1993)
20. A.V. Baranov, K. Inoue, K. Toba, A.Yamanaka, V.I. Petrov, A.V. Fedorov: Phys. Rev. B **53**, 1721 (1996)
21. K. Inoue, A.V. Baranov, A. Yamanaka: Physica B **219–220**, 508 (1996)
22. *Excitons, Topics in Current Physics Vol. 14* ed by K. Cho (Springer, Heidelberg, Berlin 1979); *Excitons*, ed. by E.I. Rashba, M.D. Sturge (North-Holland, Amsterdam 1982)
23. A.I. Ekimov, A.A. Onushchenko: JETP Lett. **40** 1136 (1984)
24. M. Ueta, H. Kanzaki, M. Kobayashi, Y. Toyozawa, E. Hanamura: *Excitonic Processes in Solids* (Springer, Heidelberg Berlin 1986) p. 212
25. E.O. Kane: J. Phys. Chem. Solids **1**, 249 (1957)
26. I.M. Cidil'kovsky: *Electrons and Holes in Semiconductors* (Nauka, Moscow 1972)
27. M.A. Chamarro, C. Gordon, P. Lavallard, A.I. Ekimov: Jpn. J. Appl. Phys. Suppl. **34-1**, 12 (1995)
28. M. Nirmal, D.J. Norris, M. Kuno, M.G. Bawendi, Al.L. Efros, M. Rosen: Phys. Rev. Lett. **75**, 3728 (1995)
29. U. Woggon, F. Gindele, O. Wind, C. Klingshirn: Phys. Rev. B **54**, 1506 (1996)
30. T. Takagahara: Phys. Rev. B **47**, 4569 (1993)
31. T. Itoh, M. Nishijima, A.I. Ekimov, C. Gourdon, Al.L. Efros, M. Rosen: Phys. Rev. Lett. **74**, 1645 (1995)
32. S. Okamoto, Y. Masumoto: J. Lumin. **64**, 253 (1995)
33. A.V. Baranov, Y. Masumoto, K. Inoue, A.V. Fedorov, A.A. Onushchenko: Phys. Rev. B **55**, 15675 (1997)
34. T. Kawazoe, Y. Masumoto: Meeting Abstracts of the Phys. Soc. Japan **52**, 186, (1997)
35. M.M. Denisov, V.P. Makarov: J. Phys. C **5**, 2651 (1972)
36. *Semiconductors, Landolt-Börnstein, New Series, Group III, Vol. 17a* ed. by O. Madelung, M. Schultz, H. Weiss (Springer, Heidelberg, Berlin 1982)
37. T. Mita, K. Sôtome, M. Ueta: J. Phys. Soc. Jpn. **48**, 486 (1980); Solid State Commun. **33**, 1135 (1980)
38. K. Cho: J. Lumin. **87–89**, 7 (2000); H. Ajiki, K. Takizawa, K. Cho: J. Lumin. **87–89**, 341 (2000)
39. A.V. Fedorov, A.V. Baranov, K. Inoue: Phys. Rev. B **56**, 7491 (1997)
40. S. Schmitt-Rink, D.A.B. Miller, D.S. Chemla: Phys. Rev. B **35**, 8113 (1987)
41. A.P. Alivisatos, T.D Harris, P.J. Carroll, M.L. Steigerwald, L.E. Brus: J. Chem. Phys. **90**, 3463 (1989)
42. M.C. Klein, F. Hache, D. Ricard, C. Flitzanis: Phys. Rev. B **42**, 11123 (1990)
43. S. Nomura, T. Kobayashi: Phys. Rev. B **45**, 1305 (1992)
44. J.C. Marini, B. Stebe, E. Kartheuser: Phys. Rev. B **50**, 14302 (1994)
45. R. Englman, R. Ruppin: J. Phys. C **1**, 614 (1968); R. Englman, R. Ruppin: Rep. Prog. Phys. **33**, 149 (1970)

46. K. Inoue, A. Yamanaka, K. Toba, A.V. Baranov, A.V. Fedorov, A.A. Onushchenko: Phys. Rev. B **54**, 8321 (1996)
47. A.V. Baranov, S. Yamauchi, Y. Masumoto: Phys. Rev. B **56**, 10332 (1997)
48. A.V. Baranov, S. Yamauchi, Y. Masumoto: J. Lumin. **87–89**, 500 (2000)
49. W. Siebrand, M.Z. Zgierski: In: *Excited States, Vol. 4* ed. by E. C.Lim (Academic Press, New York 1979) p. 1
50. R.C.C. Leite, J.F. Scott, T.C. Damen: Phys. Rev. Lett. **22**, 780 (1969); M.V. Klein, S.P.S. Porto: Phys. Rev. Lett. **22**, 782 (1969)
51. V.I. Belitsky, A. Cantarero, S.T. Pavlov: Phys. Rev. B **52**, 11920 (1995), and references therein
52. L. Zimin, S.V. Nair, Y. Masumoto: Phys. Rev. Lett. **80**, 3105 (1998)
53. S.V. Nair, Y. Masumoto: Jpn. J. Appl. Phys. **38**, 581 (1999)
54. H. Ohmura, A. Nakamura: Phys. Rev. B **59**, 12216 (1999)
55. D. Fröhlich, M. Haselhoff, K. Reinmann, T. Itoh: Solid State Commun. **94**, 189 (1995)
56. K. Edamatsu, M. Tsukii, K. Hayashibe, M. Nishijima, T. Itoh, B.P. Zhang, Y. Segawa, A.I. Ekimov: 'Resonant one- and two-photon excitation microspectroscopy of CuCl quantum dots'. In: *Nonlinear Optics, Vol. 18(2–4)* (OPA, Amsterdam 1997) pp. 295-302
57. K. Edamatsu, T. Itoh, S. Hashimoto, B.P. Zhang, Y. Segawa: J. Lumin. **87–89**, 387 (2000)

# 8 Homogeneous Width of Confined Excitons in Quantum Dots – Experimental

Yasuaki Masumoto

## 8.1 Introduction

A quantum dot has quantized energy levels that show a size-dependent blue-shift as a result of the quantum confinement effect. The quantized levels have been generally believed to be as sharp as the $\delta$-function, reflecting its atomic character and size. Other inhomogeneities in the ensemble of quantum dots are considered to make the optical spectrum of the levels broad. However, this consideration is oversimplified. As a result of unique dephasing mechanisms in quantum dots, the homogeneous width of quantum dots is not so sharp as the $\delta$-function, but is found to be finite and is sometimes broader than that of the bulk crystals. It is given by the inverse of dephasing time consisting of radiative lifetime, impurity or defect scattering time, surface or interface scattering time, phonon scattering time, and carrier–carrier scattering time at the excited states. Because electrons, excitons, and phonons are size-quantized in quantum dots, the electron–phonon and exciton–phonon interactions in quantum dots are unique. Temperature dependence of the homogeneous width reflects the unique temperature-dependent dephasing mechanisms of quantum states by phonons; their study is very important. Therefore the homogeneous linewidth of the optical spectra of quantum dots is one of the most important characters of quantum dots. Homogeneous linewidth is important not only from the basic understanding of quantum dots but also from the application point of view, because it is inversely proportional to the optical nonlinearity. Very narrow linewidth of the homogeneous optical spectra of quantum dots means a long dephasing time, which enables the multiform coherent control of the quantum state suitable for quantum computation. Because of thisimportance, the homogeneous optical spectrum of quantum dots should be studied.

So far, persistent spectral hole burning [1], photoluminescence blinking [2,3], and spectral diffusion [4] of quantum dots are observed. These phenomena are not observed in bulk semiconductors or even quantum wells or quantum wires, and therefore are considered unique for quantum dots. Mechanisms of these phenomena are considered to arise from the presence of traps nearby quantum dots; the energy of quantum dots is affected by the carriers captured by traps. The interaction between quantum dots and the surrounding matrix can work as a dephasing mechanism of quantum dots. Its contribution to the dephasing mechanism should be studied.

To observe the homogeneous linewidth of quantum dots in inhomogeneously broadened optical spectra, site-selective spectroscopy, such as hole burning and fluorescence line narrowing, and single quantum dot spectroscopy are used. Another method is coherence measurement in the time domain, such as photon echo, which can measure dephasing relaxation time $T_2$ and hence give the homogeneous linewidth. In this chapter, the experimental studies of the homogeneous width of quantum dots are reviewed. Especially, temperature-dependent homogeneous width of confined excitons in quantum dots investigated at very low temperatures by means of accumulated photon echo is discussed.

## 8.2   Spectral Hole Burning and Fluorescence Line Narrowing

Site-selective spectroscopy such as spectral hole burning and fluorescence line narrowing has been generally used to measure the homogeneous width from the inhomogeneously broadened optical spectra of atoms, ions, and molecules in the gas phase, solutions, and solids. They have also been used to measure the homogeneous width from the inhomogeneously broadened optical spectra of quantum dots [5,6]. A burned hole has a full-width at the half-maximum (FWHM) of $4\hbar/T_2$. Two absorption steps, burning and detection, are needed for the observation of spectral hole burning and, therefore, the burned hole is twice as broad as the homogeneous spectrum.

In the strong-confinement regime, hole burning was used for the study of CdSe quantum dots embedded in glass. A rather broad hole width (FWHM) of 100 meV was observed for the $1S_{3/2}1S_e$ absorption resonance at 10 K, reflecting a very fast dephasing time of 25 fs [7]. On the other hand, for CdSe quantum dots capped by organic groups whose radii range from 1.8 nm to 2.5 nm, a burned hole with 30 meV width (FWHM) was observed at 7 K [8]. Later, for 1.6-nm-radius CdSe quantum dots capped by organic groups, a burned hole with 16 meV width (FWHM) was observed for the $1S_{3/2}1S_e$ absorption resonance [9]. These different results are understood if the homogeneous width of quantum dots is dominated by the surface or surroundings of quantum dots. Further, the broad hole width of 90 meV at 50 K for 2.0-nm-radius CdSe quantum dots embedded in glass was found to be reduced to 10 meV after the sample was exposed to strong laser illumination [10]. This change was attributed to changes in the interface charge state or interface polarization under high excitation. Therefore it is considered that the homogeneous width is affected by the surface and surroundings of quantum dots.

Very recently, high-resolution spectral hole-burning measurement of chemically synthesized CdSe/ZnS quantum dots ($R$=4.5 nm) was done for the

$1S_{3/2}1S_e$ absorption resonance by means of two tunable diode lasers with the spectral linewidth of 0.4 μeV [11]. The spectral hole observed at 10 K consisted of a narrow zero-phonon line and confined acoustic phonon wings with energes of 0.67 meV and 1.5 meV. The narrowest homogeneous width was observed at 2 K and was 32 μeV. The nonlinear temperature-dependent homogeneous width was explained by the two-phonon process involving simultaneous absorption and emission of confined acoustic phonons. This measurement suggested that many preceding hole-burning measurements of chemically synthesized CdSe/ZnS quantum dots could not dissolve the zero-phonon lines from the acoustic phonon wings.

In the weak-confinement regime, the homogeneous width of $Z_3$ exciton of CuCl quantum dots embedded in NaCl crystals was investigated by hole burning and fluorescence line narrowing. The observed homogeneous width of CuCl quantum dots was much narrower than that of CdSe quantum dots and was less than 1 meV for 6.1-nm-radius CuCl quantum dots at 77 K [12]. Hole-burning spectra of CuCl quantum dots were measured by means of a nanosecond pump and probe, and temperature and size dependences of the homogeneous width were obtained [13]. In the low-temperature limit the hole width (FWHM) was 0.59 meV for 4.1-nm-radius CuCl quantum dots. Quadratic temperature dependence and inverse quadratic size dependence were obtained. After persistent spectral hole burning was known to take place in CuCl quantum dots embedded in NaCl crystals, persistent spectral hole burning was used to observe the homogeneous width [14]. At 2 K the observed hole width (FWHM) was 0.14 meV for 3.5-nm-radius dots and the zero-phonon hole was well separated from the wing of the confined acoustic phonon. Its temperature dependence was expressed by the activation of a confined acoustic phonon of 2.2 meV.

Fluorescence line narrowing was used to observe the homogeneous linewidth of CuCl quantum dots embedded in a NaCl crystal [15]. The site-selective excited luminescence of CuCl quantum dots showed a sharp linewidth of 0.4 meV at 4.2 K for 5.7-nm-radius dots. The linewidth increased with the decrease of the radius of the dots. Recently, more precise measurement of fluorescence line narrowing was done, and acoustic phonon wings were removed from the zero-phonon line [16]. The linewidth was estimated to be as sharp as 50 μeV at 10 K for 2.4-nm- and 3.2-nm-radius CuCl dots. Quadratic temperature dependence was observed and was ascribed to the two-phonon Raman process by confined acoustic phonons, as reported on the basis of the temperature-dependent accumulated photon echo data [17].

## 8.3   Single Quantum Dot Spectroscopy

Quantum dots are different from each other in size, shape, orientation, and surrounding environment. An optical spectrum of a single quantum dot was expected to be free from inhomogeneous broadening and was used to evaluate homogeneous broadening. To observe a single quantum dot, microphotoluminescence of dots by means of far-field microscopic optics, near-field optics, or cathodephotoluminescence of dots is used. The detailed descriptions of the methods are given in Chap. 4.

Well-width fluctuations form quantum-dot-like states in a GaAs/AlGaAs quantum well. A microphotoluminescence spectrum of a single GaAs/$Al_{0.35}$ $Ga_{0.66}$As quantum well grown by molecular beam epitaxy with growth interruption showed sharp luminescence lines at low temperature [18]. During the growth interruption, Ga and Al surface diffusion increases the lateral correlation length and makes interface roughness with a larger lateral correlation length. An exciton is localized in a small, atomically flat island formed in a GaAs well. A single island can be observed by a microscope with a spatial resolution of 1.5 μm. At 5 K, the observed sharp luminescence line was less than 0.1 meV, which is close to the spectrometer resolution of 0.07 meV.

Photoluminescence of a GaAs/$Al_{0.3}$$Ga_{0.7}$As single quantum well was investigated through an Al metal aperture having a small hole [19]. By means of deconvolution analysis of the spectrometer resolution of 30 μeV, the homogeneous width of a single quantum-dot-like state in a single GaAs quantum well was estimated to be 23±10 μeV below 10 K. The temperature dependence of the homogeneous width was interpreted by taking account of one phonon-assisted excitation to the upper quantum states and phonon-assisted relaxation to the lower quantum states.

A sharp microphotoluminescence line from a single GaAs/$Al_{0.4}$$Ga_{0.6}$As quantum dot of dimension 190×160×12 nm was reported [20]. It was made by using the selective growth technique and metal organic chemical-vapor deposition. The photoluminescence linewidth of a single dot was 0.9 meV at 15 K.

A single InAs quantum dot made by self-organized growth on a GaAs substrate showed very sharp photoluminescence at low temperature [21]. When the 500 nm mesa structure was formed on the sample, the number of quantum dots was reduced, and the luminescence spectrum became a set of sharp lines whose widths were less than 0.1 meV (the resolution of the measurement sytem) at 10 K. Cathodoluminescence was used to excite a single InAs self-assembled quantum dot and showed a set of lines whose linewidths were less than the 0.15 meV resolution of the measurement system below 50 K [22].

A single InP/$In_{0.5}$$Ga_{0.5}$As quantum dot made by self-organized growth was studied by microphotoluminescence [23]. It was observed through a 1-μm-diameter hole made on a metal mask evaporated on the sample. The luminescence spectrum showed sharp lines narrower than 40 μeV of the resolution of the spectrometer at 5 K.

A chemically synthesized single CdSe quantum dot showed sharp luminescence lines whose widths were less than 0.12 meV of the spectral resolution of the measurement sysytem at 10 K [24]. The microphotoluminescence spectrum of a single quantum dot showed spectral diffusion that depended on the laser power. These observations show that observation of the homogeneous width of quantum dots needs not only high spectral resolution but also time-dependent spectral information.

In summary, single quantum dot spectroscopy can study the homogeneous width of the quantum dot if the homogeneous width is broader than the resolution of the spectrometer. At low temperatures, optical spectra of single dots of many kinds are too sharp for the spectral domain measurement by means of grating monochromators.

## 8.4   Photon Echo

Photon echo is a standard measurement technique of dephasing relaxation time in time domain. Ensembles of inhomogeneously broadened transitions are coherently excited by the first short pulse, their optical dipoles dephase at the dephasing time constant of $T_2$, and the second short pulse rephases their polarization. Then macroscopic polarization is built up and produces a photon echo. Although coherently photoexcited polarizations oscillate at their own frequency and macroscopic polarization decreases in proportion to photoexcited inhomogeneous broadening, the total polarization is built up as a result of a rephasing process. Therefore, the photon echo can observe the pure dephasing time constant $T_2$ independently of inhomogeneous broadening. The photon echo technique has been used to measure the dephasing time constant $T_2$ of electronic excitations in atoms, ions, molecules, and semiconductors. Between the dephasing relaxation time $T_2$ and the corresponding homogeneous linewidth (FWHM) $\Gamma_h$, the relation $\Gamma_h=4\hbar/T_2$ holds.

When inhomogeneous broadening dominates the optical spectrum, the four-wave mixing signal decays exponentially with a time constant of $T_2/4$ [25]. Chemically synthesized 1.1-nm-radius CdSe quantum dots embedded in a polymer film were studied by the three-pulse photon echo separating the oscillation of the coherently excited longitudinal optical phonon [26]. Nearly transform-limited pulses of 15 fs duration were used to observe the photon echo. The dephasing time was 85 fs at 15 K, which means a homogeneous width of 31 meV. The homogeneous width increased nearly in proportion to the temperature.

Detailed three-pulse photon echo measurements of CdSe quantum dots were reported as a function of both the dot size and temperature [27]. A linearly temperature-dependent increase of the homogeneous width was observed for dots whose radii range from 1 nm to 2 nm. The size-dependent homogeneous width was divided into three processes: temperature-dependent phonon broadening due to low-frequency modes, temperature-insensitive lifetime broadening, and temperature-independent elastic scattering by impurities or defects. Lifetimes ranged from 400 fs for the 1-nm-radius dot to 8 ps for the 2-nm-radius dot. The lifetime broadening was considered to be proportional to $1/R$, because it comes from the process of trapping at the surface.

Elastic scattering by impurities or defects was considered to be proportional to $1/R$, reflecting the surface-to-volume ratio.

Three-pulse photon echo measurement was applied to chemically synthesized InP quantum dots embedded in a polymer film [28]. For 1.5-nm-radius InP quantum dots, the dephasing time was 280 fs around 20 K. The corresponding homogeneous width was 9.4 meV and increased linearly with increasing temperature. The temperature dependence was interpreted by the dephasing of the low-frequency acoustic phonons.

## 8.5   Accumulated Photon Echo

### 8.5.1   Accumulated Photon Echo and Persistent Hole Burning

In Chap. 5, persistent spectral hole-burning phenomena of many kinds of quantum dots are reviewed. It is shown that the phenomena are observed widely in many kinds of quantum dots. The persistent spectral hole-burning phenomena make the accumulated photon echo measurements possible under an ideally weak excitation. As is described in Chap. 5, a very sharp optical spectrum whose width is comparable to or less than 0.1 meV has been reported for several kinds of quantum dots by means of single quantum dot spectroscopy and site-selective laser spectroscopy, but the linewidth is too sharp to be studied with sufficient spectral resolution. This experimental difficulty has masked the intrinsic physical properties of quantum dots. An alternative approach in the time domain has potential to solve these problems. A very sharp homogeneous width indicates a long dephasing time, which is easily accessible by the time-domain measurements, such as photon echo. Fourier transform of the time trace observed by the photon echo gives a homogeneous optical spectrum. A possible weak point of the photon echo measurement is the necessity of intense pulses for echo detection because of the third-order nonlinear optical process of the photon echo. Intense excitation creates many carriers and increases the dephasing rate due to the mutual collisions of carriers.

The heterodyne-detected accumulated photon echo measurement is known as one of the most sensitive methods of photon echo measurement. It enables us to observe the photon echo signal under the very low excitation condition free from the dephasing due to the mutual collisions of carriers. The unique requirement for its applicability is the presence of a longer absorption recovery time compared with the laser repetition period [29,30]. Therefore, persistent spectral hole-burning phenomena in quantum dots automatically satisfy the requirement for the applicability of the accumulated photon echo [1].

Let us consider the homogeneous spectrum in the inhomogeneously broadened optical spectrum shown in Fig. 8.1. When two identical coherent light pulses, $P_1$ and $P_2$, excite the optical spectrum, interference between polarizations excited by two succesive coherent pulses takes place in the inhomogeneously broadened optical spectrum during the coherence time of the

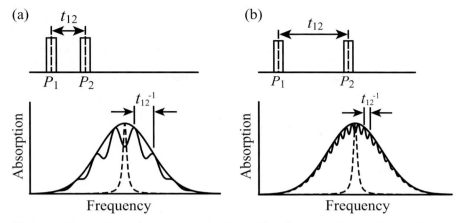

**Fig. 8.1.** Spectral interference pattern formed by the successive excitation by two identical coherent light pulses in the inhomogeneously broadened optical spectrum for (**a**) short time separation and (**b**) long time separation. *Dashed lines* show the homogeneous spectrum in the inhomogeneously broadened optical spectrum

polarization. The interference period in the frequency spectrum is expressed by $1/t_{12}$, where $t_{12}$ is the time separation between the two pulses. The interference makes a population grating, but the grating pitch can not be narrower than the homogeneous width. Therefore, the modulation depth of the population grating depends on the ratio of the homogeneous width and the interference period. If the interference period is narrower than the homogeneous width, the modulation depth is largely reduced. The longer absorption recovery time compared with the laser repetition period accumulates the interference and increases the modulation depth. As a result, the photon echo signal decays at the dephasing time of the polarization, which is the inverse of the homogeneous width.

For the photon echo measurement, the phase-matching condition between the excitation pulses and the echo signal should be satisfied. Here we use wavevectors of 3 excitation pulses, $\boldsymbol{k}_1$, $\boldsymbol{k}_2$, and $\boldsymbol{k}_3$, and a wavevector of the echo pulse $\boldsymbol{k}_e$ to describe the propagation direction of the pulses. To satisfy the phase-matching condition, collinear $(\boldsymbol{k}_1=\boldsymbol{k}_2=\boldsymbol{k}_3=\boldsymbol{k}_e)$, noncollinear $(\boldsymbol{k}_1\neq\boldsymbol{k}_2, \boldsymbol{k}_3=\boldsymbol{k}_2, \boldsymbol{k}_e=2\boldsymbol{k}_2-\boldsymbol{k}_1$ or $\boldsymbol{k}_1\neq\boldsymbol{k}_2, \boldsymbol{k}_3=\boldsymbol{k}_1, \boldsymbol{k}_e=\boldsymbol{k}_2)$, phase conjugated $(\boldsymbol{k}_1\neq\boldsymbol{k}_2, \boldsymbol{k}_3=-\boldsymbol{k}_2, \boldsymbol{k}_e=-\boldsymbol{k}_1)$, or boxcar $(\boldsymbol{k}_1=(k_x,k_y,k_z), \boldsymbol{k}_2=(-k_x,k_y,k_z), \boldsymbol{k}_3=(k_x,-k_y,k_z), \boldsymbol{k}_e=(-k_x,-k_y,k_z))$ beam configurations are used. For the accumulated photon echo measurement, the two-pulse noncollinear configuration $(\boldsymbol{k}_1\neq\boldsymbol{k}_2, \boldsymbol{k}_3=\boldsymbol{k}_1)$ with the echo direct ion of $\boldsymbol{k}_e=\boldsymbol{k}_2$ is frequently used. In this case, the echo pulse $\boldsymbol{k}_e$ is in phase with the probe pulse $\boldsymbol{k}_2$ and constructively interferes with the probe pulse at the photodetector. Because the photodetector signal is proportional to the square of the electric field, it is proportional to the product of the electric fields of the echo pulse and the

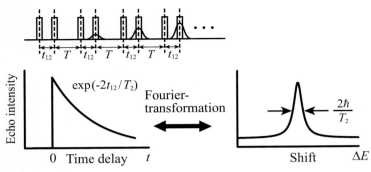

**Fig. 8.2.** The Fourier-cosine transform of the time trace observed by the heterodyne-detected accumulated photon echo gives the persistent hole-burning spectrum and vice versa

probe pulse under the condition that the echo pulse is much weaker than the probe pulse. Therefore, the echo intensity detected in this configuration decays with $\exp(-2t_{12}/T_2)$, in contrast to the decay of $\exp(-4t_{12}/T_2)$ in case of the phase-matching condition of $\boldsymbol{k}_e=2\boldsymbol{k}_2-\boldsymbol{k}_1$ [29]. It is rigorouly verified that the Fourier-cosine transform of the time trace observed by the heterodyne-detected accumulated photon echo gives the persistent hole burning spectrum and vice versa [31]. The relation is schematically displayed in Fig. 8.2. The heterodyne-detected accumulated photon echo signal decays at a decay time constant of $T_2/2$. From the measurement of $T_2$, the homogeneous width was given by $2\hbar/T_2$ for the full-width at the half-maximum or $\hbar/T_2$ for the half-width at the half-maximum.

The accumulated photon echo of quantum dots was recognized by the observation of the linear power dependence of the echo intensity and by the slow rise and slow decay of the echo signal with the increase of the irradiation time of the laser pulses for the $Z_3$ exciton in CuCl quantum dots [32] and for the $Z_{1,2}$ exciton in CuBr quantum dots [33]. The demonstrative data is shown in Fig. 8.3. Time-integrated and time-resolved intensities of the two-pulse photon echo signal evolve with the increase of the irradiation time of the pulses. The irradiation-time-dependent signal and the linear power dependence of the signal were explained by the accumulation effect of the population grating due to the bottleneck state for the relaxation. The accumulated grating is formed by the longer recovery time than the pulse repetition time $T$. The long recovery time forms the persistently burned hole simultaneously.

The excitation intensity for the observation of the accumulated photon echo for the $Z_3$ exciton in CuCl quantum dots [34] was 2 nJ/cm$^2$ [27], which was four orders of magnitude less than that for the photon echo measurement of CdSe quantum dots. The reduction of the excitation power led to the reliable dephasing time and made possible the observation of the rise of

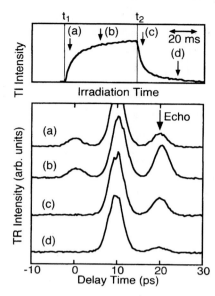

**Fig. 8.3.** *Upper figure*: Irradiation-time-dependent time-integrated intensity of the two-pulse photon echo signal of CuCl quantum dots embedded in glass with the time separation between two pulses of 10 ps. *Lower figure*: The time-resolved intensity of the two-pulse photon echo signal (from Kuroda et al. [32])

the photon echo signal coming from the phonon sideband. At 2 K the observed dephasing time was 130 ps, corresponding to the homogeneous width of 10 µeV.

### 8.5.2 Phase-Modulation Technique of the Accumulated Photon Echo – Application to Quantum Dots

To observe the photon echo signal masked by scattering of the stronger-intensity excitation pulse or the probe pulse itself, the modulation of intensity or phase and phase-sensitive detection are used. In particular, phase-sensitive detection with phase modulation is a very sensitive way to obtain the accumulated photon echo signal. The phase-modulation technique with an optical delay driven by a piezoelectric actuator is especially effective to observe the accumulated photon echo signal, because the modulation is pure phase modulation and does not contain amplitude modulation [35]. The phase-modulated echo signals were detected by a photodiode and then fed into a lock-in amplifier. The experimental setup is shown in Fig. 8.4.

The heterodyne-detected accumulated photon echo signal was observed in this experimental setup in quantum dots such as CuCl quantum dots in glass and NaCl crystals, CuBr quantum dots in glass, and CdSe quantum dots in glass. The excitation sources of 82 MHz picosecond and femtosecond Ti:sapphire laser systems, including an optical parametric generator pumped by a 200 kHz Ti:sapphire regeneratively amplifier, were used. The second harmonics of the Ti:sapphire laser were used for the excitation of CuCl and CuBr quantum dots, while the output of the optical parametric generator

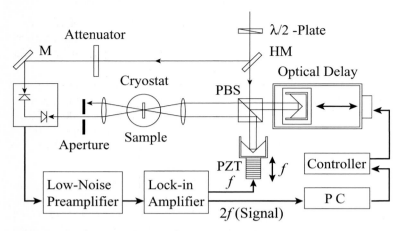

**Fig. 8.4.** Experimental setup for the accumulated photon echo measurement. Balanced photodiodes were used for the compensation of the fluctuation of the laser power. PBS: polarizing beam splitter; PZT: piezoelectric transducer driving a corner cube reflector

was used for the excitation of CdSe quantum dots. The heterodyne-detected accumulated photon echo enables us to observe long dephasing time, i.e., ultranarrow homogeneous width, under weakest excitation from the time domain, and therefore is an ideal method to study the homogeneous width of quantum dots.

The CuCl quantum dot is typical of a weak-confinement regime, the CdSe quantum dot belongs to a strong-confinement regime, and the CuBr quantum dot belongs to an intermediate-confinement regime. The samples studied were CuCl quantum dots (average radius $R=2.0$ nm) embedded in NaCl crystals and aluminoborosilicate glass, referred to as CuCl-QD/NaCl and CuCl-QD/glass, CuBr quantum dots ($R=2.9$, 4.8 and 6.5 nm) in potassium aluminoborosilicate glass referred to as CuBr-QD/glass, and CdSe quantum dots ($R=2.3$, 2.5, 2.7 and 3.6 nm) in $GeO_2$:$Na_2O$ glass, referred to as CdSe-QD/glass. Very low temperature measurements below 1.5 K were done by using a helium-3 cryostat with a calibrated $RuO_2$ thermometer, while above 1.5 K a helium-4 immersion-type cryostat and a temperature-variable continuous-flow cryostat were used.

### 8.5.3 Accumulated Photon Echo Signal and the Homogeneous Width of CuCl Quantum Dots

When the repetition time period of the excitation light is shorter than the absorption recovery time of the material, accumulation of the photon echo signal takes place. The persistent spectral hole-burning process produces an accumulated population grating and largely enhances the amplitude of the grating as a result of accumulation. The heterodyne detection of the accumulated photon echo is very sensitive as a result of both accumulation of the grating and heterodyne detection, and therefore is detected far below the excitation level corresponding to one photon per dot.

The accumulated photon echo signals for the lowest $Z_3$ exciton of CuCl quantum dots are shown at the lower part of Fig. 8.5 for three temperatures. The photon energy and the excitation density was 3.23 eV and 100 pJ/cm$^2$. Absorption spectra of the samples at 2 K are shown at the upper part of Fig. 8.5 by solid lines. At low temperatures, the echo signal showed double

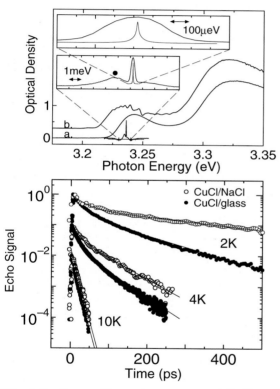

**Fig. 8.5.** *Upper*: Absorption spectra of CuCl quantum dots at 2 K: (**a**) CuCl-QD/glass and (**b**) CuCl-QD/NaCl. In the *insets*, a *solid curve* shows the persistent hole-burning spectrum of CuCl-QD/glass at 2 K with expanded scale. The very sharp Lorentzian curve drawn by a *thin line* represents the Fourier transform of the echo decay curve. *Lower*: Temperature-dependent accumulated photon echo signal of CuCl quantum dots. The *solid circles* and the *open circles* correspond to data for CuCl-QD/glass and CuCl-QD/NaCl, respectively. *Solid lines* represent double exponential decay (single exponential decay for 10 K data). The decay time constants of the slow component were 310, 117, 65, 55, 9.5, and 9.5 ps from top to bottom (from Ikezawa and Masumoto [17])

exponential decay. Then $T_2$ was obtained from the slower decay component. The fast component whose decay time is 10~30 ps is caused by a phonon sideband and does not correspond to the broadening of the zero-phonon line. The first point to notice is that the homogeneous linewidth obtained was extremely sharp. At the upper part of Fig. 8.5 and its insets, the persistent hole-burning spectrum of CuCl-QD/glass is shown by a solid line. The homogeneous linewidth is evaluated from half of the FWHM of the spectral hole. The Fourier transformed curve of the echo data at 2 K is represented in the inset by a fine solid line. The echo signal gives much sharper linewidth than that obtained by persistent spectral hole burning. Another important point is that the echo signal is different for the two samples at low temperature. At 10 K, the two signals are almost the same, but the difference becomes clear with the decrease in temperature. At 2 K, the echo signal of CuCl-QD/NaCl shows a more decay time more than twice as long as that of CuCl-QD/glass.

Homogeneous linewidth obtained from these data are plotted in Fig. 8.6 as a function of temperature. In this figure, solid (open) circles represent the homogeneous width of CuCl-QD/glass (CuCl-QD/NaCl). In comparison to previous work in the spectral domain [13,15,36,14], the data gives a very small homogeneous width at low temperatures. For example, the linewidth of CuCl-QD/NaCl measured by persistent spectral hole burning was 70 μeV

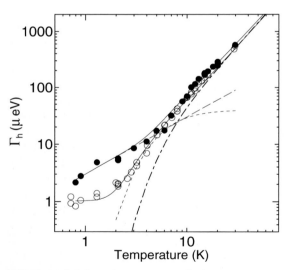

**Fig. 8.6.** *Solid circles* represent the homogeneous width of the $Z_3$ exciton of CuCl-QD/glass as a function of temperature. *Open circles* show the homogeneous width of CuCl-QD/NaCl. Above 5 K, both samples show a similar temperature dependence, but a remarkable difference can be seen below 5 K. *Long dashed* and *short dashed* curves represent the temperature-dependent TLS contribution of CuCl-QD/glass and CuCl-QD/NaCl, respectively (from Ikezawa and Masumoto [17])

at 2 K [14]. However, it was found to be 2 μeV in the accumulated photon echo measurement, which is smaller by an order of magnitude. In the persistent spectral hole-burning case, the spectral resolution of the spectrometer is comparable to the homogeneous width. Furthermore, the spectral diffusion occurring in the long interval from the hole-burning process to the absorption measurement may affect the experimental results.

The observed homogeneous width reached 1 μeV at low temperature. Even for accumulated photon echo, possible spectral diffusion during the accumulation process (several seconds) may broaden the homogeneous spectrum. Therefore, the homogeneous width is narrower than the value obtained. This is the smallest homogeneous width of quantum dot system. The value is narrower than the homogeneous width of excitonic polaritons in the bulk crystal by an order of magnitude [37,38]. This extreme narrowing of the homogeneous width of confined excitons in quantum dots is considered to come from the spatial confinement of the exciton wavefunction and that of the acoustic phonons in quantum dots.

The temperature dependences of the homogeneous widths of both samples were similar above 5 K, but an obvious difference was seen between them below this temperature. For CuCl-QD/NaCl, the homogeneous width continues to decrease rapidly below 5 K and approaches a lower limit. On the other hand, the homogeneous width of the CuCl-QD/glass changes its slope at 5 K and shows almost linear temperature dependence in the low-temperature region. It indicates that the matrix affects the dephasing rate of the exciton confined in quantum dots in the low-temperature region.

Let us discuss the origin of the temperature-dependent homogeneous broadening of confined excitons. As one mechanism that determines the homogeneous broadening at elevated temperatures, interaction of the confined exciton and the confined phonon in quantum dots is appropriate. LO phonon scattering is not important in the low-temperature range, but increases rapidly above 50 K [14]. Confined acoustic phonons are experimentally observed for CuCl quantum dots by persistent spectral hole burning [39,40]. As seen in the inset of Fig. 8.5, a sideband due to the confined acoustic phonon marked by a solid circle was clearly observed, which was separated from the zero-phonon line by 1.6 meV. For the high-temperature region, experimental data for both samples having a NaCl and a glass matrix can be fitted by assuming that the homogeneous width is determined by a two-phonon Raman process, where absorption and emission of a phonon simultaneously occur and whose probability is proportional to $n(n+1)$, where $n$ represents the occupation number of the confined acoustic phonon. Then the temperature dependence of the homogeneous broadening is given by $\Gamma_{\mathrm{h}}(T) \propto \sinh^{-2}(\hbar\omega/2kT)$, where $\hbar\omega$ is the energy of the lowest confined phonon mode. A dot-dashed line in Fig. 8.6 shows the contribution of the confined acoustic phonon for CuCl-QD/glass.

One cannot explain the observed matrix-dependent temperature dependence in the low-temperature region by this phonon mode. Another contribution that is related to the matrix of the quantum dots should be considered. One possible dephasing process is coupling to a two-level system (TLS) [41], which had been used to explain the anomalous temperature dependence of the homogeneous linewidth of impurity atoms in a glass matrix or dye molecules in a polymer. The basic concept of TLS is that there are double potential well systems in glass where quantum mechanical tunneling can occur. Temperature-dependent line broadening by the interaction with TLS is expressed in the form of $\cosh^{-2}(\hbar\omega/2kT)$, where $\hbar\omega$ is the characteristic energy of TLS. The randomness in glass is thought to distribute the energy of the TLS over a wide range. Therefore, broadening due to TLS is often approximated as $T^\alpha$ ($1<\alpha<2$) [41]. The experimental data for CuCl-QD/glass was reproduced by the sum of the contributions from the confined acoustic phonon and the $T$-linear TLS processes, as shown in Fig. 8.6 by a solid line on the open circles. On the other hand, the experimental data for CuCl-QD/NaCl can be well reproduced if the energy of TLS is concentrated on a characteristic energy $\hbar\omega'$. They exist not only in glassy systems but also in crystals [42]. Then the contribution of TLS is directly proportional to $\cosh^{-2}(\hbar\omega'/kT)$. Finally, the temperature dependence of the homogeneous width of CuCl-QD/NaCl is represented by

$$A\sinh^{-2}\left(\hbar\omega/2kT\right) + B\cosh^{-2}\left(\hbar\omega'/2kT\right) + \Gamma_0 \ .$$

A solid curve on the open circles in Fig. 8.6 shows this temperature dependence in the case of $\hbar\omega'=1$ meV. The fitting by the TLS model agrees well with the experimental results for both CuCl-QD/glass and CuCl-QD/NaCl. These satisfactory results lead to the conclusion that the matrix makes a difference in the temperature dependence of the homogeneous width of excitons in quantum dots through the difference of the energy distribution of TLS.

It is important to compare $T_2$ at the low-temperature limit and the longitudinal relaxation time $T_1$. If the pure dephasing, which does not involve population relaxation of excitons, disappears at low temperature, the transverse relaxation rate should equal half the longitudinal relaxation rate. Experimentally, $T_1$ was obtained by the luminescence decay time at very low excitation power. The luminescence showed single-exponential decay with a decay time of 3.56 ns for CuCl-QD/NaCl. The excitation density was the same as that in the photon echo measurements. In CuCl-QD/glass, on the other hand, the luminescence lifetime was 600 ps. The natural widths corresponding to these $T_1$ are 0.19 μeV (CuCl-QD/NaCl) and 1.1 μeV (CuCl-QD/glass), respectively. The observed lower limit of the homogeneous width of CuCl-QD/glass is 2 μeV at 0.7 K, almost the same as the natural width. On the other hand, the lowest value of the homogeneous width of the CuCl-QD/NaCl is 1 μeV, which is five times larger than the natural width. Therefore another low energy dephasing mechanism may exist in CuCl-QD/NaCl. The origin of this process is not clear, but an absence of temperature dependence indicates that it is impurity or surface scattering. For CuCl-QD/glass, this unknown mechanism is hidden by the fast longitudinal relaxation rate due to nonradiative population relaxation. Luminescence decay of CuCl-QD/glass did not change in the temperature range of this study. Therefore it is sure that the linear temperature dependence of the homogeneous width is not due to the change in $T_1$. At the low-temperature limit, the confined acoustic phonon and the TLS are frozen because of their finite energy and do not contribute the dephasing of confined exitons in CuCl quantum dots. The remaining dephasing processes are impurity or surface scattering and/or nonradiative relaxation, which can be less efficient in quantum dots than bulk crystals, because excitons ideally confined in a small space have less probability to interact with impurities and surfaces than excitonic polaritons propagating in the bulk crystal.

### 8.5.4 Accumulated Photon Echo Signal and the Homogeneous Width of CdSe Quantum Dots

Accumulated photon echo was observed for the lowest energy $1S_{3/2}1S_e$ transition in CdSe quantum dots ($R=3.6$ nm) displayed in Fig. 8.7b under very low excitation density ranging from 40 nJ/cm$^2$ to 1.8 μJ/cm$^2$ [43,44]. Figure 8.7a displays the temperature-dependent time traces of the accumulated photon echo signal under excitation density of 100 nJ/cm$^2$ (solid circles). No power dependence of the time trace was observed in this range, indicating that carrier–carrier scattering does not contribute to the measured dephasing time. The excitation density of 40 nJ/cm$^2$ corresponds to that of $10^{-3}$ photons per dot. The trace was fitted by two exponential decays, $a_f\exp(-t/t_f)+a_s\exp(-t/t_s)$. The fast decay time $t_f$ was 380 fs, corresponding to 1.7 meV in the spectral domain at 7 K. The slow decay time $t_s$ was 2.7 ps, corresponding to 0.25 meV at 7 K.

The fast decay component of $T_1=100$ ps observed in the pump-and-probe and luminescence time traces was much longer than $T_2$ measured by the accumulated photon echo [44], indicating that the localization of electrons and/or holes at the surface of the quantum dots contributes little to $T_2$. This suggests that the observed $T_2$ comes from the intrinsic dephasing mechanism of carriers in CdSe quantum dots. Fourier transform of the time trace showed its homogeneous spectrum is a 0.25 meV sharp line superposed on a 1.7 meV broad band, as is shown in Fig. 8.7c.

As is seen in Fig. 8.7a, the decay time of the slow decay component becomes faster with the increase of temperature. The fast decay component becomes dominant with the temperature rise. Fourier transform of the accumulated photon echo at a higher temperature is also displayed in Fig. 8.7c. The Fourier-transformed spectrum is much narrower than the persistent hole-burning spectrum and the absorption spectrum. The temperature dependence of the slow component contribution $[a_s/(a_f+a_s)]^{1/2}$ can be well fitted by the

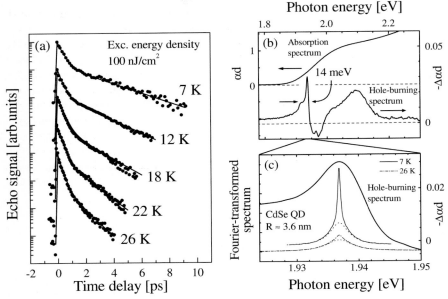

**Fig. 8.7.** (**a**) The accumulated photon echo signal from CdSe quantum dots ($R$=3.6 nm) under excitation density of 100 nJ/cm$^2$ at various temperatures (*solid circles*). *Solid lines* are fitted by two exponential decays. The dephasing time $T_2$ is twice the decay time constant. (**b**) The absorption spectrum and the persistent hole-burning spectrum of the sample. (**c**) Fourier transformed spectrum of the time traces of the accumulated photon echo signal at 7 K and 26 K. The broad components are described by *dotted lines*. The persistent hole-burning spectrum is replotted by a *thick solid line* (from Masumoto et al. [43])

expression of a Debye–Waller factor [43,45]. This means that the fast component is ascribed to the phonon sideband. On the other hand, the slow component in the echo decay, i.e., the sharp spectrum, is ascribed to the zero-phonon line.

Acoustic phonons are confined in quantum dots and their energies are inversely proportional to radii of quantum dots [5]. The energies of the confined acoustic spherical and ellipsoidal modes of CdSe quantum dots observed by Raman scattering are represented by 6.58, 11.0, and 18.8 meV/$R$ (nm) [46,47]. For $R$=3.6 nm, they are 1.8, 3.1, and 5.2 meV, respectively. The two smaller modes are observed to be stronger than the largest mode. They give the phonon sideband. Therefore the bandwidth 1.7 meV is reasonably understood as the confined acoustic phonon. Moreover, the fast decay time constant is inversely proportional to the radius of the CdSe quantum dots. This observation justifies the supposition that the fast decay time component is due to the confined acoustic phonon sideband. The sideband is considered to arise from the absorption or emission of a confined acoustic phonon.

The temperature dependence of the linewidth of the zero-phonon transition was extracted from the slow exponential decay component of the accumulated photon echo signal in Fig. 8.7a. Figure 8.8a exhibits the temperature dependence of the homogeneous linewidth of the zero-phonon line calculated from the slow decay component in the accumulated photon echo signal. The solid line is a linear least-squares fit of these data. The good agreement between the experimental data and the linear fit shows that the homogeneous width increases linearly with the increase of temperature. The linear temperature coefficient is plotted as a function of $1/R^2$ in Fig. 8.8b. The size dependence of the deformation potential coupling has been theoretically investigated for the energetically lowest pair state in CdSe quantum dots, which

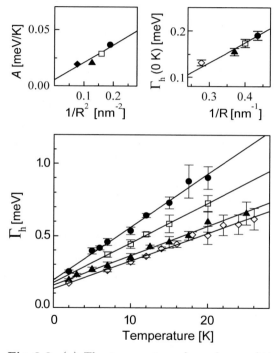

**Fig. 8.8.** (a) The temperature dependence of the homogeneous linewidth of the zero-phonon line $\Gamma_h$ estimated from the slow accumulated photon echo decay component for CdSe quantum dots. The average radii of the dots are 2.3 nm (*closed circles*), 2.5 nm (*open squares*), 2.7 nm (*closed triangles*), and 3.6 nm (*open diamonds*), respectively. The *solid lines* indicate their linear least-squares fit with temperature coefficients depicted in (**b**); (**b**) A linear temperature coefficient of the homogeneous width for four different samples as a function of the inverse square of the radius. The *solid line* exhibits the theoretical estimation for CdSe quantum dots; (**c**) Homogeneous width at 0 K for four samples as a function of the inverse of the radius (from Takemoto et al. [44])

predicts the $1/R^2$ dependence of the interaction strength on size based on the deformation coupling between confined carriers and confined acoustic phonons [48]. For comparison with the experimental data, the theoretical estimation of the linear temperature coefficient for CdSe dots is also depicted in Fig. 8.8b [48], which agrees with the experimental data quite well. This fact also serves as evidence of the deformation potential coupling between confined carriers and confined acoustic phonons in the CdSe quantum dots. Temperature broadening of the zero-phonon line is considered to arise from two phonon processes, which are both absorption and emission of confined acoustic phonons. Although the contribution of the interactions with the glassy host to the dephasing processes can be considered, it is reasonable to say that the homogeneous broadening mainly comes from the deformation potential coupling, because the deformation potential of CdSe is very large (3 eV) [49].

As shown in Fig. 8.8a, the linewidth is broadened between 2 and 26 K as the dot size decreases. The homogeneous linewidth extrapolated to 0 K is plotted in Fig. 8.8c as a function of the inverse of the radius of dots. It is inversely proportional to the radius, suggesting the dominant contribution of surface scattering. This feature qualitatively agrees with the experimental results of photon echo [27], where the dephasing mechanisms are divided into three dynamical processes: the coupling to vibrations of the nanocrystals, the lifetime broadening, and the surface scattering from impurity and defects. However, accumulated photon echo measurement can deduce the width of the sharp line extrapolated to 0 K to be $\Gamma_h(0 \text{ K})=0.13$ meV for $R=3.6$ nm, which is much narrower than those so far reported in CdSe quantum dots embedded in glass [7,10]. This value is comparable to that measured in the single dot spectroscopy of chemically grown CdSe quantum dots [24]. The dephasing time observed in this accumulated photon echo measurement is much longer than that observed in photon echo measurements [26]. This measurement was done for good-quality samples whose shortest $T_1$ is 100 ps. In this case, phonon broadening is the major broadening mechanism to determine the homogeneous linewidth above 10 K.

### 8.5.5  Lowest-Temperature Accumulated Photon Echo Signal and Homogeneous Width

The accumulated photon echo signals from CuCl-QD/NaCl, CuCl-QD/glass, CuBr-QD/glass, and CdSe-QD/glass at lowest temperatures are displayed in Fig. 8.9.

At the lowest temperature, fast and slow decay components are observed in CuCl-QD/NaCl, CuCl-QD/glass, and CdSe-QD/glass, while a single exponential decay is observed in CuBr-QD/glass. The Fourier-cosine transformation of two-exponential decay in the accumulated photon echo signal gives a narrow Lorentzian superposed on the broader Lorentzian [30]. The temperature dependence of the relative contribution of the narrower Lorentzian

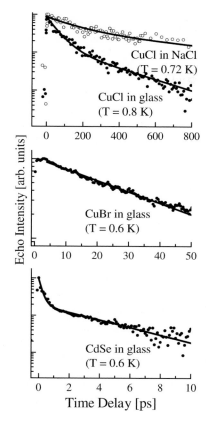

**Fig. 8.9.** Low-temperature accumulated photon echo signal for CuCl-QD/NaCl ($R$=2.0 nm), CuCl-QD/glass ($R$=2.0 nm), CuBr-QD/glass ($R$=6.5 nm) and CdSe-QD/glass ($R$=3.6 nm). *Solid lines* are fitted by 0.56exp(−t/170 ps)+0.44exp(−t/925 ps) for CuCl-QD/NaCl, by 0.83exp(−t/60 ps)+0.17exp(−t/310 ps) for CuCl-QD/glass, by exp(−t/30.5 ps) for CuBr-QD/glass and by 0.80exp(−t/0.23 ps)+0.20exp(−t/4.3 ps) for CdSe-QD/glass (from Masumoto et al. [51])

corresponding to the slower decay is explained by the Debye–Waller factor, and therefore the faster decay is ascribed to the confined-acoustic-phonon sideband [44,17]. The slow decay time strongly depends on the temperature and gives the inverse of the width of the zero-phonon line. With the decrease of temperature, decay time increases. This means the homogeneous width of the zero-phonon line becomes narrow with the decrease of temperature. The temperature-dependent homogeneous widths of CuCl-QD/NaCl, CuCl-QD/glass, CuBr-QD/glass, and CdSe-QD/glass are displayed in Fig. 8.10.

For CuCl-QD/NaCl and CuCl-QD/glass, the ultranarrow homogeneous width and its temperature dependence are discussed in Sect. 8.5.3.

For CuBr-QD/glass, another temperature broadening proportional to $T^\alpha$ ($\alpha$=1.3) is observed below 30 K [50]. This broadening is also dominated by distributed small energy excitations in the two-level system. The homogeneous width extrapolated to 0 K ranges from 23 µeV to 59 µeV, depending on the radius of quantum dots. The homogeneous width extrapolated to 0 K, and temperature-broadening coefficient are inversely proportional to the ra-

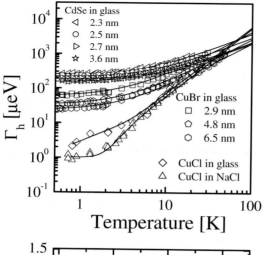

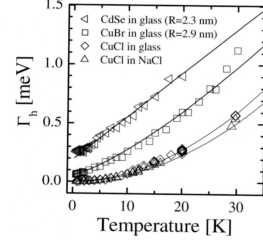

**Fig. 8.10.** The temperature-dependent homogeneous linewidth $\Gamma_h$ (FWHM) of CuCl-QD/NaCl, CuCl-QD/glass, CuBr-QD/glass, and CdSe-QD/glass. The average radius of both CuCl-QD/NaCl and CuCl-QD/glass is 2.0 nm. *Solid lines* are fitted by expressions $A+B/[\exp(\delta/kT)-1]$ for CdSe-QD/glass, and $A+BT^{1.3}$ for CuBr-QD/glass, where $A$, $B$, and $\delta$ are fitting parameters. For fitting of CuCl-QD/NaCl and CuCl-QD/glass, see Sect. 8.5.3 (from Masumoto et al. [52])

dius of the dots, which suggests the dominant contribution of surface or surroundings to the dephasing.

For CdSe-QD/glass, on the other hand, linear temperature dependence is observed between 25 K and 2 K [44]. However, the temperature dependence deviates from linear around 2 K, and the homogeneous width stops decreasing around 200 μeV. As is shown in Fig. 8.10, the whole temperature dependence is nicely fitted phenomenologically by the Bose factor $n=1/[\exp(\delta/kT)-1]$, where $\delta$ decreases with the increase of the size and is 0.2∼0.4 meV. This means very small excitation, like acoustic phonons whose energy is smaller than 0.2∼0.4 meV, is absent in CdSe quantum dots. As is discussed in Sect. 8.5.4, the linear temperature coefficient observed above 2 K is inversely

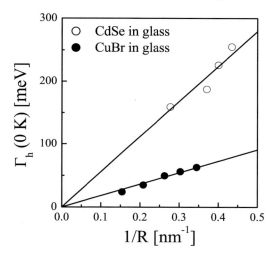

**Fig. 8.11.** The homogeneous width $\Gamma_h$ (0 K) of CuBr and CdSe quantum dots extrapolated to 0 K is proportional to inverse of $R$. *Solid lines* show the least square fitting (from Masumoto et al. [51])

proportional to square of the radius. The absolute value of the linear temperature coefficient and its radius dependence are well explained by the deformation potential interaction between the confined exciton and confined acoustic phonons in the high-temperature approximation. These experimental observations suggest that the phonon spectrum has broad peaks at 2∼4 meV, with a low-energy tail extending to 0.2∼0.4 meV. The homogeneous width extrapolated to 0 K is inversely proportional to the radius also in the case of CdSe-QD/glass, as is shown in Fig. 8.11. It is broader than the inverse of the nonradiative lifetime (100 ps) measured by luminescence decay and by pump-and-probe decay as the fast component decay.

The homogeneous width extrapolated to 0 K is broader than the the inverse of the nonradiative lifetime measured by the luminescence decay and pump-and-probe decay in all cases of CuCl, CuBr, and CdSe quantum dots [17,44,53]. Therefore, a broadening mechanism other than lifetime broadening dominates the broadening process even at the lowest temperatures. Comparing the homogeneous width of CuCl, CuBr, and CdSe quantum dots extrapolated to 0 K, it is noted that the width depends on the exciton Bohr radius in bulk crystals. In other words, it is small in the case of weak confinement, while it is large in the case of strong confinement. The homogeneous width extrapolated to 0 K is almost proportional to $1/R$ for CuBr and CdSe quantum dots, as is shown in Fig. 8.11.

Exciton–acoustic-phonon interaction is characterized by the square of the deformation potential for the exciton. The deformation potential is 0.45 eV for CuCl, 0.30 eV for CuBr, and 3.0 eV for CdSe [49,54]. Therefore, the exciton–acoustic-phonon interaction for CdSe is larger than that for CuCl and CuBr quantum dots by two orders of magnitude. It is understandable that exciton–acoustic-phonon interaction dominates the dephasing of CdSe quantum dots, but does not dominate the dephasing of CuCl and CuBr quantum dots at very low temperatures. Instead, very small energy excitations in the two-level system dominate the temperature-dependent dephasing of CuCl and CuBr quantum dots at very low temperatures.

### 8.5.6  Summary of the Accumulated Photon Echo of Quantum Dots

The homogeneous width of confined excitons in CuCl, CuBr, and CdSe quantum dots was investigated at low temperatures by means of accumulated photon echo. The homogeneous width of CuCl quantum dots becomes as narrow as 1∼2 μeV and depends on the surroundings of quantum dots at the lowest temperatures. The homogeneous width of CuBr and CdSe quantum dots becomes 23∼59 μeV and 0.16∼0.25 meV, respectively, and is proportional to the inverse of the radius of the quantum dots at 0 K. This suggests that the homogeneous width depends on the surroundings of quantum dots. Based on these results, it is considered that surroundings dominate the low-temperature dephasing mechanism of confined excitons in quantum dots, i.e., the homogeneous width of the optical spectra, of real quantum dots. Therefore the surroundings must be taken into account to understand the basic character of quantum dots.

## 8.6   Coherency Measurements

To observe the very narrow homogeneous width of the quantum dots, high-resolution spectral domain measurements of the width or time-domain measurements of the coherency are needed. For the high-resolution spectral domain measurement, monochromators are not sufficient for the observation of the ultranarrow homogeneous width of the quantum dots. Detection as well as excitation by very narrow linewidth lasers are essential, and the spectral resolution becomes a convolution of linewidths of the two lasers. Single frequency tunable lasers as narrow as 4 neV or even 200 peV are available, but 4 neV is enough for the observation of quantum dots made from direct-transition semiconductors, because the lifetime broadening is on the order of 1 μeV. For the time domain measurement of the dephasing time of quantum dots, not only photon echo and accumulated photon echo but also coherency measurements utilizing phase-locked pulses were used. Successive pulses with precise time control on the order of 1/100 fs are called phase-locked pulses. The second pulse was used to stimulate the oscillation of optical dipoles constructively or destructively, corresponding to in-phase excitation and out-of-phase excitation, respectively [55]. The phase-lock technique was also used for the observation of phase coherence of optical transitions in quantum dots.

Coherent nonlinear optical spectroscopy, degenerate and nondegenerate four-wave mixing, was applied to a single GaAs quantum dot formed by a well fluctuation in a GaAs quantum well 4.2 nm thick [56]. A metal mask with a 0.2–2.5 μm aperture was used to observe a single quantum dot. Two continuous wave single-frequency dye lasers having a bandwidth less than 4 neV were used for the four-wave mixing at 6 K, and the mixing signal was selectively obtained by homodyne detection. The observed spectrum of the mixing signal was analyzed by an expression for the nonlinear polarization including $\Gamma_{rel}=\hbar/T_1$ and $\gamma=\hbar/T_2$. The analysis gave values of $\Gamma_{rel}=35$ μeV and $\gamma=20$ $\mu$eV. The energy relaxation and dephasing relaxation rates are comparable, reflecting the absence of significant pure dephasing.

Coherent optical control of the quantum state of a single quantum dot was demonstrated by observation of its oscillatory luminescence intensity in the time trace [57]. Quantum dots investigated were naturally formed well fluctuations in a 4 nm GaAs quantum well surrounded by $Al_{0.3}Ga_{0.7}As$ barriers. Elliptical disks were oriented along the [110] axes. A metal mask with a 500-nm-diameter aperture was used to observe a single quantum dot. The photoluminescence excitation spectrum is as sharp as 17 μeV, suggesting $T_2$ of 39 ps at 6 K. A pair of phase-locked pulses excited the single quantum dot and the luminescence intensity was detected as a function of the optical delay between two pulses. The luminescence intensity showed the oscillation coming from the interference between the polarization of the lowest quantum state excited by the first pulse and that excited by the second pulse. The envelope of the oscillation decayed at a time constant of 40 ps, which agrees with $T_2$ estimated by the linewidth of the photoluminescence excitation spectrum.

A very narrow optical phonon sideband of resonantly excited stress-induced quantum dots formed in an $In_{0.1}Ga_{0.9}As/GaAs$ quantum well was used to evaluate the homogeneous width of the resonantly excited quantum state [58]. Two trains of 2 ps pulses passing through a Michelson interferometer were used to excite the sample with precise knowledge of the optical delay between the pulses. Spectral filtering of the narrow optical phonon sideband of resonant luminescence made the correlation trace; analysis gave the dephasing time of the resonantly excited state of 18.5 ps at 2 K, corresponding to homogeneous width of 36 μeV.

As is discussed in this chapter, quantum states in quantum dots sometimes have very sharp optical spectra, in other words, long coherence times reflecting their "atomic" character. Site-selective spectroscopy, such as hole burning, persistent spectral hole burning, and fluorescence line narrowing, are sometimes useful to see this by eliminating the inhomogeneous broadening. Single quantum dot spectroscopy is an excellent technique to observe the homogeneous width by eliminating the inhomogeneous broadening. Photon echo and accumulated photon echo are able to overcome the spectral resolution of the monochromators and hence are very efficient to observe the dephasing time in the time domain. Coherency measurements utilizing phase-locked pulses was recently demonstrated. Coherent control of the quantum states becomes a possible technique for ultrafast optical switches and quantum computation. Development in this direction is possible and expected.

## Note Added in Proof

Very recently, long coherence in quantum dots at low temperatures has been reported in several works and has become one of the hottest topics of quantum dots.

A heterodyne-detected photon-echo technique was successfully used to observe a $T_2$ of 630 ps as a slower decay component at 7 K in three layers of $In_{0.7}Ga_{0.3}As$ self-assembled quantum dots separated by GaAs spacers [59]. Four-wave mixing was done in an optical waveguide made of quantum dot layers. The homogeneous linewidth corresponding to $T_2$ of 630 ps is 2 μeV. The Fourier transform of the two decay components was represented by a sum of a sharp zero-phonon Lorentzian line and a strongly temperature-dependent broad wing coming from the elastic exciton-acoustic phonon interaction. Nonlinearly temperature dependent broadening of the zero-phonon line was ascribed to a sum of lifetime broadening, linear temperature broadening coming from inelastic scattering by acoustic phonons, and the activation process to the hole excited state.

By means of a photon echo, a $T_2$ of 372 ps at 5 K was observed as a slower decay component in the biexponential decay for 10 layers of $In_{0.5}Al_{0.04}Ga_{0.46}As$ self-assembled quantum dots embedded in $Al_{0.08}Ga_{0.92}As$ [60]. The homogeneous linewidth corresponding to $T_2$ of 372 ps is 3.5 μeV and was significantly narrower than the linewidth of 45 μeV observed on average by the single dot luminescence spectrum of a single layer of dots of the same composite. The difference in values of the linewidth shows that the linewidth of photoluminescence observed by the single dot spectroscopy can generally not be taken as the homogeneous linewidth as a result of spectral diffusion. The observed homogeneous linewidth of 3.5 μeV was concluded to be lifetime broadening of the ground state of the dots.

Precise fluorescence line narrowing was used to observe the homogeneous linewidth of CuCl quantum dots embedded in NaCl crystals and glass at low temperatures [61]. A very narrow-band pulsed Ti:sapphire laser and a 1 m double-monochromator were used together with a deconvolution process. Resonantly excited secondary emission consists of a sharp zero-phonon line, Stokes and anti-Stokes confined acoustic phonon sidebands and a Rayleigh scattering line. These three components were carefully decomposed from the temperature-dependent resonantly excited secondary emission spectrum. The temperature dependence of the homogeneous width was well fitted by the expression $\sinh^{-2}(\hbar\omega/2kT)$, where $\hbar\omega$ is the energy of the lowest confined phonon mode. This result fully supports the content of Sect. 8.5.3 and the result of the accumulated photon echo experiment [17].

# References

1. Y. Masumoto: J. Lumin. **70**, 386 (1996)
2. M. Nirmal, B.O. Dabbousi, M.G. Bawendi, J.J. Macklin, J.K. Trautman, T.D. Harris, L.E. Brus: Nature **383**, 802 (1996)
3. M. Sugisaki, H.-R. Ren, S.V. Nair, J.-S. Lee, S. Sugou, T. Okuno, Y. Masumoto: J. Lumin. **87–89**, 40 (2000)
4. S.A. Empedocles, M.G. Bawendi: Science **278**, 2114 (1997)
5. U. Woggon: *Optical Properties of Semiconductor Quantum Dots* (Springer, Berlin 1997)
6. S.V. Gaponenko: *Optical Properties of Semiconductor Nanocrystals* (Cambridge Univ. Press, Cambridge 1998)
7. N. Peyghambarian, B. Fluegel, D. Hulin, A. Migus, M. Joffre, A. Antonetti, S.W. Koch, M. Lindberg: IEEE J. Quant. Electron. **25**, 2516 (1989)
8. A.P. Alivisatos, A.L. Harris, N.J. Levinos, M.L. Steigerwald, L.E. Brus: J. Chem. Phys. **89**, 4001 (1988)
9. M.G. Bawendi, W.L. Wilson, L. Rothberg, P.J. Carroll, T.M. Jedju, M.L. Steigerwald, L.E. Brus: Phys. Rev. Lett. **65**, 1623 (1990)
10. U. Woggon, S. Gaponenko, W. Langbein, A. Uhrig, C. Klingshirn: Phys. Rev. B **47**, 3684 (1993)
11. P. Palinginis, H. Wang: Appl. Phys. Lett. **78**, 1541 (2001)
12. Y. Masumoto, T. Wamura, A. Iwaki: Appl. Phys. Lett. **55**, 2535 (1989)
13. T. Wamura, Y. Masumoto, T. Kawamura: Appl. Phys. Lett. **59**, 1758 (1991)
14. Y. Masumoto, T. Kawazoe, N. Matsuura: J. Lumin. **76–77**, 189 (1998)
15. T. Itoh, M. Furumiya: J. Lumin. **48–49**, (1991) 704
16. K. Edamatsu, T. Itoh, K. Matsuda, S. Saikan: Phys. Status Solidi B **224**, 629 (2001)
17. M. Ikezawa, Y. Masumoto: Phys. Rev. B **61**, 12662 (2000)
18. K. Brunner, G. Abstreiter, G. Böhm, G. Tränkle, G. Weimann: Appl. Phys. Lett. **64**, 3320 (1994)
19. D. Gammon, E.S. Snow, B.V. Shanabrook, D.S. Katzer, D. Park: Science **273**, 87 (1996)
20. Y. Nagamune, H. Watabe, M. Nishioka, Y. Arakawa: Appl. Phys. Lett. **67**, 3257 (1995)
21. J.-Y. Marzin, J.-M. Gérard, A. Izraël, D. Barrier, G. Bastard: Phys. Rev. Lett. **73**, 716 (1994)
22. M. Grundmann, J. Christen, N.N. Ledentsov, J. Böhrer, D. Bimberg, S.S. Ruvimov, P. Werner, U. Richter, U. Gösele, J. Heydenreich, V.M. Ustinov, A.Y. Egorov, A.E. Zhukov, P.S. Kop'ev, Z.I. Alferov: Phys. Rev. Lett. **74**, 4043 (1995)
23. L. Samuelson, N. Carlsson, P. Castrillo, A. Gustafsson, D. Hessman, J. Lindahl, L. Montelius, A. Petersson, M.-E. Pstol, W. Seifert: Jpn. J. Appl. Phys. **34**, 4392 (1995)
24. S.A. Empedocles, D.J. Norris, M.G. Bawendi: Phys. Rev. Lett. **77**, 3873 (1996)
25. T. Yajima, Y. Taira: J. Phys. Soc. Jpn. **47**, 1620 (1979)
26. R.W. Schoenlein, D.M. Mittleman, J.J. Shiang, A.P. Alivisatos, C.V. Shank: Phys. Rev. Lett. **70**, 1014 (1993)
27. D.M. Mittleman, R.W. Schoenlein, J.J. Shiang, V.L. Colvin, A.P. Alivisatos, C.V. Shank: Phys. Rev. B **49**, 14435 (1994)

28. U. Banin, G. Cerullo, A.A. Guzelian, C.J. Bardeen, A.P. Alivisatos, C.V. Shank: Phys. Rev. B **55**, 7059 (1997)
29. W.H. Hesselink, D.A. Wiersma: Phys. Rev. Lett. **43**, 1991 (1979)
30. S. Saikan, T. Nakabayashi, Y. Kanematsu, N. Tato: Phys. Rev. B **38**, 7777 (1988)
31. S. Saikan, H. Miyamoto, Y. Tosaki, A. Fujiwara: Phys. Rev. B **36**, 5074 (1987)
32. T. Kuroda, F. Minami, K. Inoue, A.V. Baranov: Phys. Rev. B **57**, 2077 (1998)
33. T. Kuroda, F. Minami, K. Inoue, A.V. Baranov: Phys. Status Solidi **164**, 287 (1997)
34. R. Kuribayashi, K. Inoue, K. Sakoda, V.A. Tsekhomskii, A.V. Baranov: Phys. Rev. B **57**, R15084 (1998)
35. S. Saikan, K. Uchikawa, H. Ohsawa: Opt. Lett. **16**, 10 (1991)
36. T. Kuroda, S. Matsushita, F. Minami, K. Inoue, A.V. Baranov: Phys. Rev. B **55**, R16041 (1997)
37. Y. Masumoto, S. Shionoya, T. Takagahara: Phys. Rev. Lett. **51**, 923 (1983)
38. F. Vallée, F. Bogani, C. Flytzanis: Phys. Rev. Lett. **66**, 1509 (1991)
39. S. Okamoto, Y. Masumoto: J. Lumin. **64**, 253 (1995)
40. J. Zhao, Y. Masumoto: Phys. Rev. B **60**, 4481 (1999)
41. See, for example, R.M. Macfarlane, R.M. Shelby: In: *Optical Linewidths in Glasses*, ed. by M.J. Weber [J. Lumin. **36**, 179 (1987)]; D.L. Huber: In: *Dynamical Processes in Disordered Systems*, ed. by W.M. Yen (Trans Tech, Aedermannsdorf, Switzerland 1989)
42. G.P. Flinn, K.W. Jang, J. Ganem, M.L. Jones, R.S. Meltzer, R.M. Macfarlane: Phy. Rev. B **49**, 5821 (1994)
43. Y. Masumoto, K. Takemoto, T. Shoji, B.-R. Hyun: *Proc. 24th Int. Conf. Physics of Semiconductors* (Jerusalem, 1998) VII-B-15, pdf no.1058
44. K. Takemoto, B.-R. Hyun, Y. Masumoto: Solid State Commun. **114**, 521 (2000)
45. S. Saikan, A. Imaoka, Y. Kanematsu, K. Sakota, K. Kominami, M. Iwamoto: Phys. Rev. B **41**, 3185 (1990)
46. L. Saviot, B. Champagnon, E. Duval, I.A. Kudriavtsev, A.I. Ekimov: J. Non-Cryst. Solids **197**, 238 (1996)
47. U. Woggon, F. Gindele, O. Wind, C. Klingshirn: Phys. Rev. B **54**, 1506 (1996)
48. T. Takagahara: Phys. Rev. Lett. **71**, 3577 (1993); T. Takagahara: J. Lumin. **70**, 129 (1996)
49. A. Blacha, H. Presting, M. Cardona: Phys. Status Solidi B **126**, 11 (1984)
50. B.-R. Hyun, M. Furuya, K. Takemoto, Y. Masumoto: J. Lumin. **87–89**, 302 (2000)
51. Y. Masumoto, M. Ikezawa, B.-R. Hyun, K. Takemoto, M. Furuya: Phys. Status Solidi B **224**, 613 (2001)
52. Y. Masumoto, M. Ikezawa, B.-R. Hyun, K. Takemoto, M. Furuya: In: *Proc. 25th Int. Conf. Physics of Semiconductors* (Osaka, 2000) p. 1271
53. T. Okuno, H. Miyajima, A. Satake, Y. Masumoto: Phys. Rev. B **54**, 16952 (1996)
54. A. Blacha, S. Ves, M. Cardona: Phys. Rev. B **27**, 6346 (1983)
55. A.P. Heberle, J.J. Baumberg, K. Köhler: Phys. Rev. Lett. **75**, 2598 (1995)
56. N.H. Bonadeo, G. Chen, D. Gammon, D.S. Katzer, D. Park, D.G. Steel: Phys. Rev. Lett. **81**, 2759 (1998)
57. N.H. Bonadeo, J. Erland, D. Gammon, D. Park, D.S. Katzer, D.G. Steel: Science **282**, 1473 (1998)

58. A.V Baranov, V. Davydov, A.V. Fedorov, H.-W. Ren, S. Sugou, Y. Masumoto:
    Phys. Status Solidi B **224**, 461 (2001)
59. P. Borri, W. Langbein, S. Schneider, U. Woggon, R.L. Sellin, D. Ouyang,
    D. Bimberg: Phys. Rev. Lett. **87**, 157401 (2001)
60. D. Birkedal, K. Leosson, J.M. Hvam: Phys. Rev. Lett. **87**, 227401 (2001)
61. K. Edamatsu, T. Itoh, K. Matsuda, S. Saikan: Phys. Rev. B **64**, 195317 (2001)

# 9 Theory of Exciton Dephasing in Semiconductor Quantum Dots

Toshihide Takagahara

## 9.1 Introduction

A resonant optical excitation creates an excited state population and also induces an optical polarization. Dynamics of this optical excitation is characterized by relaxation of the population as well as decay of the induced optical polarization. In lower dimensional semiconductors, electronic confinement leads to qualitative changes in population relaxation, including spontaneous emission and exciton–phonon scattering, as shown in extensive recent studies [1]. These population relaxation processes are expected to contribute to dephasing with a dephasing rate given by half the population decay rate. Pure dephasing processes that do not involve population or energy relaxation of excitons can also contribute to dephasing. Pure dephasing, which is a well-established concept for atomic systems, remains yet to be investigated in lower-dimensional semiconductors due to a lack of direct comparison between dephasing and population relaxation and between theory and experiment. Studies of pure dephasing processes in lower-dimensional semiconductors will renew and deepen our understanding of dephasing of collective excitations in solids, although several seminal studies were done on the exciton dephasing in quantum well (QW) structures [2–6].

Narrow GaAs QWs grown by molecular beam epitaxy (MBE) and with growth interruptions have provided a model system for investigating dephasing processes in lower-dimensional semiconductors. In these narrow QWs, fluctuations at the interface between GaAs and AlGaAs lead to localization of excitons at monolayer-high islands. These localized states can be regarded effectively as weakly confined quantum dot (QD)-like states. One dimension of the confinement is defined by the width of the QW, while the other two lateral dimensions are defined by the effective size of the islands. To avoid inhomogeneous broadening due to well-width and island-size fluctuations, earlier studies have used photoluminescence (PL) and PL excitation with high spatial resolution to probe excitons in individual islands [7–9]. As a result, a very narrow linewidth of about several tens of μeV was observed. Without additional information on population relaxation, it was suggested that at very low temperatures dephasing of excitons in these structures is due to radiative recombination, while at elevated temperatures dephasing is mainly caused by thermal activation of excitons to higher excited states [9]. However, this interpretation is not complete since both of the suggested processes belong to the longitudinal decay processes and the dephasing rate is, in general, composed of half the longitudinal decay rate and the pure dephasing rate. In order to examine the presence of pure dephasing in this system, we carried out nonlinear optical measurements of the exciton dephasing in GaAs QD-like islands based on the three-pulse stimulated photon echo method [10]. This method enables the simultaneous measurement of the dephasing rate and the population decay rate. At very low temperatures the observed dephasing rate $\Gamma_\perp$ is very close to half the population decay rate $\Gamma_\parallel/2$, suggesting that dephasing is caused mainly by the population decay. With increasing temperature, the dephasing rate increases much faster than the population decay rate. At elevated temperatures (>30 K), dephasing rates become much greater than $\Gamma_\parallel/2$, indicating a dominant contribution of pure dephasing. Thus our measurements revealed convincingly the presence of a pure dephasing process that dominates excitonic dephasing at elevated temperatures.

The observed strong increase of the pure dephasing rate above 30 K suggests that interactions between excitons and low-energy acoustic phonons play an essential role in the pure dephasing process. In this chapter, we present a theoretical model that takes into account the interaction between excitons and acoustic phonons and can explain satisfactorily the magnitude as well as the temperature dependence of the dephasing rate. Our model generalizes the Huang–Rhys theory of $F$-centers [11–13] to include mixing among the ground and excited exciton states through exciton–phonon interactions, and as a result enables us to identify the elementary processes of exciton pure dephasing.

Comparison of the observed temperature dependence of the exciton dephasing rate in GaAs QD-like islands with the previously reported temperature dependence for CdSe [14] and CuCl [15] nanocrystals suggests qualitative differences that depend on the strength of quantum confinement. In the strong-confinement regime as in CdSe nanocrystals, linear temperature dependence prevails up to high temperatures (~200 K). On the other hand, in the weak-confinement regime as in GaAs islands and CuCl nanocrystals, the linear temperature dependence is dominated by the nonlinear temperature dependence with increasing temperature. It is very important to clarify the underlying physics of these differences.

## 9.2   Green Function Formalism of Exciton Dephasing Rate

As discussed in Sect. 9.1, the strong increase of the exciton dephasing rate above 30 K suggests the important role of the low-energy acoustic phonons in determining the dephasing rate. The exciton dephasing rate can be estimated most directly from the half-width at half-maximum (HWHM) of the absorption spectrum, which can be calculated from the Fermi golden rule as

$$I(\hbar\omega) = \frac{2\pi}{\hbar} \text{ Av.} \sum_f |\langle f|V_R|g\rangle|^2 \, \delta(\hbar\omega + E_g - E_f) \, , \tag{9.1}$$

where $V_R$ is the electromagnetic interaction, $|f\rangle$ and $|g\rangle$ denote the final exciton state and the initial ground state, respectively, including the phonon degrees of freedom, and Av. means the average over the thermal equilibrium state of phonons. This expression can be rewritten as

$$I(\hbar\omega) \propto \text{Re} \left[ \int_0^\infty dt e^{-i\omega t} \text{ Av.} \langle g|V_R \exp\left[\frac{i}{\hbar}H_e t\right] V_R \exp\left[-\frac{i}{\hbar}H_g t\right] |g\rangle \right] , \tag{9.2}$$

where $H_e$ and $H_g$ are the Hamiltonians of the excited state and the ground state, respectively. This can be confirmed by inserting a closure relation between two values of $V_R$. Equation (9.2) is a Fourier–Laplace transform of a correlation function. In order to proceed further, the Hamiltonians will be specified as

$$H_g = \sum_\alpha \hbar\omega_\alpha b_\alpha^\dagger b_\alpha \, , \tag{9.3}$$

$$H_e = \sum_i E_i|i\rangle\langle i| + \sum_\alpha \hbar\omega_\alpha b_\alpha^\dagger b_\alpha + \sum_\alpha M_\alpha(b_\alpha + b_\alpha^\dagger) \, , \tag{9.4}$$

where the index $\alpha$ denotes the acoustic phonon mode, $E_i$ the energy of the exciton states, and $M_\alpha$ is the exciton–phonon coupling matrix within the exciton state manifold. This is a generalization of the Huang–Rhys model of $F$-centers [11] to include mixing among the exciton state manifold, which is reflected in the matrix form of $M_\alpha$. In the following, we take into account only the exciton–phonon coupling, which is linear with respect to the phonon coordinates. Even within this range, however, the well-known deformation potential coupling and the piezoelectric coupling are included. Thus our Hamiltonian is sufficiently general. We note that in the elementary processes of the exciton–phonon interaction, the crystal momentum conservation needs to be satisfied in directions where the translational invariance holds. The dephasing process becomes prominent in systems with three-dimensional (3D) electronic confinement because the 3D confinement relaxes the crystal momentum conservation and also suppresses exciton population relaxation [16,17] due to the exciton–phonon interactions.

Hereafter the three terms of $H_e$ in (9.4) will be denoted, respectively as

$$H_e = H_e^0 + H_g + V \, . \tag{9.5}$$

Then the Laplace transform of $\exp[(\mathrm{i}/\hbar)H_{\mathrm{e}}t]$ can be expanded as

$$\int_0^\infty \mathrm{d}t\, \mathrm{e}^{-st} \exp[(\mathrm{i}/\hbar)H_{\mathrm{e}}t] = \frac{1}{s - \frac{\mathrm{i}}{\hbar}H_{\mathrm{e}}} = \frac{1}{s - \frac{\mathrm{i}}{\hbar}(H_{\mathrm{e}}^0 + H_{\mathrm{g}})}$$

$$+ \frac{\mathrm{i}}{\hbar} \frac{1}{s - \frac{\mathrm{i}}{\hbar}(H_{\mathrm{e}}^0 + H_{\mathrm{g}})} V \frac{1}{s - \frac{\mathrm{i}}{\hbar}(H_{\mathrm{e}}^0 + H_{\mathrm{g}})} + \left(\frac{\mathrm{i}}{\hbar}\right)^2 \frac{1}{s - \frac{\mathrm{i}}{\hbar}(H_{\mathrm{e}}^0 + H_{\mathrm{g}})}$$

$$V \frac{1}{s - \frac{\mathrm{i}}{\hbar}(H_{\mathrm{e}}^0 + H_{\mathrm{g}})} V \frac{1}{s - \frac{\mathrm{i}}{\hbar}(H_{\mathrm{e}}^0 + H_{\mathrm{g}})} + \cdots . \tag{9.6}$$

For example, the third term on the right hand side of (9.6) can be expressed in the convolution form as

$$\left(\frac{\mathrm{i}}{\hbar}\right)^2 \int_0^t \mathrm{d}t_1 \int_0^{t_1} \mathrm{d}t_2\, \mathrm{e}^{(\mathrm{i}/\hbar)(H_{\mathrm{e}}^0 + H_{\mathrm{g}})(t-t_1)}$$

$$V \mathrm{e}^{(\mathrm{i}/\hbar)(H_{\mathrm{e}}^0 + H_{\mathrm{g}})(t_1-t_2)} V \mathrm{e}^{(\mathrm{i}/\hbar)(H_{\mathrm{e}}^0 + H_{\mathrm{g}})t_2} . \tag{9.7}$$

Substituting this term for $\exp[(\mathrm{i}/\hbar)H_{\mathrm{e}}t]$ in (9.2) and noting the commutability between $H_{\mathrm{e}}^0$ and $H_{\mathrm{g}}$ and between $M_\alpha$ and $H_{\mathrm{g}}$, we have

$$\langle g|V_{\mathrm{R}} \exp\left[\frac{\mathrm{i}}{\hbar}H_{\mathrm{e}}t\right] V_{\mathrm{R}} \exp\left[-\frac{\mathrm{i}}{\hbar}H_{\mathrm{g}}t\right]|g\rangle \rightarrow \left(\frac{\mathrm{i}}{\hbar}\right)^2 \int_0^t \mathrm{d}t_1 \int_0^{t_1} \mathrm{d}t_2 \sum_{\alpha,\beta}$$

$$\langle g|V_{\mathrm{R}} \exp\left[\frac{\mathrm{i}}{\hbar}H_{\mathrm{e}}^0(t - t_1)\right] M_\alpha \exp\left[\frac{\mathrm{i}}{\hbar}H_{\mathrm{e}}^0(t_1 - t_2)\right] \times M_\beta \exp\left[\frac{\mathrm{i}}{\hbar}H_{\mathrm{e}}^0 t_2\right]$$

$$V_{\mathrm{R}}|g\rangle\langle 0|\exp\left[\frac{\mathrm{i}}{\hbar}H_{\mathrm{g}}(t - t_1)\right] \left(b_\alpha + b_\alpha^\dagger\right) \exp\left[\frac{\mathrm{i}}{\hbar}H_{\mathrm{g}}(t_1 - t_2)\right]$$

$$\times \left(b_\beta + b_\beta^\dagger\right) \exp\left[\frac{\mathrm{i}}{\hbar}H_{\mathrm{g}}t_2\right] \exp\left[-\frac{\mathrm{i}}{\hbar}H_{\mathrm{g}}t\right]|0\rangle , \tag{9.8}$$

where $|0\rangle$ denotes symbolically the thermal equilibrium state of phonons. The phonon part of (9.8) can be written as

$$\langle 0|\exp\left[\frac{\mathrm{i}}{\hbar}H_{\mathrm{g}}(t - t_1)\right] \left(b_\alpha + b_\alpha^\dagger\right) \exp\left[\frac{\mathrm{i}}{\hbar}H_{\mathrm{g}}(t_1 - t_2)\right] \left(b_\beta + b_\beta^\dagger\right)$$

$$\cdot \exp\left[\frac{\mathrm{i}}{\hbar}H_{\mathrm{g}}t_2\right] \exp\left[-\frac{\mathrm{i}}{\hbar}H_{\mathrm{g}}t\right]|0\rangle$$

$$= \mathrm{Tr}\rho_0 \exp\left[-\frac{\mathrm{i}}{\hbar}H_{\mathrm{g}}t_1\right] (b_\alpha + b_\alpha^\dagger) \exp\left[\frac{\mathrm{i}}{\hbar}H_{\mathrm{g}}(t_1 - t_2)\right] (b_\beta + b_\beta^\dagger) \exp\left[\frac{\mathrm{i}}{\hbar}H_{\mathrm{g}}t_2\right]$$

$$= \mathrm{Tr}\rho_0 \left[b_\alpha(-t_1) + b_\alpha^\dagger(-t_1)\right] \left[b_\beta(-t_2) + b_\beta^\dagger(-t_2)\right]$$

$$= \delta_{\alpha\beta} \left[N_\alpha \mathrm{e}^{-\mathrm{i}\omega_\alpha(t_1 - t_2)} + (1 + N_\alpha)\mathrm{e}^{\mathrm{i}\omega_\alpha(t_1 - t_2)}\right] \tag{9.9}$$

with

$$\rho_0 = e^{-\beta H_g}/\text{Tr}\, e^{-\beta H_g} \ ,$$

$$b_\alpha(t) = \exp\left[\frac{i}{\hbar}H_g t\right] b_\alpha \exp\left[-\frac{i}{\hbar}H_g t\right] = e^{-i\omega_\alpha t}b_\alpha \ ,$$

and

$$N_\alpha = 1/[\exp(\beta\hbar\omega_\alpha) - 1] \quad \text{with} \ \beta = 1/(k_B T) \ .$$

Then again making a Laplace transform, we have

$$\left(\frac{i}{\hbar}\right)^2 \langle g|V_R \frac{1}{s - \frac{i}{\hbar}H_e^0} \sum_\alpha M_\alpha \left\{ \frac{N_\alpha}{s - \frac{i}{\hbar}H_e^0 + i\omega_\alpha} + \frac{1 + N_\alpha}{s - \frac{i}{\hbar}H_e^0 - i\omega_\alpha} \right\}$$

$$\times M_\alpha \frac{1}{s - \frac{i}{\hbar}H_e^0} V_R|g\rangle$$

$$= \langle g|V_R \frac{1}{s - \frac{i}{\hbar}H_e^0} \Sigma_0^{(2)}(s) \frac{1}{s - \frac{i}{\hbar}H_e^0} V_R|g\rangle \ , \tag{9.10}$$

with

$$\Sigma_0^{(2)}(s) = \left(\frac{i}{\hbar}\right)^2 \sum_\alpha M_\alpha \left\{ \frac{N_\alpha}{s - \frac{i}{\hbar}H_e^0 + i\omega_\alpha} + \frac{1 + N_\alpha}{s - \frac{i}{\hbar}H_e^0 - i\omega_\alpha} \right\} M_\alpha. \tag{9.11}$$

Since the exciton–phonon interaction Hamiltonian $V$ is linear with respect to the phonon coordinates, the terms of odd powers in $V$ in (9.6) vanish in the final expression. The next non-vanishing term is the fourth-order term in $V$, which has three contraction diagrams as depicted in Fig. 9.1.

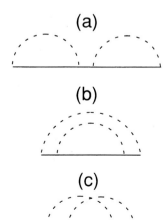

**(a)**

**(b)**

**(c)**

**Fig. 9.1.** Possible diagrams of the fourth-order term in (9.6). A *solid (dashed)* line denotes the exciton (phonon) propagator

For example, the term corresponding to Fig. 9.1c can be written as

$$
\Sigma_0^{(4)}(s) = \left(\frac{i}{\hbar}\right)^4 \sum_{\alpha,\beta} \left\{ M_\alpha \frac{N_\alpha}{s - \frac{i}{\hbar}H_e^0 + i\omega_\alpha} M_\beta \frac{1}{s - \frac{i}{\hbar}H_e^0 + i(\omega_\alpha + \omega_\beta)} \right.
$$
$$
\times M_\alpha \frac{N_\beta}{s - \frac{i}{\hbar}H_e^0 + i\omega_\beta} M_\beta + M_\alpha \frac{N_\alpha}{s - \frac{i}{\hbar}H_e^0 + i\omega_\alpha}
$$
$$
M_\beta \frac{1}{s - \frac{i}{\hbar}H_e^0 + i(\omega_\alpha - \omega_\beta)} M_\alpha \frac{1 + N_\beta}{s - \frac{i}{\hbar}H_e^0 - i\omega_\beta} M_\beta
$$
$$
+ M_\alpha \frac{1 + N_\alpha}{s - \frac{i}{\hbar}H_e^0 - i\omega_\alpha} M_\beta \frac{1}{s - \frac{i}{\hbar}H_e^0 - i(\omega_\alpha - \omega_\beta)}
$$
$$
M_\alpha \frac{N_\beta}{s - \frac{i}{\hbar}H_e^0 + i\omega_\beta} M_\beta + M_\alpha \frac{1 + N_\alpha}{s - \frac{i}{\hbar}H_e^0 - i\omega_\alpha}
$$
$$
M_\beta \frac{1}{s - \frac{i}{\hbar}H_e^0 - i(\omega_\alpha + \omega_\beta)} M_\alpha \frac{1 + N_\beta}{s - \frac{i}{\hbar}H_e^0 - i\omega_\beta} M_\beta \left. \right\} . \tag{9.12}
$$

On the other hand, the diagrams in Fig. 9.1a,b can be incorporated into the zeroth-order and second-order terms, respectively, by renormalizing the exciton propagator and the self-energy, as will be shown below.

First of all, we introduce the Green function defined by

$$
G_0^{(2)}(s) = \frac{1}{s - \frac{i}{\hbar}H_e^0 - \Sigma_0^{(2)}(s)}
$$
$$
= \frac{1}{s - \frac{i}{\hbar}H_e^0} + \frac{1}{s - \frac{i}{\hbar}H_e^0} \Sigma_0^{(2)}(s) \frac{1}{s - \frac{i}{\hbar}H_e^0}
$$
$$
+ \frac{1}{s - \frac{i}{\hbar}H_e^0} \Sigma_0^{(2)}(s) \frac{1}{s - \frac{i}{\hbar}H_e^0} \Sigma_0^{(2)}(s) \frac{1}{s - \frac{i}{\hbar}H_e^0} + \cdots , \tag{9.13}
$$

where $\Sigma_0^{(2)}$ has a meaning of the second-order self-energy and $G_0^{(2)}$ represents the exciton propagator in the phonon field. The diagram in Fig. 9.1a is included in this Green function. Using this propagator, we improve the second-order self-energy as

$$
\Sigma^{(2)}(s) = \left(\frac{i}{\hbar}\right)^2 \sum_\alpha M_\alpha \left[ N_\alpha G_0^{(2)}(s + i\omega_\alpha) + (1 + N_\alpha)G_0^{(2)}(s - i\omega_\alpha) \right] M_\alpha ,
$$

$$
\tag{9.14}
$$

and including this self-energy into the denominator, we obtain the improved Green function as

$$
G^{(2)}(s) = \frac{1}{s - \frac{i}{\hbar}H_e^0 - \Sigma^{(2)}(s)} . \tag{9.15}
$$

This Green function incorporates the diagram in Fig. 9.1b. Thus only the diagram in Fig. 9.1c, namely an irreducible diagram, should be included in the fourth order.

Now, using the exciton propagator in (9.15), we calculate the self-energy including the fourth-order irreducible diagram, namely

$$\Sigma^{(2)}(s) + \Sigma^{(4)}(s)$$

$$= \left(\frac{\mathrm{i}}{\hbar}\right)^2 \sum_\alpha M_\alpha \left[ N_\alpha G^{(2)}(s + \mathrm{i}\omega_\alpha) + (1 + N_\alpha)G^{(2)}(s - \mathrm{i}\omega_\alpha) \right] M_\alpha$$

$$+ \left(\frac{\mathrm{i}}{\hbar}\right)^4 \sum_{\alpha,\beta} \Big\{ N_\alpha N_\beta M_\alpha G^{(2)}(s + \mathrm{i}\omega_\alpha) M_\beta G^{(2)}(s + \mathrm{i}(\omega_\alpha + \omega_\beta))$$

$$\times M_\alpha G^{(2)}(s + \mathrm{i}\omega_\beta) M_\beta$$

$$+ N_\alpha(1 + N_\beta) M_\alpha G^{(2)}(s + \mathrm{i}\omega_\alpha) M_\beta G^{(2)}(s + \mathrm{i}(\omega_\alpha - \omega_\beta))$$

$$\times M_\alpha G^{(2)}(s - \mathrm{i}\omega_\beta) M_\beta)$$

$$+ (1 + N_\alpha) N_\beta M_\alpha G^{(2)}(s - \mathrm{i}\omega_\alpha) M_\beta G^{(2)}(s - \mathrm{i}(\omega_\alpha - \omega_\beta))$$

$$\times M_\alpha G^{(2)}(s + \mathrm{i}\omega_\beta) M_\beta$$

$$+ (1 + N_\alpha)(1 + N_\beta) M_\alpha G^{(2)}(s - \mathrm{i}\omega_\alpha) M_\beta G^{(2)}(s - \mathrm{i}(\omega_\alpha + \omega_\beta))$$

$$\times M_\alpha G^{(2)}(s - \mathrm{i}\omega_\beta) M_\beta \Big\} . \tag{9.16}$$

Incorporating this self-energy into the denominator, we have the improved Green function as

$$G^{(4)}(s) = \frac{1}{s - \frac{\mathrm{i}}{\hbar} H_\mathrm{e}^0 - \Sigma^{(2)}(s) - \Sigma^{(4)}(s)} . \tag{9.17}$$

We can extend this procedure up to the higher orders iteratively.

As a consequence, we find

$$\int_0^\infty \mathrm{d}t \, \mathrm{e}^{-st} \mathrm{Av.}\langle g | V_\mathrm{R} \exp\left[\frac{\mathrm{i}}{\hbar} H_\mathrm{e} t\right] V_\mathrm{R} \exp\left[-\frac{\mathrm{i}}{\hbar} H_\mathrm{g} t\right] | g \rangle$$

$$= \langle g | V_\mathrm{R} \frac{1}{s - \frac{\mathrm{i}}{\hbar} H_\mathrm{e}^0 - \Sigma(s)} V_\mathrm{R} | g \rangle , \tag{9.18}$$

with

$$\Sigma(s) = \Sigma^{(2)}(s) + \Sigma^{(4)}(s) + \Sigma^{(6)}(s) + \cdots ,$$

where the self-energy parts can be represented by irreducible diagrams as shown in Fig. 9.2.

The optical absorption spectrum is calculated from (9.18) by putting $s$ as $s \longrightarrow \mathrm{i}\omega + \delta$, where $\delta$ is half of the population decay rate of the exciton state excluding the contribution from the acoustic-phonon-mediated relaxation, because such contribution is automatically included in the self-energy part $\Sigma(s)$. More concretely, $\delta$ should include the radiative decay rate, the trapping rate to some defects, and the rate of exciton migration to neighboring islands. The latter two processes are phonon-mediated but should be included in

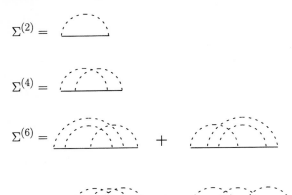

$$\Sigma^{(2)} =$$

$$\Sigma^{(4)} =$$

$$\Sigma^{(6)} = \quad + \quad$$

$$+ \quad + \quad$$

**Fig. 9.2.** Irreducible diagrams corresponding to the self-energy terms of the second, fourth, and sixth order with respect to the exciton–phonon interaction

$\delta$ because they are not taken into account in the present Green function formalism. The quantity $\delta$ can be estimated from the observed population decay rate by subtracting the phonon-assisted population decay rate within an island, which can be calculated theoretically as given in Sect. 9.7.

In order to examine the convergence of the above procedure, we have estimated the Green function up to the sixth order and compared the optical

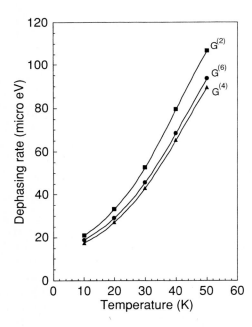

**Fig. 9.3.** Temperature dependence of exciton dephasing rates calculated by including the self-energy terms up to the second, fourth, and sixth order denoted as $G^{(2)}$, $G^{(4)}$, and $G^{(6)}$, respectively. The size parameters of the quantum disk are the same as in Fig. 9.4 (from T. Takagahara [30])

absorption spectrum and its HWHM by calculating

$$I^{(2)}(\omega) = \mathrm{Re}\left[\mathrm{Av} \cdot \langle g|V_\mathrm{R}G^{(2)}(s = \mathrm{i}\omega + \delta)V_\mathrm{R}|g\rangle\right] ,$$

$$I^{(4)}(\omega) = \mathrm{Re}\left[\mathrm{Av} \cdot \langle g|V_\mathrm{R}G^{(4)}(s = \mathrm{i}\omega + \delta)V_\mathrm{R}|g\rangle\right] ,$$

$$I^{(6)}(\omega) = \mathrm{Re}\left[\mathrm{Av} \cdot \langle g|V_\mathrm{R}G^{(6)}(s = \mathrm{i}\omega + \delta)V_\mathrm{R}|g\rangle\right] . \tag{9.19}$$

Typical results are shown in Fig. 9.3. The relevant parameters are explained in Sect. 9.4.

The size parameters of a quantum disk for Fig. 9.3 are $a$=20 nm and $b$=15 nm. It is seen that the percentage difference of the dephasing rates calculated from $G^{(2)}$ and $G^{(4)}$ is about 15–20%, whereas that calculated from $G^{(4)}$ and $G^{(6)}$ is several percent. Hence we carry out the calculation up to the fourth order and estimate the dephasing rate from the HWHM of $I^{(4)}(\omega)$.

## 9.3  Exciton–Phonon Interactions

The microscopic details of the interaction Hamiltonian between the exciton and the acoustic phonons will be described. In GaAs/AlGaAs QW's, the elastic properties of both materials are not much different, and thus the bulk-like acoustic phonon modes can be assumed as the zeroth-order approximation. Hereafter the phonon modes will be specified by the wavevector $\boldsymbol{q}$. The interaction between electrons and acoustic phonons arises from the deformation potential coupling and the piezoelectric coupling [18].

The dominant interaction term of the deformation potential coupling is given as

$$H_\mathrm{DF} = \sum_{r,q} \sqrt{\frac{\hbar|q|}{2\rho u V}} \left(D_\mathrm{c}a_{cr}^\dagger a_{cr} + D_\mathrm{v}a_{vr}^\dagger a_{vr}\right) e^{\mathrm{i}qr} \left(b_q + b_{-q}^\dagger\right) , \tag{9.20}$$

where $a(a^\dagger)$ denotes the annihilation (creation) operator of the electron in the conduction (c) or valence (v) band, $b(b^\dagger)$ is the annihilation (creation) operator of the acoustic phonon, $D_\mathrm{c}(D_\mathrm{v})$ the deformation potential of the conduction (valence) band, $u$ the sound velocity of the longitudinal acoustic (LA) mode, $V$ the quantization volume, $\rho$ the mass density, and the vector symbols of $\boldsymbol{r}$ and $\boldsymbol{q}$ are dropped.

The piezoelectric coupling is given as

$$H_\mathrm{PZ} = -\sum_{r,q} \frac{8\pi e e_{14}}{\epsilon\, q^2} \sqrt{\frac{\hbar}{2\rho u|q|V}} (\xi_x q_y q_z + \xi_y q_z q_x + \xi_z q_x q_y)e^{\mathrm{i}qr}$$

$$\cdot \left(a_{cr}^\dagger a_{cr} + a_{vr}^\dagger a_{vr}\right)\left(b_q + b_{-q}^\dagger\right) , \tag{9.21}$$

where $\epsilon(e_{14})$ is the dielectric (piezoelectric) constant and $\boldsymbol{\xi}$ is the polarization vector of the acoustic phonon modes. In this case, the transverse acoustic

(TA) mode as well as the longitudinal acoustic (LA) mode contribute to the coupling. The polarization vectors for the LA mode and the two TA modes with a wavevector $\boldsymbol{q}$ are given as

$$\boldsymbol{\xi}(LA) = (q_x, q_y, q_z)/|\boldsymbol{q}| \, ,$$

$$\boldsymbol{\xi}(TA1) = (q_y, -q_x, 0)/\sqrt{q_x^2 + q_y^2} \, ,$$

$$\boldsymbol{\xi}(TA2) = (-q_x q_z, -q_y q_z, q_x^2 + q_y^2)/|\boldsymbol{q}|\sqrt{q_x^2 + q_y^2} \, . \tag{9.22}$$

Then the matrix element of the interaction Hamiltonian between two exciton states given by

$$|X_i\rangle = \sum_{r_e, r_h} F_i(r_e, r_h) a_{cr_e}^\dagger a_{vr_h} |0\rangle \, , \tag{9.23}$$

and

$$|X_f\rangle = \sum_{r_e, r_h} F_f(r_e, r_h) a_{cr_e}^\dagger a_{vr_h} |0\rangle \, , \tag{9.24}$$

is calculated as

$$\langle X_f | H_{\mathrm{DF}} | X_i \rangle = \sum_q \sqrt{\frac{\hbar |q|}{2\rho u_{\mathrm{LA}} V}} \left( D_c \langle X_f | e^{iqr_e} | X_i \rangle - D_v \langle X_f | e^{iqr_h} | X_i \rangle \right)$$
$$\times \left( \sqrt{N_q} \text{ or } \sqrt{N_q + 1} \right) \, ,$$

$$\langle X_f | H_{\mathrm{PZ}} | X_i \rangle = -\sum_{\sigma, q} \frac{8\pi e e_{14}}{\epsilon q^2} \sqrt{\frac{\hbar}{2\rho u_\sigma |q| V}} \left( \xi_x q_y q_z + \xi_y q_z q_x + \xi_z q_x q_y \right)$$
$$\times \left( \langle X_f | e^{iqr_e} | X_i \rangle - \langle X_f | e^{iqr_h} | X_i \rangle \right)$$
$$\times \left( \sqrt{N_q} \text{ or } \sqrt{N_q + 1} \right) \, , \tag{9.25}$$

with

$$\langle X_f | e^{iqr_e(r_h)} | X_i \rangle = \int \mathrm{d}^3 r_e \int \mathrm{d}^3 r_h \, F_f^*(r_e, r_h) e^{iqr_e(r_h)} F_i(r_e, r_h) \, , \tag{9.26}$$

where $\sigma = LA$, $TA1$, and $TA2$, and the factor $\sqrt{N_q}(\sqrt{N_q + 1})$ corresponds to the phonon absorption (emission) process.

The parameter values employed for GaAs are $D_c = -14.6$ eV, $D_v = -4.8$ eV [19], $u_{\mathrm{LA}} = 4.81 \times 10^5$ cm/s, $u_{\mathrm{TA}} = 3.34 \times 10^5$ cm/s, $e_{14} = 1.6 \times 10^{-5}$ C/cm$^2$ and $\epsilon = 12.56$ [20].

## 9.4   Excitons in Anisotropic Quantum Disks

Now the theoretical formulation has been completed. In a more concrete calculation, we have to specify a model for the QD-like island structures. Here we employ the simple effective mass approximation without taking into account the valence band degeneracy. The extremely narrow linewidth of exciton emission was observed for the first time in QW samples [7–9]. The lateral fluctuation of the QW thickness gives rise to an island-like structure. The localized excitons at such structures can be viewed as the zero-dimensional excitons. In these samples, the confinement in the direction of the crystal growth is strong, whereas the confinement in the lateral direction is rather weak. Furthermore, the island structures were found to be elongated along the $[1\bar{1}0]$ direction [9]. Thus these island structures can be modeled by an anisotropic quantum disk. In order to facilitate the calculation, the lateral confinement potential in the $x$- and $y$-directions is assumed to be Gaussian as

$$V_e(r) = V_e^0 \exp\left[-\left(\frac{x}{a}\right)^2 - \left(\frac{y}{b}\right)^2\right], \ V_h(r) = V_h^0 \exp\left[-\left(\frac{x}{a}\right)^2 - \left(\frac{y}{b}\right)^2\right],$$
(9.27)

where the lateral size parameters $a$ and $b$ can be fixed in principle from the measurement of morphology by, e.g., scanning tunneling microscopy, but are left as adjustable parameters. The exciton wavefunction in such an anisotropic quantum disk can be approximated as

$$F(r_e, r_h) = \sum_{l_e, l_h, m_e, m_h} C(l_e, l_h, m_e, m_h) \left(\frac{x_e}{a}\right)^{l_e} \left(\frac{x_h}{a}\right)^{l_h} \left(\frac{y_e}{b}\right)^{m_e} \left(\frac{y_h}{b}\right)^{m_h}$$
$$\cdot \exp\left\{-\frac{1}{2}\left[\left(\frac{x_e}{a}\right)^2 + \left(\frac{x_h}{a}\right)^2 + \left(\frac{y_e}{b}\right)^2 + \left(\frac{y_h}{b}\right)^2\right]\right.$$
$$\left. - \alpha_x(x_e - x_h)^2 - \alpha_y(y_e - y_h)^2\right\} \varphi_0(z_e)\varphi_0(z_h),$$
(9.28)

with

$$\varphi_0(z) = \sqrt{\frac{2}{L_z}} \cos\left(\frac{\pi z}{L_z}\right),$$
(9.29)

where $C(l_e, l_h, m_e, m_h)$ is the expansion coefficient, $L_z$ is the QW thickness, the factor $1/2$ in the exponent is attached to make the probability distribution $|F(r_e, r_h)|^2$ follow the functional form of the confining potential, and $\alpha_x$ and $\alpha_y$ indicate the degree of the electron–hole correlation and are determined variationally. The electron–hole relative motion within the exciton state is not much different from that in the bulk because the lateral confinement is rather weak. As a result, the parameters $\alpha_x$ and $\alpha_y$ are weakly dependent on the lateral size. Because of the inversion symmetry of the confining potential, the parity is a good quantum number, and the wavefunction can be classified in terms of the combination of parities of $x^{l_e+l_h}$ and $y^{m_e+m_h}$. The exciton ground state belongs to the (even, even) series, where the first (second) index indicates the parity with respect to the $x(y)$ coordinate. As can be seen easily, the optically allowed exciton states belong to the (even, even) series, and other exciton states associated with (even, odd), (odd, even), and (odd, odd) series are dark states. In actual calculations, terms up to the sixth power are included, namely, $0 \leq l_e + l_h$, $m_e + m_h \leq 6$ to ensure the convergence of the calculation.

The potential depth for the exciton lateral motion can be guessed from the splitting energy of the heavy hole excitons due to the monolayer fluctuation of the QW thickness. The value of $|V_e^0 + V_h^0|$ is typically about 10 meV for the nominal QW thickness of about 3 nm [9,10]. Of course, even if $V_e^0 + V_h^0$ is fixed, each value of $V_e^0$ and $V_h^0$ cannot be determined uniquely. Here we employ $V_e^0 = -6$ meV and $V_h^0 = -3$ meV throughout this paper, referring to the experimental results and assuming $V_e^0 : V_h^0 = 2:1$.

A typical example of the exciton level structure is shown in Fig. 9.4 for $a = 20$ nm and $b = 15$ nm.

The disk height, namely the QW thickness, is fixed at 3 nm throughout this chapter. The transition intensities of the optically active exciton states are plotted by solid lines and the corresponding radiative lifetime is given alongside. The dark exciton states are exhibited by triangles slightly above the horizontal axis to indicate the energy positions. In the calculation of the optical absorption spectrum in (9.2), the lowest 13 exciton levels are taken into account, including the dark exciton states, because this number of levels is sufficient for converged results.

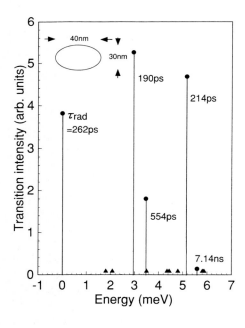

**Fig. 9.4.** Exciton energy levels in a GaAs quantum disk with parameters of $a=20$ nm, $b=15$ nm, $L_z=3$ nm, $V_e^0=6$ meV, and $V_h^0=3$ meV (see text). The origin of energy is taken at the exciton ground state. The dark exciton states are denoted by *triangles* slightly above the horizontal axis (from T. Takagahara [30])

## 9.5   Temperature–Dependence of the Exciton Dephasing Rate

First of all, we are interested in the lineshape of the calculated absorption spectrum. The lineshape of the exciton ground state is plotted in Fig. 9.5 for the quantum disk model in Sect. 9.4 at 10 K and 50 K.

The squares show the calculated spectra and the circles are the Lorentzian fit. At low temperatures, the spectra can be fitted very well by the Lorentzian, as expected. At elevated temperatures, however, the lineshape deviates from the Lorentzian and shows additional broadening. In any case, the dephasing rate, i.e., the HWHM of the absorption spectrum, can be estimated unambiguously. In addition, it is interesting to note the redshift of the exciton peak position by about several tens of microelectronvolts relative to the purely electronic transition energy indicated by the origin of energy. This is caused by the lattice relaxation energy given by (4.5) in [18].

The size dependence of the dephasing rate is shown in Fig. 9.6 for the size parameters of $(a, b)$=(12 nm,10 nm), (20 nm,15 nm), and (30 nm,20 nm). In this size range the dephasing rate is larger for smaller sizes. This can be considered to be caused by the enhanced coupling strength between the exciton and the acoustic phonons since the confinement effect on the spectral density of acoustic phonon modes is not significant in this size range. From the comparison of these results with experimental data, we can guess the likely size of the quantum disk. Hereafter we employ the size parameters of $(a, b)$=(20 nm,15 nm).

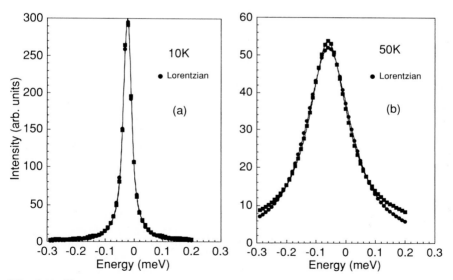

**Fig. 9.5.** Absorption spectra of the exciton ground state at (**a**) 10 K and (**b**) 50 K for the quantum disk in Fig. 9.4 (from T. Takagahara [30])

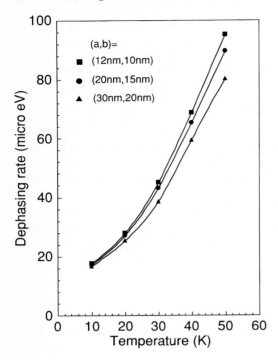

**Fig. 9.6.** Dephasing rates are plotted as a function of temperature for the exciton ground state in three quantum disks with size parameters of $(a,b)=(12$ nm, 10 nm), (20 nm, 15 nm), and (30 nm, 20 nm) and disk height of 3 nm (from T. Takagahara [30])

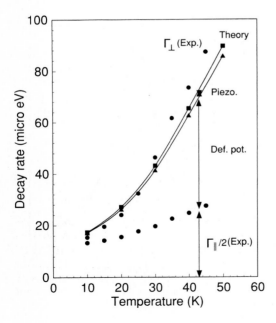

**Fig. 9.7.** Calculated dephasing rates of the exciton ground state are shown with experimental data [10] as a function of temperature. A quantum disk model is employed with the same parameters as in Fig. 9.4. The pure dephasing rate is decomposed into the contribution from the deformation potential coupling denoted as "Def. pot."and that from the piezoelectric coupling and the interference term denoted as "Piezo." (from T. Takagahara [30])

The calculated dephasing rate is plotted in Fig. 9.7 as a function of temperature with experimental data [10]. As mentioned in Sect. 9.1, the difference between the dephasing rate $\Gamma_\perp$ and half the population decay rate $\Gamma_\parallel/2$ indicates the pure dephasing rate. It is seen that the overall agreement between the theory and experiments is satisfactory concerning both the absolute magnitude and the temperature dependence. Furthermore, in the theory, we can separate out the contribution of the deformation potential coupling to the exciton dephasing; this part is shown by the arrow denoted as "Def. pot." The remaining part comes from the piezoelectric coupling and the interference term between the two couplings. This part is simply denoted as "Piezo." in Fig. 9.7. It is seen that the deformation potential coupling dominantly contributes to the pure dephasing.

In order to see the reason in more detail, we look into the matrix elements of the exciton–phonon interactions. In Fig. 9.8, we plot the angular average of the squared matrix elements of the exciton–phonon interaction defined by

$$f_{ij}^\alpha(|q|) = \int d\Omega \, |\langle i|V_\alpha(q)|j\rangle|^2 \,, \tag{9.30}$$

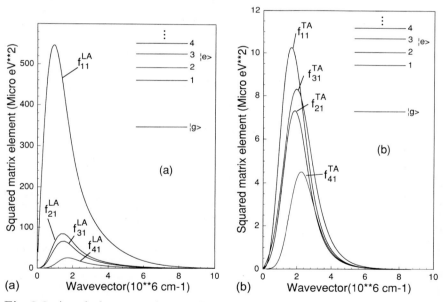

**Fig. 9.8.** Angularly averaged squared matrix elements of the exciton–phonon interaction are plotted as a function of the phonon wavevector for (**a**) the LA phonons and (**b**) the TA phonons. The *horizontal (vertical) axis* is scaled by $10^6$ cm$^{-1}$ (($\mu$eV)$^2$). The GaAs quantum disk employed is the same as in Fig. 9.4. The lowest four exciton states are numbered consecutively including the dark exciton states (from T. Takagahara [30])

where $\alpha=LA$ or $TA$, the indices $i$ and $j$ denote the exciton state, and for the case of $\alpha=TA$ the contribution from both $TA1$ and $TA2$ modes in (9.22) is combined. The quantitiy $V_{LA}$ includes the contribution from both the deformation potential coupling and the piezoelectric coupling, whereas $V_{TA}$ contains only the contribution from the piezoelectric coupling. In the inset, the energy level scheme is shown for the lowest four exciton states including the dark exciton states. From the comparison between Fig. 9.8a and b, we see that the contribution from $LA$ phonons is more than one order of magnitude larger than that from $TA$ phonons. Furthermore, it is important to note that the wavevector of the most efficiently coupled phonons is roughly determined by max.$(1/a_B, 1/L)$, where $a_B$ is the exciton Bohr radius and $L$ is the typical size of the lateral confinement. The vanishing $z$-component of the wavevector is favored because the common envelope function in (9.29) is assumed for both the electron and the hole, and they are uncorrelated in the $z$-direction. The relevant wavevector is about $10^6$ cm$^{-1}$ for GaAs islands. Hence the corresponding phonon energy is rather small ($<1$ meV), since the phonon energy versus wavevector ($|q|$) relation is $\hbar\omega=0.32(0.22)|q|$ (meV) for the $LA(TA)$ modes with $|q|$ scaled in units of $10^6$ cm$^{-1}$. This property will be invoked later in discussing the correlation between the temperature dependence of the exciton dephasing rate and the strength of the quantum confinement.

We have also calculated the dephasing rate of the excited exciton states. The results are shown in Fig. 9.9 for the lowest four optically active exciton states. In the calculation for the excited exciton states, the value of $\delta$ in (9.19) is assumed to be the same as for the exciton ground state because the relevant relaxation processes mentioned prior to (9.19) may be dependent

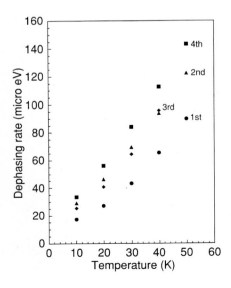

**Fig. 9.9.** Calculated dephasing rates of the lowest four optically active exciton states are shown as a function of temperature for the same quantum disk as in Fig. 9.4 (from T. Takagahara [30])

on the exciton state, but the absolute magnitude of $\delta$ is rather small. The dephasing rate is, in general, larger for the higher-lying exciton states. But this tendency is not monotonic, as seen by the reversed order of magnitude between the second and third exciton states.

## 9.6   Elementary Processes of Exciton Pure Dephasing

Now we discuss the mechanism of exciton pure dephasing. Generally speaking, pure dephasing means the decay of the dipole coherence without change in the state of the system. Any real transition to other states leads to the population decay. Thus the pure dephasing is caused by virtual processes that start from a relevant state, and through some excursion in the intermediate states, return to the same initial state. These virtual processes give rise to the temporal fluctuation of the phase of the wavefunction. Previously, this kind of temporal phase fluctuation was treated by a stochastic model of random frequency modulation [21], and the resulting pure dephasing was discussed in the context of resonant secondary emission [22].

Here we treat these processes microscopically. There are two kinds of these virtual processes which contribute to the pure dephasing. The first type of processes is induced by the off-diagonal exciton–phonon interaction. Those processes start from the exciton ground state, pass through excited exciton states, and return to the exciton ground state. The second type of processes is induced by the diagonal exciton–phonon interaction, and the relevant state

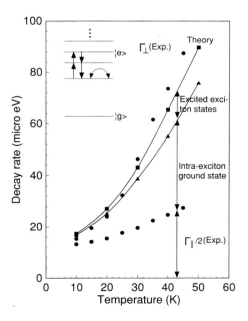

**Fig. 9.10.** Calculated dephasing rates of the exciton ground state are shown with experimental data [10] as a function of temperature for the same quantum disk as in Fig. 9.4. The pure dephasing rate is decomposed into the contribution from the diagonal exciton–phonon interaction denoted as "Intra-exciton ground state" and that from the off-diagonal interaction and the interference term denoted as "Excited exciton states" (from T. Takagahara [30])

always remains within the exciton ground state. These processes are shown schematically in the inset of Fig. 9.10.

The contribution to the pure dephasing from the second kind of processes can be singled out theoretically by carrying out the calculation, which includes only the exciton ground state. That contribution is denoted by "Intra-exciton ground state" in Fig. 9.10. The remaining part denoted as "Excited exciton states" comes from the first kind of processes and the interference between the two kinds of processes. It is seen that the "intra-exciton ground state" (diagonal) processes contribute substantially to the pure dephasing, but the contribution from the off-diagonal processes is not negligible. This feature can be understood from Fig. 9.8a,b since the squared matrix element within the exciton ground state denoted by $f_{11}^\alpha$ is much larger than other squared matrix elements. As a result, the "intra-exciton ground state" processes contribute significantly to the pure dephasing.

## 9.7    Mechanisms of Population Decay of Excitons

The possible mechanisms of the population decay will be discussed. Experimentally, two decay time constants were observed [10]. The slow time constant ($\sim$200 ps) is almost independent of temperature, suggesting radiative decay as its mechanism. In fact, the calculated radiative lifetime of the exciton ground state is around 200 ps, as shown in Fig. 9.4. On the other hand, the fast time constant ($\sim$30 ps) is weakly dependent on temperature. The likely mechanisms are thermal activation to excited exciton states and phonon-assisted migration to neighboring islands. In this section we present detailed calculation of these relaxation rates and examine the significance of these mechanisms.

### 9.7.1    Phonon-Assisted Population Relaxation

The phonon-assisted transition rate to other exciton states is calculated as

$$P_i = \sum_{j \neq i} w_{ij} \; , \tag{9.31}$$

where $w_{ij}$ is the transition rate from the exciton state $i$ to another exciton state $j$. Here the values of $P_i$ are calculated for the lowest four optically active exciton states and are plotted in Fig. 9.11 as a function of temperature.

The same quantum disk model as in Fig. 9.4 is employed and the same 13 exciton levels are included in the calculation. Since the energy difference between exciton levels is less than several meV, it is sufficient to take into account only the one-phonon processes. For the exciton ground state, the transition rate is several μeV. For the excited exciton states, the transition rates are about one order of magnitude larger than that of the exciton ground state. In general, the higher-lying exciton states have a larger population

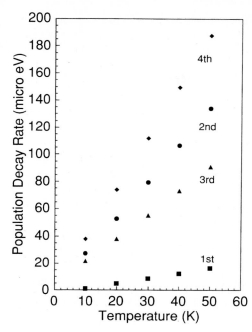

**Fig. 9.11.** Phonon-assisted population decay rates of the lowest four optically active exciton states are plotted as a function of temperature for the same quantum disk as in Fig. 9.4 (from T. Takagahara [30])

decay rate. But this trend is not monotonic, as seen by the reversed order of magnitude between the second and third exciton states.

It is interesting to note that a linear temperature dependence is clearly seen. This indicates that the energy of the relevant acoustic phonons is rather small, as shown in Fig. 9.8a,b, and the high temperature approximation holds as

$$\frac{1}{e^{\hbar\omega/k_{\mathrm{B}}T} - 1} \approx \frac{k_{\mathrm{B}}T}{\hbar\omega} . \tag{9.32}$$

This is the origin of the linear temperature dependence.

### 9.7.2   Phonon-Assisted Exciton Migration

The excitons localized at island structures can migrate among the structures, accompanying phonon absorption or emission to compensate for the energy mismatch. Since the energy mismatch is typically a few meV, the acoustic phonons dominantly contributes to the exciton migration process. Now let us consider two island sites at $R_a$ and $R_b$, and assume that the island at $R_b$ is larger in size and has a localized exciton state of lower energy than in the island at $R_a$. Then we consider the phonon-assisted exciton migration from the site $R_a$ to the site $R_b$, namely a transition of

$$|R_a ; n_q\rangle \longrightarrow |R_b ; n_q + 1\rangle ,$$

where $n_q$ indicates the occupation number of an acoustic phonon mode with wavevector $\boldsymbol{q}$. As discussed previously [23], there are three elementary processes of this transition:

(1) $|R_a ; n_q\rangle \xrightarrow{H_{\mathrm{ep}}} |R_b ; n_q + 1\rangle$ ,

(2) $|R_a ; n_q\rangle \xrightarrow{H_{\mathrm{ep}}} |R_a ; n_q + 1\rangle \xrightarrow{H_{\mathrm{ss}}} |R_b ; n_q + 1\rangle$ ,

(3) $|R_a ; n_q\rangle \xrightarrow{H_{\mathrm{ss}}} |R_b ; n_q\rangle \xrightarrow{H_{\mathrm{ep}}} |R_b ; n_q + 1\rangle$ ,

where $H_{\mathrm{ep}}(H_{\mathrm{ss}})$ represents the exciton–phonon interaction (inter-site transfer) Hamiltonian. The first process is a direct process through the overlap between exciton wavefunctions at two island sites, which is strongly dependent on the distance between the two islands. Thus this process contributes only for the case of short distance. On the other hand, the second and third processes are indirect ones that can contribute to the exciton transfer even for the case of long distance.

The inter-site transfer Hamiltonian $H_{\mathrm{ss}}$ is caused by the electron–electron interaction and is calculated as

$$
\begin{aligned}
J(R_a, R_b) &= \langle R_a | H_{\mathrm{ss}} | R_b \rangle \\
&= \sum_{r_e, r_h} \sum_{r'_e, r'_h} F^*_{c\tau, v\sigma}(r_e, r_h; R_a) F_{c\tau', v\sigma'}(r'_e, r'_h; R_b) \\
&\quad \cdot [V(c\tau r_e, v\sigma' r'_h; c\tau' r'_e, v\sigma r_h) - V(c\tau r_e, v\sigma' r'_h; v\sigma r_h, c\tau' r'_e)] ,
\end{aligned}
\tag{9.33}
$$

where the localized exciton at the site $R_i$ is described as

$$
|R_i\rangle = \sum_{\tau, r_e; \sigma, r_h} F_{c\tau, v\sigma}(r_e, r_h; R_i) a^\dagger_{c\tau r_e} a_{v\sigma r_h} |0\rangle ,
\tag{9.34}
$$

with an envelope function $F$ corresponding to (9.28). The suffix $\tau(\sigma)$ denotes the Wannier function index of the conduction (valence) band, e.g., the total angular momentum. The first term in (9.33) represents the electron–hole exchange interaction, and the second term corresponds to the Coulomb interaction. Because of the localized nature of the Wannier functions, we can approximate as

$$
V(c\tau r_e, v\sigma' r'_h; v\sigma r_h, c\tau' r'_e) \approx \delta_{r_e r'_e} \delta_{r_h r'_h} \delta_{\tau\tau'} \delta_{\sigma\sigma'} \frac{e^2}{\epsilon |\boldsymbol{r}_e - \boldsymbol{r}_h|} ,
$$

$$
V(c\tau r_e, v\sigma' r'_h; c\tau' r'_e, v\sigma r_h) \approx \delta_{r_e r_h} \delta_{r'_e r'_h} \left[ \delta_{r_e r'_e} V(c\tau r_e, v\sigma' r_e; c\tau' r_e, v\sigma r_e) \right.
$$

$$
\left. + (1 - \delta_{r_e r'_e}) \boldsymbol{\mu}_{c\tau v\sigma} \frac{[1 - 3\boldsymbol{n} \cdot^t \boldsymbol{n}]}{|\boldsymbol{r}_e - \boldsymbol{r}'_e|^3} \boldsymbol{\mu}_{v\sigma' c\tau'} \right], \tag{9.35}
$$

with

$$n = \frac{r_{\mathrm{e}} - r_{\mathrm{e}}'}{|r_{\mathrm{e}} - r_{\mathrm{e}}'|} \; , \tag{9.36}$$

$$\mu_{c\tau v\sigma} = \int \mathrm{d}^3 r \; \phi_{c\tau R}^*(r)(r - R)\phi_{v\sigma R}(r) \; , \tag{9.37}$$

where $\phi_{c\tau(v\sigma)R}(r)$ is a Wannier function localized at the site $R$. Hereafter the vector symbols will be dropped because there is no fear of confusion. The Coulomb term decreases rapidly when $|R_a - R_b|$ exceeds the lateral size of the exciton wavefunction. On the other hand, the exchange term contains the dipole–dipole interaction and has a long-range character that decreases as $|R_a - R_b|^{-3}$. Thus the exchange term contributes dominantly to the inter-site exciton transfer when $|R_a - R_b|$ is larger than the lateral size of the confining potential.

The exciton transfer probability is calculated as

$$
\begin{aligned}
\omega(R_a \longrightarrow R_b) = \frac{2\pi}{\hbar} \sum_\sigma \Big| & \langle R_b \,; n_q + 1|H_{\mathrm{ep}}^\sigma|R_a \,; n_q\rangle \\
&+ \frac{\langle R_b \,; n_q + 1|H_{\mathrm{ss}}|R_a \,; n_q + 1\rangle\langle R_a \,; n_q + 1|H_{\mathrm{ep}}^\sigma|R_a \,; n_q\rangle}{-\hbar\omega_q^\sigma} \\
&+ \frac{\langle R_b \,; n_q + 1|H_{\mathrm{ep}}^\sigma|R_b \,; n_q\rangle\langle R_b \,; n_q|H_{\mathrm{ss}}|R_a \,; n_q\rangle}{E_{\mathrm{X}}(R_a) - E_{\mathrm{X}}(R_b)} \Big|^2 \\
&\cdot \delta\big(E_{\mathrm{X}}(R_a) - E_{\mathrm{X}}(R_b) - \hbar\omega_q^\sigma\big) \; ,
\end{aligned}
\tag{9.38}
$$

where $E_{\mathrm{X}}(R_i)$ is the energy of a localized exciton at the site $R_i$ and the summation concerning the acoustic phonon mode $\sigma$ is taken over the $LA$, $TA1$, and $TA2$ modes in (9.22).

In the exchange matrix element in (9.35), the first (second) term is usually called the short- (long-)range part of the exchange interaction. The contribution from the long-range part can be rewritten into a more tractable form as [24]

$$
\begin{aligned}
\sum_{r_{\mathrm{e}} \neq r_{\mathrm{e}}'} & F_{c\tau,v\sigma}^*(r_{\mathrm{e}}, r_{\mathrm{e}}; R_a) F_{c\tau',v\sigma'}(r_{\mathrm{e}}', r_{\mathrm{e}}'; R_b)\mu_{c\tau,v\sigma} \frac{[1 - 3n \cdot^t n]}{|r_{\mathrm{e}} - r_{\mathrm{e}}'|^3}\mu_{v\sigma',c\tau'} \\
&= -\sum_{r_{\mathrm{e}} \neq r_{\mathrm{e}}'} F_{c\tau,v\sigma}^*(r_{\mathrm{e}}, r_{\mathrm{e}}; R_a)\mu_{c\tau,v\sigma}\,\mathrm{grad}_{r_{\mathrm{e}}}\mathrm{div}_{r_{\mathrm{e}}}\left[\frac{\mu_{v\sigma',c\tau'}}{|r_{\mathrm{e}} - r_{\mathrm{e}}'|}F_{c\tau',v\sigma'}(r_{\mathrm{e}}', r_{\mathrm{e}}'; R_b)\right] \\
&= -\int \mathrm{d}^3 r_{\mathrm{e}} \int' \mathrm{d}^3 r_{\mathrm{e}}' F_{c\tau,v\sigma}^*(r_{\mathrm{e}}, r_{\mathrm{e}}; R_a)\mu_{c\tau,v\sigma} \\
&\qquad \cdot \,\mathrm{grad}_{r_{\mathrm{e}}}\mathrm{div}_{r_{\mathrm{e}}}\left[\frac{\mu_{v\sigma',c\tau'}}{|r_{\mathrm{e}} - r_{\mathrm{e}}'|}F_{c\tau',v\sigma'}(r_{\mathrm{e}}', r_{\mathrm{e}}'; R_b)\right] \; ,
\end{aligned}
\tag{9.39}
$$

where the integration with respect to $r'_e$ is carried out over the whole space excluding a small sphere around the point $r_e$, which is indicated by a primed integral symbol. Then making use of a relation for an arbitrary vector field $Q(r)$

$$\int' d^3r' \, \text{grad}_r \text{div}_r \frac{Q(r')}{|r - r'|} = \frac{4\pi}{3} Q(r) + \text{grad}_r \text{div}_r \int' d^3r' \, \frac{Q(r')}{|r - r'|} \, , \quad (9.40)$$

we can rewrite (9.39) as

$$- \int d^3r_e \, F^*_{c\tau,v\sigma}(r_e, r_e; R_a) \mu_{c\tau,v\sigma} \left[ \frac{4\pi}{3} \mu_{v\sigma',c\tau'} F_{c\tau',v\sigma'}(r_e, r_e; R_b) \right.$$

$$\left. + \text{grad}_{r_e} \text{div}_{r_e} \int d^3r'_e \frac{\mu_{v\sigma',c\tau'}}{|r_e - r'_e|} F_{c\tau',v\sigma'}(r'_e, r'_e; R_b) \right] , \quad (9.41)$$

where, in the second term within the parentheses, the integration can be performed over the whole space because the singularity of $|r_e - r'_e|^{-1}$ is integrable. By a partial integration, we have the expression of the long-range exchange term as

$$- \frac{4\pi}{3} (\mu_{c\tau,v\sigma} \cdot \mu_{v\sigma',c\tau'}) \int d^3r \, F^*_{c\tau,v\sigma}(r, r; R_a) F_{c\tau',v\sigma'}(r, r; R_b)$$

$$+ \int d^3r \, \text{div}_r (F^*_{c\tau,v\sigma}(r, r; R_a) \mu_{c\tau,v\sigma})$$

$$\cdot \text{div}_r \left[ \int d^3r' \frac{\mu_{v\sigma',c\tau'}}{|r - r'|} F_{c\tau',v\sigma'}(r', r'; R_b) \right] . \quad (9.42)$$

The short-range exchange term is simply written as

$$V(c\tau r_0, v\sigma' r_0 \, ; c\tau' r_0, v\sigma r_0) \int d^3r \, F^*_{c\tau,v\sigma}(r, r; R_a) F_{c\tau',v\sigma'}(r, r; R_b) \, , \quad (9.43)$$

and the Coulomb term is calculated by

$$- \int d^3r_e \int d^3r_h \, F^*_{c\tau,v\sigma}(r_e, r_h; R_a) \frac{e^2}{\epsilon|r_e - r_h|} F_{c\tau,v\sigma}(r_e, r_h; R_b) \, . \quad (9.44)$$

Combining three terms (9.42), (9.43), and (9.44), we can estimate the exciton transfer matrix element in (9.33).

It is to be noted that when the Coulomb term and/or the exchange term are of comparable magnitude to the energy difference between localized exciton states at $R_a$ and $R_b$, the eigenstates should be mixed states of two localized excitons and a simple picture of exciton transfer between two sites cannot be applied. Thus we have to check the inequality

$$|J(R_a, R_b)| \ll |E_X(R_a) - E_X(R_b)| \, ,$$

before we apply the exciton transfer model. We can check numerically that the matrix element $|J(R_a, R_b)|$ is typically about several tens of $\mu$eV except for a very close pair of islands, whereas $|E_X(R_a)-E_X(R_b)|$ is about a few meV for typical sizes of islands. Thus the simple picture of exciton transfer can be applied safely.

In order to see the typical behavior of the exciton migration rate, we employ two quantum disk islands characterized by $L_z=3$ nm and the lateral size parameters of $(a, b)=(20$ nm, $15$ nm$)$ and $(30$ nm, $20$ nm$)$ and consider the migration between two exciton ground states whose energy difference is 0.68 meV. The exciton migration rate depends on the distance between two islands, the geometrical configuration of two islands, and on the direction of exciton polarizations.

In Fig. 9.12a,b, the migration rate is plotted as a function of the center-to-center distance between two islands. In Fig. 9.12a, two islands are aligned such that the longer axes of two ellipses are coincident with each other. In this configuration, the migration rate is larger for the exciton polarization along the longer axis than for the exciton polarization along the shorter axis because of the larger interaction through the surface charges. In Fig. 9.12b, two islands are aligned such that the shorter axes of two ellipses are coincident

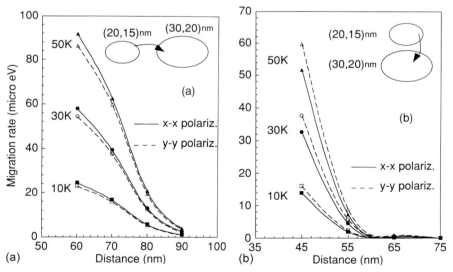

**Fig. 9.12.** Phonon-assisted migration rate from the exciton ground state in a quantum disk of $(a, b)=(20$ nm, $15$ nm$)$ to the exciton ground state in a quantum disk of $(a, b)=(30$ nm, $20$ nm$)$ is plotted as a function of the center-to-center distance between two quantum disks at temperatures of 10 K, 30 K, and 50 K. In (**a,b**), two disks are aligned such that the *longer* (*shorter*) *axes* of two ellipses are coincident with each other and the distance is measured along the *longer* (*shorter*) axis. The exciton polarization in two disks is aligned along the $x$- or $y$-direction, and this is indicated by "$x$–$x$" or "$y$–$y$ polariz.", where $x(y)$ denotes the direction of the *longer* (*shorter*) *axis* of the ellipse (from T. Takagahara [30])

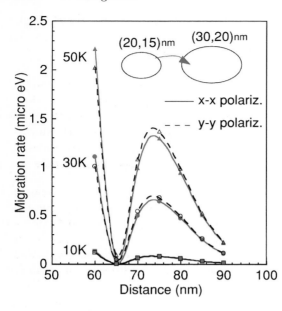

**Fig. 9.13.** Phonon-assisted migration rate from the exciton ground state in a quantum disk of $(a, b) = (20$ nm, 15 nm$)$ to the second lowest optically active exciton state in a quantum disk of $(a, b) = (30$ nm, 20 nm$)$ is plotted as a function of the center-to-center distance between two quantum disks at temperatures of 10 K, 30 K, and 50 K. The configuration of the two disks is the same as in Fig. 9.12a (from T. Takagahara [30])

with each other. In this configuration, the migration rate is larger for the exciton polarization along the shorter axis than for the exciton polarization along the longer axis. The absolute magnitude of the exciton migration rate is several tens of μeV.

We have also estimated the migration rate between the exciton ground state and the excited exciton states. The results are shown in Fig. 9.13 for the transition from the exciton ground state of an island with $(a, b) = (20$ nm, 15 nm$)$ to the second optically active exciton state in an island with $(a, b) = (30$ nm, 20 nm$)$. The configuration of two islands is the same as in Fig. 9.12a. This migration process is associated with phonon absorption ($\sim$2.11 meV), in contrast to the case in Fig. 9.12. As a result, the migration rate is several tens of times smaller than in Fig. 9.12a. In addition, the distance dependence is not monotonic. This feature arises from the interference among three terms in (9.38) and may be highly dependent on the spatial profile of the exciton wavefunction.

At present, the distance between neighboring islands cannot be determined precisely. However, from consideration of the order of magnitude, we can identify the likely mechanism of the population decay of the exciton ground state as the combination of thermal activation to the excited exciton states within an island and phonon-assisted exciton migration to neighboring islands.

## 9.8 Correlation Between Temperature Dependence of Exciton Dephasing Rate and Strength of Quantum Confinement

We have so far discussed the dephasing rate of excitons in GaAs QD-like islands. It is important to note that the temperature dependence of exciton dephasing rates can differ qualitatively for semiconductors with strong and weak quantum confinement. For quantum dots belonging to the strong-confinement regime, such as II–VI nanocrystals, a linear temperature dependence was observed up to 200 K [14]. Recent more careful measurements also show linear temperature dependence, although the measured temperature range is limited below 30 K [25]. On the other hand, for CuCl nanocrystals belonging to the weak-confinement regime, since the exciton Bohr radius is rather small, a temperature dependence similar to that in Fig. 9.7 was observed [15,26], although strongly nonlinear temperature dependence occurs above a higher temperature than in Fig. 9.7. These different temperature dependencies can be explained by the present theory if we take into account the different exciton level structures involved. For GaAs islands with lateral size about 40 nm, the energy separation among the lowest exciton levels is a few meV, and for CuCl nanocrystals with 4 nm radius, the energy separation is typically several meV. On the other hand, in CdSe nanocrystals with radii smaller than 2 nm, the relevant energy spacing is determined by the A- and B-exciton splitting ($\sim$26 meV) instead of the inter-sublevel energy of the A-exciton series, which is much larger [27].

The key feature in considering the temperature dependence of the exciton dephasing is the interplay among virtual processes within the exciton ground state, those which pass through excited exciton states, and real transitions to excited exciton states. Another important feature is the phonon mode spectrum. In this chapter, we have dealt with the continuous spectra of acoustic phonons, which is justified because of the small difference in the elastic properties between the QW material and the barrier material. The situation is slightly different in the case of CuCl and CdSe nanocrystals. They are embedded in materials with much different elastic properties. Here the acoustic phonons are affected by the elastic confinement [28,29] but are intrinsically delocalized into the surrounding medium. As a result, the mode spectra of acoustic phonons have a continuous background in addition to discrete peaks corresponding to size-quantized modes. Furthermore, as discussed in Sect. 9.5, it is important to note that the wavevector of phonons that couple most strongly with excitons is roughly given by $\max.(1/a_B, 1/L)$, where $L$ is the typical size of nanocrystals.

In the case of small CdSe nanocrystals, the energy separation between the exciton ground state and the first excited exciton state is about 20 meV, and the acoustic phonon energy corresponding to the maximum coupling with excitons is about 1 meV. Thus only the virtual processes within the exciton ground state contribute to the exciton dephasing up to around 200 K. Then the high-temperature approximation in (9.32) holds, leading to the linear temperature dependence of the dephasing rate. Above 200 K, the multi-acoustic-phonon and LO phonon assisted activation to the excited exciton states becomes possible, yielding the nonlinear temperature dependence of the dephasing rate.

In the case of GaAs islands, the energy of phonons that couple strongly with excitons is less than 1 meV. Thus at low temperatures ($\leq$10 K), the virtual processes within the exciton ground state contribute dominantly to the exciton dephasing. At elevated temperatures ($\sim$30 K), the virtual processes via excited exciton states begin to contribute to the exciton dephasing, because the energy separation between the ground and first excited exciton states is a few meV. Thus the range of the linear temperature dependence is rather narrow, and the nonlinear temperature dependence prevails.

In the case of CuCl nanocrystals, the phonon energy corresponding to the maximum coupling with excitons is estimated as about 3–4 meV. Also the energy separation between exciton levels is several meV. Thus the linear temperature dependence is not manifest, and the nonlinear temperature dependence becomes prominent above 50 K.

We see that the temperature dependence of the exciton dephasing rate is highly dependent on the exciton level structures, which reflect the strength of quantum confinement, and on the energy of phonons, which couple with excitons strongly. We can summarize as follows: in the strong-confinement regime, the diagonal virtual processes dominate the pure dephasing, in which the relevant state remains always within the exciton ground state, and the electron–phonon interactions associated with small energy acoustic phonons mainly contribute to the pure dephasing. Thus the high temperature approximation holds for the phonon occupation number, leading to the linear temperature dependence of the dephasing rate up to the temperature corresponding to the energy spacing between the ground and first excited exciton states. On the other hand, in the weak-confinement regime, the exciton levels are rather densely distributed and the off-diagonal virtual processes contribute substantially to the pure dephasing, in which the virtual transition to excited exciton states occurs effectively when the thermal energy $k_{\mathrm{B}}T$ approaches the relevant energy level separation. In this case, acoustic phonons with thermal energy contribute significantly to the exciton dephasing. As a result, the linear temperature dependence is not manifest and the nonlinear temperature dependence dominantly appears.

## 9.9    Polarization Relaxation of Excitons

Now we discuss polarization relaxation by the electron–phonon interaction among exciton doublet structures and within an exciton doublet, which are calculated in Chap. 2, taking into account the valence band structure. In this

system the relevant energy level spacings are about a few meV, and thus
we take into account only the interactions with acoustic phonons through
the deformation potential coupling and the piezoelectric coupling [30,18].
The electron–phonon interaction itself does not flip the electron spins. The
exciton eigenstates are mixed states of all the eight combinations of the Bloch
functions, as discussed in Chap. 2. Thus it is possible to make a transition
between the dominantly $x$-polarized states and the dominantly $y$-polarized
states through electron–phonon interactions.

For the inter-doublet transitions, it is sufficient to consider only the one-
phonon processes because the acoustic phonons resonant with the relevant
transition energy have a wavelength comparable to the quantum disk size, and
the matrix element of the electron–phonon interaction has a sizable magnitude. On the other hand, for the intra-doublet transitions, we have to consider
multi-phonon, or at least two-phonon processes, because the contribution
from the one-phonon processes is vanishingly small. In general, the matrix
element of the electron–phonon interaction is proportional to

$$\int d^3 r_e \int d^3 r_h F_f^*(r_e, r_h) \, e^{i\boldsymbol{q}\boldsymbol{r_e}} (\text{or } e^{i\boldsymbol{q}\boldsymbol{r_h}}) \, F_i(r_e, r_h) \,, \tag{9.45}$$

where $\boldsymbol{q}$ is the phonon wavevector, and $F_i (F_f)$ is the envelope function of the
initial (final) exciton state. Since the doublet splitting energy is several tens of
$\mu$eV, the corresponding wavevector of phonons is rather small ($\sim 10^5$ cm$^{-1}$),
and we can approximate as

$$\int d^3 r_e \int d^3 r_h F_f^*(r_e, r_h)$$

$$\left( 1 + i\boldsymbol{q} \cdot \boldsymbol{r_e}(\text{or } i\boldsymbol{q} \cdot \boldsymbol{r_h}) + \frac{1}{2}(i\boldsymbol{q} \cdot \boldsymbol{r_e})^2 \left(\text{or } (i\boldsymbol{q} \cdot \boldsymbol{r_h})^2\right) + \cdots \right)$$

$$F_i(r_e, r_h) = \int d^3 r_e \int d^3 r_h F_f^*(r_e, r_h)$$

$$\left( \frac{1}{2}(i\boldsymbol{q} \cdot \boldsymbol{r_e})^2 \left(\text{or } (i\boldsymbol{q} \cdot \boldsymbol{r_h})^2\right) + \cdots \right) F_i(r_e, r_h) \,, \tag{9.46}$$

where the first term in the parentheses of the first line vanishes due to the orthogonality of the eigenfunctions, and the second term also vanishes because
the parity is a good quantum number for the confinement potential with
inversion symmetry. As a result, the probability of one-phonon processes is
rather small, and we have to consider two-phonon processes. The probability
of the Raman-type two-phonon processes is given by

$$\sum_{\omega_1, \omega_2} \left| \sum_j \frac{\langle f|H_{\text{e-ph}}(\omega_2)|j\rangle \langle j|H_{\text{e-ph}}(\omega_1)|i\rangle}{E_i - E_j \mp \hbar\omega_1} \right|^2 \delta\left(E_f - E_i \pm \hbar\omega_1 \mp \hbar\omega_2\right), \tag{9.47}$$

where $H_{\text{e-ph}}$ is the electron–phonon interaction, $j$ denotes the intermediate exciton states, and the summation over the energies $(\omega_1, \omega_2)$ and the
wavevectors of two phonons is carried out.

**Table 9.1.** Comparison between the rates of the one-phonon and the two-phonon processes at 30 K for the intra-doublet transitions of the lowest four exciton doublets in a GaAs elliptical quantum disk with the size parameters $a=20$ nm, $b=15$ nm, and $L_z$(disk height)$=3$ nm (from T. Takagahara [30])

|       | 1-phonon | 2-phonon |
|-------|----------|----------|
| 1–1′  | $1.42 \cdot 10^{-8}$ µeV | 0.071 µeV |
| 2–2′  | 0.017 | 0.156 |
| 3–3′  | 0.068 | 0.418 |
| 4–4′  | 0.021 | 0.227 |

In Table 9.1, the calculated transition rates at 30 K of the one-phonon and the two-phonon processes are compared for the intra-doublet transitions of the lowest four exciton doublets. It can be confirmed that the two-phonon processes are dominantly contributing.

Combining the one-phonon and the two-phonon processes, the calculated relaxation times at 30 K are shown in Fig. 9.14 for the lowest four exciton doublet states having even parity. In this figure the dark exciton states consisting of the even-parity dark states and the odd-parity states are not included. The relaxation time within each exciton doublet state is rather long (several nanoseconds) because of the small energy difference. These relax-

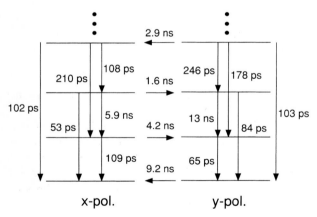

**Fig. 9.14.** Population relaxation times at 30 K for the polarization-conserving processes within the lowest four optically active exciton doublet states and for the intra-doublet relaxation processes in a GaAs elliptical quantum disk with the size parameters $a=20$ nm, $b=15$ nm, and $L_z$(disk height)$=3$ nm. It is to be noted that the energy level structures are only schematic and do not reflect the actual energy spacings (from T. Takagahara [30])

ation times are not inconsistent with a recent measurement on InGaAs QD [31]. On the other hand, the typical relaxation times within exciton states having the same polarization (x- or y-polarization) are about 100 ps. These relaxation times are in good agreement with the recently measured values [32] using single-dot pump-probe spectroscopy.

The relaxation time from the third exciton level to the second exciton level is rather long for both the x- and y-polarized states, reflecting the small energy difference between the second and the third exciton levels (<1 meV). This is explained by the same argument given by (9.46). It should be noted that the exciton level structures in Fig. 9.14 are only schematic and do not reflect the actual level spacings accurately. The relaxation time from the second exciton level to the exciton ground state is about 100 ps, whereas that from the third exciton level to the exciton ground state is several tens of picoseconds. Thus the relaxation from the third exciton level to the exciton ground state occurs directly instead of the two-step sequence through the second exciton state. In general, the transition between exciton levels whose energy difference is small would often be bypassed by some other efficient relaxation paths. This kind of phenomenon seems to have been observed experimentally [33].

Now we discuss the cross-relaxation or polarization relaxation between exciton states having mutually orthogonal polarizations. The calculated results at 30 K are given in Fig. 9.15. The cross-relaxation times are of the same order of magnitude as the relaxation times within the exciton states having the same polarization. Thus the polarization relaxation of excitons occurs rather efficiently. At the same time, it should be noted that the cross-relaxation rate is rather small for the transition between the third and the second exciton states having the orthogonal polarizations. This is also caused by the small energy difference.

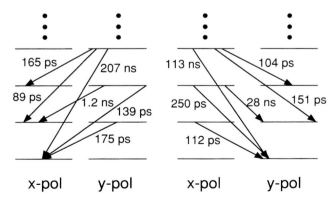

**Fig. 9.15.** Polarization relaxation times at 30 K within the lowest four optically active exciton doublet states in the same quantum disk as in Fig. 9.14 (from T. Takagahara [30])

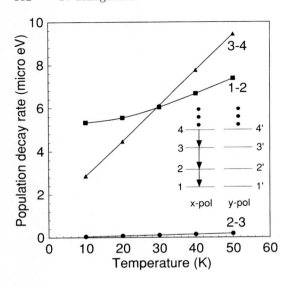

**Fig. 9.16.** Population relaxation rates are plotted as a function of temperature for the successive transitions within the $x$-polarized exciton states in the same quantum disk as in Fig. 9.14 (from T. Takagahara [30])

The temperature dependence of the relaxation rate is related to the number of phonons involved in the transition. In the case of one-phonon processes, the relaxation rate is proportional to the phonon occupation number, which can be approximated as

$$n(\omega) = \frac{1}{e^{\hbar\omega/k_{\mathrm{B}}T} - 1} \sim \frac{k_{\mathrm{B}}T}{\hbar\omega} \ , \tag{9.48}$$

at high temperatures, such that $\hbar\omega \ll k_{\mathrm{B}}T$.

In Fig. 9.16, the relaxation rates for a few transitions within the $x$-polarized exciton levels are shown as a function of temperature. Since the energy difference between the second and the third exciton states is rather small, the linear temperature dependence is clearly seen. On the other hand, as mentioned before, the relaxation within the exciton doublet states occurs mainly through two-phonon processes and the rate is typically proportional to a factor given by

$$\sum_{\omega_1,\omega_2} M(\omega_1,\omega_2)n(\omega_1)(1 + n(\omega_2)) \ , \tag{9.49}$$

where $M$ contains the matrix elements of the electron–phonon interaction. According to (9.48), the rate in (9.49) is expected to be proportional to $T^2$ at high temperatures.

In Fig. 9.17, the temperature dependence of relaxation rates is shown for the intra-doublet transitions of the lowest four exciton doublets. The $T^2$-dependence can be confirmed at high temperatures. In fact, in the temperature range from 30 K to 50 K, the calculated results can be fit well by $T^n$ ($n=1.98\sim2.24$).

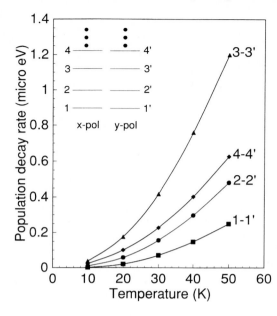

**Fig. 9.17.** Polarization relaxation rates are plotted as a function of temperature for the intra-doublet transitions in the same quantum disk as in Fig. 9.14. The *solid lines* show a dependence of $T^n$ ($n$=1.98~2.24) (from T. Takagahara [30])

## 9.10    Photoluminescence Spectrum under Selective Excitation

In Sect. 9.9, it was found that polarization relaxation occurs efficiently in our system. In order to see this effect more clearly, we have calculated the photoluminescence spectra under selective excitation. So far we have discussed the even-parity exciton states because the optically active exciton states have even parity. There are, however, odd-parity exciton states that are optically dark, and their energy positions are shown by diamonds in Fig. 2.11 for a GaAs elliptical quantum disk with the size parameters $a$=20 nm, $b$=15 nm, and $L_z$(disk height)=3 nm. The relaxation times within the odd-parity exciton states and those between the even-parity exciton states and the odd-parity exciton states are of the same order of magnitude as the relaxation times within the even-parity exciton states. This suggests that the dark exciton states possibly play important roles as the intermediate states of relaxation processes. We have formulated a set of rate equations for the occupation probability of relevant exciton levels, which include 12 odd-parity and 14 even-parity exciton levels in ascending order of energy up to 7 meV from the exciton ground state.

The photoluminescence spectrum under resonant cw excitation is obtained by a stationary solution of the set of rate equations. The results for the $x$-polarized excitation at the fourth exciton level are shown in Fig. 9.18 for 30 K.

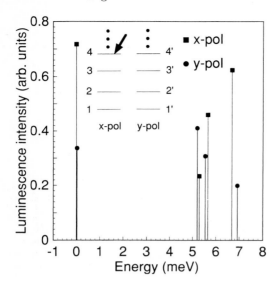

**Fig. 9.18.** The photoluminescence spectrum under resonant cw excitation at the $x$-polarized fourth exciton level at 30 K in a GaAs elliptical quantum disk with the size parameters $a$=20 nm, $b$=15 nm, and $L_z$(disk height)=3 nm (from T. Takagahara [30])

The polarization-conserving luminescence from the $x$-polarized exciton ground state is strong, but the luminescence from the $y$-polarized exciton ground state has also substantial intensity. The situation is similar also for resonant cw excitation of the $y$-polarized fourth exciton level, as shown in Fig. 9.19.

In this case, however, the luminescence from the $x$-polarized exciton ground state is stronger than that from the $y$-polarized exciton ground state, indicating more efficient relaxation from the $y$-polarized states to the $x$-

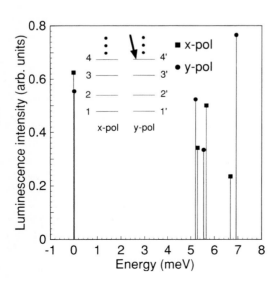

**Fig. 9.19.** The photoluminescence spectrum under resonant cw excitation at the $y$-polarized fourth exciton level at 30 K in the same quantum disk as in Fig. 9.18 (from T. Takagahara [30])

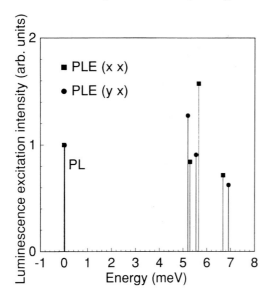

**Fig. 9.20.** Photoluminescence excitation spectrum under the detection of emission from the $x$-polarized exciton ground state in the same quantum disk as in Fig. 9.18 (from T. Takagahara [30])

polarized states than the reverse process. These results clearly indicate the presence of efficient paths of polarization relaxation.

In the same way, we can calculate the excitation spectra of the exciton photoluminescence. The excitation spectra for the detection of luminescence from the $x$-polarized exciton ground state are shown in Fig. 9.20 for both the $x$- and $y$-polarized excitation. This spectrum is very similar to the absorption

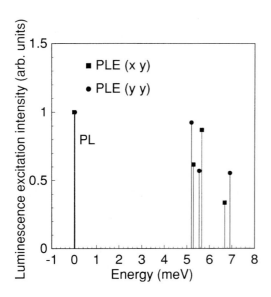

**Fig. 9.21.** Photoluminescence excitation spectrum under the detection of emission from the $y$-polarized exciton ground state in the same quantum disk as in Fig. 9.18 (from T. Takagahara [30])

spectrum in Fig. 2.11, although the intensity ratio between the $x$- and the $y$-polarized components is different for the fourth exciton doublet. Similar behaviors can be seen in the excitation spectra for the detection of luminescence from the $y$-polarized exciton ground state, as shown in Fig. 9.21.

This indicates that the relaxation processes after photoexcitation are not sensitive to the initial polarization direction because the presence of efficient polarization relaxation through the excited exciton states, and thus the emission intensity from the exciton ground state, is grossly determined by the absorption intensity of the exciting light.

## 9.11    Summary and Discussion

We have formulated a theory of exciton dephasing in semiconductor quantum dots extending the Huang–Rhys theory of $F$-centers to include the mixing among the exciton state manifold through the exciton–phonon interaction and identified the mechanisms of pure dephasing. We have reproduced quantitatively the magnitude as well as the temperature dependence of the exciton dephasing rate observed in GaAs QD-like islands. In this system, it has been found that both the diagonal and off-diagonal exciton–phonon interactions contribute to the exciton pure dephasing on the same order of magnitude. Examining the previous data of the exciton dephasing rate in GaAs islands, CuCl, and CdSe nanocrystals, we have pointed out the correlation between the temperature dependence of the dephasing rate and the strength of the quantum confinement, and explained gross features of the temperature dependence in quantum dots of various materials.

Very recently, a detailed study of the exciton dephasing rate in CuCl nanocrystals was carried out down to much lower temperatures, and an unusual temperature dependence was reported [34]. A two-level system (TLS) with a very small energy splitting was proposed as a possible mechanism of this unusual temperature dependence. At very low temperatures, the acoustic phonons are frozen out and the only remaining degrees of freedom may be the TLS, which acts as an agent of the exciton dephasing [35]. The two-level system is considered as some local modification of atomic configuration, but more details are yet to be clarified. At even lower temperatures, all the degrees of freedom are frozen out and the exciton dephasing is mainly caused by the radiative decay, approaching the limiting situation that $1/T_2 = 1/(2T_1)$.

Furthermore, we studied the population relaxation among exciton doublet structures and predicted a long relaxation time for the intra-doublet transitions of about several nanoseconds. We also found that the cross-relaxation between orthogonally polarized exciton states occurs, in general, as efficiently as the population relaxation without a change in the polarization direction.

# References

1. See, for example, *Confined Electrons and Photons, New Physics and Devices*, ed. by E. Burstein, C. Weisbuch (Plenum, New York, 1995); *Microcrystalline and Nanocrystalline Semiconductors, Materials Res. Soc. Symp. Proc. Vol. 358*, eds. R.W. Collins, C.C. Tsai, M. Hirose, F. Koch, L. Brus (Materials Research Society, Pittsburgh 1995)
2. J. Hegarty, M.D. Sturge: J. Opt. Soc. Am. B **2**, 1143 (1985)
3. M.D. Webb, S.T. Cundiff, D.G. Steel: Phys. Rev. B **43**, 12658 (1991)
4. L. Schultheis, A. Honold, J. Kuhl, K. Köhler, C.W. Tu: Phys. Rev. B **34**, 9027 (1986)
5. A. Honold, L. Schultheis, J. Kuhl, C.W. Tu: J. Opt. Soc. Am. B **4**, 210 (1987)
6. T. Takagahara: Phys. Rev. B **32**, 7013 (1985); J. Lumin. **44**, 347 (1989)
7. H.F. Hess, E. Betzig, T.D. Harris, L.N. Pfeiffer, K.W. West: Science **264**, 1740 (1994)
8. K. Brunner, G. Abstreiter, G. Böhm, G. Tränkle, G. Weimann: Phys. Rev. Lett., **73**, 1138 (1994)
9. D. Gammon, E.S. Snow, B.V. Shanabrook, D.S. Katzer, D. Park: Science **273**, 87 (1996); D. Gammon, E.S. Snow, B.V. Shanabrook, D.S. Katzer, D. Park: Phys. Rev. Lett. **76**, 3005 (1996)
10. X. Fan, T. Takagahara, J.E. Cunningham, H. Wang: Solid State Commun. **108**, 857 (1998)
11. K. Huang, A. Rhys: Proc. Roy. Soc. (London) A **204**, 406 (1950)
12. C.B. Duke, G.D. Mahan: Phys. Rev. **139**, A1965 (1965)
13. R. Kubo, Y. Toyozawa: Prog. Theor. Phys. **13**, 160 (1955)
14. R.W. Schoenlein, D.M. Mittleman, J.J. Shiang, A.P. Alivisatos, C.V. Shank: Phys. Rev. Lett. **70**, 1014 (1993)
15. T. Itoh, M. Furumiya: J. Lumin. **48–49**, 704 (1991)
16. H. Benisty, C.M. Sotomayor-Torres, C. Weisbuch: Phys. Rev. B **44**, 10945 (1991)
17. U. Bockelmann: Phys. Rev. B **48**, 17637 (1993)
18. T. Takagahara: J. Lumin. **70**, 129 (1996); T. Takagahara: Phys. Rev. Lett. **71**, 3577 (1993), and references therein
19. A. Blacha, H. Presting, M. Cardona: Phys. Status Solidi B **126**, 11 (1984)
20. *Physics of Group IV Elements and III-V Compounds, Landolt-Börnstein, Vol. 17a*, ed. by O. Madelung, M. Schulz, H. Weiss (Springer, Berlin 1982)
21. R. Kubo: In: *Fluctuation, Relaxation and Resonance in Magnetic Systems*, ed. by D. ter Haar, (Oliver and Boyd, Edinburgh 1962)
22. T. Takagahara, E. Hanamura, R. Kubo: J. Phys. Soc. Japan **43**, 802, 811, 1522 (1977)
23. T. Takagahara: Phys. Rev. B **31**, 6552 (1985)
24. T. Takagahara: Phys. Rev. B **47**, 4569 (1993)
25. K. Takemoto, B.-R. Hyun, Y. Masumoto: Solid State Commun. **144**, 521 (2000)
26. R. Kuribayashi, K. Inoue, K. Sakoda, V.A. Tsekhomskii, A.V. Baranov: Phys. Rev. B **57**, R15084 (1998)
27. A.L. Efros, M. Rosen, M. Kuno, M. Nirmal, D.J. Norris, M. Bawendi: Phys. Rev. B **54**, 4843 (1996); Detailed calculation was done here, including the electron–hole exchange interaction and the effect of non-spherical shape. The typical energy splitting is about 20 meV for a 20-Å-radius spherical CdSe quantum dot

28. A.E.H. Love: *A Treatise on the Mathematical Theory of Elasticity* (Dover, New York 1944)
29. A. Tamura, K. Higeta, T. Ichinokawa: J. Phys. C **15**, 4975 (1982)
30. T. Takagahara: Phys. Rev. B **60**, 2638 (1999)
31. H. Gotoh, H. Ando, H. Kamada, A. Chavez-Pirson, J. Temmyo: Appl. Phys. Lett. **72**, 1341 (1998)
32. N.H. Bonadeo, G. Chen, D. Gammon, D. Park, D.S. Katzer, D.G. Steel: *Tech. Digest of Quantum Electronics and Laser Science Conf.* (QELS '99, Baltimore 1999) QTuC5, p. 48
33. H. Kamada, H. Gotoh, H. Ando, J. Temmyo, T. Tamamura: Phys. Rev. B **60**, 5791 (1999)
34. M. Ikezawa, Y. Masumoto: Phys. Rev. B **61**, 12662 (2000)
35. See, for example, *Dynamical Processes in Disordered Systems, Materials Science Forum Vol. 50*, ed. by W.M. Yen (Trans Tech Publications, Switzerland 1989); *Optical Spectroscopy of Glasses*, ed. by I. Zschokke (D. Reidel Publishing Company, Dordrecht 1986)

# 10 Excitonic Optical Nonlinearities and Weakly Correlated Exciton-Pair States

Selvakumar V. Nair and Toshihide Takagahara

## 10.1 Introduction

In this chapter we discuss the optical nonlinearities associated with the confined excitonic state in quantum dots. Interest in the physics of quantum dots has been largely fuelled by their potential as efficient nonlinear optical and laser materials [1–11]. The atom-like discrete energy level structure and the consequent concentration of the oscillator strength into well-defined energies makes quantum dots (QDs) very attractive for electro-optic and nonlinear optical applications.

In addition to the spatial confinement, the Coulomb interaction between the excited electrons and holes also plays an important role in determining the excitation spectra of QDs. This is especially true in most of the currently studied crystallites of II–VI and I–VII semiconductors like CdS, CdSe, and CuCl, owing to the large exciton binding energy in these materials. The formation of excitons and biexcitons in quantum dots of radius $R$ larger than several times the exciton Bohr radius $a_{ex}$ leads to strong optical response at the exciton resonance. In fact, in this weak-confinement regime $(R \gg a_{ex})$ [12], the exciton oscillator strength is proportional to the volume of the quantum dot. Consequent super-radiant decay of the exciton has been experimentally observed [13,14] with the lifetime decreasing inversely as the volume of the QD. The mesoscopic enhancement of the exciton oscillator strength would lead to, for example, a nonlinear optical susceptibility increasing with the size of the QD [15].

The nonlinear response is, however, determined not only by the exciton states but also by multiple-exciton states, and especially by the biexcitonic excitations. Many experiments on CuCl QDs in the weak-confinement regime have revealed distinctly non-bulk-like features, including enhanced nonlinear optical susceptibility with a nonmonotonic size dependence [9], very large gain for biexcitonic lasing [8], and a blueshift of the excitonic absorption under a strong pump beam [16]. These observations point to the significance of the interplay of excitonic and biexcitonic contributions to the nonlinearities.

During the early days of quantum dot research, considerable progress was achieved in the theoretical description of the single particle electronic structure, providing a satisfactory framework for describing the optical response of QDs of radius comparable to or smaller than $a_{ex}$ [17–22]. In larger crystallites, reliable theoretical calculations of the excitonic states were developed in Refs. 23–27. Biexciton calculations [2,28] were initially restricted to QDs whose radius was smaller than a few times $a_{ex}$ due to the numerical complexity of the problem when $R \gg a_{ex}$ [29].

Nair and Takagahara studied the size dependence of the biexcitonic states and of the nonlinear optical response of semiconductor QDs of radii up to $10a_{ex}$ using a novel approach based on an exciton–exciton product state basis [30,31]. They identified the presence of weakly correlated exciton pair states that play a crucial role in determining the excitonic nonlinear optical response and provided a consistent understanding of the experimental results mentioned earlier. In this chapter we provide detailed description of the developments that led to this understanding with a theoretical emphasis.

## 10.2   Exciton States

The most widely used method for the study of electronic states in low-dimensional semiconductor structures is the envelope function method based on the effective mass approximation (EMA) [32]. Although the applicability of the EMA is limited to energy levels that lie close to points of high symmetry of the bulk band structure, this method is extensively used due to its simplicity and versatility. Further, all parameters used in the EMA have a direct physical meaning and can be experimentally determined.

Early calculations of the size quantization of the electron and the hole in spherical QDs used a single-band EMA model for both the electron and the hole [12,24,26,28,33,34]. Such a description for electrons is appropriate for most semiconductors, which have an isolated conduction band well removed in energy from all other bands. The valence bands of holes is degenerate or nearly degenerate in many semiconductors at the $\Gamma$-point ($k$=0). For such cases the multiband EMA formalism has been effectively used by several authors. See Fig. 10.1 for a schematic description of EMA in various levels of sophistication and [32] for more details.

During the last decade, more microscopic approaches such as tight-binding and pseudopotential methods have been developed and applied to QDs [17,18,21]. These calculations have shown the qualitative soundness of EMA as applied to nanocrystal QDs and also provided a quantitatively more accurate description of the single particle (electron and hole) states.

In addition to the quantum confinement of the carriers, it is necessary to consider the electron–hole Coulomb interaction. In small QDs, the Coulomb interaction may be treated perturbatively. Such a procedure works as long as the kinetic energy of confinement dominates over the Coulomb binding energy. When the Coulomb interaction between the electron and hole is strong, as is the case in the so-called weak-confinement regime ($R \gg a_{ex}$), their motion becomes correlated, and it is appropriate to refer to the electron–hole pairs

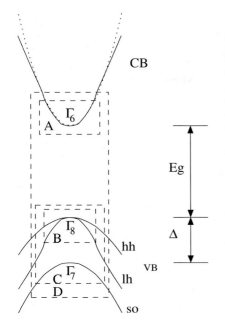

**Fig. 10.1.** A schematic showing the band structure close to the $\Gamma$-point of a typical semiconductor with a direct band gap $E_{\mathrm{g}}$. The *dashed boxes* indicate the bands included in (A) a single-band EMA for the conduction band (effectively replacing the dispersion by a parabolic approximation, as indicated by the *dotted lines*); (B) a four-band EMA for the heavy-hole (hh) and light-hole (lh) bands (typically used for materials with large spin-orbit splitting($\Delta$)); (C) a six-band EMA that includes the spin-orbit split-off band (so) as well; and (D) an eight-band EMA typically used for small band gap materials for which the conduction–valence band coupling could be important. In some materials (e.g., CuCl), $\Delta$ is negative so that the top-most valence band is non-degenerate and a single band EMA could be used for the holes as well

as excitons. In semiconductors we have Wannier excitons, and the Coulomb attraction between the electron and the hole may be regarded as screened by the macroscopic dielectric constant.

Although tight-binding and pseudopotential methods are more accurate than the EMA, it is very difficult to treat the two-particle problem of correlated electron–hole pairs within these approaches. In contrast, the two-particle problem within EMA is more amenable for numerical calculation. Fortunately, the validity of the EMA is at its best in large QDs, and it is in this regime that the electron–hole correlation is most important. Consequently, all non-perturbative calculations of the exciton states in QDs have been based on the EMA.

### 10.2.1 Formulation

The optical properties of exciton states of CuCl, CdS, and CdSe QDs of a wide range of sizes have been studied by various groups (see [35] for a review). For CuCl, the exciton Bohr radius $a_{\mathrm{ex}}$ is only $\approx 7$ Å so that the QD sizes are almost always greater than $a_{\mathrm{ex}}$. In the case of CdS QDs, although the interest has shifted to smaller QDs in recent years, a lot of earlier work was concentrated on QDs in the weak-confinement regime. For such large-size QDs, the Coulomb interaction has to be considered on an equal footing with the kinetic energy, i.e., the electron and hole confinement cannot be treated independently. This is a rather serious complication, which makes it difficult to obtain accurate results for the exciton levels. One simplification we shall resort to here is to treat the valence band, like the conduction band, in a single-band model. This is justifiable for CuCl and hexagonal CdS because of the presence of nondegenerate valence bands, as long as the confinement kinetic energies of the relevant hole levels of interest are small compared to the band edge energy difference between the two topmost valence bands.

Within the EMA, the wave function $\phi(\boldsymbol{r}_e, \boldsymbol{r}_h)$ for a single electron–hole pair is determined by the effective Schrödinger equation

$$\left[ -\frac{\hbar^2}{2m_e}\nabla_e^2 - \frac{\hbar^2}{2m_h}\nabla_h^2 - \frac{e^2}{\epsilon r_{eh}} + V \right] \phi(\boldsymbol{r}_e, \boldsymbol{r}_h) = (E - E_g)\phi(\boldsymbol{r}_e, \boldsymbol{r}_h) \,, \quad (10.1)$$

where $m_e$ ($m_h$) and $\boldsymbol{r}_e$ ($\boldsymbol{r}_h$), respectively denote the effective mass and the position vector of the electron (hole), $r_{eh}=|\boldsymbol{r}_e-\boldsymbol{r}_h|$, $\epsilon$ is the bulk dielectric constant, and $E_g$ is the bulk band gap. The confining potential $V$ is assumed to be zero inside the QD of radius $R$ and infinite outside, with the corresponding boundary condition

$$\phi(\boldsymbol{r}_e, \boldsymbol{r}_h) = 0 \quad \text{for} \quad r_e \quad \text{or} \quad r_h \geq R \,. \quad (10.2)$$

Here we have neglected image charge effects that arise from the dielectric mismatch between the QD and the host material.

The confinement kinetic energy scales as $1/R^2$, while the Coulomb binding energy scales as $1/R$, so that for small enough sizes of the QD, the Coulomb interaction can be treated as a perturbation. In fact, two limiting situations are determined by the ratio of the radius $R$ of the QD to the effective Bohr radius $a_{ex}$ of the exciton in the bulk material [12]. In the limit $R/a_{ex}\ll1$, the electron and the hole are weakly correlated as mentioned earlier, and the independent particle-in-a-sphere wave functions provide a good description of their states. In this limit of strong confinement one can obtain [26,33] an expression for the lowest eigenstate by treating the Coulomb interaction in the first order perturbation theory

$$E = E_g + \frac{\hbar^2\pi^2}{2\mu R^2} - 1.786\frac{e^2}{\epsilon R} - 0.248\frac{\mu e^4}{2\epsilon^2\hbar^2} \,, \quad (10.3)$$

where $\mu$ is the reduced mass of the electron–hole pair. The last term arises from the residual electron–hole correlation that contributes a finite amount to the energy, even in the limit $R\to0$ [26]. This expression is of limited practical value since it is valid only for $R<a_{ex}$, and in this size range a single-band description of the valence band is not appropriate in most cases. The opposite limit is $R/a_{ex}\gg1$ in which the character of the exciton as a bound system is preserved, and its translational motion is spatially confined. Efros and Efros [12] argued that in this limit the confinement energy is just that of a particle of mass equal to that of the exciton, ie., $M=m_e+m_h$. Then the lowest exciton state is given by

$$\phi(\boldsymbol{r}_e, \boldsymbol{r}_h) = N\frac{\sin \pi r_{cm}}{r_{cm}} \exp(-r_{eh}/a_{ex}) \,, \quad (10.4)$$

$$E = E_{\text{g}} + E_{\text{ex}} + \frac{\hbar^2 \pi^2}{2MR^2},$$ (10.5)

where $r_{\text{cm}}$ is the coordinate of the center of mass of the electron–hole pair, and $E_{\text{ex}}$ is the exciton binding energy of the bulk material. This proposition is especially interesting because the energy levels are determined by the total mass of the exciton, which is sensitive to the heavier particle mass, usually the hole, while the reduced mass involved in the bulk exciton binding energy is sensitive to the lighter particle mass, usually the electron. However, experimental results for CuCl particles [36] give an exciton mass much smaller than generally accepted. This was explained by Nair et al. [24], who performed a variational calculation valid over a wide size range extending from the strong- to weak-confinement regimes, and showed that the above limiting behavior is not reached in sizes practically studied. They identified the reason for this to be the finite size of the exciton and proposed that the center-of-mass motion is essentially confined to a sphere of radius $R$–$\eta a_{\text{ex}}$, where $\eta$ is expected to be of the order of unity and to depend on the electron–hole mass ratio. This so called "dead-layer" correction was investigated in more detail by Kayanuma [26] and is widely used in the literature, mainly for its simplicity.

While the qualitative models described above help understanding, modelling of the exciton and biexciton optical response calls for more precise calculations of the energy levels and wave functions. Now we discuss of some of the widely used calculational approaches for the exciton states in QDs.

### 10.2.2   Configuration Interaction in a Truncated Basis

As mentioned earlier, in QDs of radius $R$ comparable to or larger than $a_{\text{ex}}$, the Coulomb interaction is not a weak perturbation. Nevertheless, one can think of constructing the exciton state as a linear combination of electron and hole single particle product states as in a configuration interaction (CI) approach. Here we may take the Hamiltonian in (10.1) without the Coulomb interaction term as the unperturbed Hamiltonian. Then the electron and hole motions are decoupled, and the wave functions and energies for the electron are given by

$$\phi_{nlm}(r_{\text{e}}) = \left(\frac{2}{R^3}\right)^{1/2} \frac{j_l(\xi_{nl} r_{\text{e}}/R)}{j_{l+1}(\xi_{nl})} Y_{lm}(\theta_{\text{e}}, \phi_{\text{e}}),$$ (10.6)

and

$$E_{nlm} = E_{\text{g}} + \frac{\hbar^2 \xi_{nl}^2}{2m_{\text{e}} R^2},$$ (10.7)

where $j_l$ is the spherical Bessel function of order $l$, $\xi_{nl}$ is its $n$th zero, $Y_{lm}$ values are spherical harmonics, and $r_{\text{e}}$, $\theta_{\text{e}}$, and $\phi_{\text{e}}$ denote the spherical polar coordinates of the electron. The corresponding expressions for the hole are obtained by replacing the electron coordinates and mass by those of the hole and setting $E_g=0$ in (10.7).

When the Coulomb interaction is introduced, the angular momenta of the electron and hole individually are no longer conserved quantities, but their total angular momentum $L$ is conserved. Further, only the $L=0$ states (i.e., the S states) are optically excited. Here we restrict our analysis to the $L=0$ states, but generalization to higher values of $L$ is straightforward. The unperturbed electron–hole pair states with a specified value of $L$ may be constructed as linear combinations of products of single particle states following the standard angular momentum addition rules. For $L=0$ we have [37]

$$\psi_{n_1 n_2 l}(\boldsymbol{r}_{\mathrm{e}}, \boldsymbol{r}_{\mathrm{h}}) = \frac{1}{\sqrt{2l+1}} \sum_{m=-l}^{l} (-1)^{l-m} \phi_{n_1 l m}(\boldsymbol{r}_{\mathrm{e}}) \phi_{n_2 l, -m}(\boldsymbol{r}_{\mathrm{h}}) . \qquad (10.8)$$

The corresponding energies are given by

$$E_{n_1 n_2 l} = E_{\mathrm{g}} + \frac{\hbar^2}{2R^2} \left( \frac{\xi_{n_1 l}^2}{m_{\mathrm{e}}} + \frac{\xi_{n_2 l}^2}{m_{\mathrm{h}}} \right) , \qquad (10.9)$$

As the Hamiltonian is spin independent (see Sect. 10.3.3 for a discussion of the electron–hole exchange effect), we have ignored the spin of the electron. However, we note that only singlet excitons (total spin=0) are optically excited as the electromagnetic interaction is spin independent.

Using a finite set of $\psi_{n_1 n_2 l}$ values as the basis set, we can construct the matrix representation of the Hamiltonian (10.1) and diagonalize it to get the exciton eigenstates. The calculation of the matrix elements of the Coulomb potential is greatly simplified by expanding $1/r_{\mathrm{eh}}$ in spherical harmonics so that the angular integrals can be analytically performed. The radial integrals are done numerically. It is easy to see that the matrix elements of the Hamiltonian have the form

$$H_{ij} = \frac{D_i}{R^2} \delta_{ij} + \frac{C_{ij}}{R}, \qquad (10.10)$$

where $D_i$ and $C_{ij}$ are independent of $R$. The unperturbed level spacings scale as $1/R^2$, while the Coulomb term scales as $1/R$. Therefore, as the QD size increases, the CI calculation requires a larger and larger matrix to be diagonalized. In one of the early calculations of this kind, Nair et al. [24] reported satisfactory convergence using a 20-dimensional space for $R<1.5a_{\mathrm{ex}}$. A much larger basis set would yield convergent results for larger QDs, as has been demonstrated by Hu et al. [28], who also calculated biexcitonic states using a four-particle CI basis. However, for a given accuracy, the size of the basis set increases rather prohibitively with the QD size. So this method is generally limited to the strong- and intermediate-confinement regimes. It may be noted that this CI approach is widely used to describe the multiple electron–hole pairs in self-assembled QDs, since in these systems the exciton size is comparable or smaller than the QD size.

### 10.2.3   Variational Approach

A variational calculation for the exciton states in QDs was first attempted by Brus [33] by using a linear combination of the three lowest-energy free electron–hole pair states as the trial wave function. Such an approach works only when the Coulomb interaction is weak, and reliable results are obtained in a more limited size range than the CI approach described in Sect. 10.2.2.

Kayanuma [23] and Nair et al. [24] presented more accurate variational calculations of the exciton ground state using a trial wave function of the form

$$\phi(\boldsymbol{r}_\mathrm{e}, \boldsymbol{r}_\mathrm{h}) = N \left( \frac{\sin(\pi r_\mathrm{e}/R)}{r_\mathrm{e}/R} \right)^{\alpha_1} \left( \frac{\sin(\pi r_\mathrm{h}/R)}{r_\mathrm{h}/R} \right)^{\alpha_2} \exp(-\beta r_\mathrm{eh}/R) , \quad (10.11)$$

where $N$ is the normalization constant, and $\alpha_1$, $\alpha_2$, and $\beta$ are variational parameters (a single variational parameter $\beta$ was used in [23]). This form of the trial wave function is chosen to satisfy the boundary condition (10.2) and other known limiting forms. In particular, for $\alpha_1 = \alpha_2 = 1/2$ and $\beta = R/a_\mathrm{ex}$, (10.11) approaches the limiting form (10.4) for large QDs.

The $L=0$ exciton states, which include the ground state, depend only on the three Hylleraas [38] coordinates $r_\mathrm{e}$, $r_\mathrm{h}$, and $r_\mathrm{eh}$. The kinetic energy terms involving $\nabla_\mathrm{e}^2$ and $\nabla_\mathrm{h}^2$ can be expressed in terms of these three coordinates so that additional angular variables needed to specify $\boldsymbol{r}_\mathrm{e}$, $\boldsymbol{r}_\mathrm{h}$ do not appear in the evaluation of the expectation value of the Hamiltonian.

One of the striking findings of this calculation was that although the exciton energy approaches the bulk value as $R$ increases and is within 2% at $R=10a_\mathrm{ex}$ for CuCl, the size dependence of energy deviates significantly from the limiting behaviour (10.5) expected from the center-of-mass confinement picture. In particular, the exciton mass estimated from the slope of the energy versus $R^2$ curve was found to be as small as half the actual value $m_\mathrm{e}+m_\mathrm{h}$. It was found that as $R$ is increased, while $\alpha_1$ approaches the limiting value of $1/2$ rather quickly, $\alpha_2$ remains much larger than $1/2$ for both CuCl and CdS, even for $R=10a_\mathrm{ex}$. This is because when the electron and hole motion are strongly correlated, the heavier hole cannot approach the boundary as it should remain closer to the center of mass, while the confinement of the electron, which is energetically less favourable, should be minimized. It was suggested [24] that considering an exciton as a rigid particle with a finite size, the center of mass motion is confined to a sphere of radius $R-\eta a_\mathrm{ex}$, where the factor $\eta$ is of the order of unity and depends on the electron–hole mass ratio. Thus (10.5) based on the center-of-mass confinement picture can be expected to be valid only for very large sizes for which the exciton energy would be practically indistinguishable from the bulk value. Thus, in the practically accessible weak-confinement regime, the exciton ground state energy is better described by

$$E = E_\mathrm{g} + E_\mathrm{ex} + \frac{\hbar^2 \pi^2}{2M(R - \eta a_\mathrm{ex})^2} , \quad (10.12)$$

and the factor $\eta$ varies from about 1 for equal electron and hole masses to about 2 for an infinitely large mass ratio. For a more detailed analysis of this so-called "dead-layer" correction, see [26].

### 10.2.4    Kayanuma's Correlated Basis Set

Although the variational approach described in Sect. 10.2.3 can yield a very accurate description of the exciton ground state, extension to excited states is difficult. The CI approach, on the other hand, can provide several excited states but the convergence rate is rather poor. What is needed is a basis set in which electron–hole correlation is built-in so that a subspace diagonalization can give several excited states with high accuracy, and at the same time does not require a prohibitively large basis size. Kayanuma [26] achieved this for the $L=0$ subspace using a basis set of correlated electron–hole functions written in terms of polynomials and exponentials.

If we expand the $L=0$ exciton envelope function into a set of nonorthogonal basis functions [26]

$$\phi_0(\boldsymbol{r}_{\rm e}, \boldsymbol{r}_{\rm h}) = \sum_{l=0}^{l_{\max}} \sum_{m=0}^{m_{\max}} \sum_{n=0}^{n_{\max}} c_{lmn} w_m(r_{\rm e}) w_n(r_{\rm h}) r_{\rm eh}^l \exp(-r_{\rm eh}/a_{\rm ex}) , \quad (10.13)$$

with

$$w_m(r) = \prod_{k=1}^{m} \left[ r^2 - \left( \frac{k}{m} R \right)^2 \right] , \quad (10.14)$$

which explicitly satisfies the boundary condition given by (10.2).

One of the greatest advantages of using polynomials and exponentials is that the kinetic energy and Coulomb matrix elements can be analytically evaluated. The price paid is the nonorthogonality of the basis functions so that the matrix representation of the Schrödinger equation yields a generalized eigenvalue problem. Further, the analytical expressions for the Hamiltonian matrix elements involve numerically unstable summations, so care should be taken in the numerical implementation to avoid instabilities and reduce round-off errors. If values of $n_{\max}$ or $m_{\max}$ exceeding 4 are used, it is advisable to calculate the matrix elements by numerical integration rather than using analytical expressions.

The coefficients $c_{lmn}$ are determined by a generalised eigenvalue equation, which may be solved by standard numerical techniques. Truncating the expansion (10.13), with $m_{\max}=n_{\max}=3$ and $l_{\max}=2$, several exciton energy levels are obtained with high accuracy [26].

Although only the $L=0$ excitons are optically excited in a direct-gap semiconductor, the $L>0$ states are also of interest. First, the $L=1$ states can be excited by intraband excitation in the presence of an exciton [39]. Further, as we will see in Sect. 10.3.2, our own approach to biexciton calculation involves the use of $L>0$ exciton states. A straightforward generalization of the above approach leads to the expansion

$$\phi_{LM}(\boldsymbol{r}_{\mathrm{e}},\boldsymbol{r}_{\mathrm{h}}) = \sum_{l_1=0}^{l_{\max}} \sum_{l_2=|l_1-L|}^{l_1+L} F_{l_1 l_2}(r_{\mathrm{e}},r_{\mathrm{h}})$$

$$\times \sum_{m=-l_1}^{l_1} C_{m,M-m,M}^{l_1,l_2,L} Y_{l_1 m}(\Omega_{\mathrm{e}}) Y_{l_2,M-m}(\Omega_{\mathrm{h}}) , \qquad (10.15)$$

where $C_{m,M-m,M}^{l_1,l_2,L}$'s denote the Clebsch–Gordan coefficients and $Y_{lm}$ are the spherical harmonics. However, the use of such an expansion is computationally intensive because of the need to keep a large number of terms in the sum over the angular functions.

Instead, we can extend Kayanuma's approach to $L=1$ states by noting that, for two particles, any odd-parity, $L=1$ state can be expressed in the form [31] (see Appendix A)

$$\phi_{1M}^i(\boldsymbol{r}_{\mathrm{e}},\boldsymbol{r}_{\mathrm{h}}) = f_{\mathrm{e}}^i(r_{\mathrm{e}},r_{\mathrm{h}},r_{\mathrm{eh}}) Y_{1M}(\Omega_{\mathrm{e}})$$
$$+ f_{\mathrm{h}}^i(r_{\mathrm{e}},r_{\mathrm{h}},r_{\mathrm{eh}}) Y_{1M}(\Omega_{\mathrm{h}}) , \qquad (10.16)$$

where $i$ denotes the radial quantum number. Thus the $L=1$ states can be described in terms of two functions, $f_{\mathrm{e}}$ and $f_{\mathrm{h}}$, of the Hylleraas coordinates, and the same basis set used for the $L=0$ case can be used to expand $f_{\mathrm{e}}$ and $f_{\mathrm{h}}$.

A similar generalization, but more general (valid for both even and odd parity states), has been reported by Uozumi and Kayanuma [40].

For $L>1$ states, no such simple form appears to exist. Any $L=2$, even-parity state can be written as [31] (see Appendix A)

$$\phi_{2M}^i(\boldsymbol{r}_{\mathrm{e}},\boldsymbol{r}_{\mathrm{h}}) = g_{\mathrm{e}}^i(r_{\mathrm{e}},r_{\mathrm{h}},r_{\mathrm{eh}}) Y_{2M}(\Omega_{\mathrm{e}}) + g_{\mathrm{h}}^i(r_{\mathrm{e}},r_{\mathrm{h}},r_{\mathrm{eh}}) Y_{2M}(\Omega_{\mathrm{h}})$$

$$+ \sum_{l\geq 5}^{\infty} \tilde{g}_l^i(r_{\mathrm{e}},r_{\mathrm{h}}) \sum_m C_{m,M-m,M}^{l,l,2} Y_{lm}(\Omega_{\mathrm{e}}) Y_{l,M-m}(\Omega_{\mathrm{h}}) , \quad (10.17)$$

Note that the sum in the third term in (10.17) starts at $l=5$, and so the relatively slowly varying envelope of the low-energy states would be well described by the first two terms. This expectation is borne out by numerical results and thus, to a good approximation, the $L=2$ exciton states also may be written in a form identical to the $L=1$ states.

## 10.3   Biexciton States

The EMA Hamiltonian for two electrons and two holes is given by

$$H_{XX} = -\frac{\hbar^2}{2m_{\mathrm{e}}}\left(\nabla_{\mathrm{e}_1}^2 + \nabla_{\mathrm{e}_2}^2\right) - \frac{\hbar^2}{2m_{\mathrm{h}}}\left(\nabla_{\mathrm{h}_1}^2 + \nabla_{\mathrm{h}_2}^2\right) + V$$

$$- \frac{e^2}{\epsilon|\boldsymbol{r}_{\mathrm{e_1}} - \boldsymbol{r}_{\mathrm{h_1}}|} - \frac{e^2}{\epsilon|\boldsymbol{r}_{\mathrm{e_1}} - \boldsymbol{r}_{\mathrm{h_2}}|} - \frac{e^2}{\epsilon|\boldsymbol{r}_{\mathrm{e_2}} - \boldsymbol{r}_{\mathrm{h_1}}|}$$
$$- \frac{e^2}{\epsilon|\boldsymbol{r}_{\mathrm{e_2}} - \boldsymbol{r}_{\mathrm{h_2}}|} + \frac{e^2}{\epsilon|\boldsymbol{r}_{\mathrm{e_1}} - \boldsymbol{r}_{\mathrm{e_2}}|} + \frac{e^2}{\epsilon|\boldsymbol{r}_{\mathrm{h_1}} - \boldsymbol{r}_{\mathrm{h_2}}|} \ . \tag{10.18}$$

Again we neglect the image charge effects as in the case for the exciton states.

### 10.3.1   Variational Approach

The biexciton ground state in a quantum dot was calculated variationally for the first time including the image charge effect and by employing the four-particle envelope function [2] given as

$$\begin{aligned}
\varPhi(\boldsymbol{r}_{\mathrm{e_1}}, \boldsymbol{r}_{\mathrm{h_1}}, \boldsymbol{r}_{\mathrm{e_2}}, \boldsymbol{r}_{\mathrm{h_2}}) = {}& C_{\mathrm{m}} j_0(\pi r_{\mathrm{e_1}}/R) j_0(\pi r_{\mathrm{e_2}}/R) j_0(\pi r_{\mathrm{h_1}}/R) j_0(\pi r_{\mathrm{h_2}}/R) \\
& \times r_{\mathrm{h_1 h_2}}^{\gamma} \exp(-\delta r_{\mathrm{h_1 h_2}}) \\
& \times \{ \exp[-\alpha(r_{\mathrm{e_1 h_1}} + r_{\mathrm{e_2 h_2}}) - \beta(r_{\mathrm{e_1 h_2}} + r_{\mathrm{e_2 h_1}})] \\
& + \exp[-\beta(r_{\mathrm{e_1 h_1}} + r_{\mathrm{e_2 h_2}}) - \alpha(r_{\mathrm{e_1 h_2}} + r_{\mathrm{e_2 h_1}})] \} \ , \tag{10.19}
\end{aligned}$$

where $C_{\mathrm{m}}$ is the normalization constant, $r_{ij}=|\boldsymbol{r}_i-\boldsymbol{r}_j|$ and $j_0$ is the zeroth-order spherical Bessel function and represents the lowest sub-band state of the electron and the hole. In this calculation the variational parameter $\gamma$, which describes the hole–hole repulsion was fixed to be 1; unfortunately, a negative biexciton binding energy was obtained for small sizes of the quantum dot. This point was later remedied by a perturbational calculation [42], and now a monotonically increasing biexciton binding energy as $R\longrightarrow 0$ is established for the strong-confinement regime. All these calculations are concerned only with the biexciton ground state; the spin structure and the electron–hole exchange interaction are neglected. These subjects will be discussed in Sects. 10.3.2–10.3.3.

### 10.3.2   Exciton–Exciton Product State Basis

The approaches described in Sect. 10.3.1 are either limited to the biexciton ground state or to QDs comparable in size or smaller than the exciton. A reliable description of the optical nonlinearity near the exciton resonance, however, requires the knowledge of several excited states of the biexciton as well. Nair and Takagahara used an exciton–exciton product state basis to calculate several biexciton states over a wide size range in the weak-confinement regime. This approach is described below.

As all the optically excited states have a vanishing angular momentum ($L{=}0$) for the envelope function, here we consider only such biexciton states. Biexciton states $\varPhi(\boldsymbol{r}_{\mathrm{e_1}}, \boldsymbol{r}_{\mathrm{h_1}}, \boldsymbol{r}_{\mathrm{e_2}}, \boldsymbol{r}_{\mathrm{h_2}})$ with $L{=}0$ may be expanded into the exciton–exciton product states:

$$\varPhi(\boldsymbol{r}_{\mathrm{e_1}}, \boldsymbol{r}_{\mathrm{h_1}}, \boldsymbol{r}_{\mathrm{e_2}}, \boldsymbol{r}_{\mathrm{h_2}}) = \sum_{ijL} C_{ijL} G_{XX}^{ijL} + \tilde{C}_{ijL} \tilde{G}_{XX}^{ijL} \ , \tag{10.20}$$

with $G$ and $\tilde{G}$ given by

$$G_{XX}^{ijL} = \sum_{M=-L}^{L} \phi_{LM}^{i}(\boldsymbol{r}_{e_1}, \boldsymbol{r}_{h_1}) \phi_{LM}^{j*}(\boldsymbol{r}_{e_2}, \boldsymbol{r}_{h_2}) , \tag{10.21}$$

$$\tilde{G}_{XX}^{ijL} = \sum_{M=-L}^{L} \phi_{LM}^{i}(\boldsymbol{r}_{e_1}, \boldsymbol{r}_{h_2}) \phi_{LM}^{j*}(\boldsymbol{r}_{e_2}, \boldsymbol{r}_{h_1}) . \tag{10.22}$$

Here $i, j$ denote the radial quantum numbers of the exciton eigenstates. Since the Hamiltonian is invariant under exchange of the electron or hole coordinates, the biexciton wave function can be labelled by symmetry under permutations of the coordinates. We denote states that are symmetric (antisymmetric) under electron exchange by a superscript $+(-)$. A second superscript of $\pm$ is used to denote the symmetry under hole exchange. Thus, $\Phi^{\pm\pm}$ denotes states with $C_{ijL} = C_{jiL} = \pm \tilde{C}_{ijL}$, while $\Phi^{\mp\pm}$ denotes states with $C_{ijL} = -C_{jiL} = \pm \tilde{C}_{ijL}$.

Although a general two-particle state of angular momentum $L$ can have either parity, all the low-energy states in relatively large QDs will have parity $(-1)^L$. This becomes clear by noting that in the size range being considered, the exciton envelope function is approximately given by the product of the bulk exciton wave function for the relative coordinate and a particle-in-a-sphere wave function for the confinement of the center-of-mass motion [12]. As the first excited state of a hydrogenic system has a binding energy of only $1/4$ that of the ground state, it follows that all the low-energy states involve the s-like relative coordinate wave function so that the angular momentum is determined by that of the center-of-mass motion alone. Such states will have a parity $(-1)^L$. As explained in the Sect. 10.3.1, we have an efficient scheme for calculating such "natural" parity states and include only these in (10.21, 10.22).

For materials such as CuCl, the electron–hole exchange interaction modifies the exciton spectrum in a significant way. Now we discuss a scheme for including the interband exchange terms in exciton and biexciton calculation in QDs.

### 10.3.3   Electron–Hole Exchange Interaction

The exciton and the biexciton Hamiltonian including the electron–hole exchange interaction within the EMA are derived in Appendix B. For the exciton we have,

$$\left[ -\frac{\hbar^2}{2m_e} \nabla_e^2 - \frac{\hbar^2}{2m_h} \nabla_h^2 - \frac{e^2}{\epsilon r_{eh}} + \delta_{I,1} \Delta E_{exch}^0 \pi a_{ex}^3 \delta(\boldsymbol{r}_e - \boldsymbol{r}_h) \right] \phi(\boldsymbol{r}_e, \boldsymbol{r}_h)$$
$$-(E - E_g) \phi(\boldsymbol{r}_e, \boldsymbol{r}_h) = 0 , \tag{10.23}$$

where $\Delta E^0_{\text{exch}}$ is the bulk exciton exchange splitting energy, and $I$ is the sum of the electron and hole Bloch function angular momenta. For the case of the $\Gamma_6$ conduction band and the $\Gamma_7$ valence band of cubic materials that we consider (see Appendix B), the $I=1$ state is three-fold degenerate and is a mixture of spin-singlet and spin-triplet electron–hole pair states. The $I=0$ state, which has no exchange contribution to the energy, is purely spin-triplet [43].

At this point, it is useful to review some details of the symmetry of the electron and hole Bloch functions and that of the exciton. We use the case of CuCl QDs to illustrate the details. As we consider spherical QDs and use spherical band dispersion for the EMA, the band edge Bloch function of $\Gamma_6$ symmetry transforms like an $l=0$, $s=1/2$ orbital, while the $\Gamma_7$ Bloch function transforms like an $l=1$, $s=1/2$, $l+s=1/2$ orbital [44]. The four exciton states that may be formed from these two-fold degenerate electron and hole states split into a non-degenerate state of $\Gamma_2$ symmetry and a three-fold degenerate state of $\Gamma_5$ symmetry. In the present case, these states may also be labeled by their total Bloch function angular momentum, $I=0$ and $1$, respectively. The corresponding products of the electron and hole Bloch functions are

$$\psi_{00} = \frac{1}{\sqrt{2}}\left(u_{\text{c},1/2}u^*_{\text{v},1/2} + u_{\text{c},-1/2}u^*_{\text{v},-1/2}\right), \tag{10.24}$$

$$\psi_{10} = \frac{1}{\sqrt{2}}\left(u_{\text{c},1/2}u^*_{\text{v},1/2} - u_{\text{c},-1/2}u^*_{\text{v},-1/2}\right), \tag{10.25}$$

$$\psi_{11} = -u_{\text{c},1/2}u^*_{\text{v},-1/2}, \tag{10.26}$$

and

$$\psi_{1,-1} = u_{\text{c},-1/2}u^*_{\text{v},1/2}, \tag{10.27}$$

where the functions $u_{\text{c(v)}}$ are the conduction (valence) band Bloch functions defined in Appendix B. It is the $I=1$ exciton state that is optically excited as it contains the spin-singlet component. As all the optically excited states have a zero angular momentum ($L$) for the envelope function, for such states $I$ also equals the total angular momentum.

Equation (10.23) for the exciton, including the electron–hole exchange interaction, can be solved by expanding the exciton wave function as described in Sect. 10.2.4. However, the exchange energy can be obtained to a very good approximation within the first order perturbation theory as

$$\Delta E_{\text{exch}} = \Delta E^0_{\text{exch}}\pi a^3_{\text{ex}}\int |\phi(\boldsymbol{r},\boldsymbol{r})|^2\,\mathrm{d}^3 r, \tag{10.28}$$

for the $I=1$ states, and zero otherwise. Here $\phi(\boldsymbol{r}_{\text{e}},\boldsymbol{r}_{\text{h}})$ is the exciton envelope function calculated without including the exchange interaction. The wave function of the exciton is then given by the envelope function $\phi$ times the Bloch function product $\psi_{II_z}$ given by (10.24–10.27).

Now we consider the biexciton states. From $I=0$ and $I=1$ exciton states, we may generate biexciton states with the Bloch function angular momentum $J=0, 1$, or 2:

$$0 \otimes 0 = 0 \text{ or } \Gamma_2 \otimes \Gamma_2 = \Gamma_1 \,,$$
$$1 \otimes 0 = 1 \text{ or } \Gamma_5 \otimes \Gamma_2 = \Gamma_4 \,,$$
$$1 \otimes 1 = 0 \oplus 1 \oplus 2 \text{ or } \Gamma_5 \otimes \Gamma_5 = \Gamma_1 \oplus \Gamma_4 \oplus (\Gamma_3 \oplus \Gamma_5) \,.$$

Thus an $I=0$ exciton pair will get mixed with an $I=1$ pair to give $J=0$ biexciton states, while $I=1$ pairs will form $J=2$ biexcitons; a pair made up of $I=0$ and $I=1$ excitons will mix with an $I=1$ pair to give $J=1$ biexcitons.

For those biexciton states with the envelope function angular momentum $L=0$, the case that we consider, the Bloch function angular momentum $J$ completely determines the symmetry of the biexciton wave functions. However, the requirement of the antisymmetry of the wave function under the electron–electron or the hole–hole interchange puts some restrictions on the form of the envelope functions. As discussed in Sect. 10.3.2, we can have four kinds of exciton–exciton product states: $\Phi^{++}$, $\Phi^{--}$, $\Phi^{+-}$, and $\Phi^{-+}$. The $J=0$ biexciton state may be written as

$$\Psi_{00} = \Phi_0^{++}\chi_{00}^{00} + \Phi_0^{--}\chi_{00}^{11} \,, \tag{10.29}$$

where $\chi_{JJ_z}^{ss'}$ is the product of two-electron and two-hole Bloch function products with the total electron spin equal to $s$ and the total hole angular momentum equal to $s'$, and $J=s+s'$, $s+s'-1$, $\ldots |s-s'|$. Then, the EMA equations satisfied by $\Phi_0^{++}$ and $\Phi_0^{--}$ are (see Appendix B)

$$(H_{XX} - E)\Phi_0^{++} + \pi a_{\text{ex}}^3 \Delta E_{\text{exch}}^0 \left( \frac{3}{4}(\delta_1 + \delta_2 + \delta_3 + \delta_4)\Phi_0^{++} \right.$$
$$\left. - \frac{\sqrt{3}}{4}(\delta_1 + \delta_2 - \delta_3 - \delta_4)\Phi_0^{--} \right) = 0 \,, \tag{10.30a}$$

$$(H_{XX} - E)\Phi_0^{--} + \pi a_{\text{ex}}^3 \Delta E_{\text{exch}}^0 \left( \frac{-\sqrt{3}}{4}(\delta_1 + \delta_2 - \delta_3 - \delta_4)\Phi_0^{++} \right.$$
$$\left. + \frac{1}{4}(\delta_1 + \delta_2 + \delta_3 + \delta_4)\Phi_0^{--} \right) = 0 \,, \tag{10.30b}$$

where $H_{XX}$ is the Hamiltonian given by (10.18), $\delta_1 = \delta(\boldsymbol{r}_{e_1} - \boldsymbol{r}_{h_1})$, $\delta_2 = \delta(\boldsymbol{r}_{e_2} - \boldsymbol{r}_{h_2})$, $\delta_3 = \delta(\boldsymbol{r}_{e_1} - \boldsymbol{r}_{h_2})$, and $\delta_4 = \delta(\boldsymbol{r}_{e_2} - \boldsymbol{r}_{h_1})$.

The $J=1$ biexciton state with $J_z = M$ may be written as

$$\Psi_{1M} = \Phi_1^{--}\chi_{1M}^{11} + \Phi_1^{+-}\chi_{1M}^{01} + \Phi_1^{-+}\chi_{1M}^{10} \tag{10.31}$$

and the corresponding EMA equations are

$$(H_{XX} - E)\Phi_1^{--} + \pi a_{ex}^3 \Delta E_{exch}^0 \left( \frac{1}{2}(\delta_1 + \delta_2 + \delta_3 + \delta_4)\Phi_1^{--} \right.$$

$$- \frac{1}{2\sqrt{2}}(\delta_1 + \delta_2 - \delta_3 - \delta_4)\Phi_1^{+-}$$

$$\left. + \frac{1}{2\sqrt{2}}(\delta_1 - \delta_2 + \delta_3 - \delta_4)\Phi_1^{-+} \right) = 0 , \qquad (10.32a)$$

$$(H_{XX} - E)\Phi_1^{+-} + \pi a_{ex}^3 \Delta E_{exch}^0 \left( -\frac{1}{2\sqrt{2}}(\delta_1 + \delta_2 - \delta_3 - \delta_4)\Phi_1^{--} \right.$$

$$+ \frac{3}{4}(\delta_1 + \delta_2 + \delta_3 + \delta_4)\Phi_1^{+-}$$

$$\left. + \frac{1}{2}(\delta_1 - \delta_2 - \delta_3 + \delta_4)\Phi_1^{-+} \right) = 0 , \qquad (10.32b)$$

$$(H_{XX} - E)\Phi_1^{-+} + \pi a_{ex}^3 \Delta E_{exch}^0 \left( -\frac{1}{2\sqrt{2}}(\delta_1 - \delta_2 + \delta_3 - \delta_4)\Phi_1^{--} \right.$$

$$+ \frac{1}{2}(\delta_1 - \delta_2 - \delta_3 + \delta_4)\Phi_1^{+-}$$

$$\left. + \frac{3}{4}(\delta_1 + \delta_2 + \delta_3 + \delta_4)\Phi_1^{-+} \right) = 0 . \qquad (10.32c)$$

The $J=2$, $J_z=M$ biexciton state has the form

$$\Psi_{2M} = \Phi_2^{--} \chi_{2M}^{11} , \qquad (10.33)$$

and $\Phi_2^{--}$ satisfy

$$(H_{XX} - E)\Phi_2^{--} + \pi a_{ex}^3 \Delta E_{exch}^0 (\delta_1 + \delta_2 + \delta_3 + \delta_4)\Phi_2^{--} = 0 . \qquad (10.34)$$

Equations (10.30a–10.34) may be solved by expanding the functions $\Phi$ into exciton–exciton product states as in the case without the exchange interaction described in Sect. 10.3.1.

## 10.4   Exciton and Biexciton Energy Levels: The Case of CuCl

We consider CuCl as a prototype material to discuss the exciton and biexciton energy levels calculated by the approach in Sect. 10.3. The calculated excitonic energy levels are plotted in Fig. 10.2. Although the $L=0$ excitons have been extensively discussed in the literature, there exists little experimental or theoretical information on $L>1$ exciton states. The $L=1$ excitons are especially interesting since these may be excited in infrared spectroscopy as well as in two-photon spectroscopy. Both these phenomena have received some experimental attention [39,45].

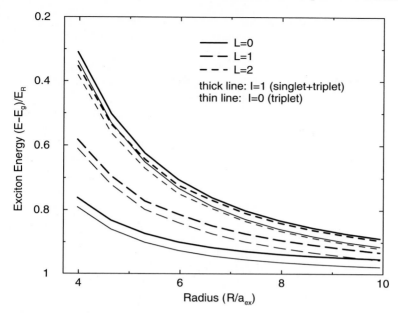

**Fig. 10.2.** Calculated energy levels of the $L=0$, 1, and 2 exciton states in semiconductor QDs. $E_R$ is the exciton Rydberg, $a_{ex}$ the exciton Bohr radius, and $E_g$ is the bulk band gap energy

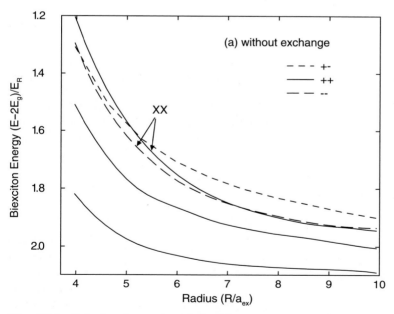

**Fig. 10.3.** Calculated energies of the biexciton states without the electron–hole exchange interaction included. Notations are the same as in Fig. 10.2

The results presented, though scaled by the exciton Rydberg $E_R$, correspond to an electron–hole mass ratio $m_e/m_h=0.28$ appropriate for CuCl [46] with $m_e=0.5$ and $m_h=1.8$.

The $L=0$ biexciton energy levels, with and without the exchange interaction included, are shown in Fig. 10.3 and Fig. 10.4. Here we have used four $L=0$, three $L=1$, and two $L=2$ states, giving a total of 58 product states forming a nonorthogonal basis. In the absence of the exchange interaction, the states are labeled by the symmetry under exchange of the electrons and holes as described in Sect. 10.3. These states have more degeneracy than expected from the conservation of the total angular momentum $J$. For example, the $--$ states with $J=0$, 1, and 2 all have the same energy, giving a nine-fold degenerate state. The exchange interaction mixes states of the same total angular momentum $J$ as well as lifts the degeneracy of the $--$ states with $J=0$, 1, and 2.

We note that in the bulk semiconductors there is a large discrepancy between the best variational estimates of the biexciton binding energy and the experimental results. For example, in CuCl the variational calculation of Akimoto and Hanamura [47] and Brinkman, Rice, and Bell [48] gives a very low value of 11 meV [46] compared to the experimental value of 32 meV. This large discrepancy was not noticed at the time these calculations were reported, since the electron–hole mass ratio known at that time was substantially smaller than the presently accepted values, which led to a fortuitous agreement with experiments.

We find that a small part of this discrepancy may be attributed to the neglect of the electron–hole exchange interaction. To see this, we note that the lowest biexciton state is of $\Gamma_1$ symmetry because this state has its envelope function symmetric under exciton exchange and hence has a bonding character, like the bonding orbital of the hydrogen molecule, forming a bound state. As discussed in Sect. 10.3.3, the electron–hole exchange interaction mixes the $\Gamma_2$–$\Gamma_2$ and the $\Gamma_5$–$\Gamma_5$ exciton product states that contribute to the $\Gamma_1$ biexciton states, and hence the biexciton ground state has an exchange contribution less than twice the exciton exchange energy. The experimentally quoted biexciton binding energy is the difference between twice the energy of the $\Gamma_5$ exciton and the biexciton ground state energy, while the theoretical value is calculated as the difference between twice the energy of the $\Gamma_2$ exciton and the biexciton ground state energy [47,48]. Thus, inclusion of the electron–hole exchange interaction increases the biexciton binding energy. In the following we employ the former definition of the biexciton binding energy.

For the largest size for which calculations have been done ($10a_{ex}$), the biexciton energy is increased by about $1.3\Delta E^0_{exch}$, while in bulk CuCl the exchange correction is quoted to be $1.6\Delta E^0_{exch}$, obtained as a first-order perturbative estimate using an explicit variational wave function by Bassani et al. [49]. We use $\Delta E^0_{exch}=4.4$ meV for CuCl [50].

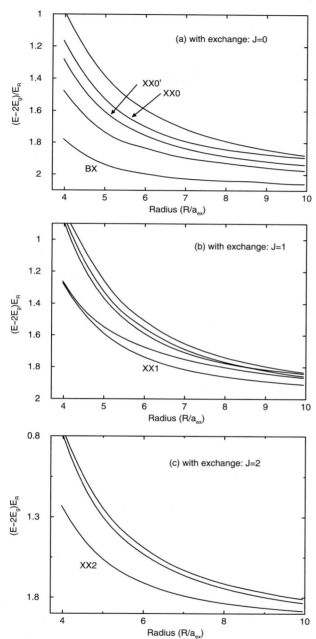

**Fig. 10.4.** Calculated energies of the biexciton states including the electron–hole exchange interaction. BX, XX0 (XX0′), XX1, and XX2, respectively, denote the biexciton ground state and the weakly correlated exciton pair states with (**a**) $J=0$; (**b**) $J=1$; and (**c**) $J=2$

Nonetheless, the variationally calculated bulk biexciton binding energy is still substantially smaller than the experimental result. On the other hand, our calculation gives a biexciton binding energy of 30.3 meV for a CuCl QD of $R$=70 Å. Although this is slightly smaller than the bulk value of 32 meV, we can say that there is substantial improvement over the older calculations in the bulk material.

Figure 10.5 shows the size dependence of the calculated biexciton binding energy. The biexciton binding energy in CuCl QDs was measured by Masumoto et al. [51]. This experimental result is also shown in Fig. 10.5. As the radius of the QD increases from 28 Å to 70 Å, the biexciton binding energy decreases from $0.257E_R$ (50 meV) to $0.156E_R$ (30.3 meV), while the experimental result in the same size range varies from $0.33E_R$ (64 meV) to $0.216E_R$ (42 meV), which is somewhat larger than the calculated result. These numbers correspond to $E_R$=194.4 meV [52] ($a_{ex}$=7.07 Å). A proper comparison with the experiment is, however, made difficult by the lack of precise knowledge of the exciton Rydberg to be used and the size of the crystallites in the experimental sample. Figure 10.5 shows calculated results corresponding to $E_R$=194.4 meV as well as $E_R$=213 meV, where the latter value is widely used in the literature on CuCl QDs (see, e.g., [51]). As the experimentally estimated size [51] corresponds to matching the exciton energy to that predicted by the center-of-mass confinement picture, it is very sensitive to the values of

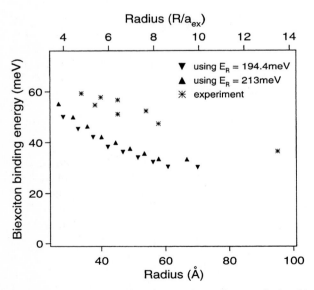

**Fig. 10.5.** The calculated size dependence of the biexciton binding energy in CuCl quantum dots. Two sets of results corresponding to the exciton Rydberg $E_R$=194.4 meV and 213 meV are shown. The experimental results of [51] are also shown

the exciton mass, exciton Rydberg, and bulk exciton energy used. Keeping these reservations in mind, there is reasonable agreement between the experiment and the theory. The theory somewhat underestimates the biexciton binding energy, possibly due to insufficient exciton–exciton correlation built into the wave function by the truncated basis set. However, the reasonable agreement with experiments indicates that the limited number of basis states used to make the problem numerically tractable does provide reliable results. The discrepancy between the theory and the experiment may also be partly attributed to the fact that the experimental sample contains somewhat flattened (platelet-shaped) crystallites [53] compared to the spherical shape that we consider.

Among the excited states of the biexciton, the most interesting ones are the nearly degenerate states occurring slightly above twice the ground state exciton energy. These states, marked XX in Fig. 10.3 and XX0, XX1, and XX2 in Fig. 10.4a–c, have an envelope function well approximated by the product of two ground state excitons. Consideration of the oscillator strengths for the exciton to biexciton transitions offers further important insights into the nature of these weakly correlated exciton pair states.

## 10.5  Transition Dipole Moments

### 10.5.1  Formulation

Now we discuss the calculation of transition dipole moments for excitation of the exciton states and the biexciton states. We consider only the interband transitions between the $\Gamma_7$ valence band and the $\Gamma_6$ conduction band. Then the polarization operator, in the second quantized site representation, is given by

$$P_z^+ = \mu_{\mathrm{cv}} \sum_\sigma \int \hat{\psi}_{\mathrm{c}\sigma}^\dagger(\boldsymbol{r}) \hat{\psi}_{\mathrm{v},-\sigma}^\dagger(\boldsymbol{r}) \, \mathrm{d}^3 r \,, \tag{10.35}$$

where $\mu_{\mathrm{cv}}$ is the interband transition dipole moment, and the light polarization is assumed to be along the $z$-axis. In terms of the conduction and valence band edge Bloch functions given by (10.71–10.74) in Appendix B, $\mu_{\mathrm{cv}} = -\mathrm{i}e/(\sqrt{3}\Omega_{\mathrm{cell}}) \int \zeta_z(\boldsymbol{r}) z \zeta_0(\boldsymbol{r}) \mathrm{d}^3 r$, where the integral is over a unit cell of volume $\Omega_{\mathrm{cell}}$. In this case, only the $I=1$ exciton with $I_z=0$ is excited from the ground state. Using the exciton state for $I=1$, $I_z=0$ given by

$$|X\rangle_{10} = \frac{1}{\sqrt{2}} \sum_\sigma \int \mathrm{d}^3 r_{\mathrm{e}} \mathrm{d}^3 r_{\mathrm{h}} \, \phi(\boldsymbol{r}_{\mathrm{e}}, \boldsymbol{r}_{\mathrm{h}}) \hat{\psi}_{\mathrm{c}\sigma}^\dagger(\boldsymbol{r}_{\mathrm{e}}) \hat{\psi}_{\mathrm{v},-\sigma}^\dagger(\boldsymbol{r}_{\mathrm{h}})|0\rangle \,, \tag{10.36}$$

we obtain the corresponding dipole moment to be

$$\mu_{10}^X = {}_{10}\langle X|P_z^+|0\rangle = \sqrt{2}\mu_{\mathrm{cv}} \int \phi(\boldsymbol{r}, \boldsymbol{r}) \mathrm{d}^3 r \,. \tag{10.37}$$

Now we consider the exciton to biexciton transitions. Again taking the light polarization to be along the $z$-axis, only transitions which conserve the $z$-component of the angular momentum are allowed. In addition, transitions that would cause the total spin to change are forbidden. These selection rules lead to the following restrictions: the $J=0$ biexciton can be excited only from the $I=1$, $I_z=0$ exciton states, while the $J=1$ biexciton states are excited from the $I=0$ (if $J_z=0$) and $I=1$, $I_z=J_z$ (if $J_z\neq0$) exciton states and the $J=2$ biexciton states are excited only from the $I=1$, $I_z=J_z$ exciton states.

Using the second quantized form of the $J=0$ biexciton state $|XX\rangle_{00}$ corresponding to the wave function given by (10.29), we obtain the dipole moment for its excitation from $|X\rangle_{10}$ to be

$$
\mu_{00}^{XX} = {}_{00}\langle XX|P_z^+|X\rangle_{10}
$$

$$
= \sqrt{2}\mu_{cv} \int \left( \Phi_0^{++}(\boldsymbol{r}_e, \boldsymbol{r}_h, \boldsymbol{r}, \boldsymbol{r}) - \sqrt{\frac{1}{3}}\Phi_0^{--}(\boldsymbol{r}_e, \boldsymbol{r}_h, \boldsymbol{r}, \boldsymbol{r}) \right)
$$

$$
\cdot \phi(\boldsymbol{r}_e, \boldsymbol{r}_h)\mathrm{d}^3r_e\mathrm{d}^3r_h\mathrm{d}^3r \ . \tag{10.38}
$$

In general, we may write the biexciton state of Bloch function angular momentum $J$ and its $z$-component $M$ in a concise notation (cf. (10.29), (10.31), and (10.33)):

$$
|XX\rangle_{JM} = a_J^{++}\Phi_J^{++}|\chi\rangle_{JM}^{00} + a_J^{--}\Phi_{JM}^{--}|\chi\rangle_{JM}^{11}
$$

$$
+ a_J^{+-}\Phi_{JM}^{+-}|\chi\rangle_{JM}^{01} + a_J^{-+}\Phi_{JM}^{-+}|\chi\rangle_{JM}^{10} \ , \tag{10.39}
$$

where $a_J^{pp'}=1$ if $J=0$, $p=p'$ or $J=1$, $p\neq p'$ or $J=1$ or 2, $p=p'=-$, and vanishes otherwise. Then the transition dipole moment from the exciton state $|X\rangle_{IM}$ to the biexciton state $|XX\rangle_{JM}$ may be expressed as

$$
{}_{JM}\langle XX|P_z^+|X\rangle_{IM} = \sqrt{2}\mu_{cv} \sum_{pp'} a_J^{pp'} I_J^{pp'}
$$

$$
\times \mathcal{M}^{pp'}(IM; JM) \ , \tag{10.40}
$$

where

$$
I_J^{pp'} = \int \Phi_J^{pp'}(\boldsymbol{r}_e, \boldsymbol{r}_h, \boldsymbol{r}, \boldsymbol{r})\phi(\boldsymbol{r}_e, \boldsymbol{r}_h)\mathrm{d}^3r_e\mathrm{d}^3r_h\mathrm{d}^3r \ . \tag{10.41}
$$

The values of $\mathcal{M}^{pp'}(IM; JM)$ are tabulated in Table 10.1. The corresponding result for other polarizations can be obtained by invoking symmetry. We find that for $x$- and $y$-polarizations,

$$
{}_{JM'}\langle XX|P_x^+|X\rangle_{IM} = {}_{J1}\langle XX|P_z^+|X\rangle_{I1}\mathcal{N}(IM; JM') \ \text{if} \ J = I = 1
$$

$$
= {}_{J0}\langle XX|P_z^+|X\rangle_{I0}\mathcal{N}(IM; JM') \ \text{otherwise} \ , \tag{10.42}
$$

and

$$
{}_{JM'}\langle XX|P_y^+|X\rangle_{IM} = {}_{JM'}\langle XX|P_x^+|X\rangle_{IM} \exp(\mathrm{i}(M - M')\pi/2) \ . \tag{10.43}
$$

The values of $\mathcal{N}(IM; JM')$ are tabulated in Table 10.2.

**Table 10.1.** $\mathcal{M}^{pp'}(IM;JM)$ appearing in (10.40) for the transition dipole moment from the exciton state $|X\rangle_{IM}$ to the biexciton state $|XX\rangle_{JM}$ for the $z$-polarization

| $J,pp' \rightarrow$<br>(I,M) | 0,++ | 0,-- | 1,-- | 1,+- | 1,-+ | 2,-- |
|---|---|---|---|---|---|---|
| 0,0 | 0 | 0 | $-\sqrt{2}$ | $-1$ | 1 | 0 |
| 1,1 | 0 | 0 | 0 | 1 | 1 | $\sqrt{2}$ |
| 1,0 | 1 | $-1/\sqrt{3}$ | 0 | 0 | 0 | $2\sqrt{2/3}$ |
| 1,$-1$ | 0 | 0 | 0 | $-1$ | $-1$ | $\sqrt{2}$ |

**Table 10.2.** $\mathcal{N}(IM;JM')$ appearing in (10.42) for the transition dipole moment from the exciton state $|X\rangle_{IM}$ to the biexciton state $|XX\rangle_{JM'}$ for the $x$-polarization

| $J,M' \rightarrow$<br>(I,M) | 0,0 | 1,0 | 1,1 | 1,$-1$ | 2,0 | 2,$\pm 1$ | 2,2 | 2,$-2$ |
|---|---|---|---|---|---|---|---|---|
| 0,0 | 0 | 0 | $-1/\sqrt{2}$ | $1/\sqrt{2}$ | 0 | 0 | 0 | 0 |
| 1,1 | $-1/\sqrt{2}$ | $1/\sqrt{2}$ | 0 | 0 | $1/\sqrt{8}$ | 0 | $-\sqrt{3}/2$ | 0 |
| 1,0 | 0 | 0 | $1/\sqrt{2}$ | $1/\sqrt{2}$ | 0 | 0 | $-\sqrt{3/8}$ | $\sqrt{3/8}$ |
| 1,$-1$ | $1/\sqrt{2}$ | $1/\sqrt{2}$ | 0 | 0 | $-1/\sqrt{8}$ | 0 | 0 | $\sqrt{3}/2$ |

### 10.5.2   Results for CuCl

The physical nature of the biexcitonic states and their relevance to optical response becomes clearer on considering the oscillator strengths for their excitation from the excitonic states. In this section we actually discuss only the transition dipole moments. The oscillator strength $f$ of a transition is related to the transition dipole moment $\mu$ through $f=2m_0E|\mu|^2/e^2\hbar$, where $E$ is the energy of the transition.

Only the $\Gamma_5$ ($I=1$) excitons are optically excited from the ground state. As the $\Gamma_5$ exciton state is threefold degenerate, subsequent excitation of the biexciton states will be dependent on the polarization of the exciton state. Figure 10.6 shows the dipole moments for transitions from the $\Gamma_5$ exciton ground state with $I_z=0, \pm 1$ to the biexciton states. Only a few dominant transitions are shown, and the dipole moment for excitation of the exciton ground state is also shown for comparison. The light polarization is taken to be along the $z$-axis. Figures 10.7 and 10.8 show the squared dipole moments for the exciton-to-biexciton transitions as a function of the transition frequency for a few different values of the radius. While the limited data shown in Fig. 10.6 illustrate the size dependence of the dipole moments discussed in

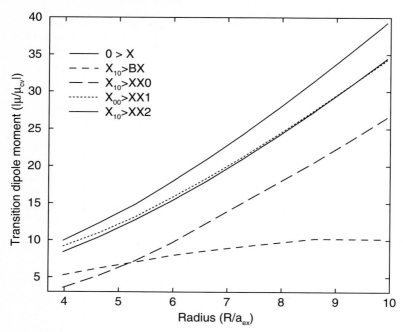

**Fig. 10.6.** Transition dipole moments for excitation of the lowest $I=1$ exciton state ($X_{10}$) and for the dominant transitions from the lowest exciton to the biexciton states. The biexciton states involved in the transitions are indicated by the energy level labels used in Fig. 10.4; 0 denotes the ground state and $X_{II_z}$ denotes the lowest $I=0$ or $I=1$ exciton state

detail below, Figs. 10.7 and 10.8 provide a complementary picture suitable for describing the excited state absorption, discussed in Sect. 10.7.2.

As shown in Sect. 10.5.1, for the $z$-polarization, only the $I=1$, $I_z=0$ exciton states can be excited by one-photon absorption from the ground state, while only the $J=0$, $J_z=0$ and $J=2$, $J_z=0$ biexciton states are excited by a subsequent one-photon absorption. We note that this process is sufficient to discuss excitation of biexciton states by a two-step absorption of linearly polarized photons by the ground state. While experiments measuring the coherent nonlinear optical susceptibility, for example, are described by such processes, a pump–probe experiment could probe exciton and biexciton excitations by different polarizations. The latter case is discussed in Sect. 10.7.3.

The dipole moment for the transition from the biexciton ground state to the exciton, commonly referred to as the M-line emission, increases with the radius of the QD at small sizes, but saturates toward the bulk value at larger sizes. For $R=10a_{ex}$, the M-line dipole moment of $10.2\mu_{cv}$ corresponds to an oscillator strength of $\approx 1900 f_{bulk}$ for CuCl, which may be compared with the measured bulk value of $2500 f_{bulk}$ [54]. Here $f_{bulk}$ is the oscillator strength per unit cell of the bulk $\Gamma_5$ exciton.

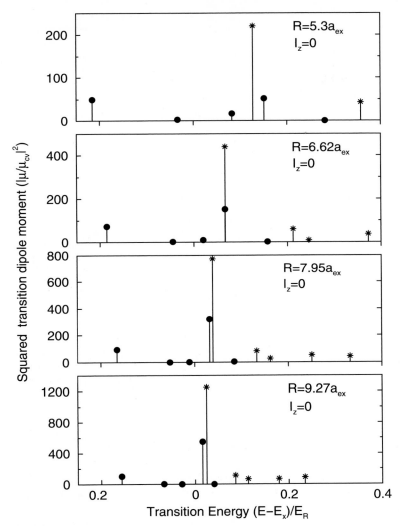

**Fig. 10.7.** Squared transition dipole moments for transitions from the lowest $I=1$, $I_z=0$ exciton to the $J=0$, 1, and 2 biexciton states, respectively, marked by •, ○, and *. The polarization of light is taken to be along the $z$-axis

While the exciton oscillator strength is proportional to the volume of the crystallite, provided the exciton envelope function is coherent over the whole crystallite, the biexciton oscillator strength tends towards a constant value in the bulk limit. This behaviour may be understood by the following simple physical argument. The creation of a biexciton from an exciton state involves creation of a second exciton spatially close to the first one within the volume of the biexciton. Thus in the bulk limit, the M-line oscillator strength is of the

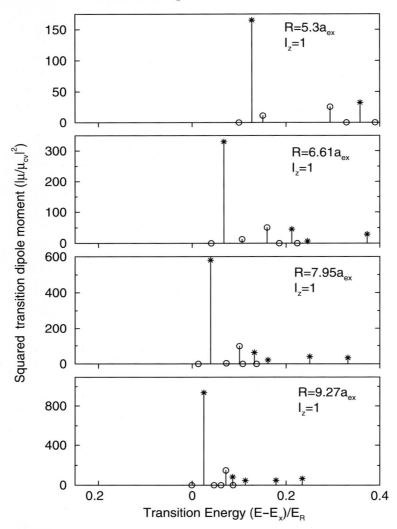

**Fig. 10.8.** Squared transition dipole moments for transitions from the lowest $I=1$, $I_z=1$ exciton to the $J=0$, 1, and 2 biexciton states, respectively, marked by $\bullet$, $\circ$, and $*$. The polarization of light is taken to be along the $z$-axis

order of the exciton oscillator strength corresponding to a coherence volume equal to the volume of the biexciton. This is a constant, dictated by the size of the biexciton.

The calculated M-line oscillator strength corresponds to a radiative decay time of 660 ps for $R=28$ Å, gradually decreasing to 175 ps as the radius of the QD increases to 70 Å. This may be compared with the measured biexciton decay time of 50 ps in bulk CuCl [55] and 70 ps in CuCl nanocrystals of

42 Å radius [56,57]. These decay times for the QD are calculated assuming a host dielectric constant of 2.25, appropriate for glass. Even for the largest size considered by us, the radiative decay time is significantly larger than the bulk value because of the smaller dielectric constant of the host material compared to that of bulk CuCl, and also because our calculated biexciton oscillator strength in the large-$R$ limit is somewhat smaller than the bulk value. The above discrepancy between the theory and experiments may also be attributed to the participation of nonradiative processes in actual samples. On the other hand, the mesoscopic enhancement of the exciton oscillator strength implies that the radiative decay time of the exciton is inversely proportional to the volume of the QD. For the exciton ground state, we find a decay time of 740 ps for $R=28$ Å, decreasing to 45 ps for $R=70$ Å with good correspondence to experimental data [14].

A very interesting result of this calculation [30,31] is the existence of the two nearly degenerate excited biexciton states (labelled XX0 and XX2 in Fig. 10.4) with a large oscillator strength, as is evident from Fig. 10.6. These states have oscillator strengths that increase in proportion to the QD volume, and the sum of their oscillator strengths approximately equals twice that of the exciton, especially at large sizes. For linearly polarized excitation, the states that share such a large oscillator strength have $J=0$ and $J=2$, $J_z=0$. Interestingly, we find that the wave functions of these states are well approximated by a product of two independent ground state exciton states, especially at larger sizes. Because of their large oscillator strength, these states dominate the excited state absorption as well as crucially influence the excitonic optical nonlinearity. Before entering into a discussion of optical nonlinearities it is, therefore, instructive to take a closer look at these weakly correlated exciton pair states.

## 10.6   Weakly Correlated Exciton Pair States

Let us consider the creation of a second exciton in a QD much larger in size than the exciton. Such a process will be most efficient when the second exciton is created uncorrelated with the first one, as it then would have an oscillator strength of the same order as that of creating a single exciton. Such an uncorrelated exciton pair can be an approximate eigenstate of a large QD because the exciton–exciton interaction is short ranged (dipole–dipole-like), unlike the electron–hole interaction in an exciton. We may, in fact, construct two such excited states with almost the same energy

$$
\begin{aligned}
\Phi_{XX}^{\pm\pm} = (1/\sqrt{2})\,[&\phi_X^{\mathrm{g}}(\boldsymbol{r}_{\mathrm{e1}},\boldsymbol{r}_{\mathrm{h1}})\phi_X^{\mathrm{g}}(\boldsymbol{r}_{\mathrm{e2}},\boldsymbol{r}_{\mathrm{h2}}) \\
\pm\, &\phi_X^{\mathrm{g}}(\boldsymbol{r}_{\mathrm{e1}},\boldsymbol{r}_{\mathrm{h2}})\phi_X^{\mathrm{g}}(\boldsymbol{r}_{\mathrm{e2}},\boldsymbol{r}_{\mathrm{h1}})]\ ,
\end{aligned}
\tag{10.44}
$$

where $\phi_X^{\mathrm{g}}$ is the envelope function of the exciton ground state.

In the limit of large $R$ these two states ($++$ and $--$) have a combined oscillator strength and energy twice those of the exciton ground state. The exchange interaction splits these into four states, two with $J=0$, and one each with $J=1$ and $J=2$. The corresponding wave functions are given by

$$\Psi_{XX0} = \frac{\sqrt{3}}{2}\Phi_{XX}^{++}\chi_{00}^{00} - \frac{1}{2}\Phi_{XX}^{--}\chi_{00}^{11} , \tag{10.45}$$

$$\Psi_{XX0'} = \frac{1}{2}\Phi_{XX}^{++}\chi_{00}^{00} + \frac{\sqrt{3}}{2}\Phi_{XX}^{--}\chi_{00}^{11} , \tag{10.46}$$

$$\Psi_{XX1} = \Phi_{XX}^{--}\chi_{1M}^{11} , \tag{10.47}$$

$$\Psi_{XX2} = \Phi_{XX}^{--}\chi_{2M}^{11} , \tag{10.48}$$

where (10.45–10.46) are obtained by diagonalizing (10.30a, 10.30b) in the subspace of the two functions given by (10.44). The diagonalization is achieved by noting that integrals involving cross terms like $\phi_X^g(r, r)\phi_X^g(r_{e2}, r_{h2})$ $\phi_X^g(r, r_{h2})\phi_X^g(r_{e2}, r)$ tend to zero as $R\to\infty$. Only two of these, XX0 and XX2 (respectively with $J=0$ and $J=2$), are excited by multistep excitation via the $I=1$ exciton ground state. Both these states have an exchange energy of twice that of the exciton ground state. The dipole moments for excitation of these states may be calculated using (10.40), (10.41), and Table 10.1. Noting that the second term in (10.44) makes a negligible contribution to the integral in (10.41), it follows that the transition dipole moments for excitation of the states XX0 and XX2, respectively, equal to $\sqrt{2/3}$ and $\sqrt{4/3}$ times that of the exciton ground state. Thus, in the limit of large $R$, the states XX0 and XX2 will have a combined oscillator strength of twice that of the exciton ground state. For finite $R$, the exciton–exciton interaction would modify this picture, but our numerical results agree with the above description, to a good approximation, especially at larger sizes. The four weakly correlated exciton pair states XX0, XX0', XX1, and XX2 described above are shown in Fig. 10.4a–c.

Interestingly, the factor of two in the oscillator strength may also be understood as the bosonic enhancement factor corresponding to the creation of a second identical exciton. It would be interesting to extend this picture to the creation of multiple exciton states in large QDs. We note that the independent boson picture implicit in this argument is reasonable as long as the QD is large enough to accommodate the excitons without considerable overlap. Experimentally, QDs provide a unique opportunity to create a definite number of excitons in a small and well-defined volume allowing observation of the bosonic enhancement in the exciton creation.

As the size of the QD is reduced, the two excitons overlap with each other, the state corresponding to $\Phi_{XX}^{--}$ acquires a repulsive energy, as is well known with the case of the antibonding state of the hydrogen molecule. On the other hand, the $\Phi_{XX}^{++}$ state gets more and more mixed with and repelled by the biexciton ground state. The net effect of this size-dependent evolution of the weakly correlated exciton pair states is a weakening of their oscillator strength as well as a blueshift of the corresponding exciton–biexciton transition, as the QD size is reduced.

In addition, there is a $J=1$ weakly correlated product state (labeled XX1 in Fig. 10.4b), which corresponds to the product of the $I=0$ ($\Gamma_2$) and $I=1$ ($\Gamma_5$) exciton ground states. As $\Gamma_2$ excitons are not optically excited, this state cannot be excited by absorption of two identically polarized photons. However, it can be excited by absorption of, for example, a $z$-polarized photon from the $I=1$, $I_z=\pm 1$, or the $I=0$ states. The latter process has a dipole moment comparable to that of the exciton and is also shown in Fig. 10.6.

## 10.7   Nonlinear Optical Properties

### 10.7.1   Size Dependence of the Third-Order Nonlinear Susceptibility

As discussed in Sect. 10.6, the weakly correlated exciton pair states have a large oscillator strength. As the excitonic and two-excitonic contributions to the third-order nonlinear susceptibility have opposite signs, the weakly correlated pair states would play a crucial role in determining the resonant excitonic nonlinearity in large QDs. We shall now investigate this in detail.

The third order nonlinear susceptibility, $\chi^{(3)}(-\omega;\omega,\omega,-\omega)$ may be obtained from perturbation theory as [58]

$$
\chi^{(3)}(-\omega;\omega,\omega,-\omega) = -\mathrm{i}N \sum_{e,b} \frac{|\mu_{eg}|^2}{2\hbar^3 \gamma_\parallel^e} \frac{\gamma_{eg}}{(\omega_{eg}-\omega)^2 + \gamma_{eg}^2}
$$
$$
\times \left[ \frac{2|\mu_{eg}|^2}{\mathrm{i}(\omega_{eg}-\omega) + \gamma_{eg}} - \frac{|\mu_{be}|^2}{\mathrm{i}(\omega_{be}-\omega) + \gamma_{be}} \right]
$$
$$
-\mathrm{i}N \sum_{e,b} \frac{|\mu_{eg}|^2 |\mu_{be}|^2}{4\hbar^3}
$$
$$
\times \frac{1}{\mathrm{i}(\omega_{eg}-\omega) + \gamma_{eg}} \frac{1}{\mathrm{i}(\omega_{bg}-2\omega) + \gamma_{bg}}
$$
$$
\times \left[ \frac{1}{\mathrm{i}(\omega_{eg}-\omega) + \gamma_{eg}} - \frac{1}{\mathrm{i}(\omega_{be}-\omega) + \gamma_{be}} \right] , \quad (10.49)
$$

where we have retained only the near-resonant terms. Here $\hbar\omega_{ij}$, $\mu_{ij}$ and $\gamma_{ij}$, respectively, denote the energy, dipole moment and dephasing rate corresponding to a transition between the states $i$ and $j$. The subscripts g, e, and b denote the ground state, the exciton states and the biexciton states, respectively, $\gamma_\parallel^e$ denotes the exciton population decay rate, and $N$ is the number density of the quantum dots.

The first two terms in (10.49) arise from the saturation of the exciton population while the last two terms arise from the two-photon coherence of the biexciton state. Thus there will be resonant enhancement of $\chi^{(3)}$ at the exciton-to-biexciton transition energy as well as at the exciton energy. The former case is especially interesting as the increase in $\chi^{(3)}$ is not accompanied by an increase in absorption, unlike at the exciton resonance. Consequently, the dynamics at this two-photon resonance would be governed by the dephasing time of the biexciton and thus promise a fast response time. However, as the QD size increases, the oscillator strength of the bound biexciton saturates towards a constant value and shows no mesoscopic enhancement. On the other hand, in the weak-confinement regime that we consider, the mesoscopically enhanced exciton oscillator strength would lead to mesoscopic enhancement of $\chi^{(3)}$.

In fact, the resonant excitonic $\chi^{(3)}$ of CuCl QDs has been observed [9] to increase with the radius of the QD, exhibiting such a mesoscopic enhancement. But as $R$ is increased to about 50 Å (at 77 K), $\chi^{(3)}$ was seen to saturate and then to rather abruptly decrease with further increase in $R$ (see Fig. 10.9). We shall see that this saturation of the excitonic contribution to $\chi^{(3)}$ and the reversal of its size dependence arise from competing contributions from the weakly correlated exciton pair states and from the exciton ground state. The weakly correlated exciton pair states also have mesoscopically enhanced oscillator strengths, and a proper consideration of the size dependence of $\chi^{(3)}$ should include contribution from such states, as described by (10.49).

Now we consider the size dependence of the mesoscopically enhanced $\chi^{(3)}$ at the lowest exciton resonance. Since there will be considerable linear absorption at the exciton resonance, it would be appropriate to consider the

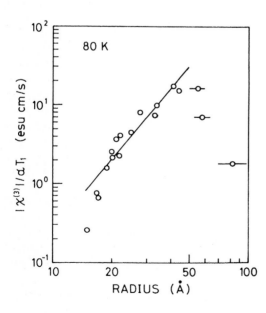

**Fig. 10.9.** The figure of merit $|\chi^{(3)}|/\alpha T_1$ as a function of QD radius. The *straight line* indicates the $R^3$ dependence. Here $T_1$ is the population decay time referred to as $\tau$ in the text (from [9])

figure of merit $\eta=|\chi^{(3)}|/\alpha$, where $\alpha$ is the linear absorption coefficient given by

$$\alpha(\omega) = N \frac{\omega}{nc} \frac{4\pi}{\hbar} \sum_{e} \frac{|\mu_{eg}|^2 \gamma_{eg}}{(\omega - \omega_{eg})^2 + \gamma_{eg}^2} , \tag{10.50}$$

where $n$ is the refractive index of the sample, which, for a dilute collection of QDs, may be approximated by that of the host material. Following the experimental results of [9,13] on CuCl QDs, the size dependence of $\gamma_{\parallel}^{e}$ and $\gamma_{eg}$ are fitted as,

$$\gamma_{eg} = \begin{cases} 0.04(45 - R) + \gamma_h & \text{if } R \leq 45\,\text{Å} \\ \gamma_h & \text{if } R > 45\,\text{Å} \end{cases} \tag{10.51}$$

$$\frac{1}{\gamma_{\parallel}^{e}} = \tau = 1.07 \times 10^6 R^{-2.26} \tag{10.52}$$

where $R$ is in angstroms and $1/\gamma_{\parallel}^{e}$ is in ps, and this form is assumed for all the exciton levels. Although $\gamma_h$ that appears in (10.51) is reported to be 0.9 meV [9], we treat it as a free parameter and discuss the dependence of $\chi^{(3)}$ on $\gamma_h$. In the absence of experimental information on the homogeneous linewidth of the biexciton states, we take it to be the same as that of the exciton.

In Fig. 10.10a the maximum value of $|\chi^{(3)}|/\alpha$ is plotted as a function of the radius of the QD. In the calculation of $\chi^{(3)}$, the lowest four $L=0$ exciton levels and five lowest-energy biexciton states each with $J=0$, 1, and 2 were included. The calculation used material parameters appropriate for CuCl [52] and $n=2.25$, appropriate for a glass matrix. In the size range considered, the peak value of $\chi^{(3)}$ occurs almost exactly at the exciton resonance frequency. Several values of $\gamma_h$ are considered. For a small value of $\gamma_h$ (less than 1 meV), $\eta$ increases at small sizes sublinearly with $R$, and the rate of increase slightly decreases as $R$ increases to $10a_{ex}$. This is easily understood as the size dependence of the dominating resonant excitonic contribution to $\chi^{(3)}$, determined by the exciton oscillator strength, which increases as $R^3$ and the population decay time $\tau$, which decreases as $R^{-2.26}$. But as $\gamma_h$ is increased, this behaviour dramatically changes. At the radius for which $\gamma_h$ becomes comparable to the energy difference $\delta E = E_{XX0} - 2E_X$ or $E_{XX2} - 2E_X$, the size dependence of $\chi^{(3)}$ tends to saturate and, interestingly, $\chi^{(3)}$ decreases with further increases in $R$. This correspondence between $\gamma_h$ and the size dependence of the energy difference ($\delta E$) is illustrated in Fig. 10.10b. For the case of $\gamma_h=3$ meV, the size at which $\delta E \approx \gamma_h$ is estimated to be $R=68$ Å and this value is in good agreement with the radius at which $\chi^{(3)}/\alpha$ shows a maximum in Fig. 10.10a.

This behaviour may be easily understood as arising from the weakly correlated exciton pair state, which makes a competing contribution to $\chi^{(3)}$ and tends to cancel the strong excitonic contribution. In very small QDs where the electron–hole correlation is negligible, $\chi^{(3)}$ arises from the atom-like level

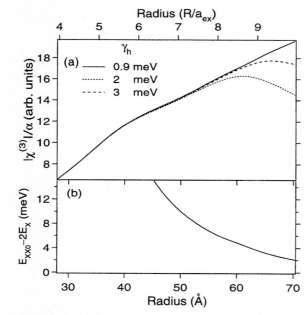

**Fig. 10.10.** (a) Calculated size dependence of the peak value of $|\chi^{(3)}/\alpha|$ near the exciton resonance in CuCl QDs. All the curves are scaled to the same value at $R=28$ Å. The hump seen at $R\approx40$ Å arises from the assumed size dependence of $\gamma_{\mathrm{h}}$ and has no special physical significance. (b) The size dependence of the energy difference between the weakly correlated exciton pair state ($E_{XX0}$) and twice the exciton ground state energy ($E_X$)

filling mechanism, while in the bulk semiconductor excitons behave like independent bosons, strongly suppressing the excitonic contribution to $\chi^{(3)}$. The presence of the weakly correlated exciton pair states with nearly twice the exciton energy and with nearly twice the exciton oscillator strength that we have identified in large QDs implies an approach to such a bulk-like behaviour.

An explicit demonstration of this cancellation between excitonic and two-excitonic contributions may be presented using a three-level model. In the present case of a three-level model consisting of the ground state, the lowest exciton state and the weakly correlated two-exciton state, there are two competing contributions to $\chi^{(3)}/\alpha$ proportional to[1]

$$\frac{2|\mu_X|^2}{(\omega_X - \omega) - \mathrm{i}\Gamma} - \frac{|\mu_{XX}|^2}{(\omega_{XX} - \omega) - \mathrm{i}\Gamma} , \tag{10.53}$$

[1] These two terms arise from the first term in (10.49) which in the present case dominates over the second term because $\gamma_{\parallel}^{\mathrm{e}} \ll \gamma_{\mathrm{eg}}$.

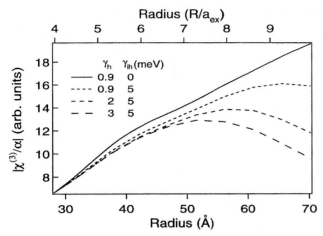

**Fig. 10.11.** Calculated size dependence of the peak value of $|\chi^{(3)}/\alpha|$ near the exciton resonance in CuCl QDs, but with inhomogeneous broadening $\gamma_{\mathrm{ih}}$ included

where the three levels are labelled by 0, X, and XX, and $\omega_X$ and $\omega_{XX}$ denote the transition frequencies and $\mu_X$ and $\mu_{XX}$ the dipole moments for the transitions $0{\to}X$ and $X{\to}XX$, respectively. $\Gamma$ denotes the homogeneous widths of these transitions. The two terms in (10.53) exactly cancel when $|\mu_{XX}|^2{=}2|\mu_X|^2$ and $\omega_X{=}\omega_{XX}$, a situation which the QD level structure approaches as $R$ increases.

In actual samples, there is also inhomogeneous broadening, probably due to size and shape inhomogeneities of the microcrystals. Figure 10.11 shows the size dependence of $|\chi^{(3)}|/\alpha$ for different values of homogeneous and inhomogeneous broadening of the exciton and biexciton states. These results were obtained by assuming a phenemenological Gaussian inhomogeneous broadening with a common width $\gamma_{ih}$ for all the one-photon transition frequencies, $\omega_{\mathrm{eg}}$ and $\omega_{\mathrm{be}}$. Then the average over the inhomogeneous broadening is equivalent to that over the excitation photon energy and average $\chi^{(3)}$ is given by

$$\int \chi^{(3)}(-\omega';\omega',\omega',-\omega')\exp(-(\omega-\omega')^2/\gamma_{ih}^2)\,\mathrm{d}\omega'\ . \tag{10.54}$$

A similar averaging has been done for $\alpha$. Increasing the inhomogeneous width causes the saturation radius to shift to lower values.

Note that the above discussion was in connection with experimental measurements at 77 K, and the observed saturation of $\chi_3$ is determined by the homogeneous linewidth of the exciton, in addition to the inhomogeneous broadening. At lower temperatures, the homogeneous linewidth could be much smaller than the 0.9 meV observed at 77 K. In fact, recent experiments [59,60] have shown that the homogeneous width of the exciton in QDs at 2 K is of the order of a few microelectronvolts. It follows that at very low temperatures, the single QD nonlinearity would continue to increase with size much beyond the threshold of 50 Å observed in [9] before the cancellation effect due to the weakly correlated pair states sets in. It would be very interesting to do single QD measurements of the excitonic nonlinearity in larger QDs of CuCl and similar materials. Considering the experimental progress in single quantum dots made during the last few years, it is hoped that such measurements would soon become a reality.

While the data shown in Figs. 10.10 and 10.11 are given in arbitrary units, it is interesting to compare the absolute value of $\chi^{(3)}$ with experiments. Using the bulk exciton oscillator strength for CuCl to be $5.85 \times 10^{-3}$ per unit cell [55], $|\chi^{(3)}|/\alpha = 2.7 \times 10^{-9}$ esu cm in a crystallite of 37.4 Å radius. We have used $\gamma_h = 0.9$ meV. This is in close agreement with the measured value of $3.4 \times 10^{-9}$ esu cm [9].

Although most experiments in the weak-confinement regime are done on CuCl quantum dots, optical properties of QDs of II–VI semiconductors like CdS and CdSe have been widely studied since the early days of quantum dot research. It is therefore interesting to apply our results to materials like CdS and CdSe. Although the present calculation uses an electron–hole mass ratio of 0.28, appropriate for CuCl, the results would be expected to be applicable to many other materials because of the weak dependence of exciton and biexciton states on the electron–hole mass ratio [2]. In fact, the electron–hole mass ratios of 0.23 for CdS [52] and 0.28 for CdSe [52] are very close to that of CuCl.

For CdS, taking $a_{ex} = 30$ Å and $E_R = 29$ meV [52], we note that a homogeneous width $\gamma_h = 2$ meV $(=0.067 E_R)$ would lead to saturation of $\chi^{(3)}$ at a size corresponding to $\delta E \approx 0.067 E_R$. Referring to Fig. 10.10b and scaling the energies by $E_R = 194.4$ meV, we find that this corresponds to $R \approx 6.5 a_{ex}$ or about 200 Å. For CdSe, taking $E_R = 15.7$ meV [52], we get the same value of saturation radius $6.5 a_{ex}$ with a homogeneous broadening of only 1 meV. Thus, in materials with smaller exciton binding energy, the effect of the weakly correlated exciton pair states becomes important at smaller values of $R/a_{ex}$, unless the exciton linewidth is also correspondingly smaller.

Different experimental measurements of the size dependence of $\chi^{(3)}$ in CdSe and CdS$_x$Se$_{1-x}$ QDs have reported conflicting results [61]. A careful analysis of these results by Schanne-Klein et al. [62] has related this behaviour to the difference between fresh and photodarkened samples and has shown that the excitonic contribution to $\chi^{(3)}$ is an increasing function of $R$. They observed that the figure of merit $\chi^{(3)}/\alpha\tau$ in CdSe QDs is enhanced by a factor of 4.4 as the radius increases from 27 Å to 44 Å. The experimentally studied size range is much smaller than the size at which we expect the weakly correlated exciton pair states to suppress the mesoscopic enhancement of $\chi^{(3)}$. Therefore, we may expect further enhancement of the figure of merit in larger QDs. It is, however, difficult to make a quantitative prediction of the optimal size at which the figure of merit is maximized, because of insufficient knowledge of the homogeneous linewidths of the exciton and biexciton in these materials. Further experimental investigation of CdS and CdSe crystallites of larger sizes would be interesting.

It is important to note the physics of the weakly correlated exciton pair states discussed here is quite general. But the details of the valence band structure seem to play a rather important role. Recently, the size dependence of the figure-of-merit of the optical nonlinearity was investigated for CuBr nanocrystals. In sharp contrast to the case of CuCl nanocrystals, that quantity continues to increase with the nanocrystal size up to $R/a_B^* \sim 30$ [63]. Both materials have similar values of the exciton binding energy and the exciton Bohr radius. Thus this striking difference between CuCl and CuBr nanocrystals may be explained most likely in terms of the difference in the valence band structure. In CuBr the topmost valence band is four-fold degenerate ($j=3/2$) in contrast to the doubly degenerate valence band ($j=1/2$) in CuCl. Thus the biexciton energy level structure would be widely spread and the oscillator strength of transition from the exciton to the biexciton would also be widely distributed. This results in such a situation that there is no prominent biexciton state having a large oscillator strength of transition from the exciton state and leads to the absence of strong cancellation between two contributions in the $\chi^{(3)}$ expression. It is an intriguing theoretical challenge to confirm this conjecture.

## 10.7.2    Excited State Absorption from the Exciton Ground State

Recent progress in experiments has revealed the discrete energy level structures not only in the excitation spectrum [64,65] but also in the excited state absorption spectra by resonant pump–probe technique [66]. Using the exciton and biexcitonic states calculated above, we can theoretically predict the absorption spectra of excited crystallites in which one exciton has already been created.

We consider a pump–probe experiment in which a linearly polarized pump pulse excites a crystallite into the exciton ground state and a collinear probe pulse that follows probes the absorption spectra of this excited crystallite. We take the pump–probe propagation direction to be the $x$-axis and the pump-polarization to be along the $z$-axis, without loss of generality. The created exciton is in the $I=1$ ($\Gamma_5$) state with $I_z=0$. Subsequent absorption of a probe photon can then excite the $J=0$, and $J=2$, $J_z=0$ two-exciton states if the probe is $z$-polarized, and the $J=1$, $J_z=\pm1$, $J=2$ and $J_z=\pm1$ two-exciton states if the probe is $y$-polarized. The oscillator strengths for all these processes can be calculated using the expressions given in Sect. 10.5.1.

Figure 10.12 shows the oscillator strengths for transitions from the exciton ground state assuming the probe to be unpolarized. As expected from the large oscillator strength of the weakly correlated exciton pair states, these dominate the excited state absorption. The lowest energy absorption peak redshifted from the exciton energy corresponds to the biexciton ground state.

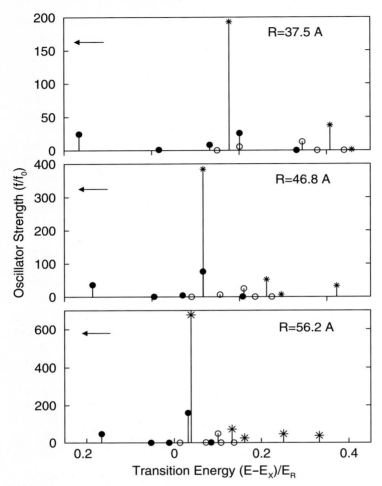

**Fig. 10.12.** Oscillator strengths $f$ for the induced absorption from the $I=1$, $I_z=0$ exciton ground state by an unpolarized probe beam. Symbols •, ∘, and *, respectively, denote transitions to the $J=0$, $J=1$, and $J=2$ biexciton states. $f_0=2m_0|\mu_{cv}|^2E/e^2\hbar$, where $E$ is the energy of the transition and $E_X$ is the exciton ground state energy. The *arrows* indicate the oscillator strength of the exciton ground state

There are a few weak transitions to the excited, but bound, biexciton states occurring below the exciton energy. The strong absorption peaks due to excitation of the $J=0$ and $J=2$ weakly correlated pair states (XX0 and XX2) occur blueshifted from the exciton. It is reasonable to argue that the experimentally observed blueshift of the probe absorption in the presence of a strong pump beam [16] involves excitation of such exciton pair states. For $R=47$ Å,

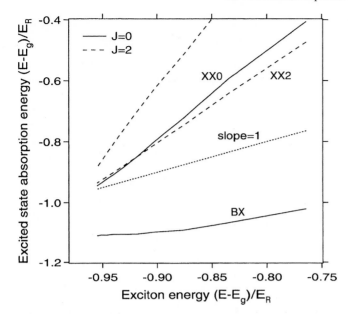

**Fig. 10.13.** Energies of the dominant excited state absorption peaks appearing in Fig. 10.12 as a function of the exciton ground state energy. BX, XX0, and XX2 denote transitions to the biexciton ground state and to the weakly correlated exciton pair states with $J=0$ and 2, respectively. The *dotted line* is of slope 1 and is only shown for reference

the blueshift is found to be 11.5 meV, and these transitions have a combined oscillator strength of about 1.3 times that of the exciton ground state. This result agrees very well with the measured blueshift of about 10 meV for $R \approx 45$ Å.

In Fig. 10.13. we plot the energies of a few dominant excited state absorption peaks as a function of the exciton ground state energy, i.e., the pump photon energy. It is interesting to note that the strongest excited state absorption to the $J=2$ two-exciton state shown in Fig. 10.13 has a linear dependence on the exciton ground state energy with a slope of about 2.4. Although the origin for this rather simple relationship is not clear, it is interesting to speculate, invoking the center-of-mass confinement picture, that the weakly correlated exciton pair has an energy equal to that of two excitons independently confined in a region of half the volume of the QD. Such a picture gives the confinement kinetic energy of the weakly correlated pair to be $2\sqrt[3]{4}=3.174$ times that of a single exciton. Consequently, the corresponding excited state absorption energy will be linearly dependent on the exciton energy with a slope of about 2.2, in close agreement with the actual value.

Much of the behaviour described above has been experimentally observed by Ikezawa et al. [66]. This is the subject of Sect. 10.7.3.

### 10.7.3    Experimental Observation
of the Weakly Correlated Exciton Pair States

The first experimental observation of the weakly correlated exciton pair state was reported in CuCl nanocrystals [66]. Prior studies of excited biexciton states were limited to the strong-confinement regime where the biexciton state is blurred by the size inhomogeneity and the ground state biexciton is masked by broad exciton bleaching [28,14]. In contrast to these reports, Ikezawa et al. reported observation of exciton addition spectra by a size-selective time-resolved pump–probe technique. An excited biexciton state and a triexciton state were clearly identified.

The experimental sample was CuCl QDs embedded in a NaCl crystal. The absorption band of the sample was inhomogeneously broadened because of the size distribution of the QDs. The second harmonics of 300 fs pulses from a Ti:sapphire laser and a Ti:sapphire regenerative amplifier were used as pump pulses for size-selective excitation after spectral filtering by a grating. The typical spectral width of the pump was 1.7 meV. A part of the amplified laser pulse was focused on pure water to produce a white continuum that served as the probe beam. The transient absorption spectra were recorded by a spectrometer and a charge-coupled device multichannel detector.

Figure 10.14 shows the absorption spectrum of the sample (solid line in the upper panel). The $Z_3$ exciton is shifted to the higher-energy side compared to the bulk material due to quantum confinement. The inset shows the spectrum of the spectrally filtered pump pulse (solid line) along with that of the unfiltered laser pulse (broken line). From the calculated size-dependence of the confined exciton energy, we estimate that the pump pulse energy of 3.259 eV excites QDs of 23 Å radius. The dashed line shows the absorption spectrum 10 ps after the excitation. The solid line in the lower part of Fig. 10.14 shows the difference between the two spectra in the upper part. The difference spectrum consists of a spectral hole at the pump photon energy and two induced absorption structures (3.180 eV and 3.296 eV) on each side of the spectral hole. Some additional bleaching around 3.335 eV due to the $Z_{1,2}$ exciton is also seen.

By scanning the pump photon energy, QDs of different sizes can be excited. The differential absorption spectra for four different pump energies are shown in the lower part of Fig. 10.14. The prominent induced absorption bands shift with the pump photon energy. The energies of the spectral hole and the induced absorption peaks are plotted in Fig. 10.15 as a function of the excitation photon energy. The spectral hole coincides with the pump energy, so the corresponding points (solid circles) lie on a line of slope 1. Open (solid) triangles show the energies of the induced absorption peaks on the higher (lower) energy side of the spectral hole.

The solid line through the open triangles is a straight line fit with slope 2.0. This line crosses the spectral hole near the $Z_3$ exciton energy in bulk CuCl. Further, temporal measurements have shown that the spectral hole

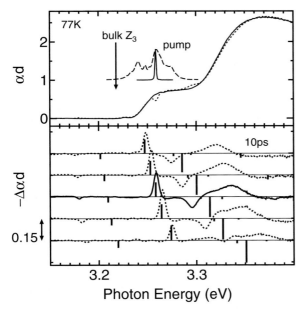

**Fig. 10.14.** *Upper panel*: The *solid line* shows the absorption spectrum of CuCl QDs in a NaCl crystal at 77 K, while the *dashed line* shows the spectrum 10 ps after excitation. *Lower panel*: The *solid line* shows the differential absorption spectrum corresponding to the upper figure, while the *dashed lines* are those for different excitation photon energies. The radii of the QDs excited are 26.5, 24.6, 23.2, 21.9, and 20.3 Å from top to bottom, respectively. The *thick solid bars* are theoretical results normalized at the spectral hole. In the *inset*, the filtered and unfiltered pump spectra are shown by a *solid line* and a *dashed line*, respectively (from Ref. [66])

and the high-energy-side induced absorption band exhibit a similar temporal evolution with an exponential decay with a decay constant of 480 ps, which is very close to the luminescence decay time of excitons in CuCl QDs [14]. These observations indicate that the higher-energy-side induced absorption arises from the excitons pre-excited by the pump beam. Furthermore, this induced absorption is comparable to or somewhat stronger than the excitonic absorption. All these features agree well with the strong transition associated with the induced absorption into the weakly correlated exciton pair state.

The calculated induced absorption spectrum from the exciton ground state is shown by thick solid bars in Fig. 10.14. The energy shifts of the strong transitions show good correspondence with the experiment. The relative strengths of the induced absorption lines and the spectral holes are well reproduced by the theory, although there is some discrepancy in the energy positions of the induced absorption lines. The strong induced absorption peak above the pump energy is thus clearly attributed to the excitation of the XX2 weakly correlated pair states.

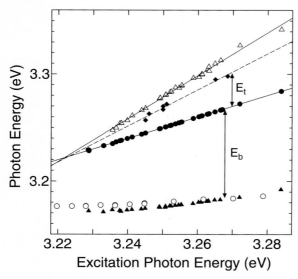

**Fig. 10.15.** The excitation energy dependence of the structures appearing in the differential absorption spectrum. The *solid circles* representing spectral hole energies lie on a line of slope 1.0. *Open* and *solid triangles* show the energies of the induced absorption peaks located at the higher and lower energy sides, respectively, of the spectral hole. *Solid diamonds* show the additional induced absorption observed under high-density excitation, $E_b$ denotes the biexciton binding energy defined as $2E_X-E_{XX}$, and $E_t$ is $E_{XXX}-E_{XX}-E_X$. The triexciton binding energy defined as $3E_X-E_{XXX}$ is given by $E_b-E_t$ (from [66])

At higher pump intensities it is possible to induce further absorption from the biexciton state to the triexciton manifold. The observed excitation power dependence of the differential absorption spectrum is shown in Fig. 10.16a. Under low-density excitation, these spectra are similar to those in Fig. 10.14, and no luminescence from biexcitons is observed. At higher excitation powers, additional induced absorption, indicated by an arrow, is seen. Simultaneously, the biexciton luminescence also became observable.

The time evolution of the additional induced absorption is shown in Fig. 10.16. The temporal evolution of the spectral hole and the induced absorption to the XX2 states are also shown for comparison. The solid lines through the solid circles and open triangles show exponential decay with a time constant of 480 ps. On the other hand, the decay constant of the additional induced absorption is less than 100 ps and corresponds to the radiative lifetime of the biexciton, which is known to be 70 ps for CuCl QDs [56]. These facts suggest that the additional induced absorption arises from a transition from the biexciton ground state to a triexciton state.

A clear experimental test of this assignment was performed by a two-color pump–probe experiment. The energy of the second pump pulse is tuned to

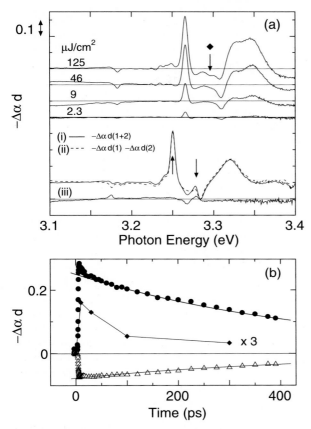

**Fig. 10.16.** (a) *Upper panel*: Excitation density dependence of the differential absorption spectrum at 10 ps after excitation for a fixed pump photon energy of 3.265 eV. Additional induced absorption shown by the *arrow* appears with increasing excitation. *Lower panel*: Results of the two-color pump–probe measurement. The photon energy of the first (second) pump pulse is indicated by an *up* (*down*) *arrow*: (i) shows the differential absorption spectrum induced by both the first and the second pump pulses; (ii) is the sum of the differential absorption spectra for each pump pulse acting alone; (iii)=(i)−(ii). (b) Time evolution of the additional induced absorption under high-density excitation together with the spectral hole and the higher-energy induced absorption. Symbols are the same as in Fig. 10.15 (from Ref. [66])

the induced absorption caused by the first pulse. This combination leads to an efficient generation of biexcitons in QDs of a particular size selected by the first pump pulse, and the probe beam measures the induced absorption to the triexciton state. The two-color differential absorption spectra is show in the lower part of Fig. 10.16a. The photon energy of the first and second pump pulses are indicated by up and down arrows, respectively. The two pump pulses were separated by a delay of 5 ps, and the probe pulse delay with respect to the second pump pulse was also 5 ps. The solid line (i) shows the differential absorption spectrum induced by the two successive pump pulses, while the dotted line (ii) shows the sum of the differential spectra induced by either pump pulse acting alone. The difference between these two spectra is labeled (iii). An increased absorption region is observed between the spectral hole and the induced absorption in spectrum (iii). The relative peak position is similar to that seen in the additional induced absorption in the upper part of Fig. 10.16a. Both of these are of the same origin, namely excitation of a triexciton from the biexciton ground state. The excitation energy dependence of this triexciton absorption is shown in Fig. 10.15 by solid diamonds. The fitted line has a slope of 1.7, and this line also crosses the other two lines near the bulk $Z_3$ exciton energy.

The linear proportionality of the induced absorption energy to the exciton confinement energy suggests that the exciton addition energy is primarily determined by the increase of the kinetic energy, which is inversely proportional to the QD radius. For the X to XX2 transition, the observed slope of 2.0 is very close to the calculated value and supports the simple physical picture of the XX2 state as two excitons independently confined in a region of half the volume of the QD (antibonding excitons). The observed triexciton state has an energy greater than that of an independent biexciton-exciton pair, but smaller than that of three excitons. Thus, while this state is a bound state, it may be regarded as an antibonding combination of a biexciton and an exciton. Assuming the added exciton to occupy half the volume of the QD without affecting the biexciton ground state, we may write $E_{XXX}-E_{XXg}=\sqrt[3]{4}(E_X-E_{bk})+E_{bk}$ with $E_{XXX(XXg)}$ being the energy of the triexciton (biexciton ground) state. This gives the slope in Fig. 10.15 as $\sqrt[3]{4}\approx1.6$, again in good agreement with the experimental value of 1.7.

### 10.7.4   Recent Progress in Nonlinear Nano-Optics

Recently, the nonlinear optical experiments on a single quantum dot, referred to as nonlinear nano-optics, has been successfully demonstrated [67]. The degenerate nonlinear spectra show almost the same features as the photoluminescence (PL) spectra. On the other hand, the nondegenerate nonlinear spectroscopy can identify the correlation among particular sets of sharp lines in the PL spectra and can reveal the underlying physics. This type of nondegenerate nonlinear spectroscopy has led to the observation of the exciton entanglement in a single quantum dot [68]. In this experiment, each of the exciton doublet states split by an external magnetic field is excited resonantly under the condition that the combination of the two photon energies is well off-resonance with the two-exciton states, and thus the contribution from the two-exciton states is negligible. This is reasonable because the biexciton binding energy ($\sim$1 meV in a GaAs quantum dot) is much larger than the doublet splitting energy, and the excited exciton pair states are blueshifted by

several meV from twice the exciton energy. Thus the detuning of the exciton
to biexciton transitions is much larger than the spectral bandwidth of the
excitation lasers and the homogeneous linewidth of excitons and biexciton
states. Therefore the strong resonant nonlinearity responsible for generation
of the entangled state owes its origin to avoidance of the excited state absorption to two exciton states of the kind discussed at length in this chapter.
However, when short pulses are used, their spectral bandwidth would cover
the bound biexciton state and the excited two-exciton (bound and possibly
scattering) states. It is very interesting and important to investigate the exciton entanglement under these situations because the actual application to
quantum information processing would require the use of optical pulses.

## 10.8   Summary and Conclusions

We have presented a theory of the excitonic and biexcitonic states in semiconductor QDs in the weak-confinement regime and discussed recent experimental results. The most important finding discussed here is the presence of
excited biexciton states with large oscillator strengths, which play a crucial
role in determining the nonlinear optical properties of QDs in the weak-
confinement regime. These states consist of two weakly correlated ground
state excitons and, consequently, have oscillator strengths for excitation from
the exciton ground state that increase proportionally to the volume of the QD.
In fact, the combined oscillator strength of the nearly degenerate $J=0$ and
$J=2$ weakly correlated two-exciton states is nearly twice that of the exciton
ground state. These states also have energies close to twice that of the exciton ground state. Consequently, the two-pair states give rise to a competing
contribution to the mesoscopically enhanced $\chi^{(3)}$ at the exciton resonance,
leading to a saturation and reversal of the size dependence of $\chi^{(3)}$. These
results provide a consistent understanding of the experimentally observed [9]
size dependence of $\chi^{(3)}$ in CuCl QDs. This mechanism of the saturation of
the mesoscopically enhanced $\chi^{(3)}$ is of quite fundamental character and is
applicable to other materials.

The excited state absorption from the exciton ground state is also discussed. Again, the excitation of the weakly correlated exciton pair state dominates the spectrum. As the size of the QD is reduced, these two-exciton states
acquire a repulsive energy, and we argue that the experimentally observed
blueshift of the exciton absorption peak in the pump–probe experiment of
[16] corresponds to excitation of the weakly correlated exciton pair states. Direct experimental observation [66] of the weakly correlated exciton pair states
and triexciton states using a two-color pump–probe technique provides clear
experimental support for the theoretical ideas developed here.

The cancellation effect in $\chi^{(3)}$ that begins as the size of the QD increases
indicates the approach towards a bosonic (harmonic) behaviour of excitons
in the low-density regime. In the bulk limit, one would expect an exact can-

cellation of the resonant one-exciton and two-exciton contributions so that $\chi^{(3)}$ is determined by nonresonant contributions from the bound biexciton and other excited two-exciton states. However, this cancellation may not be complete even in a harmonic approximation, because of possible differences in the dephasing rates of the one- and two-exciton states. Consequently, a calculation of the bulk limit of $\chi^{(3)}$ at the exciton resonance requires careful consideration of the size dependence of the relaxation rates as well as of the off-resonant contribution.

Finally, we note that the weakly correlated exciton pair states, discussed here in the context of quantum dots, are of quite general nature and would also exist in other semiconductor structures like quantum wells and wires. In fact, analysis of lasing in II–VI quantum well structures has indicated that such antibonding exciton pair states do play an important role in the lasing mechanism [69].

## Appendix A: Two-Particle States with $L=1,2$

Here we derive (10.16) and (10.17). We may write a general two-particle wave function of angular momentum $L$ and its $z$-component $M$ as

$$\phi_{LM}(\boldsymbol{r}_1, \boldsymbol{r}_2) = F(r_1, r_2, r_{12})Y_{LM}(\Omega_1) + G(r_1, r_2, r_{12})Y_{LM}(\Omega_2)$$
$$+ \sum_{ll'}^{\infty} g_{ll'}(r_1, r_2) \sum_{m} C^{l,l',L}_{m,M-m,M} Y_{lm}(\Omega_1)Y_{l',M-m}(\Omega_2) , \quad (10.55)$$

where $r_{12}=|\boldsymbol{r}_1-\boldsymbol{r}_2|$. Expanding $F$ and $G$ as

$$F(r_1, r_2, r_{12}) = \sum_{l} F_l(r_1, r_2) \sum_{m}(-1)^m Y_{lm}(\Omega_1)Y_{l,-m}(\Omega_2) , \quad (10.56)$$

and similarly for $G$, and using

$$(-1)^m Y_{lm}(\Omega_1)Y_{L,M}(\Omega_1) = \sum_{l'm'} C^{ll'L}_{m,-m',-M} C^{ll'L}_{000}$$
$$\sqrt{\frac{(2l+1)(2l'+1)}{4\pi(2L+1)}}Y_{l'm'}(\Omega_2) , \quad (10.57)$$

we rewrite (10.55) as

$$\phi_{LM}(\boldsymbol{r}_1, \boldsymbol{r}_2) = \sum_{ll'} \left[ (F_{l'}(r_1, r_2) + G_l(r_1, r_2)) A_{ll'L} + g_{ll'}(r_1, r_2) \right]$$
$$\times \sum_{m} C^{l,l',L}_{m,M-m,M} Y_{lm}(\Omega_1)Y_{l',M-m}(\Omega_2) , \quad (10.58)$$

where $A_{ll'L} = C_{000}^{l,l',L} \sqrt{\frac{(2l+1)(2l'+1)}{4\pi(2L+1)}}$ is nonzero for all $l, l'$ such that $|l-l'| \leq L \leq$ $(l+l')$ with $l+l'+L$ even. Noting that only such $l, l'$ values appear in the general form given by (10.15) for states with parity $(-1)^L$, we restrict what follows to those states.

Comparing (10.58) with (10.15), we have

$$(F_{l'}(r_1, r_2) + G_l(r_1, r_2))A_{ll'L} + g_{ll'}(r_1, r_2) = f_{ll'}(r_1, r_2) \,. \tag{10.59}$$

For $L=1$, (10.59) may be satisfied with $g_{ll'}=0$, by choosing

$$G_{l+2} = G_l + \frac{f_{l+2,l+1}}{A_{l+2,l+1,1}} - \frac{f_{l,l+1}}{A_{l,l+1,1}} \,, \tag{10.60}$$

$$F_{l-1} = \frac{f_{l,l-1}}{A_{l,l-1,1}} - G_l \,. \tag{10.61}$$

Then (10.55) reduces to (10.16).

For $L=2$, (10.59) gives,

$$(G_l + F_{l+2})A_{l,l+2,2} + g_{l,l+2} = f_{l,l+2} \,, \tag{10.62}$$

$$(G_l + F_{l-2})A_{l,l-2,2} + g_{l,l-2} = f_{l,l-2} \,, \tag{10.63}$$

$$(G_l + F_l)A_{ll2} + g_{ll} = f_{ll} \,. \tag{10.64}$$

(10.62) and (10.63) can be satisfied with $g_{l,l+2}=g_{l,l-2}=0$ by choosing

$$G_{l+4} = G_l + \frac{f_{l+4,l+2}}{A_{l+4,l+2,2}} - \frac{f_{l,l+2}}{A_{l,l+2,2}} \,, \tag{10.65}$$

and

$$F_{l-2} = \frac{f_{l,l-2}}{A_{l,l-2,2}} - G_l \,, \tag{10.66}$$

which leaves $G_0, G_1, G_2, G_3$ arbitrary. Using (10.8), (10.64) may be written as

$$\left(G_l + \frac{f_{l+2,l}}{A_{l+2,l,2}} - G_{l+2}\right) A_{ll2} + g_{ll} = f_{ll} \,. \tag{10.67}$$

Now we may choose $G_0, G_1, G_2, G_3$ such that $g_{ll}=0$ for $l=1,2,3,4$ ($g_{00}$ is identically zero), but $g_{ll}$ for $l \geq 5$ cannot be made to vanish, in general. Thus we get (10.17) for the $L=2$ state with even parity.

## Appendix B: Electron–Hole Exchange Interaction

We derive the exciton and biexciton EMA equations including the electron–hole exchange interaction. We follow the treatment of [70] and start from an effective 2-band Hamiltonian describing electrons and holes in a semiconductor:[2]

$$
\begin{aligned}
H = & \sum_{k\sigma} E_c(\boldsymbol{k}) a^\dagger_{\boldsymbol{k}\sigma} a_{\boldsymbol{k}\sigma} - \sum_{k\sigma} E_v(\boldsymbol{k}) b^\dagger_{\boldsymbol{k}\sigma} b_{\boldsymbol{k}\sigma} \\
& + \frac{1}{2} \sum_{\boldsymbol{k}_1 \boldsymbol{k}_2 \boldsymbol{k}_3 \boldsymbol{k}_4} \sum_{\sigma_1 \sigma_2} V^{cccc}_{\sigma_1 \sigma_2 \sigma_2 \sigma_1} (\boldsymbol{k}_1 \boldsymbol{k}_2 \boldsymbol{k}_3 \boldsymbol{k}_4) a^\dagger_{\boldsymbol{k}_1 \sigma_1} a^\dagger_{\boldsymbol{k}_2 \sigma_2} a_{\boldsymbol{k}_3 \sigma_2} a_{\boldsymbol{k}_4 \sigma_1} \\
& + \frac{1}{2} \sum_{\boldsymbol{k}_1 \boldsymbol{k}_2 \boldsymbol{k}_3 \boldsymbol{k}_4} \sum_{\sigma_1 \sigma_2} V^{vvvv}_{\sigma_1 \sigma_2 \sigma_2 \sigma_1} (\boldsymbol{k}_1 \boldsymbol{k}_2 \boldsymbol{k}_3 \boldsymbol{k}_4) b^\dagger_{-\boldsymbol{k}_1 -\sigma_1} b^\dagger_{-\boldsymbol{k}_2 -\sigma_2} b_{-\boldsymbol{k}_3 -\sigma_2} b_{-\boldsymbol{k}_4 -\sigma_1} \\
& - \sum_{\boldsymbol{k}_1 \boldsymbol{k}_2 \boldsymbol{k}_3 \boldsymbol{k}_4} \sum_{\sigma_1 \sigma_2} V^{cvvc}_{\sigma_1 \sigma_2 \sigma_2 \sigma_1} (\boldsymbol{k}_1 \boldsymbol{k}_2 \boldsymbol{k}_3 \boldsymbol{k}_4) a^\dagger_{\boldsymbol{k}_1 \sigma_1} b^\dagger_{-\boldsymbol{k}_3 -\sigma_2} b_{-\boldsymbol{k}_2 -\sigma_2} a_{\boldsymbol{k}_4 \sigma_1} \\
& - \sum_{\boldsymbol{k}_1 \boldsymbol{k}_2 \boldsymbol{k}_3 \boldsymbol{k}_4} \sum_{\sigma_1 \sigma_2 \sigma_3 \sigma_4} \overline{V}^{cvcv}_{\sigma_1 \sigma_2 \sigma_3 \sigma_4} (\boldsymbol{k}_1 \boldsymbol{k}_2 \boldsymbol{k}_3 \boldsymbol{k}_4) a^\dagger_{\boldsymbol{k}_1 \sigma_1} b^\dagger_{-\boldsymbol{k}_4 -\sigma_4} b_{-\boldsymbol{k}_2 -\sigma_2} a_{\boldsymbol{k}_3 \sigma_3} ,
\end{aligned}
$$

$$(10.68)$$

where $a^\dagger_{\boldsymbol{k}\sigma}$ ($b^\dagger_{\boldsymbol{k}\sigma}$) is the creation operator for an electron (hole) with a wave function $\psi^{c(v)}_{\boldsymbol{k}\sigma}$ of the Bloch form

$$
\psi^i_{\boldsymbol{k}\sigma}(\boldsymbol{r}) = \frac{1}{\sqrt{\Omega}} u^i_{\boldsymbol{k}\sigma}(\boldsymbol{r}) \exp(i\boldsymbol{k} \cdot \boldsymbol{r}) , \tag{10.69}
$$

where $\Omega$ is the normalization volume. $E_c(\boldsymbol{k})$ and $E_v(\boldsymbol{k})$ denote the band dispersions of the conduction and valence bands, and the Coulomb matrix elements $V$ are given by

$$
\begin{aligned}
V^{ijkl}_{\sigma_1 \sigma_2 \sigma_3 \sigma_4} (\boldsymbol{k}_1 \boldsymbol{k}_2 \boldsymbol{k}_3 \boldsymbol{k}_4) = & \int \mathrm{d}^3 x \mathrm{d}^3 y \psi^{i*}_{\boldsymbol{k}_1 \sigma_1}(\boldsymbol{x}) \psi^{j*}_{\boldsymbol{k}_2 \sigma_2}(\boldsymbol{y}) v(|\boldsymbol{x} - \boldsymbol{y}|) \\
& \times \psi^k_{\boldsymbol{k}_3 \sigma_3}(\boldsymbol{y}) \psi^l_{\boldsymbol{k}_4 \sigma_4}(\boldsymbol{y}) ,
\end{aligned}
$$

$$(10.70)$$

with $v = e^2/(\epsilon|\boldsymbol{x} - \boldsymbol{y}|)$. The quantity $\overline{V}$ is given by the same expression as for $V$ but with $v$ replaced by $e^2/|\boldsymbol{x} - \boldsymbol{y}|$. The Hamiltonian (10.68) is obtained from the many-electron Hamiltonian by making a two-band approximation and keeping only those terms that conserve the number of the electron–hole pairs [70]. The effect of other excitations is phenomenologically included by screening the electron–hole Coulomb interaction. The electron–hole exchange interaction is, however, not screened [71].

We consider cubic materials with a conduction band of $\Gamma_6$ symmetry and a valence band of $\Gamma_7$ symmetry, each two-fold degenerate. We note that the

---

[2] This equation is the same as (2.7) of [70], except that we explicitly take into account the spin-orbit-coupled nature of the valence band.

index $\sigma=\pm1/2$ in the expressions above refers to the spin in the case of the electron, and to the $j_z=\pm1/2$ component of the $\Gamma_7$ band in the case of the hole. For brevity, in what follows we refer to this Bloch function angular momentum as "spin" in either case. The corresponding Bloch functions are of the form (at $\boldsymbol{k}=0$)

$$u^c_{0,1/2}(\boldsymbol{r}) = \zeta_0(\boldsymbol{r}) \uparrow\,, \tag{10.71}$$

$$u^c_{0,-1/2}(\boldsymbol{r}) = \zeta_0(\boldsymbol{r}) \downarrow\,, \tag{10.72}$$

$$u^v_{0,1/2}(\boldsymbol{r}) = \frac{-\mathrm{i}}{\sqrt{3}}(\zeta_x(\boldsymbol{r}) + \mathrm{i}\zeta_y(\boldsymbol{r})) \downarrow -\frac{\mathrm{i}}{\sqrt{3}}\zeta_z(\boldsymbol{r}) \uparrow\,, \tag{10.73}$$

$$u^v_{0,-1/2}(\boldsymbol{r}) = \frac{-\mathrm{i}}{\sqrt{3}}(\zeta_x(\boldsymbol{r}) - \mathrm{i}\zeta_y(\boldsymbol{r})) \uparrow +\frac{\mathrm{i}}{\sqrt{3}}\zeta_z(\boldsymbol{r}) \downarrow\,, \tag{10.74}$$

where $\zeta_0$ is an s-like cell-periodic function and $\zeta_x$, $\zeta_y$, $\zeta_z$ transform like $x$, $y$, and $z$. Arrows $\uparrow$ and $\downarrow$ denote the spin states.

A general electron–hole pair state may be constructed as

$$|p\rangle = \sum_{\boldsymbol{k}\boldsymbol{k}'} C^{II_z}_{\boldsymbol{k}\boldsymbol{k}'} |p^{II_z}_{\boldsymbol{k}\boldsymbol{k}'}\rangle\,, \tag{10.75}$$

where $|p^{II_z}_{\boldsymbol{k}\boldsymbol{k}'}\rangle$ is the electron–hole pair state with total "spin" $I$ and its $z$-component $I_z$

$$|p^{10}_{\boldsymbol{k}\boldsymbol{k}'}\rangle = \frac{1}{\sqrt{2}} \left(a^\dagger_{\boldsymbol{k},1/2}b^\dagger_{\boldsymbol{k}',-1/2} + a^\dagger_{\boldsymbol{k},-1/2}b^\dagger_{\boldsymbol{k}',1/2}\right)|0\rangle\,, \tag{10.76}$$

$$|p^{11}_{\boldsymbol{k}\boldsymbol{k}'}\rangle = a^\dagger_{\boldsymbol{k},1/2}b^\dagger_{\boldsymbol{k}',1/2}|0\rangle\,, \tag{10.77}$$

$$|p^{1,-1}_{\boldsymbol{k}\boldsymbol{k}'}\rangle = a^\dagger_{\boldsymbol{k},-1/2}b^\dagger_{\boldsymbol{k}',-1/2}|0\rangle\,, \tag{10.78}$$

and

$$|p^{00}_{\boldsymbol{k}\boldsymbol{k}'}\rangle = \frac{1}{\sqrt{2}} \left(a^\dagger_{\boldsymbol{k},1/2}b^\dagger_{\boldsymbol{k}',-1/2} - a^\dagger_{\boldsymbol{k},-1/2}b^\dagger_{\boldsymbol{k}',1/2}\right)|0\rangle\,. \tag{10.79}$$

Minimization of the expectation value of $H$ given by (10.68) leads to

$$(E_c(\boldsymbol{k}) - E_v(\boldsymbol{k}) - E)C^{II_z}_{\boldsymbol{k}\boldsymbol{k}'} - \sum_{\boldsymbol{l}\boldsymbol{l}'} \Big(V^{cvvc}_{1/2,-1/2,-1/2,1/2}(\boldsymbol{k}, -\boldsymbol{l}', -\boldsymbol{k}', \boldsymbol{l})$$
$$- \delta_{I,1}\overline{V}^{cvcv}_{1/2,-1/2,1/2,-1/2}(\boldsymbol{k}, -\boldsymbol{l}', \boldsymbol{l}, -\boldsymbol{k}')\Big) C^{II_z}_{\boldsymbol{l}\boldsymbol{l}'} = 0 \quad. \tag{10.80}$$

Now we make the effective mass approximation for the band dispersions and evaluate the Coulomb matrix elements in the Wannier approximation [70]:

$$V^{cvvc}_{1/2,-1/2,-1/2,1/2}(\boldsymbol{k}, -\boldsymbol{l}', -\boldsymbol{k}', \boldsymbol{l}) = \frac{1}{\Omega^2} \int \mathrm{d}^3x\,\mathrm{d}^3y\,v(|\boldsymbol{x} - \boldsymbol{y}|)$$
$$\times \exp[\mathrm{i}(\boldsymbol{l} - \boldsymbol{k})\cdot\boldsymbol{x} + \mathrm{i}(\boldsymbol{l}' - \boldsymbol{k}')\cdot\boldsymbol{y}]\,, \tag{10.81}$$

$$\overline{V}^{cvcv}_{1/2,-1/2,1/2,-1/2}(\boldsymbol{k},-\boldsymbol{l}',\boldsymbol{l},-\boldsymbol{k}') = \pi a_{\mathrm{ex}}^3 \Delta E_{\mathrm{exch}}^0 \frac{1}{\Omega^2}$$

$$\int \mathrm{d}^3x \exp[\mathrm{i}(\boldsymbol{l}-\boldsymbol{k}+\boldsymbol{l}'-\boldsymbol{k}')\cdot\boldsymbol{x}] \quad (10.82)$$

with

$$\pi a_{\mathrm{ex}}^3 \Delta E_{\mathrm{exch}}^0 = \frac{2}{3\Omega_{\mathrm{cell}}} \int \mathrm{d}^3x \mathrm{d}^3y \frac{e^2}{|\boldsymbol{x}-\boldsymbol{y}|} \zeta_0(\boldsymbol{x})\zeta_x(\boldsymbol{y})\zeta_0(\boldsymbol{y})\zeta_x(\boldsymbol{x}) , \quad (10.83)$$

where $\Omega_{\mathrm{cell}}$ is the volume of a unit cell,[3] and $\Delta E_{\mathrm{exch}}^0$ equals the bulk exciton exchange splitting within the present approximation. Invoking the above approximations on (10.80), and Fourier transforming to real space, we get (10.23) of Sect. 10.3.3. We note that $I$ is not the real spin but stands for the sum of the Bloch function angular momenta of the electron and the hole. Using the hydrogenic wave function of the bulk exciton, it is easy to verify that $\Delta E_{\mathrm{exch}}^0$ equals the bulk exciton exchange splitting energy.

Now we derive the biexciton EMA equation (10.30a,10.30b). For the total "spin" $J=0$, a general two-pair state may be written as

$$|m\rangle = \sum_{\boldsymbol{kk'll'}} \sum_{S=0,1} K^{SS}_{\boldsymbol{kk'll'}} |m^{SS}_{\boldsymbol{kk'll'}}\rangle , \quad (10.84)$$

with

$$|m^{00}_{\boldsymbol{kk'll'}}\rangle = \frac{1}{4} \sum_{\sigma_1,\sigma_2} (-1)^{(\sigma_1-\sigma_2)} a^\dagger_{\boldsymbol{k},\sigma_1} a^\dagger_{\boldsymbol{k'},-\sigma_1} b^\dagger_{\boldsymbol{l},\sigma_2} b^\dagger_{\boldsymbol{l'},-\sigma_2} |0\rangle , \quad (10.85)$$

and

$$|m^{11}_{\boldsymbol{kk'll'}}\rangle = \frac{1}{2\sqrt{3}} \left[ \sum_\sigma a^\dagger_{\boldsymbol{k},\sigma} a^\dagger_{\boldsymbol{k'},\sigma} b^\dagger_{\boldsymbol{l},-\sigma} b^\dagger_{\boldsymbol{l'},-\sigma} \right.$$
$$\left. -\frac{1}{2} \sum_{\sigma_1,\sigma_2} a^\dagger_{\boldsymbol{k},\sigma_1} a^\dagger_{\boldsymbol{k'},-\sigma_1} b^\dagger_{\boldsymbol{l},\sigma_2} b^\dagger_{\boldsymbol{l'},-\sigma_2} \right] |0\rangle . \quad (10.86)$$

The antisymmetry of the wave function under the electron–electron (hole–hole) exchange requires that $K^{00}$ be even and $K^{11}$ be odd under the interchange of $\boldsymbol{k}$, $\boldsymbol{k}'$ or $\boldsymbol{l}$, $\boldsymbol{l}'$. Proceeding as in the case of the exciton, we get (10.30a,10.30b).

To derive the EMA equations for $J=1$ and $J=2$ biexciton states, we note that a general two-pair state with total "spin" $J$ and its $z$-component $J_z$ may be written as

$$|m\rangle_{JJ_z} = \sum_{\boldsymbol{kk'll'}} \sum_{S,S'=0,1} K^{SS'}_{\boldsymbol{kk'll'}} |m^{SS'}_{\boldsymbol{kk'll'}}(JJ_z)\rangle \quad (10.87)$$

---

[3] We have neglected the long-range part of the exchange interaction, which is expected to be small (see [27]).

with

$$|m^{SS'}_{\bm{kk'll'}}(JJ_z)\rangle = \frac{1}{2} \sum_{ss'=\pm 1/2} \sum_{\sigma_1,\sigma'_1=\pm 1/2} \sum_{\sigma_2,\sigma'_2=\pm 1/2} C^{SS'J}_{ss'M} C^{\frac{1}{2}\frac{1}{2}S}_{\sigma_1\sigma'_1 s} C^{\frac{1}{2}\frac{1}{2}S'}_{\sigma_2\sigma'_2 s'}$$
$$a^{\dagger}_{\bm{k},\sigma_1} a^{\dagger}_{\bm{k'},\sigma'_1} b^{\dagger}_{\bm{l},\sigma_2} b^{\dagger}_{\bm{l'},\sigma'_2}|0\rangle \ , \quad (10.88)$$

where the values of $C$ denote the Clebsch–Gordan coefficients. The EMA equations (10.32a–c) and (10.34) for $J=1$ and $J=2$ states may now be derived as for the case of $J=0$.

# References

1. S. Schmitt-Rink, D.A.B. Miller, D.S. Chemla: Phys. Rev. B **35**, 8113 (1987)
2. T. Takagahara: Phys. Rev. B **39**, 10206 (1989)
3. E. Hanamura: Phys. Rev. B **37**, 1273 (1988)
4. S.V. Nair, K.C. Rustagi: Superlattices and Microstructures **6**, 337 (1989)
5. R.K. Jain, R.C. Lind: J. Opt. Soc. Am. **73**, 647 (1983)
6. P. Roussignol, D. Ricard, C. Flytzanis: Appl. Phys. A **44**, 285 (1987)
7. J. Yumoto, S. Fukushima, K. Kubodera: Opt. Lett. **12**, 832 (1987)
8. Y. Masumoto, T. Kawamura, K. Era: Appl. Phys. Lett. **62**, 225 (1993)
9. T. Kataoka, T. Tokizaki, A. Nakamura: Phys. Rev. B **48**, 2815 (1993)
10. V. Colvin, M. Schlamp, A.P. Alivisatos: Nature **370**, 354 (1994)
11. B.O. Dabbousi, M.G. Bawendi, O. Onitsuka, M.F. Rubner: Appl. Phys. Lett. **66**, 1316 (1995)
12. Al.L. Éfros, A.L. Éfros: Phys. Tekh. Poluprovodn. **16**, 1209 (1982) [Sov. Phys. Semicond. **16**, 772 (1982)]
13. A. Nakamura, H. Yamada, T. Tokizaki: Phys. Rev. B **40**, 8585 (1989)
14. T. Itoh, T. Ikehara, Y. Iwabuchi: J. Luminescence, **45**, 29 (1990); T. Itoh, M. Furumiya, T. Ikehara, C. Gourdon: Solid State Commun. **73**, 271 (1990)
15. E. Hanamura: Solid State Commun. **62**, 465 (1987)
16. K. Edamatsu, S. Iwai, T. Itoh, S. Yano, T. Goto: Phys. Rev. B **51**, 11 205 (1995)
17. P.E. Lippens, M. Lannoo: Phys. Rev. B **39**, 10935 (1989)
18. S.V. Nair, L.M. Ramaniah, K.C. Rustagi: Phys. Rev. B **45**, 5969 (1992)
19. L.M. Ramaniah, S.V. Nair: Phys. Rev. B **47**, 7132 (1993)
20. A.I. Ekimov, F.Hache, M.C. Schanne-Klein, D. Ricard, C. Flytzanis, I.A. Kudryavtsev, T.V. Yazeva, A.V. Rodina, Al.L. Efros: J. Opt. Soc. Am. B **10**, 100 (1993)
21. L-W. Wang, A. Zunger: J. Chem. Phys. **100**, 2394 (1994)
22. L-W. Wang, A. Zunger: Phys. Rev. B **53**, 9579 (1996)
23. Y. Kayanuma: Solid State Commun. **59**, 405 (1986)
24. S.V. Nair, S. Sinha, K.C. Rustagi: Phys. Rev. B **35**, 4098 (1987)
25. T. Takagahara: Phys. Rev. B **36**, 9293 (1987)
26. Y. Kayanuma: Phys. Rev. B **38**, 9797 (1988)
27. T. Takagahara: Phys. Rev. B **47**, 4569 (1993)
28. Y.Z. Hu, M. Lindberg, S.W. Koch: Phys. Rev. B **42**, 1713 (1990)

29. A study of the limiting behaviour of biexcitonic states in large QDs using a simplified model has been reported by L. Belleguie, L. Bányai: Phys. Rev. B **47**, 4498 (1993)

30. S.V. Nair, T. Takagahara: Phys. Rev. B **53**, R10 516 (1996)

31. S.V. Nair, T. Takagahara: Phys. Rev. B **55**, 5153 (1997)

32. G. Bastard: *Wave Mechanics Applied to Semiconductor Heterostructures* (Les Editions de Physique, Paris 1988)

33. L.E. Brus: J. Chem. Phys. **80**, 4403 (1984)

34. G.W. Bryant: Phys. Rev. B **37**, 8763 (1988)

35. U. Woggon: *Optical Properties of Semiconductor Quantum Dots*, Springer Tracts in Modern Physics Vol. 136 (Springer-Verlag, Berlin 1997)

36. A.I. Ekimov, A.A. Onushchenko, A.G. Plyukhin, Al.L. Éfros: Sh. Eskp. Teor. Fiz. **88**, 1490 (1985) [Sov. Phys. JETP 61, 891 (1985)]

37. L.D. Landua, E.M. Lifshitz: *Quantum Mechanics*, Course of Theoretical Physics Vol. 3, 3rd ed. (Butterworth-Heinemann, Oxford 1977) p. 434

38. E.A. Hylleraas: Z. Phys. **54**, 347 (1929); H.A. Bethe: *Quantum Mechanik der Ein- und Zwei-Elektronenprobleme*, Handbuch der Physik (Springer, Berlin 1933) p. 34

39. K. Edamatsu, Y. Mimura, K. Yamanaka, T. Itoh: *Proc. Japanese Physical Society, Part 2*, Osaka, Sept. 1995, p. 252

40. T. Uozumi, Y. Kayanuma, K. Yamanaka, K. Edamatsu, T. Itoh: Phys. Rev. B **59**, 9826 (1999)

41. T. Takagahara: Phys. Rev. B **39**, 10206 (1989)

42. L. Banyai: Phys. Rev. B **39**, 8022 (1989)

43. K. Cho: In: *Excitons*, ed. by K. Cho (Springer, Berlin 1979), and references therein

44. G.F. Koster, J.O. Dimmock, R.G. Wheeler, H. Satz: *Properties of Thirty-Two Point Groups* (MIT Press, Cambridge 1963)

45. T. Itoh, K. Edamatsu: *Proc. of the Japanese Physical Society, Part 2*, Kanazawa, March 1996, p. 280

46. M. Ueta, H. Kanzaki, K. Kobayashi, Y. Toyozawa, E. Hanamura: *Excitonic Processes in Solids* (Springer, Berlin 1986)

47. O. Akimoto, E. Hanamura: Solid State Commun. **10**, 253 (1972)

48. W.F. Brinkman, T.M. Rice, B. Bell: Phys. Rev. B **8**, 1570 (1973)

49. F. Bassani, J.J. Forney, A. Quattropani: Phys. Status Solidi B **65**, 591 (1974)

50. N. Nagasawa: private communication

51. Y. Masumoto, S. Okamoto, S. Katayanagi: Phys. Rev. B **50**, 18658 (1994)

52. *Physics of II–VI and I–VII compounds, Semimagnetic Semiconductors, Landolt-Börnstein, Vol. 17b*, ed. by O. Madelung, M. Schulz, H. Weiss (Springer, Berlin 1982). We obtain the exciton Rydberg by adding 4.4 meV of exchange energy to the binding energy of the $\Gamma_5$ exciton quoted here

53. T. Itoh, Y. Iwabuchi, M. Kataoka: Phys. Status Solidi B **145**, 567 (1988)

54. R. Shimano, M. Kuwata-Gonokami: Phys. Rev. Lett. **72**, 530 (1994)

55. H. Akiyama, T. Kuga, M. Matsuoka, M. Kuwata-Gonokami: Phys. Rev. B **42**, 5261 (1990)

56. Y. Masumoto, S. Katayanagi, T. Mishina: Phys. Rev. B **49**, 10782 (1994)

57. M. Ikezawa, Y. Masumoto: Phys. Rev. B **53**, 13694 (1996)

58. C. Flytzanis: In: *Quantum Electronics: A Treatise, Vol. I*, ed. by H. Rabin, C.L. Tang, (Academic, New York 1975). Also see equation (6.1) of [2]

59. Y. Masumoto, T. Kawazoe, N. Matsuura: J. Lumin. **76**, 189 (1998)
60. M. Ikezawa, Y. Masumoto: Phys. Rev. B **61**, 12662 (2000)
61. D.W. Hall, N.F. Borrelli: J. Opt. Soc. Am. B **5**, 1650 (1988); S.H. Park, R.A. Morgan, Y.Z. Hu, M. Lindberg, S.W. Koch, N. Peyghambarian: J. Opt. Soc. Am. B **7**, 2097 (1990); P. Roussignol, D. Ricard, C. Flytzanis: Appl. Phys. B **51**, 437 (1990); H. Shinojima, J. Yumoto, N. Uesugi: Appl. Phys. Lett. **60**, 298 (1992); M.C. Schanne-Klein, F. Hache, D. Ricard, C. Flytzanis: J. Opt. Soc. Am. B **9**, 2234 (1992)
62. M.C. Schanne-Klein, L. Piveteau, M. Ghanassi, D. Ricard: Appl. Phys. Lett. **67**, 579 (1995)
63. Y. Li, M. Takata, A. Nakamura: Phys. Rev. B **57**, 9193 (1998)
64. M.G. Bawendi, W.L. Wilson, L. Rothberg, P.J. Carroll, T.M. Jedju, M.L. Steigerwald, L.E. Brus: Phys. Rev. Lett. **65**, 1623 (1990)
65. D.J. Norris, A. Sacra, C.B. Murray, M.G. Bawendi: Phys. Rev. Lett. **72**, 216 (1994)
66. M. Ikezawa, Y. Masumoto, T. Takagahara, S.V. Nair: Phys. Rev. Lett. **79**, 3522 (1997)
67. N.H. Bonadeo, G. Chen, D. Gammon, D.S. Katzer, D. Park, D.G. Steel: Phys. Rev. Lett. **81**, 2759 (1998)
68. G. Chen, N.H. Bonadeo, D.G. Steel, D. Gammon, D.S. Katzer, D. Park, L.J. Sham: Science **289**, 1906 (2000)
69. F. Kreller, M. Lowisch, J. Puls, F. Henneberger: Phys. Rev. Lett. **75**, 2420 (1995)
70. E. Hanamura, H. Haug: Phys. Reports C **33**, 209 (1977)
71. L.J. Sham, T.M. Rice: Phys. Rev. **144**, 708, (1966)

# 11 Coulomb Effects in the Optical Spectra of Highly Excited Semiconductor Quantum Dots

Selvakumar V. Nair

## 11.1 Introduction

The physics of quantum dots is often likened to that of atoms and molecules because of the discrete nature of their energy levels. Although this analogy has some merits, one difference worth emphasizing is the possibility of creating multiple electron–hole pair excitations in a quantum dot (QD). Such multiparticle complexes quickly relax into a quasi-equilibrium state in which they exhibit features of a spatially confined many-particle system such as an atom, with an added advantage that the eventual radiative decay of this state offers a convenient way to probe the intricacies of multiparticle correlations. Recent advances in fabrication and spectroscopy of semiconductor QDs have indeed made it possible to make well-controlled studies of such a quantum-confined electron–hole gas.

With the advent of self-assembled QDs of III–V semiconductors and the realization of single quantum dot spectroscopy, studies of the optical properties of QDs have made great strides [1]. In recent years single quantum dot spectroscopy has revealed extremely sharp spectral lines attributed to the annihilation of single electron–hole pairs in a quantum dot [2]. Many of the single QD luminescence experiments also show multiexcitonic features in the optical spectra [3–5]. This is only to be expected as multiple electron–hole pairs can be generated inside a single QD by optical excitation at moderate laser powers.

In highly excited QDs, the interparticle Coulomb interaction is expected to modify not only the energies but the broadening of the spectral lines as well. In order to understand the physical properties of such a quantum-confined electron–hole plasma, we present a calculation of the energy levels and optical spectra of model QD systems at varying levels of sophistication.

We first consider an idealized system of a spherical QD embedded in a semiconductor matrix with square-well confining potential. The material parameters are chosen to mimic the salient features of the energy level structure of InAs/GaAs self-assembled QDs. The electrons and holes in the QD interact via the Coulomb potential $V(r)=e^2/(\epsilon_0 r)$, where $\epsilon_0$ is the background dielectric constant of the material. In QDs of experimental interest, even a single electron–hole pair corresponds to an effective carrier density of $\sim 10^{17}$ cm$^{-3}$. Thus, at high excitation powers, we have a fairly dense electron–hole plasma.

We use a density functional approach based on the local density approximation (LDA) to calculate the renormalization of the single-particle energies and the oscillator strengths of multiexciton emission lines.

We then take a more realistic model of a cylindrical InP quantum dot in a GaInP matrix, including strain effects due to the lattice mismatch, and again study the multiparticle spectra using an LDA approach. Going beyond LDA, we show that elastic Coulomb scattering can lead to dephasing of certain final state configurations, causing a significant broadening of the emission lines in highly excited quantum dots.

Finally, we present a configuration interaction calculation for cylindrical QDs for a few electron–hole pairs and elucidate the spin structures of triexciton states. We also predict weak configuration-crossing transitions from the biexciton and triexciton states.

## 11.2   Local Density Approximation for Electrons and Holes

The density functional theory (DFT) in the local density approximation (LDA) provides a computationally efficient and rather accurate scheme for the study of the electron gas. In the spirit of the effective mass approximation, the inhomogeneous electron gas in many semiconductors may be treated as composed of particles with an isotropic and parabolic dispersion. In this case, the conventional DFT applies with the electron mass replaced by the effective mass, and the Coulomb interaction screened by a macroscopic dielectric constant $\epsilon_0$. The corresponding theory for holes is formally equivalent, but complicated somewhat by the degenerate valence band structure of most semiconductors [6]. Commonly, there are several hole bands of the same (or almost the same) energy – heavy-hole, light-hole, and spin-orbit split-off bands of tetrahedral semiconductors, for example – which may mix under external perturbations or quantum confinement. Although this problem was addressed in the early days of DFT, a satisfactory theory for multicomponent hole gas was formulated only recently [7].

Here we are interested in identifying the characteristic features of many electron–hole spectra of QDs, and for reasons of clarity, we limit ourselves to a simple model of single bands for electrons and holes. This is not totally unrealistic as, in most QDs of interest, the top-most valence band is predominantly heavy-hole-like with the light-hole band split-off by strain and confinement.

The density functional theory in the local density approximation leads to single-particle Kohn–Sham equations for the electron (hole) orbitals $\psi_i^e(\mathbf{r})$ $(\psi_i^h(\mathbf{r}))$ and energies $E_i^e(E_i^h)$ given by

$$\left( -\frac{\hbar^2}{2m_e}\nabla^2 + V^e(\mathbf{r}) + V_H(\mathbf{r}) + V_{XC}^e(\mathbf{r})\psi_i^e(\mathbf{r}) \right) = E_i^e\psi_e(\mathbf{r}) \,, \qquad (11.1)$$

$$\left(-\frac{\hbar^2}{2m_{\mathrm{h}}}\nabla^2 + V^{\mathrm{h}}(\boldsymbol{r}) - V_{\mathrm{H}}(\boldsymbol{r}) + V_{\mathrm{XC}}^{\mathrm{h}}(\boldsymbol{r})\psi_i^{\mathrm{h}}(\boldsymbol{r})\right) = E_i^{\mathrm{h}}\psi_i^{\mathrm{h}}(\boldsymbol{r}) ,\qquad(11.2)$$

where $m_{\mathrm{e}}$ ($m_{\mathrm{h}}$) is the effective mass and $V^{\mathrm{e}}$ ($V^{\mathrm{h}}$) is the confining potential of the electron (hole). The Hartree potential $V_{\mathrm{H}}$ is determined by the electron and hole densities $\rho^{\mathrm{e}}$ and $\rho^{\mathrm{h}}$, respectively

$$V_{\mathrm{H}} = e^2 \int \frac{\rho^{\mathrm{e}}(\boldsymbol{r}) - \rho^{\mathrm{h}}(\boldsymbol{r})}{\epsilon_0|\boldsymbol{r} - \boldsymbol{r}'|}\mathrm{d}^3\boldsymbol{r}' .\qquad(11.3)$$

The essential input to such a calculation is the exchange correlation potentials for the electron and the hole. For the electron it may be written as

$$V_{\mathrm{XC}}^{\mathrm{e}} = \frac{\delta}{\delta\rho^{\mathrm{e}}(\boldsymbol{r})} \int \epsilon_{\mathrm{XC}}[\rho^{\mathrm{e}}, \rho^{\mathrm{h}}]\rho(\boldsymbol{r})\mathrm{d}^3 r ,\qquad(11.4)$$

where $\rho$ is the sum of the electron and hole densities, $\rho^{\mathrm{e}}$ and $\rho^{\mathrm{h}}$, respectively, and $\epsilon_{\mathrm{XC}}$ is the exchange correlation energy per particle. A similar equation holds for the hole.

To implement this scheme, we need a model for the exchange correlation energy of an electron–hole gas. Application of DFT to the electron–hole plasma has not received much attention and few parameterizations of the electron–hole exchange-correlation energy $\epsilon_{\mathrm{XC}}$ are present in the literature. Many of these are ad hoc in nature and of uncertain validity.

In a pioneering work, Vashishta and Kaliya [8] computed $\epsilon_{\mathrm{XC}}$ for a locally neutral electron–hole gas (see [6] for a review). However, in an inhomogeneous system like a QD, even for the case of equal number of electrons and holes (globally neutral), the local density will not be neutral and we need an expression for $\epsilon_{\mathrm{XC}}$ for a non-neutral case. Our approach here is to use an interpolation between the Vashishta–Kaliya potential ($\epsilon_{\mathrm{XC}}^{\mathrm{VK}}$) for the neutral case, and the well-known Ceperley–Alder potential ($\epsilon_{\mathrm{XC}}^{\mathrm{CA}}$) [9] for one component (electron or hole only) gas, as [10]

$$\epsilon_{\mathrm{XC}} = (1 - \xi)\epsilon_{\mathrm{XC}}^{\mathrm{VK}}(2^{1/3}r_{\mathrm{s}})/2 + \xi\left[\frac{m_{\mathrm{e}}}{\mu}\epsilon_{\mathrm{XC}}^{\mathrm{CA}}\left(\frac{m_{\mathrm{e}}}{\mu}r_{\mathrm{s}}\right)\theta(\xi)+\right.$$
$$\left.\frac{m_{\mathrm{h}}}{\mu}\epsilon_{\mathrm{XC}}^{\mathrm{CA}}\left(\frac{m_{\mathrm{h}}}{\mu}r_{\mathrm{s}}\right)\theta(-\xi)\right] ,\qquad(11.5)$$

in units of the exciton Rydberg. Here $\xi=|\rho^{\mathrm{e}}-\rho^{\mathrm{h}}|/\rho$ and $r_{\mathrm{s}}=(4\pi/3\rho)^{(1/3)}$.

## 11.3 Application of the Local Density Approximation to Quantum Dots

Typical self-assembled quantum dots of experimental interest have shapes of low symmetry, making a self-consistent many-body calculation of multiple electron–hole states computationally intensive. We approach this problem in terms of idealized models of increasing sophistication, viz., a spherical QD without strain effects and a cylindrical structure that includes strain.

### 11.3.1 Spherical Approximation

We consider a spherical semiconductor QD embedded in a semiconductor matrix with square-well confining potential. The parameters given in Table 11.1 are chosen to give an energy level spacing of about 60 meV (10 meV) for the lowest states of electrons (holes), in approximate agreement with those observed in typical self-assembled QDs [11–14], and calculated by more sophisticated models [15]. For spherically symmetric systems, the Kohn–Sham equations reduce to radial wave equations, which may be solved by numerical integration. LDA self-consistency is achieved by a standard iteration procedure [16].

The calculated single-particle energy level structure for up to 20 electron–hole pairs is shown in Fig. 11.1. The ground state configuration is obtained by filling the energy levels closest to the band edges subject to the constraint

**Table 11.1.** Parameters used in the LDA calculation of a spherical QD. $R$ is the radius of the QD, $m_e (m_h)$ is the effective mass, $V_e$ ($V_h$) is the confining potential for the electron (hole), and $\epsilon_0$ is the background dielectric constant

| $R(nm)$ | $m_e$ | $m_h$ | $V_e$ (eV) | $V_h$ (eV) | $\epsilon_0$ |
|---------|-------|-------|------------|------------|--------------|
| 8 | 0.06 | 0.4 | 215 | 45 | 12 |

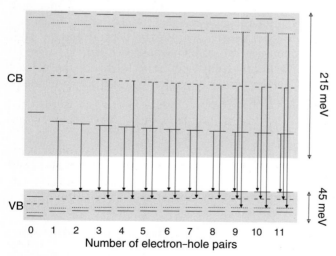

**Fig. 11.1.** Effective single-particle energy levels of many electron–hole pairs in a spherical QD calculated by a density functional approach. The material parameters used are given in Table 11.1. Only a few lowest-energy states with angular momentum $l=0$ (*solid lines*), $l=1$ (*dashed lines*), and $l=2$ (*dotted lines*) are shown. The *shaded regions* indicate the energy ranges of the confining potential for the electron and the hole. The *vertical arrows* show the possible optical emission processes

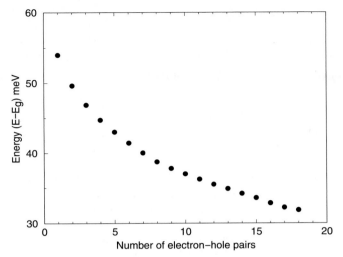

**Fig. 11.2.** The energy difference between the lowest electron–hole states as a function of the number of electron–hole pairs ($n_{\mathrm{eh}}$) in a quantum dot. The energy is measured relative to the band gap of the QD material

that each level of angular momentum $l$ can accommodate a maximum of $2(2l+1)$ particles. As the number of electron–hole pairs increases, the band gap gradually decreases (Fig. 11.2). However, this quantum-confined band gap renormalization is much weaker than that of the bulk semiconductor at the same carrier density. This is easily understood as due to the confinement induced phase-space reduction for interparticle scattering pathways. Nevertheless, the reduction in the band gap is significant, and is easily observable. Yuan et al. have recently reported observation of a dynamic bandgap renormalization by femtosecond time-resolved photoluminescence measurements [17]. The observed maximum redshift of 30 meV corresponds to a population of about 20 electron–hole pairs, which is reasonable.

The possible optical emission processes from each $n_{\mathrm{eh}}$-pair ground state configuration are also indicated in Fig. 11.1. Such single electron–hole pair annihilation processes involve transitions from the ground state of $n_{\mathrm{eh}}$-pairs to the ground or excited configurations of $n_{\mathrm{eh}}-1$ pairs. The transition dipole moment of an electron–hole recombination is given by

$$\mu_{ij} = \mu_{\mathrm{cv}} \int \psi_i^{\mathrm{e}}(\boldsymbol{r})\psi_j^{\mathrm{h}}(\boldsymbol{r})\mathrm{d}^3 r \; , \tag{11.6}$$

where $\mu_{\mathrm{cv}}$ is the interband transition dipole moment of the bulk material.

The emission spectra from the ground state configuration of $n_{\mathrm{eh}}$-pairs for several values of $n_{\mathrm{eh}}$ are shown in Fig. 11.3. As is apparent from Figs. 11.1 and 11.3, the one- and two-pair states both show a single emission energy, but the latter is shifted to the low-energy side and corresponds to the biexciton.

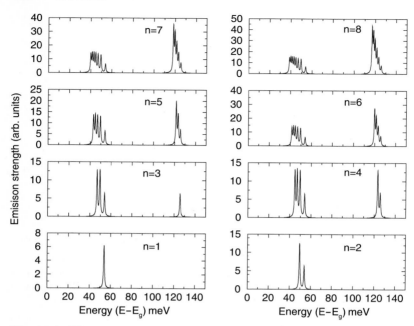

**Fig. 11.3.** The emission spectrum from the ground state configuration of $n$ pairs for $n=1$ to 8. The emission is assumed to proceed in a cascade of $n$ to various $n–1$ configurations. Any excited (intraband) configuration generated in this process is assumed to relax almost instantaneously to its ground state

Although our simplifying assumptions forbid comparison with experiments, it is interesting to note that the calculated biexciton binding energy (4.5 meV) is close to the values reported in the literature [3,5] for InAs QDs. The three-pair state involves two electrons (holes) in the lowest orbital and one in the next orbital. Consequently, an additional emission line appears on the high-energy side. More interestingly, the lowest energy transition (from the three-pair ground state to an excited configuration of two pairs) shifts further to the low-energy side.

Each emission spectrum shown in Fig 11.3 corresponds to a definite number of electron–hole pairs. Experimentally, this can be realized by pulse excitation of carriers, either in the barrier material or directly in the QD. However, most experiments are performed by cw pumping that invariably creates a distribution of carrier numbers. In this case, the emission spectrum should be calculated by averaging the spectrum over this distribution. Considering optical pumping of the barrier and subsequent capture of the electron–hole pairs by the QD, one can perform a detailed rate equation modelling for the carrier distribution. However, to a good approximation, the carrier number distribution may be taken to be the Poisson distribution. The corresponding calculated emission spectra for various pump levels (or average number of

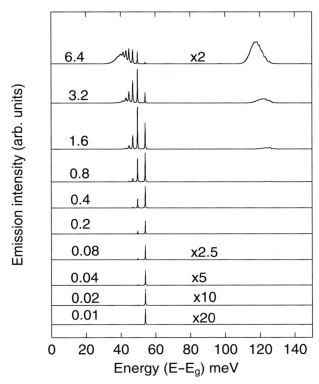

**Fig. 11.4.** Emission spectra of multiexcitons in a model spherical quantum dot described in the text. Each spectrum corresponds to a definite optical pump intensity and is labelled by the average number of electron–hole pairs $\langle n_{\mathrm{eh}} \rangle$ excited

electron–hole pairs $\langle n_{\mathrm{eh}} \rangle$) are shown in Fig. 11.4. In addition to a homogeneous broadening of 200 meV for all levels, the spectra also include density-dependent broadening due to Coulomb scattering, discussed in Sect. 11.4.

The spectra show a single emission line due to a single electron–hole pair only when the pumping is very weak. The two-pair emission (biexciton) is visible even when $\langle n_{\mathrm{eh}} \rangle$ is as small as 0.08. For $\langle n_{eh} \rangle$ close to 1, the spectrum shows comparable intensities for the one-pair and two-pair emission lines and a three-pair emission line is also visible. This is a consequence of the saturation of the one-pair state and the Poisson distribution of the carrier number and underlines the need for using extremely low optical pumping levels to study the single exciton emission line. With increasing pump intensity, the spectrum develops further peaks on the low-energy side, gradually broadening into a band a few tens of meV in width. These features are in excellent agreement with the experimentally observed multiexciton emission spectra in several self-assembled QDs [4,18,19].

## 11.3.2    Cylindrical Quantum Dots

While the spectra calculated for spherical QDs appear to capture most of the experimentally observed features of multiexciton spectra, it is instructive to consider a more realistic model of a self-assembled QD – one that includes strain effects and band offsets appropriate for experimental samples. Encouraged by the fact that several samples show nearly cylindrically symmetric QD shapes such as a lens or a truncated pyramid, we consider a cylindrically symmetric geometry.

All the results presented here are for an InP QD embedded in a $Ga_{0.5}In_{0.5}P$ matrix. The QD shape is taken to be a truncated pyramid with height 7 nm, and base and top faces 30 nm and 16 nm in diameter, respectively. For the strain calculation we consider a large cylindrical region of diameter 120 nm and height 100 nm, while the energy level calculation is done in a smaller region of diameter 60 nm and height 21 nm. We first obtain the strain distribution by minimizing the elastic energy using a finite element approach, using bulk elastic constants and bond lengths given in Table 11.2.

With the external potential (confining potential) determined by the strain distribution and the band offsets, we calculate the single-particle energies and eigenfunctions of the electron and the hole. To this end we discretize the Schrödinger equation within a cylindrical domain that encloses the QD and use an efficient sparse-matrix algorithm based on the ARPACK package [20] to obtain a few lowest-energy eigenvalues of electron and hole.

For the LDA calculation, we calculate the electron and hole densities from the single-particle wave functions and redo the single-particle calculation

**Table 11.2.** The material parameters used for InP and $Ga_{0.5}In_{0.5}P$. The conduction band offset between InP and $Ga_{0.5}In_{0.5}P$ is taken to be 0.591 eV

| Property | GaInP | InP |
|---|---|---|
| electron mass ($m_e$) | 0.077 | 0.077 |
| hole masses: | | |
| $m_{hh}^{\parallel}$ | 0.61 | 0.61 |
| $m_{hh}^{\perp}$ | 0.15 | 0.15 |
| deformation potentials (eV): | | |
| $a_c$ | −7.5 | −7.0 |
| $a_v$ | 0.4 | 0.4 |
| $b$ | −1.9 | −2.0 |
| $d$ | −4.75 | −5.0 |
| bandgap $E_g$ (eV) | 1.97 | 1.424 |
| elastic constants: | | |
| $C_{11}$ | 12.17 | 10.22 |
| $C_{12}$ | 6.01 | 5.76 |
| $C_{44}$ | 5.82 | 4.60 |
| bond length (Å) | 2.45 | 2.541 |

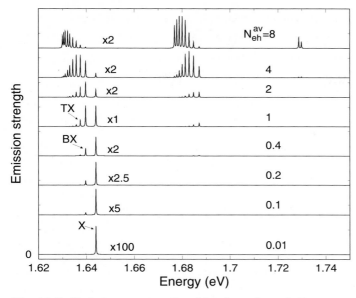

**Fig. 11.5.** Emission spectra of multiexcitons in an InP quantum dot as calculated by a density functional approach. Each spectrum corresponds to a definite optical pump intensity and is labeled by the average number of electron hole pairs $\langle n_{eh} \rangle$ excited

by adding the exchange-correlation potential to the external potential. This procedure is repeated to achieve self-consistency.

The emission spectra calculated for various pumping powers, assuming a steady state generation of carriers (cw pumping) are shown in Fig. 11.5. The average number of electron–hole pairs in the QD is indicated against each spectrum. The spectral features are very similar to those of spherical QDs discussed in Sect. 11.3.1. The spectrum experimentally measured by Sugisaki et al. [19] for a sample with structure very close to that considered here is shown in Fig. 4.15 [21]. The agreement between theory and experiment is very good, except for some spectral lines that develop at the higher-energy side of the exciton. These lines are tentatively assigned to charged excitons and call for a more detailed calculation.

The biexciton binding energy is found to be 4.3 meV, similar to the spherical case, while the triexciton state is predicted to emit 6.5 meV below the exciton state. As this calculation uses spin-averaged electron and hole densities, this triexciton emission line should be understood as an average of the spin fine-structure predicted by the CI calculation discussed in Sect. 11.5.

## 11.4   Beyond the Local Density Approximation: Spectral Broadening and Relaxation by Coulomb Scattering

In addition to energy shifts, the Coulomb interaction also leads to a broadening of the spectral lines [22]. To understand why the Coulomb scattering can lead to line broadening, consider the three-pair ground state, which emits light by recombination of the electron and hole in the lowest-energy orbitals. This leaves the system in an excited two-pair state. As the potential well for the hole is rather shallow, this three-pair state can elastically scatter to a continuum state, as illustrated in Fig. 11.6. Thus the final state of the optical emission process considered is actually a resonance in the continuum and will be broadened by elastic Coulomb scattering.

Theoretically, the broadening of the single-particle energy levels may be calculated as the imaginary part of the self-energy ($\Sigma$) of the electron and the hole. Within LDA, the single-particle self-energy is a real quantity, so we need to go beyond LDA. Our approach is to use the GW approximation [23], which gives the self-energy of a particle of spin $\sigma$ as

$$\Sigma_\sigma(\boldsymbol{r}, \boldsymbol{r}', E) = \frac{i}{2\pi} \int G_\sigma(\boldsymbol{r}, \boldsymbol{r}', E + E') W(\boldsymbol{r}, \boldsymbol{r}', E') e^{iE'\delta} \mathrm{d}E' , \qquad (11.7)$$

where $G$ is the single-particle Green function, and $W$ is the screened Coulomb interaction. Using the LDA orbitals, we may approximate the Green function of the electron as $G_\sigma(\boldsymbol{r}, \boldsymbol{r}', E) = \sum_i \psi_{i\sigma}^{\mathrm{e}}(\boldsymbol{r}) \psi_{i\sigma}^{\mathrm{e}*}(\boldsymbol{r}') / (E - E_i)$. Here $\psi_{i\sigma}^{\mathrm{e}}$ denotes the $i$th electron orbital with energy $E_i^{\mathrm{e}}$ and spin $\sigma$. A similar expression holds for the hole.

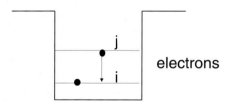

electrons

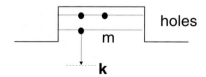

holes

**Fig. 11.6.** A schematic illustration of elastic electron–hole scattering processes that lead to a carrier-density-dependent broadening of the spectral lines

Then it is easily shown that, for occupied orbitals, the imaginary part of the self-energy arises from the screened exchange contribution

$$\mathrm{Im}\Sigma_\sigma(E = E_i^\mathrm{e}) = -\sum_j^\mathrm{occ} \mathrm{Im}W(\boldsymbol{r}, \boldsymbol{r}', E_j - E_i)\psi_{j\sigma}^\mathrm{e}(\boldsymbol{r})\psi_{j\sigma}^{\mathrm{e}*}(\boldsymbol{r}') \qquad (11.8)$$

Describing screening using the RPA response function $\chi$ and keeping terms only to the lowest order (i.e., $W=v+v\chi v$), where $v=e^2/\epsilon_0|\boldsymbol{r}-\boldsymbol{r}'|$, the imaginary part of $\Sigma$ is determined by the poles in $\chi(\omega)$ for $\omega=E_j-E_i$ in the continuum. That is, the scattering from $j$ to $i$ should be able to cause an excitation of an electron or a hole into the continuum. As the confining potential for the hole is rather shallow, scattering processes that involve the hole being pushed into the barrier layer (continuum) become feasible. In our model, if the second electron state is occupied, the lowest interband transition gets broadened. The expectation value of $2\mathrm{Im}\Sigma(E=E_i)$ gives the broadening($\Gamma_i$) of the level $i$

$$\langle i|\mathrm{Im}\Sigma_\sigma(E = E_i)|i\rangle = \pi \sum_{jm}^\mathrm{occ} \sum_{k\sigma}^\mathrm{unocc} \left|M_{ji;mk}^{\sigma\sigma'}\right|^2 \delta(E_{ji}^\mathrm{e} - E_{km}^\mathrm{h}) , \qquad (11.9)$$

where

$$M_{ji;mk}^{\sigma\sigma'} = \int \psi_{i\sigma}^\mathrm{e}(\boldsymbol{r})\psi_{j\sigma}^{\mathrm{e}*}(\boldsymbol{r}) \frac{e^2}{\epsilon_0|\boldsymbol{r} - \boldsymbol{r}'|} \psi_{m\sigma'}^{\mathrm{h}*}(\boldsymbol{r}')\psi_{k\sigma'}^\mathrm{h}(\boldsymbol{r}')\mathrm{d}^3r\mathrm{d}^3r' , \qquad (11.10)$$

and $E_{ji}^\mathrm{e}=E_j^\mathrm{e}-E_i^\mathrm{e}$.

We note that the above expression may also be derived very easily as the Fermi golden rule for the relevant scattering process involved. However, this simplification applies only when the screening is small, while the approach presented here is more general and can include higher-order corrections.

The calculated linewidth ($\Gamma_0$) of the lowest electronic orbital is shown in Fig. 11.7. It is well-approximated by $\Gamma_0=0.090(n_\mathrm{eh}-2)+0.041(n_\mathrm{eh}-2)^2$ meV, where $n_\mathrm{eh}$ is the number of electron–hole pairs in the QD. Here the two terms represent contributions from the scattering of the two holes in the lowest zero angular momentum ($l=0$) state and from the remaining $n_\mathrm{eh}-2$ holes, respectively.

The resulting optical emission spectra shown in Fig. 11.4 include the density-dependent line broadening. With increasing pump intensity, the second electron state becomes occupied, leading to the broadening of the lowest-energy transition due to the scattering process explained above. A similar broadening should also be applied to the spectra shown in Fig. 11.5 for cylindrical QDs. However, we have not performed this calculation since our numerical scheme for the cylindrical geometry does not allow us to evaluate eigenfunctions in the continuum in a reasonable way.

The Coulomb scattering process described above also provides an efficient channel for the relaxation of an electron captured into an excited state in

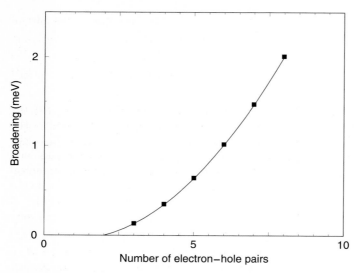

**Fig. 11.7.** The *squares* show the calculated width of the lowest-energy emission line. The *solid line* is a fit given by $0.090(n{-}2)+0.041(n{-}2)^2$ meV

the presence of a hole. The electron can relax in picosecond timescales by elastically scattering the hole out of the QD. As the closely spaced hole-energy levels can provide efficient recapture and energy relaxation of the hole by acoustic phonon emission, this process can circumvent the widely discussed phonon bottleneck in the electron relaxation by LO phonon emission.

## 11.5 Spin Fine Structure of a Few Exciton Spectra: the Configuration Interaction Approach

The LDA calculation described in Sect. 11.2 uses a spin-averaged density, and therefore misses the spin fine-structure of the multiexciton states. Also, for a few excitons, incomplete self-interaction correction in LDA could be a source of error. For small number of electron–hole pairs, however, it is easy to do a configuration interaction (CI) calculation that does not suffer from either of these weaknesses. Here we use the CI approach and study the details of the spin structure and Coulomb correlation that would appear in the optical emission spectrum for up to three electron–hole pairs [24].

All the results presented are for an InP QD embedded in a GaInP matrix with structure described in Section 11.3.2.

The hole states are assumed to be derived from the heavy-hole band. Then the electron and hole single-particle states may be written as

$$\Psi_{nm}^{\pm 1/2}(\boldsymbol{r}) = \psi_{nm}^{e}(r,z)\exp(-im\phi)\chi_{c}^{\pm 1/2} \tag{11.11}$$

$$\Psi_{nm}^{\pm 3/2}(\boldsymbol{r}) = \psi_{nm}^{h}(r,z)\exp(-im\phi)\chi_{v}^{\pm 3/2} \tag{11.12}$$

where the envelope functions $\psi_{nm}(r,z)$ are labeled by the radial $(n)$ and angular $(m)$ quantum numbers, $\chi_c^{1/2}=S\!\uparrow$, $\chi_c^{-1/2}=S\!\downarrow$ are the conduction band Bloch functions with axial angular momenta (spin) $\pm 1/2$, and $\chi_v^{3/2}=-\frac{(X+iY)}{\sqrt{2}}\uparrow$, $\chi_v^{-3/2}=\frac{(X-iY)}{\sqrt{2}}\downarrow$ are the valence band (heavy-hole) Bloch functions of axial angular momenta (spin-orbit coupled) $\pm 3/2$.

Owing to the assumed cylindrical symmetry, each multiparticle state may be characterized by the total axial angular momentum $(M)$ of the envelope wave function and the total Bloch function angular momentum of the electrons $(J_e)$ and the holes $(J_h)$. For interband optical transitions we restrict our considerations to light polarized in the $x$–$y$ plane so that the following selection rules hold

$$\Delta J = \pm 1 \quad \text{and} \quad \Delta M = 0 \, , \tag{11.13}$$

where $J = J_e + J_h$.

Multiparticle basis states for $N_e$ electrons and $N_h$ holes are constructed as a product of an $N_e$ electron slater-determinant and $N_h$ hole slater-determinant, within a restricted subspace spanned by 12 lowest single-particle states of electrons and holes. The calculated exciton (X) absorption spectrum is shown in Fig. 11.8. The dominant single-particle configurations for some representative excitonic states are indicated in the figure. Here $s$ and $p$ denote the lowest $m=0$ and $m=1$ eigenstates, the subscripts e and h refer to electron and hole states, and the superscript is the occupancy. Two features are immediately apparent. The electron–hole Coulomb interaction shifts the lowest transition energy by 21.7 meV (excitonic binding) and also causes a noticeable transfer of oscillator strengths between allowed transitions. A similar

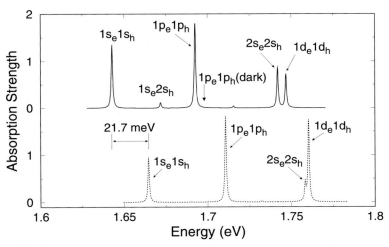

**Fig. 11.8.** Excitonic absorption by an InP quantum dot calculated with (*solid line*) and without (*dashed line*) including the electron–hole Coulomb interaction

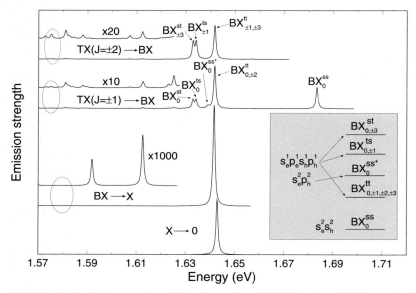

**Fig. 11.9.** Emission spectra of single exciton (X), biexciton (BX), and triexciton (TX) ground states in an InP quantum dot as calculated by a configuration interaction approach. The triexciton emission lines are labelled by the final biexciton state involved in the transition. A schematic of the relevant biexciton level structure is shown in the *inset* (*shaded box*). See the text for details

behavior has been reported in the PLE spectrum of InAs QDs [25]. The exciton binding energy is rather accurately given by the first-order perturbation theory (20.5 meV), which is equivalent to considering a single configuration consisting of the lowest electron and hole single-particle states. Inclusion of all other excited configurations contributes only 1.2 meV. It is instructive to note that the LDA calculation gives the exciton binding energy to be 17 meV, only 20% lower than the CI value.

Interestingly, the Coulomb interaction also splits the fourfold degenerate (excluding spin) $m_e=\pm1$, $m_h=\pm1$ state in to an optically active $M=0$ state separated by about 5 meV from a "dark" $M=0$ state, and two $M=\pm2$ states. While the original fourfold degeneracy is due to the assumed cylindrical symmetry, it is important to note that the Coulomb-interaction-induced splitting is comparable to that expected due the low symmetry of realistic structures [26]. In the absence of the electron–hole exchange interaction, all the optically active exciton states considered are fourfold degenerate, only two of which ($J=\pm1$) are excited by light polarized in the $x$–$y$ plane, as follows from the selection rules stated above.

In Fig. 11.9 we show the emission spectrum from single exciton (X), biexciton (BX), and triexciton (TX) ground states. The main biexciton emission (from the BX ground state with $J=0$) corresponds to the transition that leaves behind an exciton in the ground state and is redshifted from the exciton emission line by 1.4 meV. Unlike the case of the exciton, the biexciton binding energy is determined by a delicate balance of contributions from several excited configurations. In fact, the first-order perturbation theory (single configuration) gives a negative binding energy of 0.75 meV. This negative contribution arises due to the relatively strong repulsion between the two holes occupying the same single-particle orbitals, because the hole eigenfunctions in this structure tend to concentrate closer to the wider base of the pyramid compared to the electrons, which are more uniformly spread out. Inclusion of multiparticle correlations through excited configurations eventually leads to the binding energy of 1.4 meV. Experiments on QDs similar to that considered here show a rather wide scatter in the biexciton binding energy in the range 1.5–3 meV [19].

It must be noted that the energetics of biexciton binding discussed here is not universal. In structures where the electron and hole eigenfunctions differ considerably from each other, the repulsive energy of the like particles may dominate over the electron–hole attraction, leading to a negative biexciton binding energy.

A very interesting feature in the biexciton spectrum is the low-energy emission lines, though very weak, that appear several tens of meV below the main biexciton emission line. These transitions correspond to the biexciton decay leaving behind excited excitonic states. Similar to the corresponding phenomenon in bulk semiconductors, known as the inverse exciton series [27], these transitions reveal the mixing of excited configurations in the biexciton ground state. Considering the the virtual interconfiguration switching associated with these transitions, we refer to them as *configuration-crossing* transitions.

The triexciton emission spectrum (Fig. 11.9) shows two dominant sets of transitions. To understand the TX-to-BX emission process, we refer to the schematic energy level structure shown in the inset of Fig. 11.9. A few relevant low-energy BX states with $M=0$ are shown with the dominant single-particle configuration indicated against each. BX states with two electrons (holes) in the singlet combination ($J_e=0$) are labelled by a superscript s, and those with the triplet combination ($J_e=\pm1$) by t. A second superscript s or t denotes the singlet ($J_h=0$) and triplet ($J_h=\pm2$) spin states of the holes. The numeral subscript is the total $J$ value.

The main configuration in the triexciton ground state ($M=0$) involves two electrons (holes) in the lowest single-particle state and one electron (hole) in the first excited state. This is a fourfold degenerate ($J=\pm1$ and $J=\pm2$) state out of which the $J=\pm1$ states emit to the BX ground state ($J=0$), producing the high-energy emission line labeled TX→BX$_0^{ss}$. The set of emission lines appearing close to and below the BX and X emission lines correspond to the transition that involves annihilation of an electron–hole pair in the ground state, leaving behind in an excited BX state with various spin combinations as indicated in Fig. 11.9. Thus, the singlet–triplet splitting of about 10 meV of this excited BX state leads to a fine structure in the TX emission line at the band edge. As in the case of the biexciton, the triexciton emission spectrum also shows some lines much below the band edge that are configuration-crossing transitions due to the high-energy configurations weakly mixed in the TX ground state.

The LDA calculation predicts a much larger biexciton binding energy of 4.3 meV. In spite of the inadequacies of the LDA approach as applied to a few-particle system, we have reasons to believe that the LDA result for biexciton binding energy is only a slight overestimate. To test LDA, we performed an accurate calculation of the biexciton binding energy in a spherical QD using a correlated Gaussian basis [28]. For the QD structure considered in Section 11.3.1, the biexciton binding energy was found to be 3.9 meV, which compares well with the LDA result of 4.5 meV reported in Section 11.3.1. It should be noted that the CI approach for QDs appears to be seriously handicapped by the neglect of all continuum states in the barrier and the wetting layer. In view of this, the quantitative correctness of CI calculations of biexciton and multiexciton energetics is dubious. Nevertheless, CI does allow us to qualitatively identify new features like triexciton spin multiplets and weak configuration-crossing transitions that are missed by the LDA approach.

## 11.6    Conclusions

We have discussed the optical properties of multiple electron–hole pairs in a quantum dot using an idealized model of a spherical QD as well as a more realistic case of a cylindrical QD, including the effects of strain due to lattice mismatch. The interparticle Coulomb interaction is treated in an LDA approach for up to about 20 electron–hole pairs, and a CI calculation is also presented for a few electron–hole pairs.

As the number of electron–hole pairs is increased, the excitonic emission line gradually redshifts reminiscent of the band gap renormalization in bulk materials. In addition, Coulomb scattering leads to a broadening of the spectral lines, as often the final state left behind by optical emission from a many-electron–hole pair state lies in the continuum of the many-body energy spectrum. While no detailed measurements of these quantities are available at present, the predicted spectra capture most of the features of experimentally observed multiexciton emission spectra.

The Coulomb correlation between carriers shows up in the optical spectra in a variety of interesting ways as the biexciton binding energy, the spin fine structure in the triexciton emission spectrum, and as a novel class of configuration-crossing transitions. The last two features are not yet experimentally observed. We find that the LDA approach is rather reliable and computationally convenient for studying the electronic structure of QDs containing a few to several tens of electron–hole pairs. However, for effective application of this method, an improved exchange correlation potential applicable to a non-neutral, spin-polarized electron–hole gas is highly desirable.

While the CI approach appears to be conceptually well founded, our results indicate that unavoidable neglect of continuum states in the CI basis is a serious handicap of this method. Considering that CI is a very popular method for calculating the multiexciton spectra of quantum dots, further studies of its quantitative accuracy are necessary.

# References

1. For a review, see B. Bimberg, M. Grundmann, N.N. Ledenstov: *Quantum Dot Heterostructures* (Wiley, Chichester 1999)
2. K. Brunner, U. Bockelmann, G. Abstreiter, M. Walther, G. Böhm, G. Tränkle, G. Weimann: Phys. Rev. Lett. **69**, 3216 (1994)
3. K. Brunner, G. Abstreiter, G. Böhm, G. Tränkle, G. Weimann: Phys. Rev. Lett. **73**, 1138 (1994)
4. E. Dekel, D. Gershoni, E. Ehrenfreund, D. Spektor, J.M. Garcia, P.M. Petroff: Phys. Rev. Lett. **80**, 4991 (1998)
5. H. Kamada, H. Ando, J. Temmyo, T. Tamamura: Phys. Rev. B **58**, 16243 (1998)
6. P. Vashishta, R.K. Kalia, K.S. Singwi: In: *Electron–Hole Droplets in Semiconductors*, ed. by C.D. Jefries, L.V. Keldysh (North-Holland, Amsterdam 1983)
7. R. Enderlein, G.M. Sipahi, L.M.R. Scolfaro, J.R. Leite: Phys. Rev. Lett. **79**, 3712 (1997)
8. P. Vashishta, R.K. Kalia: Phys. Rev. B **25**, 6492 (1982)
9. D.M. Ceperley, B.J. Alder: Phys. Rev. Lett. **45**, 566 (1980), as parameterized by J.P. Perdew, A. Zunger: Phys. Rev. B **23**, 5048 (1981)
10. S.V. Nair, Y. Masumoto: J. Lumin., **87–89**, 438 (2000)
11. H. Drexler, D. Leonard, W. Hansen, J.P. Kotthaus, P.M. Petroff: Phys. Rev. Lett. **73**, 2252 (1994)
12. M. Fricke, A. Lorke, J.P. Kotthaus, G. Medeiros-Ribeiro, P.M. Petroff: Europhys. Lett. **36**, 197 (1996)
13. D. Pan, E. Towe, S. Kennerly: Electron. Lett. **34**, 1019 (1998); D. Pan, E. Towe, S. Kennerly: Appl. Phys. Lett. **73**, 1937 (1998)
14. Y. Tang, D. Rich, I. Mukhametzhanov, P. Chen, A. Madhukar: J. Appl. Phys. **84**, 3342 (1998)
15. A.J. Williamson, L.W. Wang, A. Zunger: Phys. Rev. B **62**, 12976 (2000)
16. G.D. Mahan, K.R. Subbaswamy: *Local Density Theory of Polarizability* (Plenum, 1990)
17. Z.L. Yuan, E.R.A.D. Foo, J.F. Ryan, D.J. Mowbray, M.S. Skolnick, M. Hopkinson: Phys. Status Solidi A **178**, 345 (2000)
18. F. Findeis, A. Zrenner, G. Böhm, G. Abstreiter: Solid State Commun. **114**, 227 (2000)
19. M. Sugisaki, H.-W. Ren, S.V. Nair, K. Nishi, Y. Masumoto: Solid State Commun. **117**, 435 (2001)
20. ARPACK is an iterative eigenvalue solver for large-scale problems based on the Implicitly Restarted Arnoldi Method. The source code developed by R. Lehoucq, K. Maschhoff, D. Sorensen, and C. Yang is freely distributed (see www.caam.rice.edu/software/ARPACK)

21. M. Sugisaki, H.-W. Ren, S.V. Nair, K. Nishi, Y. Masumoto: Solid State Commun. **117** 435 (2001)
22. S.V. Nair, Y. Masumoto: Phys. Status Solidi (a) **178**, 303 (2000)
23. L. Hedin: Phys. Rev. **139**, A796 (1965)
24. S.V. Nair, Y. Masumoto: Phys. Status Solidi (b) **224**, 739 (2001)
25. P. Hawrylak, G.A. Narvaez, M. Bayer, P. Forchel: Phys. Rev. Lett. **85**, 389 (2000)
26. C. Pryor, M-E. Pistol, L. Samuelson: Phys. Rev. B **56**, 10 404 (1997)
27. E. Tokunaga, A.L. Ivanov, S.V. Nair, Y. Masumoto: Phys. Rev. B **59**, R7837 (1999)
28. S.V. Nair, unpublished

# 12 Device Applications of Quantum Dots

Kenichi Nishi

## 12.1 Improvements of Characteristics in Quantum Dot Devices

In quantum nanostructures like quantum wires or quantum dots, where electrons are two- or three-dimensionally confined into nanometer-scale semiconductor structures, novel physical properties are expected to emerge. These expected novel properties will give rise to new semiconductor devices as well as to drastically improved device performance [1]. Optical and electrical devices with quantum dots (QDs) have already been proposed; improvements in device performance by the use of QDs have been partially proven. In this section, the origins of those improvements is studied from the viewpoint of physical property changes in the QD structure, and device performances are summarized.

### 12.1.1 Thermal Broadening in Bulk and Quantum Well Semiconductors

Most of the expected improvements in device performance originates from the change in the density of states. Figure 12.1a shows nanometer-scale semiconductor structures along with their confinement dimensions. In Fig. 12.1b, densities of states in semiconductor bulk, quantum well (QW), quantum wire (QWI), and QD structures are schematically depicted. Figure 12.1c shows the energy distributions of electrons in bulk, QW, and QD structures under a low carrier-density limit condition, where electrons obey Boltzmann's distribution. In the bulk structure, as the density of states is proportional to the square root of electron energy, the energy distribution of electrons has a width of about 1.8 $kT$, where $k$ is Boltzmann's constant and $T$ is temperature. In QWs, where electrons are one-dimensionally confined into a thin layer that forms a quantum mechanical potential well, the density of states becomes like a staircase. Thus, at the band edge the density of states becomes constant and independent of energy. There the energy distribution of electrons becomes about 0.7 $kT$, which is less than half that in the bulk structure. This reduction enables a concentration of electrons into a narrower energy distribution. Thus when the electrons are injected into such structures, a narrower emission linewidth can be expected. At present, both structures are

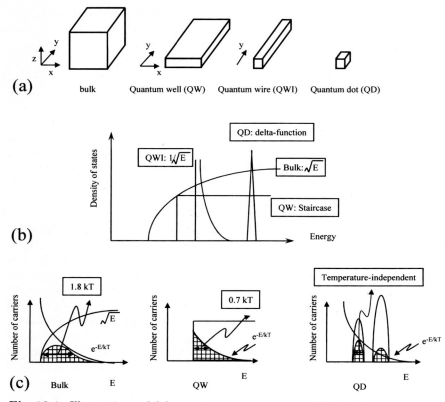

**Fig. 12.1.** Illustrations of (**a**) quantum nanostructures; (**b**) calculated density of states; (**c**) energy spreading under the low-carrier-limit condition, where electrons have the Boltzmann distribution, in bulk material, a QW, and a QD

used in semiconductor devices, such as semiconductor lasers. It should be noted that as the energy distribution width is linearly proportional to ambient temperature, basic device performance is fundamentally dependent on temperature.

### 12.1.2   Density of States in Quantum Nanostructures

When the dimension of the confinement potential increases by more than two, drastic changes in electron energy distribution become apparent. In QWIs, the density of states is inversely proportional to the square root of energy, and in QDs it becomes delta-function-like, as shown in Fig. 12.1b. Then, the width of the electron energy distribution (shown in Fig. 12.1 for the QD case) is zero as a first approximation. This means that electrons injected into those structures are distributed in certain discrete energy levels, and the energy distribution width is fundamentally independent of temperature. In real semiconductor structures, due to the many interaction processes such as electron–electron and electron–phonon scattering, certain widths in the energy distribution appear, but those widths are expected to be much smaller compared to the fundamental distribution width of bulk (1.8 kT) and QW structures (0.7 kT). In QDs, the electronic states can be understood as confined states in atoms or molecules where electrons are bound in discrete energy levels formed in a three-dimensional confinement potential. It should be noted that due to Pauli's principle, only two electrons with different spins can occupy the same energy level. Thus, in a QD, even though the density of states can be expressed like a delta-function, the integration of the electron density gives two electrons in a certain quantum level. In other words, it is Pauli's principle that determines the fundamental energy distributions of electrons in semiconductor structures.

The electron wavefunction spreads over the entire structure of a bulk structure, which results in a smaller number of electrons for a certain energy width. On the contrary, in QDs, as each QD exists in a spatially different position, each wavefunction is localized, which enables electrons to occupy the same energy level when the localized positions are different, like the impurity levels in bulk semiconductors.

The electronic states in QDs have been recognized as the main reason for the improvement of device performance in semiconductors. When the band-to-band transition is considered, its transition spectrum is mainly determined from the energy distributions of electrons. Therefore, in QDs the spectrum width should be less temperature sensitive compared to bulk structures or QWs. Thus, for example, the temperature dependence of the threshold current or output efficiency in laser diodes can be drastically improved by the use of QDs [2]. The other point is that, as the distribution width is decreased, the maximum of the gain spectrum in a semiconductor should be increased when the number of injected electrons is the same. This will reduce the threshold-current density in semiconductor lasers, resulting in reduced power consumption [3]. This effect also gives rise to an increase in the dynamic response of QD lasers [4]. Hence, the main expected device performance improvement from the use of QDs or QWIs originate from temperature-insensitive and steep (delta-function-like in QDs) densities of states.

### 12.1.3   Other Characteristic Changes of Quantum Dots for Device Applications

A recent study of QDs reported possibilities for improving other device characteristics, which had not been well investigated. One is an anisotropic physical property that originates from the anisotropic shape or alignment of the nanostructures, including the surrounding matrix. In quantum wires, a fundamental anisotropic shape has been expected to show anisotropic optical and electrical characteristics [5]. This can concentrate certain physical properties in one direction. Preferential cavity direction has been calculated in quantum wire lasers [6]. In QDs with asymmetric shapes, the same physical properties can appear. Thus, polarization control is one of the realized device performance improvements in QD lasers.

In QDs made of lattice-mismatched materials, lattice distortion inside and outside the QDs is not simple like the pseudomorphic deformations in strained QW structures, where the in-plane lattice constant of the QW matches that of the thick substrate. In QDs, because of the increased deformation ability, it is possible to deform three-dimensionally as well as in the surrounding matrix without forming dislocations [7]. In many cases the strain components of a strained QD are smaller compared to those of a strained QW having the same heterostructure. This is because strains in a matrix release those in a QD. Such reduction of the strain effect usually results in a drastic change in band gap energy. This increases the possible band gap range possible in QDs. Long-wavelength emissions from QDs formed on GaAs have been observed [8] and have been used to create vertical-cavity surface-emitting lasers (VCSELs).

In Fig. 12.2, the characteristic changes in QDs and the expected device performance improvements are summarized.

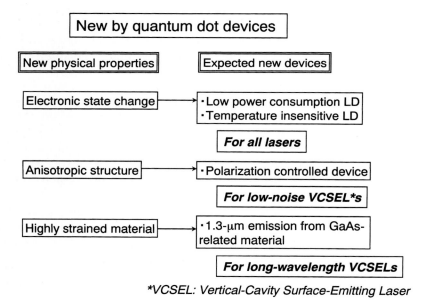

**Fig. 12.2.** Summary of characteristic changes in QDs and consequent device performance improvements

## 12.1.4   Required Quantum Dots Dimensions for Device Applications

Certain requirements must be satisfied in QDs to achieve the expected improvements. First, regarding the quantum confinement effect, the dimensions of QDs should be in the order of several to several tens of nanometers. When it is desired that electrons are to exist only in one quantum level, the energy separation between ground and the first excited state should be much larger than the thermal broadening energy $kT$, which is about 26 meV at room temperature.

Let us estimate the quantum confinement energy in a cylindrical QD structure. (Usually QDs have a relatively flat shape.) Therefore, the cylindrical or columnar shape shown in Fig. 12.3a is a good approximation for this purpose. The quantum confinement along the axial direction ($z$-direction in Fig. 12.3) can be approximately obtained from a calculation conventionally performed for a QW structure with a thickness of $L_z$. The confinement energy can become large enough to appear as a quantum confinement effect at room temperature when the dimension is about 10 nm or less in typical semiconductor heterostructures, like (In)GaAs/(Al)GaAs. This dimension along the $z$-direction can be controlled relatively easily by using well-established growth techniques. On the other hand, it is not simple to analytically estimate the lateral confinement effect. To obtain a rough image in a lateral

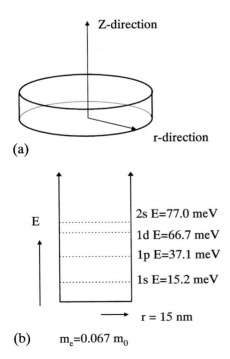

Z-direction

r-direction

**(a)**

E

2s E=77.0 meV

1d E=66.7 meV

1p E=37.1 meV

1s E=15.2 meV

r = 15 nm

**(b)**   $m_e$=0.067 $m_0$

**Fig. 12.3.** Schematic of (**a**) a cylindrical QD structure and (**b**) calculated eigenvalues due to lateral quantum confinement ($m_e$=0.067 $m_0$, $r$=15 nm)

quantum confinement, an infinitely high potential barrier is assumed for the lateral confinement. Then, electron–wave functions become Bessel functions inside the quantum column with a radius $r$. Quantum confinement energy can be calculated from the zeroth points of the Bessel functions using the effective masses of the semiconductors. An example is shown in Fig. 12.3b. Using an electron effective mass of GaAs as a typical example (0.067 $m_0$, where $m_0$ is an electron mass in a free space), a lateral dimension of less than about 30 nm is required for realistic quantum confinements. So, as a first approximation, a lateral dimension of less than 30 nm will be necessary to show a three-dimensional quantum-confinement effect in hemispherical or cylindrical QDs.

## 12.1.5    Required Characteristics for Quantum Dot Optical Devices

In devices that utilize an oscillator strength that is condensed into a narrow energy width, like QD lasers, the absolute energy level of the QDs should be the same. This means that the size, shape, and alloy composition of the QDs need to be close to identical. Then the inhomogeneous broadening of QD luminescence is eliminated, resulting in a real concentration of electron energy states. For certain types of devices that can take advantage of broad inhomogeneous broadening, one may make use of this inhomogeneous broadening with the scattered size distribution of QDs, which will not be affected or limited by ambient temperature.

Density, including the possibility of a multistack structure, is another factor for consideration in device applications. To obtain a macroscopic physical parameter, i.e., light output in laser devices, the number of interacting QDs should be as high as possible.

The reduction of nonradiative centers in QDs will inevitably be important for their use in optical devices. Nanostructures made by high-energy beam patterning cannot be used because damage is incurred from the beam around the nanostructures. This type of damage around the surface of self-assembled dots (SADs) is critical for the development of optical devices, such as laser diodes, because the surface-to-volume ratio of QDs is drastically increased compared to QWs. From this viewpoint, the self-assembling technique introduced in next section is very important to the realization of semiconductor QDs.

When QDs are incorporated into the layered structure of a semiconductor for device applications, electrical control is critical to the operation of the device. It is desirable that both an electric field can be applied to change certain physical properties of the QDs and that carriers can be injected into QDs for light emission.

It is not easy to fabricate QDs that can satisfy these requirements. This is why real QD devices have not yet been realized in the long time since the publication of several papers with fascinating predictions. The self-assembling method is definitely one of the most promising for the formation of a close-to-ideal QD structure for device applications.

### 12.1.6   Advantages of Self-Assembled Quantum Dots

The self-assembling method for fabricating QDs has been recognized as one of the most promising methods for forming QDs that can be practically incorporated into QD devices. In the crystal growth of highly lattice-mismatched material systems, self-assembling formation of nanometer-scale three-dimensional islands has been reported [9]. It is known that the considerably small lateral dimension of about 30 nm can be realized with high density just by growing a lattice-mismatched material, for example, InAs on GaAs [10], InGaAs on GaAs [11], and InP on InGaP [12]. Quantum Dots formed by the self-assembling method have high optical quality that is usually good enough for room-temperature operation.

Size dispersion is a problem that has yet to be overcome. From precise studies of dot formation, InAs self-assembled QDs were found to form with a certain aspect ratio dependency on growth temperatures. At relatively high temperatures, the QDs have a high aspect ratio with an improved uniformity [13]. Also, by using a slow growth rate, this improvement in size dispersion was found to exist. It is assumed that the enhanced surface mobility of the adatom played a role in these size-uniformity improvements by making the size and shape close to equilibrium. From optical measurements, the room-temperature luminescence linewidth had a width of about 21 meV, which is close to the ideal QW value. Thus, uniformity of size is not considered a crucial problem for device applications.

The lattice mismatch between a QD and the matrix is the fundamental driving force of self-assembling. This somewhat limits the available materials systems that can be used for QDs. However, even in the III–V semiconductor systems, by choosing an appropriate material system, visible to near infrared light emissions have been reported in InAlAs QDs, in AlGaAs [14] for visible, and in InAs QDs on InP [15] at about 2 μm.

In(Ga)As on GaAs is the most commonly used material system because lattice mismatches can be controlled by the In alloy ratio up to about 7%. The band gap of In(Ga)As QDs lies around 1.2 eV (1000 nm). This band gap is mainly determined by the strain in the QDs. By changing the strain with a strain-reducing layer while keeping the shape and size constant, the band gap can be controlled up to values as low as 0.9 eV (1350 nm) for InGaAs-covered InAs QDs on GaAs [16]. Thus, by designing the QD structure to include a covering layer, the bandgap can to some extent be independently controlled.

Many optical devices use self-assembling QDs in their active layers. Improved characteristics shown by those QD devices are explained in Sect. 12.2.

## 12.2    Optical Devices with Quantum Dots

Characteristic improvements in many optical devices have been predicted by using QDs in their core layers. Light-emitting and detecting devices, optical modulators, and optical memory using QDs have been proposed. Conventional predictions discuss the advantages that originate from QDs with steep density-of-state, while real structures inevitably possesses a certain amount of inhomogeneous broadening. In this section, QD optical devices are discussed.

### 12.2.1    Quantum Dot Lasers
### with Improved Temperature Characteristics

Semiconductor lasers are regarded as one of the most promising optical devices using QDs. As mentioned in Sect. 12.1, fascinating characteristic changes have been predicted, including reduced threshold current and the resultant low power consumption, temperature-insensitive operation, and improved modulation characteristics. In Fig. 12.4, the incorporation of QDs into semiconductor lasers is illustrated. With self-assembled QDs, the first report of improvements in characteristic temperature was made by Kristaedter et al. [17]. A laser structure with a shallow mesa waveguide was made using an InGaAs self-assembled QD active layer on a (100) GaAs substrate. A $T_0$ value of 350 K was reported at a measurement temperature of 77 K. This value was much higher than the theoretically predicted $T_0$ of QW lasers. By improving QDs to show apparent three-dimensional quantum confinement effects such as discrete energy levels, infinite $T_0$ values, as predicted by theoretical analysis, had been realized. Huffaker et al. reported lasing at relatively long wavelengths ($\sim$1200 nm) from InGaAs QDs on GaAs [18]. From 77 K to about

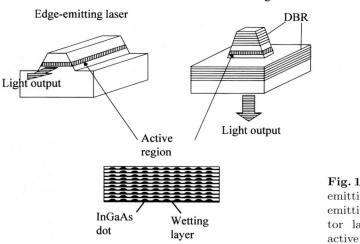

**Fig. 12.4.** Edge-emitting and surface-emitting semiconductor lasers with QD active regions

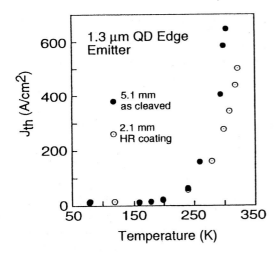

Fig. 12.5. Threshold current density of the QD laser at various temperatures reported by Huffaker et al. [18]. From 77 K to about 200 K, threshold current density did not change in QD lasers

200 K, the threshold current density did not change in QD lasers with relatively low mirror-loss structures (cavity length of 5.1 mm with cleaved facets, or 2.1 mm in a device with high-reflection coatings). The threshold current density at various temperatures is shown in Fig. 12.5. The same group also studied the relation between temperature sensitivity and sublevel energy separation in QD lasers [19]. From their results, they concluded that the energy separation of sublevels should be as large as possible to increase the characteristic temperature. By carefully controlling the growth temperature, the group succeeded in obtaining very stable laser operation. Under cw operation, $T_0$ became infinite below 300 K, above that, $T_0$ was still as large as 126 K [20]. This can be regarded as evidence of a real temperature-characteristic improvement in QD lasers.

## 12.2.2 Lasing Wavelength Control in Quantum Dot Lasers

It should be noted that, from the practical point of view, the emission wavelengths of QD lasers are important because they determine the application field. This is particularly true in the case of 1.3-µm light emission from QDs grown on GaAs, which are of interest because the wavelength dispersion of the usual silica optical fiber becomes zero. Mirin et al. reported 1.3-µm light emissions from InGaAs QDs with an In content of 0.3 [8]. By adopting alternate-beam epitaxy, where group III and group V materials were supplied alternatively to increase surface adatom migration, high-aspect-ratio QDs were formed on GaAs. It was found that the QDs were so uniform that a room-temperature PL showed a full-width at half-maximum linewidth of only 28 meV. Many papers that followed concurred with Mirin's result and reported 1.3-µm emissions from QDs on GaAs. Because the quantum confinement of those QDs becomes large due to the greater band gap difference

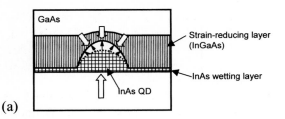

(a)

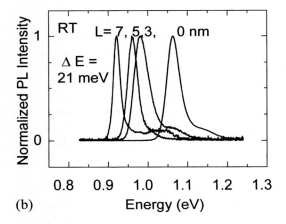

(b)

**Fig. 12.6.** (a) Schematic of the strain-reduction structure. InAs QDs on GaAs are covered by InGaAs that has an intermediate lattice constant between InAs and GaAs. Experimental results of PL spectra from InAs QDs with different covering layer thicknesses are shown in (b). The PL peak energy change corresponds to wavelength change from 1.16 to 1.35 μm

between the GaAs and the 1.3-μm emitting material, QD lasers made with those QDs succeeded in showing higher temperature characteristics. Several examples of this were discussed in Sect. 12.2.1.

By adapting other strained layers surrounding the QDs, it was found that the wavelength of emissions from QDs can be controlled. A great amount of the strain inside of the QDs usually causes an increase of its band gap by several hundreds of meV. When the strain inside the QDs can be controlled, its band gap can be changed dramatically. For the purpose of study, a strain-reduction structure, shown in Fig. 12.6a, was proposed for InAs QDs on GaAs [16]. There, strained InGaAs or InAlAs having an In content of 0.2 was used to cover the InAs QDs. The PL emission wavelength from the same QDs, but with a different covering layer thickness, changed from 1.16 to 1.35 μm, as shown in Fig. 12.6b. This proves that controlling the amount of strain inside of a QD is a good way to control the wavelength of an emission.

### 12.2.3   Reduction of Threshold Current Density in Quantum Dot Lasers

Due to the concentration of the electron energy distribution, the maximum optical gain at or under a certain carrier-injection level can be drastically increased in QD lasers compared to the gain obtained for bulk or QW lasers. This can decrease the threshold current density, resulting in lower power consumption. When a QD was fabricated using QW growth and its patterning/etching, nonradiative centers caused by damage during fabrication increased the threshold current in those lasers. After improvements were made to the surface and interface quality of self-assembling QDs, a reduction of the threshold current as well as a $T_0$ increase were reported.

In the relatively early stages of the development of self-assembled QD lasers, threshold current densities of 120 A/cm$^2$ at 77 K and 950 A/cm$^2$ at room temperature were reported for a QD laser having one sheet of an InGaAs QD active layer [17]. Molecular beam epitaxy was used to fabricate self-assembled QDs; in-situ surface monitoring was performed using reflection high-energy electron diffraction. Another QD laser also with one sheet of InGaAs QD active layer showed a threshold current density 815 A/cm$^2$ at 80 K [21]. Atomic layer epitaxy, where one monolayer of InAs and GaAs is alternatively formed, was used in the latter case. A room-temperature PL showed distinct subpeaks that originated from higher-order sub-band transitions showing three-dimensional quantum confinement. Due to the lack of a sufficient number of QDs, the laser operation achieved had a higher-order sub-band transition with a greater degeneracy, and the threshold current could not be reduced compared to QW lasers.

InGaAs quantum disks have been created using metal organic vapor-phase epitaxy with a self-organizing method [22]. Well-aligned disk-like structures were formed on a GaAs (311)B surface, as shown in Fig. 12.7. By forming two disk layers for the active region, ridge-waveguide lasers were fabricated. The threshold current of a double-quantum disk laser on (311)B was 23 mA. Figure 12.8 shows light versus current of the quantum disk and reference QW lasers. As the ridge was 2 μm wide and the cavity was 900 μm long, threshold current density was calculated to be 1.28 kA/cm$^2$. In a reference double-QW laser made on a (100) GaAs surface, threshold current density was calculated to be 1.5 kA/cm$^2$. This result shows that threshold current density was clearly reduced by quantum disks compared to QW lasers having the same device dimensions.

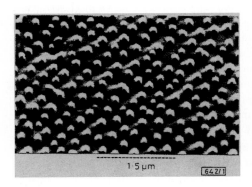

**Fig. 12.7.** A scanning electron microscope image of well-aligned quantum disk-like structures formed on GaAs (311)B surface. Growth had proceeded in a self-organizing way using MOVPE (from [22])

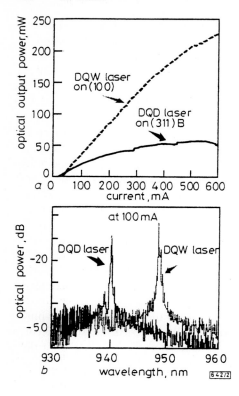

**Fig. 12.8.** Light vs. current characteristics of the quantum disk and reference QW lasers. Threshold current density of the quantum disk laser was calculated to be 1.28 kA/cm² which was better than that of reference a QW laser (1.5 kA/cm²) (from [22])

After realizing high-density QDs with uniform, high aspect-ratio QDs, lasing at ground state with low threshold current density became possible [23]. Using an alternate supply of In and As to grow nominally three-monolayer-thick InAs, in-plane density of $5.7 \times 10^{10}$ cm$^{-2}$ was achieved. Growth at a lower temperature (490°C) compared to that previously used for high-aspect-ratio formation (550°C) was effective in increasing the density. Three periods of the layers increased the total density to $1.7 \times 10^{11}$ cm$^{-2}$. The laser structure with a 30-μm-wide stripe and a 2030-μm-long cavity lased at a threshold current density of 76 A/cm². The lasing wavelength was 1.23 μm. Figure 12.9 shows the laser structure and the lasing characteristics. It should be noted that the uniform and dense QD structure having a high aspect ratio enabled the lasing at the low threshold current density.

Further improvements were reported by Liu et al. [24] with a newly designed active layer. They used molecular beam epitaxy for QD formation with a special technique called the "dot-in-a-well" (DWELL) method. InAs QDs formed with a 2.4-monolayer supply were grown on a 5-nm-thick In$_{0.15}$Ga$_{0.85}$As layer. Then InGaAs with the same alloy composition and thickness was used to cover the QDs. This type of structure can be regarded as QDs put in the middle of a 10-nm-thick InGaAs QW. When InAs QDs are

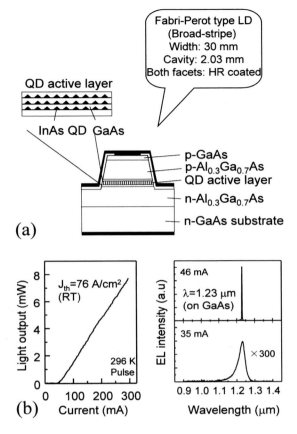

**Fig. 12.9.** (a) A laser structure made of three periods of QD layers, where total QD density is as large as $1.7 \times 10^{11}$ cm$^{-2}$; (b) Light output vs. injection current as well as lasing wavelength. Threshold current density of 76 A/cm$^2$ and lasing wavelength of 1.23 µm were obtained

formed on InGaAs instead of the usually used GaAs, the density of the QDs is drastically increased. Moreover, it has been shown that by covering InAs QDs with InGaAs, the PL linewidth can become as narrow as that of the usual QWs; this presumably originates from the improved size uniformity in the buried structure. Thus, the DWELL-type structure is advantageous for realizing the ideal QD laser.

Room temperature PL linewidth was 37 meV. In the measurements, the PL from the surrounding GaAs was not observed, indicating that a radiative transition had not occurred in the GaAs. The carrier-capture process into the QDs may become fast because 10-nm-thick InGaAs can capture carriers from surrounding GaAs, and the carriers in the QW can preferably relax into QDs rather than going up to GaAs from thermal emission.

A laser structure with 100-µm-wide stripe and 7.8-mm-long cavity was tested under a pulsed current injection. As shown in Fig. 12.10, a threshold current of 200 mA was obtained. This corresponds to a threshold current

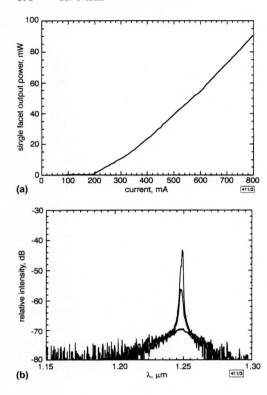

**(a)**

**(b)**

**Fig. 12.10.** (**a,b**) Laser characteristics obtained in a DWELL
(dot-in-a-well)    structure    with
100-μm-wide stripe and 7.8-mm-
long cavity. A threshold current
of 200 mA corresponds to threshold current density of 26 A/cm²
(from [24])

density of 26 A/cm². Characteristic temperature was 60 K from 280 to 320 K,
and 34.5 K above 320 K.

The use of a low-optical-loss cavity structure, such as long cavity length
and high reflection-coated facets is effective in reducing threshold current
density. Thus, the relatively long cavity length of 7.8 mm is used in the
DWELL laser. Moreover, by applying a highly reflective coating, a threshold
current density as low as 21 A/cm² was obtained. This value is almost one
order of magnitude lower than the threshold-current density of the typical
QW laser.

Owing to its long cavity length, the operating currents of the QD laser
have not been appreciably decreased, which means that commercial devices
made with QDs would not have any advantages resulting from reduced operating powers. However, by using an oxide current confinement, the threshold
current was reduced considerably to 1.2 mA [25]. By using a method such
as the above to confine current, reduced operating current in QD lasers will
become a reality.

### 12.2.4   Vertical-Cavity Surface-Emitting Lasers with Quantum Dots

Vertical-cavity surface-emitting lasers (VCSELs) are attracting much interest due to their small threshold current and circular beam pattern. The major anticipated application of VCSELs is as optical interconnections for short- and middle-range distances. By using QDs in VCSELs, some novel features have been demonstrated.

The first VCSEL with QDs was reported by Saito et al. [26]. Figure 12.11 shows a schematic view of the QD-VCSEL. By solid-source MBE, an 18-period of $n$-type doped GaAs/AlAs distributed Bragg reflector (DBR) was first grown on a GaAs (100) surface. On the DBR, a one-wavelength cavity with ten periods of a InGaAs self-assembled QD active layer was formed. This was followed by a top 14-period of p-type DBR. The QDs were grown under RHEED observation, and QDs were formed by supplying about five-monolayers of InGaAs. In the growth sequence, 0.2 monolayers of InAs and GaAs were successively supplied with a 2-second interval As beam. QD density measured by AFM on the uncovered structure was about $2 \times 10^{10} \mathrm{cm}^{-2}$. Ten sheets of QD layers were grown with 30-nm-thick AlGaAs barrier layers.

Typical device characteristics measured from the 25-μm-square device structure are shown in Fig. 12.12. The device was operated at room temperature under a continuous-current condition. The threshold current was 32 mA and the lasing wavelength was 962 nm. The PL peak was positioned at around 1000 nm, that is, 40 nm longer than the lasing wavelength. Precise PL measurement showed that a higher excited state transition exists at about

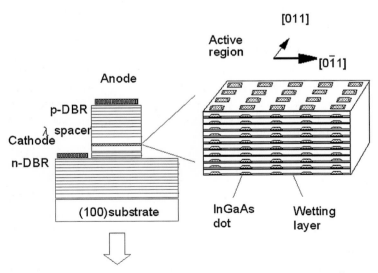

**Fig. 12.11.** Schematic view of the QD-VCSEL reported by Saito et al. [26]. Ten periods of QD layers are sandwiched by $n$- and $p$-DBRs

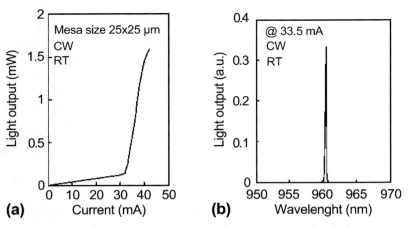

**Fig. 12.12.** Light output vs. current characteristic and a lasing spectrum of a typical QD-VCSEL with 25-μm square mesa shown in Fig. 12.11

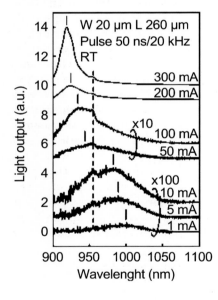

**Fig. 12.13.** Electroluminescence spectra measured from a cleaved edge of the QD-VCSEL shown in Fig. 12.11. Gain saturation according with injection current increase is obvious

960 nm. Thus, it is assumed that the lasing of the QD-VCSEL originates from the higher-order transition of InGaAs QDs, even though ten QD layers were formed to increase the optical gain.

For an understanding of the lasing mechanism of this QD-VCSEL, electroluminescence (EL) obtained from a cleaved edge of the VCSEL is shown in Fig. 12.13 because the luminescence obtained from the perpendicular direction is strongly modulated by the vertical cavity. Under a 1-mA current injection, a luminescence whose peak lay at 1000 nm was obtained. This coincides with the PL peak position obtained under low excitation, which means that the luminescence at 1000 nm originated from the ground state transition. By further increasing the injection current, the EL peak moved to a higher energy level with an increasing EL intensity, and under the lasing condition, the peak was positioned at about 930 to 940 nm. As the cavity structure determines the lasing wavelength, the lasing occurred at 962 nm. This clearly indicates a gain saturation of the ground state transition and higher gain from the excited states due to the higher degeneracy of the states. To lase at a ground state transition, it is necessary to increase the maximum gain by narrowing the inhomogeneous broadening. As another approach, it is effective to decrease optical loss by, for example, increasing the number of DBR pairs. By adding ten more DBR periods, ground-state lasing in a QD-VCSEL was achieved with almost the same active layer structure.

In QD-VCSELs, lasing operation can be modified by the anisotropic shape of the QDs. Theoretical calculation revealed that large optical anisotropy emerges in strained quantum wires, as discussed in Sect. 1.4. Self-assembled QDs contain large amounts of strain inside and outside the structure, and thus, when the shape is not completely symmetric, such optical anisotropy will exist, resulting in a change of device characteristics. From a number of structural characterizations, such as transmission electron microscopy [27] and AFM observations [8,28], self-assembling InGaAs QDs on GaAs have been found to have larger size along the [0$\bar{1}$1] compared to the [011] direction. From precise characterization with RHEED diffraction and high-resolution scanning electron microscopy (SEM), typical anisotropic shape of the lateral dimension ratio as large as 1.94 has been observed, as shown in Fig. 12.14. This anisotropy causes an in-plane optical characteristic modulation that is indicated by PL measurements. When QD-VCSELs were made with anisotropic QDs, lasing was found to be polarized along the longer direction of [0$\bar{1}$1]. Figure 12.15 shows the lasing characteristics of a QD-VCSEL with 25-μm-square mesa structure, where lasing had taken place from a ground state transition. Most of the lasing light polarized along the longer direction of the QDs, and the power suppression ratio at a 1-mW output was

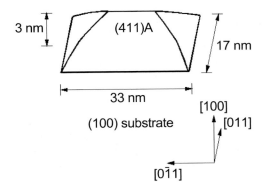

**Fig. 12.14.** Estimated shape of a self-assembled QD made of In-GaAs. RHEED diffraction, SEM, and AFM were used to characterize its shape, and the lateral dimension ratio was found to be as large as 1.94

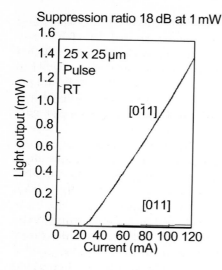

Suppression ratio 18 dB at 1 mW

**Fig. 12.15.** Light output along two major crystallographic axes from a QD-VCSEL, where lasing had taken place from ground state transition

18 dB. More than 20 devices were tested for their characterization, and all of them showed the same polarization. On the contrary, control samples, where conventional QWs were used for active layers, did not show such polarization. This indicates that QDs offer the advantage of polarization control in addition to the traditionally predicted ones, such as low power consumption and improved temperature sensitivity.

Another advantage of using QDs in VCSELs is the possibility of creating long-wavelength VCSELs. Lott et al. reported 1.3-μm VCSEL operation from three periods of InAs QDs sandwiched by DBRs [29]. Current confinement using two aluminum oxide layers enabled lasing with a low threshold current of 1.8 mA. The lasing wavelength was 1301 nm. Other characteristics such as quantum efficiency and output power were also satisfactory for practical use. Thus, the use of QDs in long-wavelength VCSELs holds promise for the realization of commercially viable devices.

## 12.2.5  Miscellaneous Improvements in Quantum Dot Lasers

QDs can vary so many physical properties in semiconductors that unpredicted characteristics of these semiconductor devices become apparent. Some examples have already been seen in semiconductor lasers. Very small current apertures made of aluminum oxide by AlAs oxidation made it possible to reduce drastically the operating current of lasers. In VCSELs where the lateral dimension of the active region is small, this type of current confinement method works very efficiently. However, as lateral size decreases, nonradiative as well as leak currents at the edge of the active region increase the operating current along with the pure radiative current. This problem can be reduced when current spreading in an active region is suppressed. The use of QDs, or three-dimensional islands, enables lateral current spreading to be reduced, resulting in a decrease of the operating current of VCSELs having oxide apertures. From the characterization of carrier diffusion in usual QW and QD layers, even at room temperature diffusion length was found to be decreased by three times in QD layers [30]. Such an effect will become important in the efforts to create lasers with very low operating currents, where even a slight nonradiative current will have a great effect on the total operating current.

The lasing wavelength of semiconductor lasers is strongly temperature dependent in usual laser structures. This phenomenon fundamentally originates from the temperature-dependent bandgap. Thus, except for VCSELs where cavity and DBR structure determine the lasing wavelength, the lasing wavelength that corresponds to the gain maximum varies as the band gap changes. One way to overcome this wavelength change is to utilize the gain saturation phenomenon of the QD structure. As the injected current increases in a QD structure, gain from the ground state saturates, and its maximum moves to the higher-energy side. Actually, this should happen in bulk and QW structures, but due to the larger integrated density of states, a very high current injection is necessary. In QD lasers, at relatively small current injection, this gain saturation occurs and can be used to compensate for the lasing wavelength change. Mirin et al. reported that from 80 K to 300 K, the lasing wavelength moved from 1010 nm to 1020 nm in QD lasers, while it changed from 890 nm to 940 nm in a usual QW laser [28]. Figure 12.16 shows the change of lasing wavelength versus the ambient temperature. The active layer of the QD laser was made of self-assembling a $In_{0.3}Ga_{0.7}As$ layer with a nominal thickness of 13.3 nm. By using alternating molecular beam epitaxy, a high-aspect-ratio structure was created having a density $2$–$3 \times 10^{10}$ cm$^{-2}$.

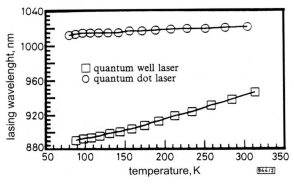

**Fig. 12.16.** Lasing wavelengths change with ambient temperatures. From 80 K to 300 K, lasing wavelengths moved from 1010 nm to 1020 nm in QD lasers, while they changed from 890 nm to 940 nm in a typical QW laser. This drastic decrease in the change in lasing wavelength of the QD laser was apparent (from [28])

Compared to other QD lasers, the total number of QDs was relatively small since only one sheet of QD layer was used for the active layer instead of multistacking layers. This limited number of QDs enhanced gain saturation due to state filling, resulting in the temperature-insensitive lasing wavelength. Thus, the choice of the appropriate number of QDs in a QD laser and the laser structure (cavity length, mirror loss, etc.) can result in improvements of characteristics in QD lasers.

### 12.2.6   Other Optical Devices

The new physical properties of QDs can be utilized in optical devices other than those already cited. Although many characteristic improvements have been shown in lasers, novel and improved characteristics are expected in other optical devices.

Semiconductor optical modulators are now used mainly in monolithic light sources for high-speed long-wavelength optical communications. Their usual use is to modulate light intensity since the quantum-confined Stark effect (QCSE), which is observed in a QW under an electric field, reduces absorption edge energy. Theoretical calculation of the QCSE has shown that this effect can be enhanced by an order of magnitude at maximum in a dense, small QD [31]. The possibility of a large change in the refractive index in QDs under an electric field was shown in the same reference. By considering such effects, it is reasonable to presume that QDs can be recognized as an appropriate material system for fabricating semiconductor optical modulators. Moreover, Sahara et al. theoretically showed that "blue chirp" can be achieved in QD modulators when the absorption peak and the signal light position are appropriately set [32]. This "blue chirp" in the modulated light wavelength is very advantageous for use in very long distance optical communication. In the typical optical fiber, the dispersion of the signal light at 1.55 μm often limits the maximum distance attainable because zero dispersion exists at 1.3 μm. But, if the signal light wavelength is modulated so as to make it shorter (become blue), pulse compression first takes place, then the pulse width spreads due to wavelength dispersion. The physical origin of this effect is the discrete optical absorption peaks in the QD structures. Unlike other semiconductor structures like bulk and QWs, it is easy to decrease optical absorption at a certain energy level when an electric field is applied. Thus optical modulators with QDs will have superior performance when ideal QDs are incorporated in them.

Usually, inhomogeneous broadening in nonideal QDs is considered an obstacle to realizing QD optical devices like lasers. This is true when the concentration of oscillator strength is a fundamental origin of characteristic improvements. However, some device applications take advantage of this inhomogeneous broadening. Jimenez et al. proposed a novel device, a "QD spectrometer" where not only light intensity but also wavelength or energy can be detected [33]. QDs with rather wide inhomogeneous broadening are thought to be a light-detecting layer. Due to discrete and size-dependent absorption peaks, carriers should be selectively generated in different QDs when the light energy is different. By adopting resonant QWs having external electric fields to remove the carriers from certain QDs, the energy of the absorbed light can be resolved from the applied electric field. Thus, a spectrometer results. The spectral resolution is determined from homogeneous broadening of QDs [34] and the size and shape distribution of the QDs. Inhomogeneous broadening will limit the complete spectral range. Therefore for this particular application, a larger inhomogeneous broadening is required for better device performance.

## 12.3   Future of Quantum Dot Devices

### 12.3.1   Ideal Quantum Dot Structures for Device Applications

As was discussed in Sect. 12.1, many device applications need uniform and dense QD structures. QD formation methods still need improvement to satisfy these requirements. When both of these necessary characteristics are realized without sacrificing each other, real commercially viable QD devices can be fabricated.

Uniform QD structures can be realized by the effective use of higher-index semiconductor surfaces. Because the Stranski–Krastanow growth mode originates from competition between the surface energy and the strain energy, just varying the surface alone will have a great effect on QD formation. It has been shown that InAs and InGaAs self-assembled QDs grown on (311)B GaAs have better uniformity compared to those grown on a (100) surface [35]. The size uniformity improvement was measured to be as large as 20%, as indicated by structural as well as optical measurements. Further study in the use other surfaces and surface treatments must be made in order to create more uniform self-assembling QD structures. Another interesting feature of the (311)B surface is the emergence of the self-alignment of QDs. When a less-strained InGaAs layer was grown on the surface, a spontaneous lateral alignment was observed along a direction that was 60° inclined from the [011] direction in a gas-source MBE-grown structure [36]. A similar alignment of InGaAs QDs was indicated in a MOVPE-grown structure [37]. The anisotropic strain field that emerges on the surface may possibly be the origin of the alignment.

As discussed in Section 1.4, QDs with a high aspect ratio tend to have a uniform size distribution. Many devices now use this kind of QD. By adopting the appropriate growth procedures, uniform QDs can be grown. However, ultimate size uniformity has not been achieved by using only this method. In combination with other techniques, the growth of more uniform QDs is expected.

For practical use, dense structures are usually desirable. As a simple technique, using migration-enhanced epitaxy or alternate-beam epitaxy is a good way to increase the QD density in a growth plane. Another effective way to increase the number of QDs is by using multistacked layers. This multilayered structure also holds promise for improving size uniformity through vertical interaction of the strain field. Tersoff et al. revealed such uniformity improvements in multilayers [38]. The first QD-VCSEL used ten periods of QD layers to increase the nominal optical gain. In that structure, the QD height was as small as 3 nm, resulting in no degradation of the PL intensity even after the formation of ten QD layers. A problem associated with high-aspect-ratio QDs is the difficulty of forming a multistacked structure, because the height of the high-aspect-ratio QDs often exceeds 10 nm and the accumulated strain in the stacked structure becomes large. Thus, when the QD layers are stacked, dislocations can be introduced into the structure. In some conventional strained QWs, positive and negative strains are introduced into QWs and barriers, respectively, to compensate for the total strain in the layers. Unfortunately, this technique is hard to incorporate in QDs because the magnitude of strain differs between the QD region and the other flat regions. The manufacture of stacked QDs with high aspect ratios will inevitably require a smart structure and a highly controlled growth process.

## 12.3.2   Ultimate Device Performances with Quantum Dots

The reported numerical parameters in recent QD devices often overcome conventional bulk or QW devices. This is especially true in many of the reports on QD lasers. However, it should be noted that, for example, threshold current density is decreased, but the current itself is not completely reduced compared to the case of the conventional QW lasers. Thus QD lasers have not been adopted for use in commercial devices. Uniformity and density must still be improved to reduce the current level itself. A few papers did report realistic improvements in QD lasers. A very low threshold current of about 1 mA was reported with a QD structure, which showed a threshold current density of as low as about 20 A/cm$^2$. Such QD lasers will be put into commercial use when output power and other characteristics for a particular use (modulation frequency, temperature characteristics, etc.) are satisfied.

If QD devices are categorized from the application point of view, band gap wavelength is the most important parameter. InAs QDs with a high aspect ratio have a bandgap of about 1.3 μm. This is particularly important for certain devices used in optical communication. Temperature-insensitive characteristics are very attractive for hostile environment communication systems because a rather wide temperature range (−20°C to 85°C) is assumed for their use. QD lasers with high or infinite $T_0$ value in these temperature ranges will make it possible to discard the temperature control or feedback circuits, thereby reducing a great amount of the module cost. Another field where this wavelength range is applicable is as the active layer for long-wavelength VCSELs. The use of 1.3-μm VCSELs as high-bit-rate and mid- or long-distance optical interconnection light sources may be possible. As shown in Section 2.5, such VCSELs have been reported. When output power and modulation bandwidth satisfy certain regulations, including operational lifetime, QDs will be incorporated into such devices.

In the 1.55-μm range, where dense wavelength-division multiplexing (DWDM) systems are used for ultrahigh-bit-rate optical communication, many novel optical devices using semiconductors are expected. Therefore, new functional devices, like semiconductor optical amplifiers, optical multiplexers, and demultiplexers, are inevitable. The discrete energy levels of QDs can be used to detect certain wavelength ranges without absorbing other wavelengths. For optical amplifiers, wide inhomogeneous broadening can be used positively for a wide-wavelength range amplifier. Those applications have not been demonstrated, partly because QDs with such a bandgap region cannot be easily formed. However, when QDs for this wavelength range are realized with satisfactory uniformity and quality, many QD devices are expected to be incorporated into DWDM applications.

It should be said that device applications to other wavelength regions will emerge when low power consumption and high temperature characteristics are realized. Even for use in blue lasers, the QD structure is attractive for reducing operating powers.

One concern about QDs in device applications is their operational lifetime. Self-assembling QDs are made of a highly strained material system. In strained QWs, this problem has been overcome. Thus, it is expected that even in self-assembled QDs, lifetime is not degraded by the strained system. However, it is still necessary to prove device lifetime before QDs can really be incorporated into commercial devices. After demonstrating those device characteristics, QDs are expected to be used in many kinds of active and passive optical devices.

# References

1. H. Sakaki: Jpn. J. Appl. Phys. **19**, L735 (1980)
2. Y. Arakawa, H. Sakaki: Appl. Phys. Lett. **40**, 939 (1982)
3. M. Asada, Y. Miyamoto, Y. Suematsu: IEEE J. Quantum Electron. QE-**22**, 1915 (1986)
4. Y. Arakawa, K. Vahara, A. Yariv: Surf. Science **174**, 155 (1986)
5. N. Notomi, J. Hammersberg, J. Zeman, H. Weman, M. Potemski, H. Sugiura, T. Tamamura: Phys. Rev. Lett. **80**, 3125 (1998)
6. A.A. Yamaguchi, A. Usui: J. Appl. Phys. **79**, 3340 (1996)
7. K. Nishi, A.A. Yamaguchi, J. Ahopelto, A. Usui, H. Sakaki: J. Appl. Phys. **76**, 7437 (1994)
8. R.P. Mirin, J.P. Ibbetson, K. Nishi, A.C. Gossard, J.E. Bowers: Appl. Phys. Lett. **67**, 3795 (1995)
9. L. Goldstein, F. Glas, J.Y. Marzin, M.N. Charasse, G. LeRoux: Appl. Phys. Lett. **47**, 1099 (1985)
10. J.M. Moison, F. Houzy, F. Barthe, L. Leprince, E. Andre, O. Vatel: Appl. Phys. Lett. **64**, 196 (1994)
11. D. Leonard, M. Krishnamurthy, C.M. Reaves, S.P. Denbaars, P.M. Petroff: Appl. Phys. Lett. **63**, 3203 (1993)

12. J. Ahopelto, A.A. Yamaguchi, K. Nishi, A. Usui, H. Sakaki: Jpn. J. Appl. Phys. **32**, L32 (1993)
13. H. Saito, K. Nishi, S. Sugou: Appl. Phys. Lett. **74**, 1224 (1999)
14. R. Leon, S. Fafard, D. Leonard, J.L. Merz, P.M. Petroff: Appl. Phys. Lett. **67**, 521 (1995)
15. V.M. Ustinov, A.E. Zhukov, A.Y. Egorov, A.R. Kovsh, S.V. Zaitsev, N.Y. Gordeev, V.I. Kopchatov, N.N. Ledentsov, A.F. Tsatsul'nikov, B.V. Volovik, P.S. Kop'ev, Z.I. Alferov, S.S. Ruvimov, Z. Liliental-Weber, D. Bimberg: Electron. Lett. **34**, 670 (1997)
16. K. Nishi, H. Saito, S. Sugou, J.-S. Lee: Appl. Phys. Lett. **74**, 1111 (1999)
17. N. Kirstaedter, N.N. Ledentsov, M. Grundmann, D. Binberg, V.M. Ustinov, S.S. Ruvimov, M.V. Maximov, P.S. Kop'ev, Z.I. Alferov, U. Richter, P. Werner, U. Gosele, J. Heydenreich: Electron. Lett. **30**, 1416 (1994)
18. D.L. Huffaker, G. Park, Z. Zou, O.B. Shchekin, D.G. Deppe: Appl. Phys. Lett. **73**, 2564 (1998)
19. O.B. Shchekin, G. Park, D.L. Huffker, D.G. Deppe: Appl. Phys. Lett. **77**, 466 (2000)
20. H. Chen, Z. Zou, O.B. Shchekin, D.G. Deppe: Electron. Lett. **36**, 1703 (2000)
21. H. Shoji, K. Mukai, N. Ohtsuka, M. Sugawara, T. Uchida, H. Ishikawa: IEEE Photonics Tech. Lett. **7**, 1385 (1995)
22. J. Temmyo, E. Kuramochi, M. Sugo, T. Nishiya, R. Noetzel, T. Tamamura: Electron. Lett. **31**, 209 (1995)
23. H. Saito, K. Nishi, Y. Sugimoto, S. Sugou: Electron. Lett. **35**, 1561(1999).
24. G.T. Liu, A. Stintz, H. Li, K.J. Malloy, L.F. Lester: Electron. Lett. **35**, 1163 (1999)
25. G. Park, O.B. Shchekin, D.L. Huffaker, D.G. Deppe: IEEE Photonics Tech. Lett. **13**, 230 (2000)
26. H. Saito, K. Nishi, I. Ogura, S. Sugou, Y. Sugimoto: Appl. Phys. Lett. **69**, 3140 (1996)
27. Y. Nabetani, T. Ishikawa, S. Noda, A. Sasaki: J. Appl. Phys. **76**, 347 (1994)
28. R. Mirin, A. Gossard, J. Bowers: Electron. Lett. **32**, 1732 (1996)
29. J.A. Lott, N.N. Ledentsov, V.M. Ustinov, N.A. Maleev, A.E. Zhukov, A.R. Kovsh, M.V. Maximov, B.V. Volovik, Zh.I. Alferov, D. Bimberg: Electron. Lett. **36**, 1384 (2000)
30. J.K. Kim, T.A. Strand, R.L. Naone, L.A. Coldren: Appl. Phys. Lett. **74**, 2752 (1999)
31. D.A.B. Miller, D.S. Chemla, S. Schmitt-Rink: Appl. Phys. Lett. **52**, 2154 (1988)
32. R. Sahara, M. Matsuda, H. Shoji, K. Morio, H. Soda: IEEE Photonics Tech. Lett. **8**, 1477 (1996)
33. J.L. Jimenez, L.R. Fonseca, D.J. Brady, J.P. Leburton, D.E. Wohlert, K.Y. Cheng: Appl. Phys. Lett. **71**, 3558 (1997)
34. K. Matsuda, T. Saiki, H. Saito, K. Nishi: Appl. Phys. Lett. **76**, 73 (2000); T. Matsumoto, M. Ohtsu, K. Matsuda, T. Saiki, H. Saito, K. Nishi: Appl. Phys. Lett. **75**, 3246 (1999)
35. K. Nishi, R. Mirin, D. Leonard, G. Medeiros-Ribeiro, R.M. Petroff, A.C. Gossard: J. Appl. Phys. **80**, 3466 (1996)
36. K. Nishi, T. Anan, A. Gomyo, S. Kohmoto, S. Sugou: Appl. Phys. Lett. **70**, 3579 (1997)
37. J.-S. Lee, S. Sugou, Y. Masumoto: J. Cryst. Growth, **205**, 467 (1999)
38. J. Tersoff, C. Teichert, M.G. Lagally: Phys. Rev. Lett. **76**, 1675 (1996)

# Index

Printing (Computer to Film): Saladruck, Berlin
Binding: Stürtz AG, Würzburg